T0321253

Oceanic Circulation Models:
Combining Data and Dynamics

NATO ASI Series

Advanced Science Institutes Series

A Series presenting the results of activities sponsored by the NATO Science Committee, which aims at the dissemination of advanced scientific and technological knowledge, with a view to strengthening links between scientific communities.

The Series is published by an international board of publishers in conjunction with the NATO Scientific Affairs Division

A	**Life Sciences**	Plenum Publishing Corporation
B	**Physics**	London and New York
C	**Mathematical**	Kluwer Academic Publishers
	and Physical Sciences	Dordrecht, Boston and London
D	**Behavioural and Social Sciences**	
E	**Applied Sciences**	
F	**Computer and Systems Sciences**	Springer-Verlag
G	**Ecological Sciences**	Berlin, Heidelberg, New York, London,
H	**Cell Biology**	Paris and Tokyo

Series C: Mathematical and Physical Sciences - Vol. 284

Oceanic Circulation Models: Combining Data and Dynamics

edited by

David L. T. Anderson

Department of Atmospheric, Oceanic and Planetary Physics,
Oxford University, Oxford, U.K.

and

Jürgen Willebrand

Institut für Meereskunde,
Universität Kiel, Kiel, F.R.G.

Kluwer Academic Publishers

Dordrecht / Boston / London

Published in cooperation with NATO Scientific Affairs Division

Proceedings of the NATO Advanced Study Institute on
Modelling the Ocean General Circulation and Geochemical Tracer Transport
Les Houches, France
February 15–26, 1988

Library of Congress Cataloging in Publication Data

```
NATO Advanced Study Institute on 'Modelling the Ocean General
   Circulation and Geochemical Tracer Transport' (1988 : Les Houches,
   Haute-Savoie, France)
   Ocean circulation models : combining data and dynamics / edited by
David L.T. Anderson and Jürgen Willebrand.
      p.   cm. -- (NATO ASI series. Series C, Mathematical and
physical sciences ; vol. 284)
      "Proceedings of the NATO Advanced Study Institute on 'Modelling
the Ocean General Circulation and Geochemical Tracer Transport' held
in Les Houches, France, February 15-26, 1988."
   Includes index.
   ISBN 978-0-7923-0394-7
   1. Ocean circulation--Computer simulation--Congresses.
2. Chemical oceanography--Computer simulation--Congresses.
I. Anderson, D. L. T. (David L. T.)  II. Willebrand, J. (Jürgen),
1941-   .  III. Title.  IV. Series: NATO ASI series.  Series C,
Mathematical and physical sciences ; no. 284.
GC228.5.N38  1988
551.47'01--dc20                                         89-15518
```

ISBN 978-0-7923-0394-7   ISBN 978-94-009-1013-3 (eBook)
DOI 10.1007/978-94-009-1013-3

Published by Kluwer Academic Publishers,
P.O. Box 17, 3300 AA Dordrecht, The Netherlands.

Kluwer Academic Publishers incorporates the publishing programmes of
D. Reidel, Martinus Nijhoff, Dr W. Junk and MTP Press.

Sold and distributed in the U.S.A. and Canada
by Kluwer Academic Publishers,
101 Philip Drive, Norwell, MA 02061, U.S.A.

In all other countries, sold and distributed
by Kluwer Academic Publishers Group,
P.O. Box 322, 3300 AH Dordrecht, The Netherlands.

Printed on acid free paper

This book contains the proceedings of a NATO Advanced Research Workshop held within the programme of activities of the NATO Special Programme on Global Transport Mechanisms in the Geo-Sciences running from 1983 to 1988 as part of the activities of the NATO Science Committee.

Other books previously published as a result of the activities of the Special Programme are:

BUAT-MENARD, P. (Ed.) – *The Role of Air-Sea Exchange in Geochemical Cycling* (C185) 1986

CAZENAVE, A. (Ed.) – *Earth Rotation: Solved and Unsolved Problems* (C187) 1986

WILLEBRAND, J. and ANDERSON, D.L.T. (Eds.) – *Large-Scale Transport Processes in Oceans and Atmosphere* (C190) 1986

NICOLIS, C. and NICOLIS, G. (Eds.) – *Irreversible Phenomena and Dynamical Systems Analysis in Geosciences* (C192) 1986

PARSONS, I. (Ed.) – *Origins of Igneous Layering* (C196) 1987

LOPER, E. (Ed.) – *Structure and Dynamics of Partially Solidified Systems* (E125) 1987

VAUGHAN, R. A. (Ed.) – *Remote Sensing Applications in Meteorology and Climatology* (C201) 1987

BERGER, W. H. and LABEYRIE, L. D. (Eds.) – *Abrupt Climatic Change – Evidence and Implications* (C216) 1987

VISCONTI, G. and GARCIA, R. (Eds.) – *Transport Processes in the Middle Atmosphere* (C213) 1987

SIMMERS, I. (Ed.) – *Estimation of Natural Recharge of Groundwater* (C222) 1987

HELGESON, H. C. (Ed.) – *Chemical Transport in Metasomatic Processes* (C218) 1987

CUSTODIO, E., GURGUI, A. and LOBO FERREIRA, J. P. (Eds.) – *Groundwater Flow and Quality Modelling* (C224) 1987

ISAKSEN, I. S. A. (Ed.) – *Tropospheric Ozone* (C227) 1988

SCHLESINGER, M.E. (Ed.) – *Physically-Based Modelling and Simulation of Climate and Climatic Change* 2 vols. (C243) 1988

UNSWORTH, M. H. and FOWLER, D. (Eds.) – *Acid Deposition at High Elevation Sites* (C252) 1988

KISSEL, C. and LAY, C. (Eds.) – *Paleomagnetic Rotations and Continental Deformation* (C254) 1988

HART, S. R. and GULEN, L. (Eds.) – *Crust/Mantle Recycling at Subduction Zones* (C258) 1989

GREGERSEN, S. and BASHAM, P. (Eds.) – *Earthquakes at North-Atlantic Passive Margins: Neotectonics and Postglacial Rebound* (C266) 1989

MOREL-SEYTOUX, H. J. (Ed.) – *Unsaturated Flow in Hydrologic Modeling* (C275) 1989

BRIDGWATER, D. (Ed.) – *Fluid Movements – Element Transport and the Composition of the Crust* (C281) 1989

ANDERSON, D.L.T. and WILLEBRAND, J. (Eds.) – *Ocean Circulation Models: Combining Data and Dynamics* (C284) 1989

CONTENTS

Tracer Inverse Problems

Carl Wunsch

A Geometrical Interpretation of Inverse Problems

Dirk Olbers

Determining Diffusivities from Hydrographic Data by Inverse Methods with Applications to the Circumpolar Current

Dirk Olbers and Manfred Wenzel

Ocean Acoustic Tomography: a Primer

Robert A. Knox

The Circulation in the Western North Atlantic Determined by a Nonlinear Inverse Method.

Herlé Mercier

Altimeter Data Assimilation into Ocean Circulation Models — Some Preliminary Results

William R. Holland

Assimilation of Data into Ocean Models

D. J. Webb

Driving of Non-linear Time-dependent Ocean Models by Observation of Transient Tracers — a Problem of Constrained Optimisation

Jens Schröter

Assimilation of XBT Data Using a Variational Technique

J. Sheinbaum and D.L.T. Anderson

The Role of Real-Time Four-Dimensional Data Assimilation in the Quality Control, Interpretation, and Synthesis of Climate Data

A. Hollingsworth

Introduction to Chemical Tracers of the Ocean Circulation

J.-F. Minster

On Oceanic Boundary Conditions for Tritium, on Tritiugenic ^{3}He, and on the Tritium-^{3}He Age Concept

Wolfgang Roether

Ocean Carbon Models and Inverse Methods

Berrien Moore III, Bert Bolin, Anders Björkström, Kim Holmén and Chris Ringo

Model of the Nutrient and Carbon Cycles in the North Atlantic. An Application of Linear Programming Methods

Reiner Schlitzer

The Design of Numerical Models of the Ocean Circulation

K. Bryan

Instabilities and Multiple Steady States of the Thermohaline Circulation

Jochem Marotzke

Subgridscale Representation

Greg Holloway

PREFACE

This book which is the outcome of a NATO-Advanced Study Institute on Modelling the Ocean Circulation and Geochemical Tracer Transport is concerned with using models to infer the ocean circulation. Understanding our climate is one of the major problems of the late twentieth century. The possible climatic changes resulting from the rise in atmospheric carbon dioxide and other trace gases are of primary interest and the ocean plays a major role in determining the magnitude, temporal evolution and regional distribution of those changes. Because of the poor observational basis the ocean general circulation is not well understood. The World Ocean Circulation Experiment (WOCE) which is now underway is an attempt to improve our knowledge of ocean dynamics and thermodynamics on global scales relevant to climate change. Despite those efforts, the oceanic data base is likely to remain scarce and it is crucial to use appropriate methods in order to extract the maximum amount of information from observations.

The book contains a thorough analysis of methods to combine data of various types with dynamical concepts, and to assimilate data directly into ocean models. The properties of geochemical tracers such as ^{14}C, He, Tritium and Freons and how they may be used to impose integral constraints on the ocean circulation are discussed. Specific applications of assimilating satellite altimeter data into models, of inferring the structure of major current systems such as in the Western North Atlantic and the Antarctic Circumpolar Current and of using acoustic tomography to remotely sense the ocean interior are given. Ocean boundary conditions for tracers and, where appropriate, the role of biology in changing their distribution are discussed.

In addition to inverse models, prognostic models of ocean circulation are needed to interpret the data. The book contains a review of numerical aspects of ocean general circulation models and their application to understanding the dynamics of the ocean mean state. One of the weak parameterisations in such models is that of subgrid scale mixing. For WOCE it is necessary to put emphasis on mixing processes which are decisive for water mass formation and large-scale heat transport. A full account of the many difficulties in parameterising mixing is given.

To promote a coherent structure to the course, four principal lecturers developed their topic in depth over a series of five lectures. Some specific applications were further developed by supporting lecturers. A number of presentations were made by students covering a wide range of topics. Regrettably although the standard of the talks was generally very high, space has prevented us from including most of these here. We have included a small selection however where the material was especially close to the general theme of the course.

Primary support for the Study Institute was provided by NATO under the auspices of the Special Programme on Global Transport Mechanisms in the Geo Sciences. Additional support for the school was provided by the Les Houches Centre de Physique Winter Programme, and by the National Science Foundation. It is a great pleasure to thank these organisations and in particular Dr. L.V. da Cunha and Prof. Nino Boccara who helped to make everything run smoothly.

We would like to thank Shona Anderson and Anu Dudhia who coped with the endless technical problems to produce this volume; and Joachim Dengg, Anu Dudhia, Jochem Marotzke and Detlef Stammer for carefully checking the manuscripts.

Finally we would like to thank the lecturers who made the school so exciting and have provided in this book a stimulating record of the events.

David Anderson

Jürgen Willebrand

Feb 1989

PARTICIPANTS

David Anderson,
University of Oxford,
Department of Atmospheric Physics,
Clarendon Laboratory,
Parks Road,
Oxford OX1 3PU,
U.K.

Sukru Turan Besiktepe,
Middle East Tech.,
University Institute of Marine Sciences,
P.K. 28,
Erdemli-Icel,
TURKEY

Frederick Bingham,
Scripps Institution of Oceanography,
A-030,
University of California, San Diego,
La Jolla, CA 92093,
U.S.A.

Philip Bogden,
Mail-code A-030,
Scripps Institution of Oceanography,
University of California, San Diego,
La Jolla, CA 92093,
U.S.A.

Bert Bolin,
Department of Meteorologie,
University of Stockholm,
Arrhenius Laboratory,
S-106 91 Stockholm,
SWEDEN

Kirk Bryan,
Geophysical Fluid Dynamics Laboratory,
Princeton University,
NOAA,
Box 308,
Princeton, NJ 08540,

Garry Budin,
Hooke Institute for Atmospheric Research,
Clarendon Laboratory,
Parks Road,
Oxford OX1 3PU,
U.K.

Eric Chassignet,
Rosenstiel School of Marine
and Atmospheric Science,
Division of Meteorology and
Physical Oceanography,
4600 Rickenbacker Causeway,
Miami, FL. 33149-1098,
U.S.A.

C.K. Chu,
Columbia University,
Deptartment of Applied Physics
and Nuclear Engineering,
Seeley W. Mudd Building,
New York,
N.Y. 10027,
U.S.A.

Joachim Dengg,
Institut für Meereskunde,
der Universität Kiel,
Düsternbrooker Weg 20,
2300 Kiel
F.R.G.

Clare Dickson,
Meteorological Office,
London Road,
Bracknell
Berkshire RG12 2SZ,
U.K.

Francois Dulac,
Centre des Faibles Radioactivités,
C.N.R.S./C.E.A,
Domaine du CNRS,
B.P. 1,
F-91190 Gif-Sur-Yvette,
FRANCE

Nick Hall,
Imperial College of Science and Technology,
The Blackett Laboratory,
Prince Consort Road,
London, SW7 2BZ,
U.K.

Christoph Heinze,
Max-Planck-Institut für Meteorologie,
Bundesstrasse 55,
2000 Hamburg 13,
F.R.G.

William Holland,
NCAR,
National Center for Atmospheric Research,
P.O.Box 3000,
Boulder, CO 80303,
U.S.A.

Tony Hollingsworth,
E.C.M.W.F.,
Shinfield Park,
Reading RG2 9AX,
U.K.

Greg Holloway,
Institute of Ocean Sciences,
P.O. Box 6000,
9860 West Saanish Road,
Sidney, British Columbia V8L 4B2,
CANADA

Daniel Jamous,
Laboratoire d'Oceanographie,
Dynamique et de Climatologie,
Tour 14, 2e étage,
4, Place Jussieu,
75252 Paris Cedex,
FRANCE

Xingjan Jiang,
Department of Atmospheric Science,
Oregon State University,
Corvallis, OR 97331,
U.S.A.

Mitsuhiro Kawase ,
Massachusetts Institute of Technology,
Cambridge, MA 02139,
U.S.A.

Robert Knox,
Scripps Institution of Oceanography,
University of California,
San Diego,
La Jolla, CA 92093,
U.S.A.

Stephanie Legutke,
Institut für Meereskunde
der Universität Hamburg,
Troplowitzstr. 7,
2000 Hamburg 54,
F.R.G.

Christian LeProvost,
Institut de Mecanique,
B.P. 53 - Centre de Tri,
38041 Grenoble Cedex,
FRANCE

Leo Maas,
Netherlands Institute for Sea Research,
Postbox 59 1790,
Ab den Burg,
Texel,
The Netherlands

Gurvan Madec,
Laboratoire D'Oceanographie,
Dynamique et de Climatologie,
L.O.D.Y.C.,
Tour 14,
Place Jussieu,
75252 Paris Cedex 05,
FRANCE

Jochem Marotzke,
Institut für Meereskunde an
der Universität Kiel,
Düsternbrooker Weg 20,
2300 Kiel,
F.R.G.

Francoise Martel,
Laboratoire D'Oceanographie,
Dynamique et de Climatologie,
Tour 14,
2e étage,
4, Place Jussieu,
75252 Paris Cedex 05,
France

Laurent Mémery,
Laboratoire D'Oceanographie,
Dynamique et de Climatologie,
Tour 14,
2e etage,
4, Place Jussieu,
75252 Paris Cedex 05,
France

Herlé Mercier,
IFREMER Centre de Brest,
B.P. 337
29273 Brest Cedex,
FRANCE

Nicolas Metzl,
Laboratoire de Physique
et Chimie Marines,
Université Pierre et Marie Curie,
4, Place Jussieu,
75230 Paris Cedex 05,
FRANCE

Elizabeth Michel,
Centre des Faibles Radioactivités,
Laboratoire Mixte,
C.N.R.S. - C.E.A.,
Avenue de la Terrasse,
91190 Gif sur Yvette,
FRANCE

Jean Francois Minster,
CNES/GRGS,
18 Av. E. Belin,
31057 Toulouse Cedex,
FRANCE

Berrien Moore III,
Institute of Earth, Ocean, and Space
Science Engeneering,
Research Building,
University of New Hampshire,
Durham, New Hampshire 03824,
U.S.A.

Dirk Olbers,
Alfred-Wegener-Institut für
Polar- und Meeresforschung,
Columbus-Center,
2850 Bremerhaven,
F.R.G.

J.D. Opsteegh,
Rijksuniversiteit Utrecht,
Instituut Meteorologie en Oceanografie,
Princetonplein 5,
3584 Utrecht,
THE NETHERLANDS

Pierre-M. Poulain,
Scripps Institution of Oceanography,
UCSD, Mail Code A-03,
La Jolla, CA 92093,
U.S.A.

Chris Ringo,
Institute of Earth, Ocean, and Space,
Science Engeneering
Research Building,
University of New Hampshire,
Durham, New Hampshire 03824,
U.S.A.

Wolfgang Roether,
Universität Bremen,
FB1,
FG Tracer Oceanographie,
Postfach 330440,
2800 Bremen 33,
F.R.G.

Hagen Ross,
Alfred-Wegener-Institut für
Polar- und Meeresforschung,
Am Handelshafen,
2850 Bremerhaven,
F.R.G.

Reiner Schlitzer,
Universität Bremen,
FB1,
FG Tracer Oceanographie,
Postfach 330440,
2800 Bremen 33,
F.R.G.

Jens Schröter,
Alfred-Wegener-Institut für
Polar- und Meeresforschung,
Columbus-Center,
2850 Bremerhaven,
F.R.G.

Julio Sheinbaum,
Hooke Institute for Atmospheric Research,
Clarendon Laboratory,
Parks Road,
Oxford OX1 3PU,
U.K.

O.M. Smedstad,
Florida State University,
Department of Meteorology,
Tallahassee,
FL. 32306,
U.S.A.

Detlef Stammer,
Institut für Meereskunde
an der Universität Kiel,
Düsternbrooker Weg 20,
2300 Kiel 1,
F.R.G.

David Stevens,
School of Mathematics and Physics,
University of East Anglia,
Norwich NR4 7TJ,
U.K.

David Straub,
3408 Woodlard PK Ave N,
Seattle,
WA 98103,
U.S.A.

H.E. de Swart,
Institute of Meteorologie and Oceanography,
Princetonplein 5,
3584 CC Utrecht,
THE NETHERLANDS

Karl Taylor,
L-262,
Lawrence Livermore National Laboratory,
P.O. Box 808,
Livermore, CA 94550,
U.S.A.

Joaquin Tintore,
Universitat de les Illes Balears,
Departament de Fiscia,
Facultat de Ciencies,
E-07071 Palma de Mallorca,
SPAIN

Jean Tournadre,
IFREMER,
Centre de Brest,
BP 337,
29273 Brest Cedex,
FRANCE

Hellen Wallace,
Maths Department,
Exeter University
Exeter,
Devon, EX4 4QE,
.U.K.

David Webb,
Institute of Oceanographic Sciences,
Brook Road,
Wormley, Godalming,
Surrey, GU8 5UB,
U.K.

Jurgen Willebrand,
Institut für Meereskunde
an der Universität Kiel,
Düsternbrooker Weg 20,
2300 Kiel 1,
F.R.G.

Joerg-Olaf Wolf,
Max-Planck-Institut für Meteorologie,
Bundesstrasse 55,
2000 Hamburg 13,
F.R.G.

Carl Wunsch,
Massachusetts Institute of Technology,
Cambridge, MA 02139,
U.S.A.

Enrico Zambianchi,
Gruppo Oceano,
Dipartemento di Fisica,
Università degli Studi di Roma 'La Sapienza',
Piazzale Aldo Moro, 2
00185 Roma,
ITALIA

Ru Ling Zhou,
Atmospheric and Planetary Science,
Columbia University and NASA Goddard,
Institute for Space Studies,
2880 Broadway
New York, N.Y. 10025

TRACER INVERSE PROBLEMS

Carl Wunsch
Center for Meteorology and Physical Oceanography
Department of Earth, Atmospheric, and Planetary Sciences
Massachusetts Institute of Technology
Cambridge, MA 02139

1. INTRODUCTION

1.1 The General Problem

Charts of tracer distributions, salt, oxygen, tritium, and potential vorticity such as those shown in Figs. 1 and 2 have long been the bedrock of the study of the oceanic general circulation. Much of what is believed well-determined about the circulation is derived from such figures. Few would dispute the argument that these figures show that water must flow at depth from the Mediterranean into the North Atlantic, or that 'overflow' water emanating from high latitudes is filling up the western North Atlantic Basin. But much confusion has been caused by failure to remember that figures such as 1 and 2 represent inventories and do not by themselves carry any information about water movement or dynamics. One might postulate that the flow which produced the Mediterranean salt tongue took place thousands of years ago, the fluid then having 'frozen' in place, no subsequent flow having taken place at all. That such a view is rejected is based upon information about the ocean not contained explicitly in such property distribution maps.

1

D. L. T. Anderson and J. Willebrand (eds.), Oceanic Circulation Models: Combining Data and Dynamics, 1–77.
© 1989 by Kluwer Academic Publishers.

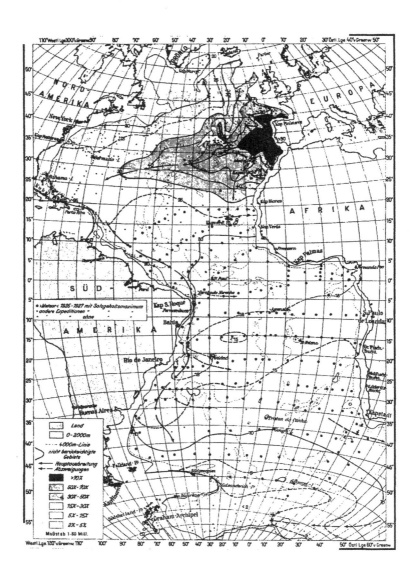

Figure 1a
Salinity at the core of the Mediterranean water (from Wüst, 1935). The
arrows show Wüst's inference of water mass movement direction (original
is in color).

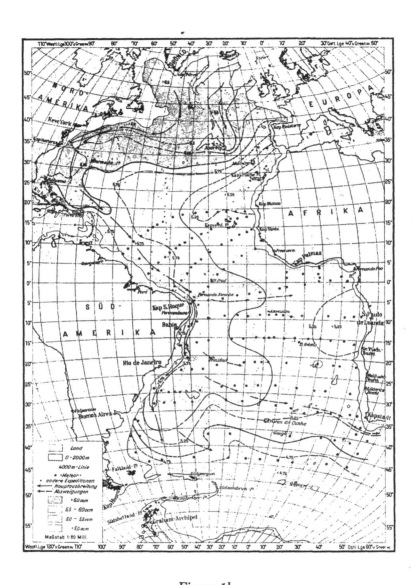

Figure 1b
Wüst's (1935) chart of oxygen concentration showing the distribution of North Atlantic Deep Water. Remember that such a chart by itself depicts only an inventory; inferences about movement require additional information.

4

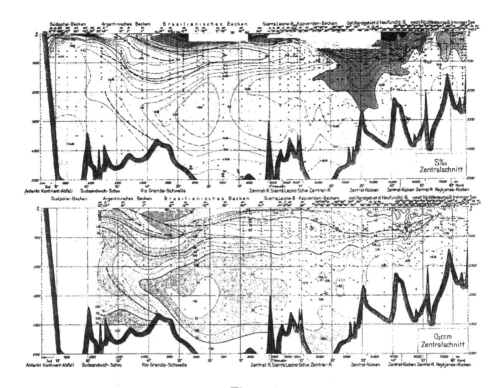

Figure 2

Meridional section down the western Atlantic ocean (also from Wüst, 1935), showing (upper panel) salinity, and (lower panel), oxygen concentration, as well as Wüst's inferred 'spreading' directions for the Antarctic Intermediate Water. As discussed in the text, the problem of steady tracer inverse methods is to extract numerical estimates of flow and mixing rates from charts such as these and those in Fig. 1, to render quantitative what Wüst and others had to leave in qualitative form.

Figs. 3 and 4 differ from those in the previous figures in displaying transient tracers. Fig. 4, from Weiss *et al.* (1985), in particular shows fluorocarbon concentrations on the equator in the North Atlantic. One is confident that a few years prior to the survey, the effective concentrations were zero there. How does one use the inferred change in concentration to make estimates about the flow or mixing field? It is not obvious that such estimates should be made in the same way, and for similar purposes as those made from the apparently steady distributions appearing in Figs. 1 and 2.

Descriptive oceanographers have cleverly exploited a variety of tracers to make intelligent inferences about how water moves around the ocean. But almost all these inferences have been strictly qualitative, almost geological in tone. What has been much less clear is how to make numerical estimates of such important quantities as flow and mixing rates. Wüst, who was one of the pioneers of the use of property charts, generally confined himself to drawing arrows of apparent 'spreading' (*Ausbreitung*), making neither numerical estimates nor separating flow and mixing. He drew plausible arrows along the axis of the tongue in Fig. 1; unfortunately direct measurements suggest that the predominant flows are more likely around the tongue rather than along it.

In these lectures, I would like to outline some of the issues involved in making inferences from tracers, but which apply to using data of any kind. One should recognize that the use of tracers for inferring flow properties is a problem with a wide-range of applications outside oceanography. Apart from the obvious meteorological analogues, such methods are widely used in medicine, hydrology, and chemical engineering, among other fields. A few references in related areas are Buffham (1985), Nauman and Buffham (1983) and Sheppard (1962), but generally speaking, I have not found any existing general mathematical framework for systematic inferences from tracers.

Later on we will set up a variety of models for using tracers for quantitative purposes; but to motivate some of the introductory material which follows, let us examine two possibilities. Consider the two-dimensional spatial grid displayed in Fig. 5; the two dimensions could be thought of as horizontal, but we do not need to be so specific. Let us suppose, following Fiadeiro and Veronis (1984), that we have been provided with observations of a steady tracer, C, at the grid points '•' and we seek to determine the flow field which gives rise to the observed distribution.

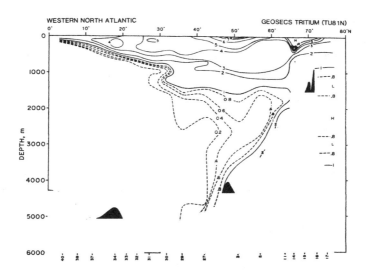

Figure 3a
Tritium distribution in the western North Atlantic at the time of t.
GEOSECS observations of 1972 (from Östlund and Fine, 1979).

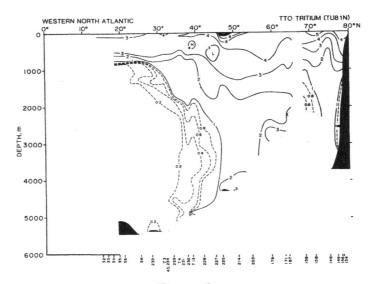

Figure 3b
Tritium distribution in nearly the same place as in Fig. 3a, but 10 yea
later at the time of the TTO expedition (from Östlund, 1983).

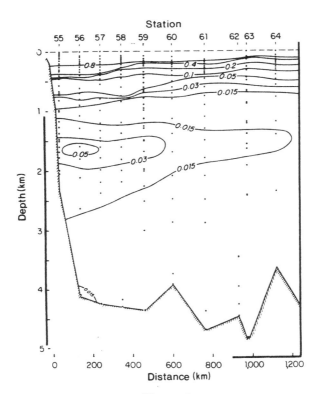

Figure 4
*Fluorocarbon distribution on the equator in 1983 (Weiss et al., 1985). It is
a reasonable supposition that the concentration was zero at this location
some 15 years before. Transient tracer modelling seeks to exploit the
information in figures such as that depicted here and in Fig. 3, to deduce
(a), the rate of uptake of various substances across the air/sea interface,
and (b), the rate of water movement and mixing. How one might do that
is the subject of part of these lectures.*

For sensible reasons we think that the equation (i.e. the model) governing the
distribution is

$$\mathbf{u}.\nabla C = K\nabla^2 C \tag{1.1}$$

where K is an eddy coefficient which we choose to assume is known. Being lucky
enough to have dense measurements, we think we can deal with the equation (1.1)
in finite difference form, and write it as

$$(C_{i,j+1}-C_{i+1,j})\psi_{i,j} + (C_{i+1,j} - C_{i,j-1})\psi_{i,j-1} + (C_{i,j-1} - C_{i-1,j})\psi_{i-1,j-1}$$

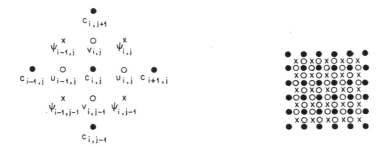

Figure 5
Grid used to discuss tracer inversions in finite difference form. In the forward problem, the stream function is assumed known at the position marked by ×, or if one uses the two velocity components, they are assumed known at the open circles. One calculates the tracer concentrations at the filled circles. In the inverse problem, the tracer concentration is assumed known at one or more of the filled circles and the problem is to find either the stream function or the velocity components on the appropriate grid locations.

$$+(C_{i-1,j}-C_{i,j+1})\psi_{i-1,j}$$
$$= -2K\left[\frac{1}{\delta}(C_{i+1,j} + C_{i-1,j} - 2C_{i,j}) + \delta(C_{i,j+1} + C_{i,j-1} - 2C_{i,j})\right] \quad (1.2)$$

where $\psi_{i,j}$ is a stream function evaluated at × on the grid, and $\delta = \Delta x/\Delta y$. The collection of finite difference equations can be written in a condensed matrix/vector notation as

$$\mathbf{A}\psi = \mathbf{d} \quad (1.3)$$

$$\psi = \begin{pmatrix} \psi_{11} \\ \psi_{12} \\ \vdots \\ \psi_{MN} \end{pmatrix}$$

$$\mathbf{d} = -2K\left[\frac{1}{\delta}(C_{i+1,j} + C_{i-1,j} - 2C_{i,j}) + \delta(\ldots)\right]$$

and $\mathbf{A}$ is the coefficient matrix of appropriate $(C_{i,j+1} - C_{i+1,j})$, etc. (see Fiadeiro and Veronis, 1984, Wunsch, 1985).[1]

---

[1] Normally, we use bold capital letters to denote matrices and bold lower case letters to denote vectors, although the usage is not wholly consistent. The context should make clear what is intended.

A count of the collection (1.3) shows that there are more unknown values of $\psi_{i,j}$, than there are equations. Deciding precisely why this underdeterminism occurs, and what we should do about it is part of the purpose of these lectures. If the tracer C is biochemically active, we would append source/sink terms $\mathbf{q}$ to these equations which could then be written

$$\mathbf{A}\psi = \mathbf{d}', \quad \mathbf{d}' = \mathbf{d} + \mathbf{q} \tag{1.4}$$

Whether $\mathbf{q}$ can be considered known and thus carrying additional information, or whether it is to be considered as unknown, and thereby contributing to the underdetermined nature of the problem has to be discussed on a case by case basis (although a safe general conclusion is that usually $\mathbf{q}$ are neither fully known nor unknown).

It is very important to notice that (1.3) formally treats the tracer distribution as known, with the streamfunction as unknown. But mere rearrangement of (1.2) allows us to write the finite difference equations as

$$\mathbf{A}_2 \mathbf{C} = 0 \tag{1.5}$$

where $\mathbf{C}$ is now the $C_{i,j}$ ordered as a vector, the problem usually treated in text-books, with the flow field as given, and the tracer distribution to be found. Experienced oceanographers will recognize that it is rare to be able to make a simple distinction between a completely unknown flow field and completely known tracer field, or vice-versa. Part of the fascination of oceanographic inverse problems is the need to grapple with such untidy situations ignored by conventional mathematics, but which themselves lead to interesting mathematical issues. However, understanding comes best by considering first either the form (1.3) or (1.5) separately.

An equation count of (1.5) shows that here too, there are fewer equations than unknowns, and indeed the homogeneous solution $\mathbf{C} = 0$ is not precluded. But the conventionality of the problem immediately suggests why: to fully pose the problem, we need to specify both initial conditions on $\mathbf{C}$, as well as the appropriate boundary conditions. Eqs. (1.3) represent a (hyperbolic) partial differential equation for the streamfunction, and one must specify its boundary conditions too. Wunsch (1985) shows that the missing equations in (1.3) are precisely those describing the appropriate boundary data.

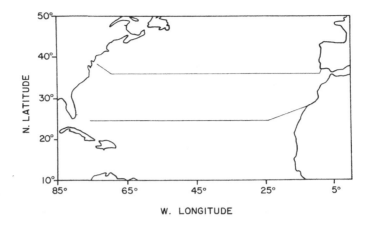

Figure 6

Closed 'box' formed by hydrographic sections (in this case the observations of Roemmich and Wunsch, 1985). One uses balance equations for various tracers entering and leaving the box to infer flow rates and mixing.

Consider now a somewhat different situation, based upon that analyzed in Wunsch (1978). We suppose that a ship has made hydrographic observations along the section lines displayed in Fig. 6. From the conventional dynamic method, we know that we can calculate the thermal wind between any pair, j, of the stations in the figure:

$$v_j(z) = \frac{g}{f} \int_{z_0}^{z} \frac{\partial \rho}{\partial s} dz + b_j \equiv v_{R_j} + b_j$$

where s is distance along the hydrographic line, f is the Coriolis parameter, z_0 is an arbitrary reference depth, and ρ is the density field. The 'classical' problem of physical oceanography has been the inability to calculate the integration constant (reference level velocity) b_j. Suppose the tracer $\mathbf{C}$ has known sources/sinks $\mathbf{Q}$ in the volume delineated by the hydrographic sections. Then we can write

$$\sum_j \int_{\text{bottom}}^{\text{top}} \rho_j(z) C_j(z) v_j(z) \Delta s_j dz = Q \tag{1.6a}$$

(where the vertical integral would, in practice, be replaced by an approximating sum) or

$$\sum_j \int_{\text{bottom}}^{\text{top}} \rho_j(z)C_j(z)b_j\,\Delta s_j dz = Q - \sum_j \int_{\text{bottom}}^{\text{top}} \rho_j C_j(z)v_{R_j}\,\Delta s_j dz$$

or,

$$\mathbf{Ab} = \mathbf{d} \tag{1.6b}$$

since the b_j are depth independent.

So far, (1.6) is one equation in many unknowns (a typical trans-Atlantic section would have $O(100)$ station pairs in it). The strategy is to add further constraints of the same form as (1.6), including tracer conservation (by 'conservation' we mean that the budget of inflow/outflow has to equal the known sources/sinks) in different layers and the addition of further tracers (including mass and salt, as well as more exotic ones such as potential vorticity and radiocarbon); the system then appears algebraically in precisely the form (1.4), albeit the specific computations involved in defining the coefficient matrix $\mathbf{A}$, and the right-hand sides are different. It remains true that any plausible bookkeeping suggests that the number of unknowns will likely always exceed the number of knowns. The model we have used, purely geostrophic, can be made much more complex, involving mixing terms, higher order dynamics, etc., without however, changing the structure of the systems (1.4 or 1.6).

Finally, as another prototype, consider a transient tracer. Let us suppose that C evolves over time, and the governing model is

$$\frac{\partial C}{\partial t} + \mathbf{u}.\nabla C = K\nabla^2 C - \lambda C \tag{1.7}$$

where λ represents a decay constant.

A prototypical finite difference scheme (upstream differencing — see Roache, 1976) is

$$C_{i,j}^{t+1} - C_{i,j}^t + \frac{U\Delta t}{\Delta x}[uC_{i,j}^t - uC_{i-1,j}^t + vC_{i,j}^t - vC_{i,j-1}^t] \tag{1.8}$$
$$- \frac{K\Delta t}{(\Delta x)^2}[C_{i+1,j}^t + C_{i-1,j}^t - 2C_{i,j}^t + C_{i,j+1}^t + C_{i,j-1}^t - 2C_{i,j}^t] - \lambda\Delta t C_{i,j}^t = 0$$

where U is a velocity scale, and $\Delta x, \Delta t$ etc. are all conventional.

Again depending upon circumstance, one may have to regard the time differences as part of the set of unknowns, or as potentially perfectly known, or somewhere in between. The collection (1.8) can be written as (1.4), i.e. simply a large linear combination of quantities assumed known, C_{ij}^t, and the unknowns (e.g. u, v).

We will later find it helpful to re-write equations (1.8) (with no loss of generality) as

$$\mathbf{C}(t+1) = \mathbf{A}_1 \mathbf{C}(t) + \mathbf{B}\mathbf{q}(t) \qquad (1.9)$$

where $\mathbf{A}_1$ is simply that matrix operator which produces the evolution of tracer distribution

$$\mathbf{C}(t) = \{C_{ij}^t\} \quad \text{to} \quad \mathbf{C}(t+1) = \{C_{ij}^{t+1}\}$$

and $\mathbf{B}\mathbf{q}(t)$ contains all sources, sinks and boundary values (see Wunsch, 1987, 1988b) where formally now $\mathbf{C}$ is the unknown rather than the known.

It is no coincidence that the three models we have written in Eqs. (1.1), (1.6), (1.8) are all sets of linear equations (linear constraints) involving combinations of things supposed known (observations of the tracer C in space and time) with various quantities such as u, K we would like to determine. Tracer inverse problems may be defined as the inference of fluid properties, especially flow and mixing rates, from systems of equations such as (1.4b).

Much of these lectures is devoted to discussion of various ways of extracting information from linear combinations of data and unknowns of the generic forms we have written. Certain extensions of the conventional inverse problem will be encountered, particularly the class of problems which are best regarded as 'regularization' problems rather than inverse ones, where we seek missing information not about the flow field, but about the tracer field itself. Such problems, e.g. finding a best estimate of the total tritium injection into the ocean are closely related to inverse problems and are easily understood in the context we develop.

In the most general terms, the model equations, be they of the linear type or some other form, provide a 'connection' between the different parts of the oceanic space/time domain, where the differences can be in time or space or both. The classical 'forward' tracer problems take equations such as (1.5 or 1.9), with u, v specified (geostrophically or otherwise), along with complete initial conditions and appropriate boundary conditions everywhere for C, and compute the field C from these 'data' everywhere in the domain. The advection/diffusion equation thus links tracer concentrations in different parts of the region under study. It is easy to show (Wunsch, 1987), that a properly specified problem (in the Cauchy/Hadamard sense), leads to a system of linear algebraic equations with the number of equations precisely equal to the number of unknowns. The general problem of using tracers to make inferences about the ocean, is in common with the use of data of any type, largely a question of whether systems of algebraic equations can be solved when they correspond to classically 'ill-posed' mathematical problems.

A number of extensions are not treated here at all. It is easy to define problems which are intrinsically non-linear; indeed as discussed by Wunsch and Minster (1982), systems (1.2, 1.3 etc.) are all implicitly non-linear because the numbers used to calculate the coefficient matrices **A** are based upon measurements, and must in all cases be regarded as being in part unknown.

Many techniques are available for tackling such problems, but their treatment would take us too far afield. As with ordinary forward problems, complete understanding of linear systems has to precede any serious approach to non-linear ones. We also do not treat methods involving an infinite number of unknowns, ruling out variational methods and techniques based more directly upon functional analysis. Intuition about infinite dimensional methods is best gained as an extension of that developed in finite dimensions. Furthermore, it appears that in practice, almost all problems end on computers and are intrinsically not only discrete, but finite.

1.2 On Determinateness

The use of tracers for studying the ocean circulation in large part falls under the general category of statistical inference (as we will see), and many of the methods best used are labeled 'inverse' ones. These methods are usually directed at solving problems like those described in section 1.1, that are in some fashion underdetermined. A lot of the confusion that exists about inverse problems, and the methods employed to solve them, derives from the widespread, but specious, belief that underderminism is a grave weakness of inverse problems, peculiar to them, and that tackling an underdetermined problem is best avoided by any sensible investigator. Usually (e.g. Fiadeiro and Veronis, 1984), avoidance is achieved by reformulating the problem so as to make it (apparently) overdetermined. In Section 3, I will demonstrate that most overdetermined least-squares problems appear overdetermined solely because of the conventional (and often unconscious) suppression of a discussion of very restrictive assumptions being imposed.

Before proceeding however, it is useful to remember that ordinary, analytical, 'forward' problems are often underdetermined, and all problems involving data, forward or inverse, are of necessity underdetermined.

Two distinct types of underdeterminancy can be distinguished: 'structural', and 'statistical'. Examination of a simple forward example helps explain the meaning of the two types of underdeterminism. Consider the Dirichlet problem:

Solve the Laplace-Poisson equation

$$\nabla^2 \varphi = \rho \qquad (1.10)$$

within a domain D, with boundary ∂D, subject to $\varphi = \varphi_0$ on ∂D, for given ρ.

With some very mild conditions on φ_0, ∂D, and ρ the solution is unique, stable, and readily computed. But if we replace the Dirichlet problem with the Neumann problem: solve (1.10) subject to $\partial \varphi / \partial n = \varphi_0'$ on ∂D, where n denotes the local normal, everyone knows that the problem is indeterminate up to an additive constant. This 'null-space' of the problem is so familiar, and so simple in structure, that no one worries that the solution is infinitely underdetermined in the sense that the additive constant can take on an arbitrary value.

A problem with a more interesting null-space comes from electromagnetic theory. One writes the magnetic field, $\mathbf{B}$, as $\mathbf{B} = \nabla \times \mathbf{A} + \nabla q$, where q is an arbitrary scalar function. q is in the operator null-space and cannot be determined. Gauge transformations make different choices of $\partial q / \partial t$.

This null-space is again so familiar, that few are troubled by the complex underdetermined nature of the problem. Lanczos (1961) discusses null spaces of differential operators and their analogues in vector spaces.

So ordinary forward problems are often what we denote as structurally underdetermined — the information available is inadequate to calculate all the structure of the solution and we 'regularize' by suppressing the calculation of the null space (the constant in the Neumann problem; the scalar field in the electromagnetic problem). Attempts to determine the null space would, of course, be futile without additional information, and lead to grossly unphysical results if tried naively.

Let us now go back to the Dirichlet problem, which does not have a solution null-space. Suppose instead that we render the problem stochastic by specifying the boundary data as

$$\varphi = \varphi_0 + n \quad \text{on} \quad \partial D \qquad (1.11)$$

where the noise, n, represents an uncertainty in the actual conditions. We cannot then solve Eq. (1.10) for the solution; we could calculate a most probable solution, a smoothest one or a maximum mean square one, etc., but none of these choices is definitive. The solution remains uncertain because of the uncertainty of the necessary data. This underdeterminancy is what we are calling statistical. Similar underdeterminancy would enter through specification of a random component in ρ in (1.10). Problems indeterminate in the statistical sense were studied at length by Schröter and Wunsch (1986) for linear and non-linear general circulation models.

Inverse problems are often (but not always) underdetermined either structurally, statistically or both. But given the appearance of these issues of ordinary forward problems, one should not be especially troubled by the uncertain nature of the solutions.

2. INTERPOLATION AND MAP MAKING

2.1 Interpolation

Charts are so fundamental to the interpretation of tracers that it is important to understand how best to make them. Because most models require gridding of some sort, and because computer drawn charts are almost always generated from interpolated data, we simultaneously address issues of sampling and interpolation. Underlying all such issues is the notion that there is some continuous function, $T(x,t)$, which we desire to reconstruct from a set of discrete observations $T(x_i,t)$, $i = 1$ to N.

To understand the issues involved it is best to begin with a very idealized problem, in one dimension only. We suppose that $T = T(x)$, and that we have available a set of observations $T(n\Delta x) = T(x_n) = T_n$ where $-\infty \leq n \leq \infty$. What is usually called the Shannon or Shannon-Whittaker sampling theorem tells us the circumstances under which the function $T(x)$ can be calculated, exactly, from these discrete observations.

In summary form, the theorem says (see for example, Freeman, 1965 or Bracewell, 1978) that if the Fourier transform of $T(x)$, defined as

$$\hat{T}(s) = \int_{-\infty}^{\infty} T(x)\exp(-2\pi i s x)dx \qquad (2.1)$$

vanishes identically for $s \geq |s_c|$, then

$$T(x) = \sum_{n=-\infty}^{\infty} T_n \frac{\sin\frac{1}{2}(2\pi x/\Delta x - 2n\pi)}{\frac{1}{2}(2\pi x/\Delta x - 2n\pi)} \qquad (2.2)$$

exactly, for any x, as long as $\Delta x \leq 1/2s_c$. ($T(x)$ is then a so-called bandlimited function).

The proof of this statement is straightforward; see the references. It is important to understand its implications. First, the sampling theorem can be regarded as a statement concerning the ability to do perfect interpolation. The expression (2.2) is exact if and only if the conditions of the theorem are met. These conditions are, in addition to the bandlimited character, that the samples T_n are perfect, containing no measurement noise, and that there are an infinite number of these samples available. Neither of these latter two conditions can ever be met in practice; it is also possible to prove that any function whose domain of definition is finite, with the function vanishing outside that domain (as temperature does outside the domain of an ocean basin) cannot be bandlimited. (The conventional statement is that a function cannot be simultaneously bandlimited and timelimited, with time representing a generic observation coordinate.)

Despite the inability to meet the conditions of the theorem in practice, it is an important reference point for understanding what happens as its various suppositions are relaxed. Furthermore, the rough rule of thumb which says that a function can be accurately reproduced anywhere, if it is essentially bandlimited, and the sampling is at least as often as $\Delta x \leq 1/2s_c$, is normally an excellent practical guide.

The subject of sampling theorems is an extensive one. We confine ourselves here to some summary remarks and references. If the assumption of infinite samples is dropped, then one can no longer do perfect interpolation; one can however do better than merely truncating (2.2). The best estimate, which we will denote $\hat{T}(x)$ is of the form

$$\hat{T}(x) = \sum_{n=-N_1}^{N_2} T(x_n)W(x - x_n) \qquad (2.3)$$

where W is no longer the $(\sin \pi x)/\pi x$ function of (2.2), but is a prolate spheroidal wave function (see Landau and Pollak, 1962).

The Shannon-Whittaker theorem can be generalized to two and more dimensions (see Freeman, 1965 and Petersen and Middleton, 1962), but we will not pursue this subject further here either. Suffice it to say that as in one dimension, one can obtain rules of thumb for adequate spatial sampling to properly represent an intrinsically continuous function. But we do need to ask what in fact happens if we strongly violate the condition $\Delta x \leq 1/2s_c$?

Undersampling as this is called, leads to the well-known phenomenon of 'aliasing'. If a field has been undersampled, then reconstruction through formulas such as (2.2) or (2.3) still leads to a nice smooth, interpolated field; unfortunately this field does not coincide with the original underlying continuous tracer distribution. Instead, undersampled high wavenumbers masquerade (i.e. appear under an alias) as low wavenumbers, so that high frequency spatial wiggles in the true tracer field appear instead as low frequency structures, with potentially disastrous results.

The precise mathematical statement is that all wavenumbers

$$s = 2ns_c \pm s_0, \quad n = 0, 1, \ldots \tag{2.4}$$

alias into the wavenumber s_0. Roemmich and Wunsch (1985) showed that the sampling density for the hydrography in the International Geophysical Year was such that there appeared to be more low wavenumber structure in temperature and salinity than in a much more densely sampled recent set of measurements. The reader should consider how the small scale features in Fig. 7 would appear if simply subsampled, and the extent to which a numerical derivative of the subsampled field could represent that of the true field. Adverse effects become exaggerated when one must differentiate the data (see below).

The Shannon-Whittaker sampling theorem is mainly of theoretical use, although as already noted, it does provide a practical rule of thumb for determining adequate sampling. To do actual interpolation in one or more dimensions, the form (2.2) has two shortcomings: (1) the 'interpolation function', $(\sin \pi x)/\pi x$, extends a long way from the position, call it x_0, for which we seek an interpolated value. Many multiplications and additions would be required to interpolate one point, leading to a heavy computational burden; (2) Given that all data are inaccurate, and real sampling schemes in two or more dimensions rarely lead to uniform data distributions, and we never have infinite amounts of data, Eq. (2.2) does not by itself provide any estimate of how accurately we can in practice expect to interpolate at a given point. To be quantitatively useful, any interpolation or gridding scheme must produce an error estimate.

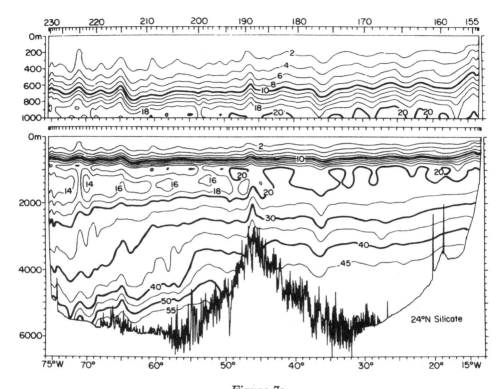

Figure 7a

Silicate distribution in a zonal section across the North Atlantic at 24°N in units of micro-gram atoms/litre (from Roemmich and Wunsch, 1985). Note the presence of much small horizontal scale structure, particularly near 1500 m. If the station spacing were much wider than that which led to this particular chart, the small scale structures would 'alias' into apparent long wavelength features, leading to an incorrect picture of the actual silicate structure, and potentially disastrous consequences if differentiated horizontally.

2.2 The Gauss-Markov Theorem

We need some elementary statistical notation. Let a bracket denote expected value, i.e. $\langle y \rangle$ denotes the mean of y, the averaging being taken in the sense of an integral over the probability density of y. If $\{y_i\} = \{y(\mathbf{r}_i)\}$ denotes a set of values (measured or theoretical), at points $\mathbf{r}_i$ and if $\langle y \rangle = 0$ (a special case), then

$$\mathbf{R} \equiv \{R_{ij}\} = \{\langle y_i y_j \rangle\} = \mathbf{R}\{\mathbf{r}_i, \mathbf{r}_j\}$$

19

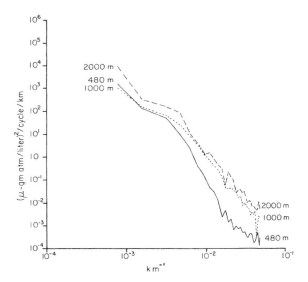

Figure 7b
Wave number spectrum of silicate distribution depicted in Fig. 7a at three different depths. The spectra are generally 'red', with most of the energy at low wavenumbers. However, as discussed in the text, differentiation of the fields suppresses the long wavelength information in favor of the high wavenumber information, thus potentially amplifying the least significant structures, including any small scale noise.

is the 'covariance' of y. If

$$R_{ij} = R(\mathbf{r}_i - \mathbf{r}_j) \tag{2.5}$$

then y is 'spatially stationary' or homogeneous (in the 'wide sense'), its second moments depending only upon the separation of the evaluation points. If

$$R_{ij} = R(|\mathbf{r}_i - \mathbf{r}_j|) \tag{2.6}$$

then y is in addition, isotropic. We now state a restricted case of a very important and general theorem, usually called the Gauss-Markov theorem:

Let $\mathbf{r}_0$ be a location where an interpolated value $y(\mathbf{r}_0)$ of a measurement set $\{\mathbf{d}\} = \{d_i = y_i + n_i\}$ is required. Suppose that the measurements are corrupted by random noise n_i of zero mean and covariance

$$\langle n(\mathbf{r}_i)n(\mathbf{r}_j)\rangle = \mathbf{N}(|\mathbf{r}_i - \mathbf{r}_j|) \tag{2.7}$$

The field we seek to map has the statistics given by $\mathbf{R}$ and suppose the noise is uncorrelated with the value of y:

$$\langle y(\mathbf{r}_i)n(\mathbf{r}_j)\rangle = 0 \quad \text{all } i,j \tag{2.8}$$

Suppose further, as is usually the case, the interpolated value is to be some weighted average of the observations:

$$\hat{y}(\mathbf{r}_0) = \sum_j B_j(\mathbf{r}_0)d(\mathbf{r}_j) = \mathbf{B}(\mathbf{r}_0)\mathbf{d} \tag{2.9}$$

$(\mathbf{B}$ is a $1 \times n$ matrix$)$

Then we can evaluate the variance of the difference between the correct value at r_0, and our interpolated value

$$E \equiv \langle(\hat{y}(\mathbf{r}_0) - y(\mathbf{r}_0))^2\rangle = \langle(\mathbf{B}\mathbf{d} - y(\mathbf{r}_0))^2\rangle \tag{2.10}$$

The Gauss-Markov theorem states that the minimum of (2.10) is reached when $\mathbf{B}$ is chosen as

$$\mathbf{B} = \mathbf{R}(\mathbf{r}_0,\mathbf{r}_j)\left[\mathbf{R}(\mathbf{r}_j,\mathbf{r}_k) + \mathbf{N}(\mathbf{r}_j,\mathbf{r}_k)\right]^{-1} \tag{2.11}$$

$$\mathbf{R}(\mathbf{r}_0,\mathbf{r}_j) \equiv \{\mathbf{R}(\mathbf{r}_0,\mathbf{r}_1),\mathbf{R}(\mathbf{r}_0,\mathbf{r}_2)...,\ \mathbf{R}(\mathbf{r}_0,\mathbf{r}_n)\}$$

with the minimum of (2.10) being

$$E_{\min} = \mathbf{R}(\mathbf{r}_0,\mathbf{r}_0) - \mathbf{R}(\mathbf{r}_0,\mathbf{r}_j)\left[\mathbf{R}(\mathbf{r}_j,\mathbf{r}_k) + \mathbf{N}(\mathbf{r}_j,\mathbf{r}_k)\right]^{-1}\mathbf{R}^T(\mathbf{r}_k,\mathbf{r}_0) \tag{2.12}$$

The proof is not difficult. Good explanations are given by Liebelt (1967) and Luenberger (1969) and almost any book on linear statistical inference. Bretherton, Davis and Fandry (1976) discuss the oceanographic use of the theorem, and in particular relax the assumption $\langle y\rangle = 0$. The usual proof of the theorem does not require many of the special assumptions, (e.g. $\langle y_i n_j\rangle = 0$) we have made. The result is closely related to ordinary weighted least-squares (but Luenberger (1969) discusses some very important differences); in one dimension for stationary $\mathbf{R}$ and $\mathbf{N}$, it leads to the Wiener 'smoothing' filter.)

The consequences of this formalism and the Gauss-Markov theorem are several. First, it is a proof that the choice (2.11) for $\mathbf{B}$ is optimum; no other linear interpolator could do better under the stated conditions. (For Gaussian processes, it is possible to even show that no non-linear interpolator could do better; see Deutsch, 1965).

A corollary is the converse, that if we choose any other **B**, we are *not* optimum.

Many who encounter this theorem for the first time object to its use on the grounds that the assumption that one knows the system covariances **R**, **N**, is unrealistic. But in practice, even extremely crude information about the covariances can be used, and is very helpful. It is also probably true, that if one has no idea about the noise or solution statistics, then one has no business trying to interpolate and map, or in conducting large-scale measurement programs.

Often one does not want to use the optimal interpolator **B**, either because one cannot afford it, or because a near-optimal operator is much more efficient or adequate (see Gelb, 1974). The expression (2.10) still determines the actual interpolation error. For example, a simple two-point linear interpolation scheme

$$y(\mathbf{r}_0) = \frac{|\mathbf{r}_2 - \mathbf{r}_0|}{|\mathbf{r}_2 - \mathbf{r}_1|} y(\mathbf{r}_1) + \frac{|\mathbf{r}_1 - \mathbf{r}_0|}{|\mathbf{r}_2 - \mathbf{r}_1|} y(\mathbf{r}_2) \tag{2.13}$$

is obviously a special case of (2.5), with

$$\mathbf{B} = \left\{ \frac{|\mathbf{r}_2 - \mathbf{r}_0|}{|\mathbf{r}_2 - \mathbf{r}_1|}, \frac{|\mathbf{r}_1 - \mathbf{r}_0|}{|\mathbf{r}_2 - \mathbf{r}_1|} \right\} \tag{2.14}$$

The expected error of the interpolation is then (2.10) with the substitution of (2.14) and is a function both of the choice of interpolation scheme, and the true covariances. Whether two-point linear interpolation is adequate for some particular purpose can only be evaluated case-by-case.

Some simple extreme cases of one-dimensional interpolation are illuminating. Obviously if the true field varies strongly with curvature between two data points used for interpolation, a linearly interpolated value may be far from correct. If the noise present in the two data points is very large, the interpolation will similarly be a poor one.

The great power and versatility of (2.10) is that it accounts in the best way for the spatial structure of both the observational noise and the expected structure of the solution. One may have to compromise optimal procedures because they are not affordable, but can still evaluate the expected error of whatever scheme is being used. *A tracer map then is, in our definition, simply a two dimensional interpolation to a grid, sufficiently regular to be contourable, of a field, initially*

22

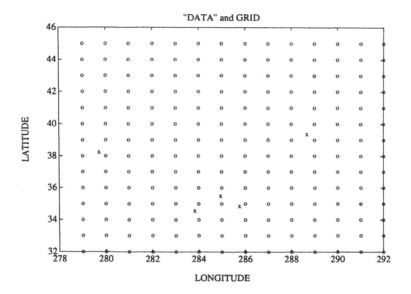

Figure 8a

'Observations' of a tracer C are supposedly made at positions denoted
×. This data is then mapped onto the uniform grid depicted for use in
a finite difference model. Although the ratio of number of grid points to
data points has been purposely exaggerated here, the general principle that
there will be a large variation on the accuracy with which this mapping
can be done is a universal one.

observed on an irregular pattern. To be useful, such a map is always accompanied by a map of the expected error (see Bretherton *et al.* (1976) for further discussion of technical details).

A map and its accompanying error are displayed in Fig. 8. The error field can easily be shown (from 2.10) to derive from two distinct sources: (1) noise in the data and (2) the absence of information.

The latter error source is best understood by supposing that the observations were all perfect, but that we were attempting to map (grid) at a point $\mathbf{r}_0$ located in a region far (as measured relative to the spatial scale of the covariance $\mathbf{R}$) from any observation. The mapped value in such a region will tend toward zero with an error approaching $\mathbf{R}(\mathbf{r}_0, \mathbf{r}_0) = \langle y(\mathbf{r}_0)^2 \rangle$ and is solely due to the absence of data. (Note that the maximum error one can have, in the mean square, is $\langle y(\mathbf{r}_0)^2 \rangle$. Any data serves to reduce this maximum possible uncertainty.)

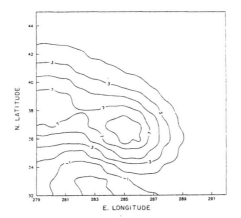

Figure 8b

Tracer concentration map constructed from Eq. (2.9) with specific choices of R and N using data points and grid as depicted in Fig. 8a. The estimated concentrations fade to zero in the upper right corner, because the spatial decorrelation scale of R was specified as short compared to the distance from this corner to the nearest data points. The estimator produces the value zero (estimated as the true mean), in the absence of any information.

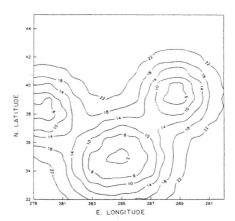

Figure 8c

Map of estimated mean-square error (diagonal elements of Eq. 2.12) corresponding to map in Fig. 8b. The asymptotic value of 25, far from the data, is the *a priori* estimate of the mean square value of C. Major point is the considerable non-uniformity of the accuracy of 8b. See Bretherton, Davis and Fandry (1976) for more details.

Roemmich (1983) used this method for mapping along hydrographic sections and Kawase and Sarmiento (1985) used it on the North Atlantic nutrient fields. One of the virtues of the ability to make maps is that one can grid arbitrarily spaced data onto uniform grids of the type appearing in finite difference formulations of the governing equation, a form with which many investigators are most comfortable. Although the procedure is practical (a good example may be seen in Tziperman, 1988) the use of finite difference equations may be ill-advised unless the error estimate accompanying the gridding is also employed. The error estimate of the Gauss-Markov theorem shows that unless the data were themselves obtained by uniform procedures on a uniform grid (Tziperman's data were almost of this form), then the gridded estimates will have potentially very uneven accuracies associated with them, leading to unpredictable results if treated as of equal accuracy.

Second, finite difference equations involve differentiation of data; indeed Laplacians, as in (1.2) necessitate second derivatives of the data. One of the early lessons of data handling is to avoid numerical differentiation if at all possible. The reasons for this stricture are easy to show. Consider the inverse form of (2.1), the field T written in terms of its Fourier transform.

$$T(x) = \int\limits_{-\infty}^{\infty} \exp(2\pi i s x)\hat{T}(s)ds$$

Then the first derivative of T is

$$T'(x) = \int\limits_{-\infty}^{\infty} (2\pi i s)\exp(2\pi i s x)\hat{T}(s)ds$$

and its second derivative is

$$T''(x) = \int\limits_{-\infty}^{\infty} (2\pi i s)^2 \exp(2\pi i s x)\hat{T}(s)ds$$

etc. for higher derivatives. The magnitude of the Fourier transform of the nth derivative is a factor $|2\pi i s|^n$ times that of the Fourier transform of T itself.

Differentiation (or its numerical analogue) thus tends to destroy information at small wavenumbers, or large scales, relative to information at high wavenumbers/small scales. Usually a mappable field will carry most of its useful information on the large scales, and tend to have increasing noise at the smallest scales (eddy noise, internal waves, measurement errors, etc.). Fig. 7b displays the horizontal wave number spectrum of the silica distribution in Fig. 7a. Differentiation reduces the energy at the low wave number end — where most of the structure lies. Thus any time one takes derivatives or differences, (analogous results can be derived for discrete derivatives), there is a strong tendency to suppress the information-carrying scales, amplifying the relative importance of the noise. The higher the derivative, the greater the effect. One is usually advised, if possible, to use the governing equations in integral form (notice that Eq. 1.6 is an integral form of the advection diffusion equation. Some further comments on this point may be found in Wunsch, 1985).

It may appear to the reader that one can arbitrarily specify the matrices $\mathbf{R}$, $\mathbf{N}$ in the mapping operator in (2.11). But these matrices must be consistent with the actual data, and if they are not, either the mapped field y or the errors as computed as the difference between the mapped and measured value, or both, will fail to have covariances consistent with the *a priori* estimates. If such an inconsistency arises, one must reject the map. A good exercise for the reader is to take a small data set and attempt to map it, then checking the consistency of the result with the prior covariances. Gaining consistency can be very difficult.

2.3 Determining a Mean Value

One of the important uses of certain transient tracers (fluorocarbons, tritium, etc.) is the determination of the total amount of the tracer in the ocean, or in a large sub-region, as a means of inferring rates of air-sea fluxes. The problem is to determine at a fixed time t, with suitable accuracy, the inventory of a tracer C within a volume V. The Gauss-Markov theorem provides appropriate machinery for a discussion of how well the estimate can be made.

Let the measurements of C within the volume be denoted y_i and suppose that each is made up of the large-scale mean, m, plus a deviation from that mean of θ_i, so that we can write

$$m + \theta_i = y_i, \quad i = 1 \text{ to } N$$

or

$$\mathbf{D}m + \theta = \mathbf{y}, \quad \mathbf{D}^T = [1, 1, .., 1] \tag{2.15}$$

We seek a best estimate, $\hat{m}$, of m.

The simplification of the Gauss-Markov theorem in Eq. (5-24) of Liebelt provides the appropriate recipe (cf. Leith, 1973). **R** is now defined as the spatial covariance of the measured tracer field about its true mean value. If one is attempting to estimate the mean in a field of statistically stationary eddies, then **R** contains the covariance of the eddy field plus the covariance of the measurement noise (including the chemical analysis errors). To emphasize this combined error rewrite $\mathbf{R} = \mathbf{R}_e + \mathbf{R}_n$ where $\mathbf{R}_e$ is the covariance of the eddy field, and $\mathbf{R}_n$ is everything else. Notice that the distinction between noise and signal is somewhat arbitrary. We are now choosing to treat the eddy field as noise. For other purposes, we may want to map on the eddy scale. Suppose an *a priori* estimate of the size of m exists, and is called m_0, i.e. $\langle m^2 \rangle = m_0^2$.

The best estimate of the mean (Liebelt, 1967, Eq. 5-26) can be written

$$\hat{m} = \left[\frac{1}{m_0^2} + \mathbf{D}^T \mathbf{R}^{-1} \mathbf{D} \right]^{-1} \mathbf{D}^T \mathbf{R}^{-1} \mathbf{y}$$

$$= \frac{1}{\dfrac{1}{m_0^2} + \mathbf{D}^T \mathbf{R}^{-1} \mathbf{D}} \mathbf{D}^T \mathbf{R}^{-1} \mathbf{y} \tag{2.16}$$

($\mathbf{D}^T \mathbf{R}^{-1} \mathbf{D}$ is a scalar). The expected error of this estimate is

$$E = \left(\frac{1}{m_0^2} + \mathbf{D}^T \mathbf{R}^{-1} \mathbf{D} \right)^{-1} = \frac{1}{\dfrac{1}{m_0^2} + \mathbf{D}^T \mathbf{R}^{-1} \mathbf{D}} \tag{2.17}$$

The expressions (2.16 and 2.17) reduce to familiar cases in certain limits. Let the θ_i be uncorrelated (independent fluctuations about the mean if the statistics are Gaussian), with variance σ^2, making **R** diagonal; then we have

$$\hat{m} = \frac{1}{\dfrac{1}{m_0^2} + \dfrac{N}{\sigma^2}} \frac{1}{\sigma^2} \sum_{i=1}^{N} y_i = \frac{m_0^2}{\sigma^2 + N m_0^2} \sum_{i=1}^{N} y_i \tag{2.18}$$

using

$$\mathbf{D}^T \mathbf{D} = N, \quad \mathbf{D}^T \mathbf{y} = \sum_{i=1}^{N} y_i$$

and,

$$E = \frac{1}{\dfrac{1}{m_0^2} + \dfrac{N}{\sigma^2}} = \frac{\sigma^2 m_0^2}{\sigma^2 + N m_0^2} \tag{2.19}$$

If we further assume that $m_0^2 \to \infty$

$$\hat{m} = \frac{1}{N} \sum_{i=1}^{N} y_i \tag{2.20}$$

and

$$E = \frac{\sigma^2}{N} \tag{2.21}$$

the ordinary average, with mean square error σ^2/N which is the usual 'square root of N rule' for the standard error of the mean. In the general case, the accuracy of the sample mean as a representation of the true mean thus depends upon the spatial characteristics of θ, the noise of measurement, any prior information on how big m might be, and the distribution of samples.

2.4 A Priori Information

The employment of the value m_0^2, formally representing $\langle m^2 \rangle$, is an example of the use of prior information to improve estimates of statistical quantities. Then if m_0 is finite, not only is the estimate of the mean different from its value otherwise, but the variance of the estimate is also reduced.

Prior information and its use raise subtle questions. When we let m_0 become arbitrarily large, this limit does not mean that we believe the true value of m is infinite, but only that it could be arbitrarily large (the conventional estimate (2.20) is finite). Taking the limit as $m_0 \to \infty$ is in effect a disclaimer of any knowledge whatever of the possible value of m, prior to the calculation. Such a situation will be perceived as highly unusual; normally we have some idea of what the range of possible values should be. The Gauss-Markov theorem says we should use that information. As N gets large, it is clear that the relative information content of the prior estimate diminishes until in the limit it becomes irrelevant — a useful result that says in the presence of infinite data, the dependence upon prior estimates disappears.

The covariance $\mathbf{R} = \mathbf{R}_e + \mathbf{R}_n$ is also the result of prior information of a statistical type. Many investigators are bothered by the notion that some form of prior information needs to be available. But it is not a question of need: if the investigator really has no idea about the noise structure, $\mathbf{R}$ can be replaced by $\delta^2 \mathbf{I}$ and δ^2 permitted to go to infinity. However, one is entitled to question the sense of an oceanographer who makes a tracer survey at sea having no idea whatever as to the error magnitudes and time/space structure, and no idea as to the size and structure of the field being mapped. One may choose to suppress this information in generating estimates from the data; but such a choice should be deliberate and not inadvertent.

Finally, note that having calculated (say) $\hat{m}$, one cannot go back and replace m_0^2 by $\hat{m}^2$ in (2.18), (2.19) thereby reducing the putative error estimate. The assumptions underlying the derivation of the Gauss-Markov theorem would then have been violated and the result would be meaningless.

3. SIMPLE ESTIMATION

3.1 Elementary Least-Squares

Most inference problems we face with tracers are formally underdetermined. There is a widespread perception that ordinary least-squares problems, which are usually formally overdetermined, are so much more powerful for inferring model unknowns, that one should render tracer (and other) problems overdetermined, at almost any cost. To understand why this impression is erroneous, it is useful to go back and consider elementary least squares.

Ordinary least-squares estimation is very familiar — so familiar that its profundity is often overlooked. Risking boredom, consider the problem motivated by the 'data' shown in Fig. 9.

We begin with a familiar simple estimation problem. Let t be an independent variable, which might be time, but could equally well be a space coordinate or simply an index. Suppose that a set of observations of a physical variable, call it $\theta(t)$, perhaps temperature at a point in the ocean, has been made at positive t_n, $n = 1$ to N, as depicted in Fig. 9a. Call the measurements $y(t_n)$.

We have reason to believe that there is a linear relationship between $\theta(t)$ and t in the form

$$\theta(t) = a + bt \tag{3.1}$$

$$y(t) = \theta(t) + n(t) = a + bt + n(t) \tag{3.2}$$

where $n(t)$ is the 'noise', at least in part due to observation inaccuracies, and we want to find a, b.

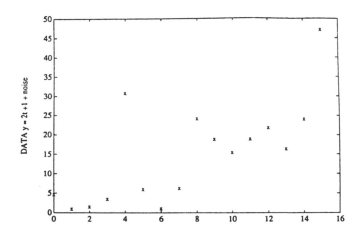

Figure 9a
'Data' generated from the rule $y = 1 + 2t +$ noise.

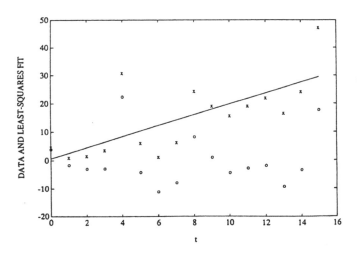

Figure 9b
Least squares fit to data of Fig. 9a, and the residuals (open circles) left
by the fit. These residuals would pass a test for being indistinguishable
from white noise of uniform variance, thus justifying the conclusion that
the linear fit is adequate to explain the data.

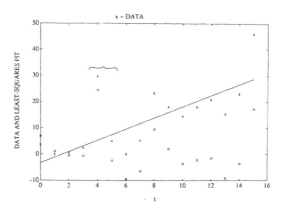

Figure 9c
Least-squares fit to data of Fig. 9a, except that 'measurements' of y at times 4–6 were supposed to be noisier than at other times. The fitted straight-line parameters are different from the values in Fig. 9b, supporting the conclusion that least-square fits are non-unique and dependent upon detailed a priori assumptions about the noise statistics. (One would necessarily have to check the a priori hypothesis that these three measurements have a greater noise level than the others. Probably one could not reject the hypothesis with only three numbers available.)

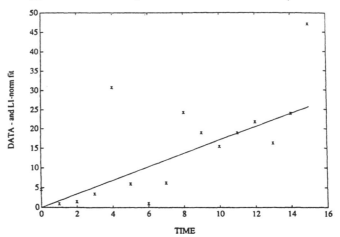

Figure 9d
Fit of the data of Fig. 9a by minimizing the average absolute deviation of the measurement rather than the average square. Such a fit is more 'robust' than ordinary least-squares, being insensitive to outliers. In effect however, a different assumption about the noise statistics has been made, and a different fit results.

One is taught at an early age to use least-squares, so let's see where that leads. The set of observations can then be written

$$a + bt_1 + n(t_1) = y(t_1)$$
$$a + bt_2 + n(t_2) = y(t_2)$$
$$\vdots$$
$$a + bt_N + n(t_N) = y(t_N) \qquad (3.3)$$

To condense the notation, define a matrix

$$\mathbf{A} = \begin{pmatrix} 1 & t_1 \\ 1 & t_2 \\ 1 & t_3 \\ \vdots & \vdots \\ 1 & t_N \end{pmatrix} \qquad (3.4a)$$

a vector of unknowns

$$\mathbf{q} = \begin{pmatrix} a \\ b \end{pmatrix} \qquad (3.4b)$$

and a data vector

$$\mathbf{y} = \begin{pmatrix} y(t_1) \\ \vdots \\ y(t_N) \end{pmatrix} \qquad (3.4c)$$

so,

$$\mathbf{Aq} = \mathbf{y} - \mathbf{n} \qquad (3.5)$$

One is taught, following Legendre and Gauss, that a sensible thing to do is to find a value $\hat{\mathbf{q}}$ that minimizes the sum of the squares of the differences between $\mathbf{Aq}$ and $\mathbf{y}$, in the form

$$\text{min}: \quad \delta = \sum_{j=1}^{N} (a + bt_j - y(t_j))^2 \qquad (3.6)$$

Using the matrix notation, (3.6) can be written

$$\text{min}: \quad (\mathbf{A\hat{q}} - \mathbf{y})^T (\mathbf{A\hat{q}} - \mathbf{y}) \qquad (3.7)$$

Differentiating (3.6) or (3.7) with respect to a, b, or q and demanding that the derivatives vanish (anticipating a minimum rather than a maximum) we have

$$\mathbf{A}^T \mathbf{A\hat{q}} = \mathbf{A}^T \mathbf{y} \qquad (3.8)$$

Assuming $(\mathbf{A}^T\mathbf{A})^{-1}$ exists,

$$\hat{\mathbf{q}} = (\mathbf{A}^T\mathbf{A})^{-1}\mathbf{A}^T\mathbf{y} \tag{3.9}$$

The fit is displayed in Fig. 9b. This classical result is deceptively simple. It is deceptive because we should ask a great many questions about its meaning.

Consider a few of them. We are looking for two unknowns $\mathbf{q}^T = [a, b]$. We therefore needed only two observations, not $N > 2$ of them, to find a, b. (1) What really was the purpose of making all the excess observations? (2) Did any of the observations count more in finding a, b than some of the others? (3) Can we be sure that there is actually a solution to (3.8), even though we were so mathematically definite in writing $(\mathbf{A}^T\mathbf{A})^{-1}(\mathbf{A}^T\mathbf{A}) = \mathbf{I}$? (4) Is there some way to use the solution to tell if the model (3.1) were 'right'? (5) Suppose before we began, we thought that it was reasonably likely that $a \sim 4$; is there some way that this belief could be used?

Before grappling with these questions, consider the specific solution in Fig. 9b. Substituting $\hat{\mathbf{q}}$ back into the original equations, and subtracting the result from the right hand side (the observations), we obtain the 'residuals', plotted in Fig. 9b. The reader may have noticed, and be troubled by the following fact: we have not only determined the two formal problem unknowns, a, b, but we have also been able to calculate N additional numbers we did not have before — the N values of the noise occurring in each observation. That is to say, we started with N equations in, apparently, 2 unknowns, but have somehow determined $N + 2$ numbers we did not have before. How is this possible?

To answer this question, notice that the formal statement of the problem was to determine a, b from the set of equations (3.3) which we could write alternatively as

$$\mathbf{A}\mathbf{q} + \mathbf{n} = \mathbf{y} \tag{3.10}$$

or

$$\mathbf{A}_1\mathbf{q}_1 = \mathbf{y} \tag{3.11}$$

$$\mathbf{A}_1 = [\mathbf{A}, \mathbf{I}_N], \quad \mathbf{q}_1^T = [a, b, \mathbf{n}^T]$$

This latter form, (3.11), makes explicit that the noise must be regarded as something we need to determine, and the superficially overdetermined least-squares equation can always be regarded as underdetermined because each observation equation always contains a minimum of one new unknown, n_i, in addition to the model

parameters. It will be shown below how to solve this underdetermined problem
directly; I am not advocating that one should actually do it this way in practice,
but it has the great advantage of making it explicit that the noise is part of the solution and must be scrutinized just as carefully as the parameters we were explicitly
seeking.

How has this underdetermined problem been apparently reduced to one yield-
ing a unique solution? The answer is simple: a very special statistical assumption
was made, in the minimum (3.6) and (3.7) — that the noise subset of the unknowns were statistically uncorrelated with each other, had a constant variance
and zero mean, and should take on the smallest possible mean square value. Stated
mathematically, we made the *a priori* assumptions

$$\langle n_i \rangle = 0, \quad \langle n_i n_j \rangle = \sigma_0^2 \delta_{ij} \tag{3.12}$$

the bracket denoting an ensemble average, and σ_0^2 is a constant.

These statistical assumptions were implicit in the minimizations (3.6). Because
the user of least-squares may be unaware of it, does not mean that the assumptions
have not been made. It should now be clear that the answer to the least-squares
problem is unique only because of *a priori* statistical assumptions. If we change the
statistical assumptions in any way, we change the answer. For example, suppose
that the three bracketed observations in Fig. 9c are thought to have a higher noise
variance than the others, by (say), a factor of 100. Then the solution changes to
that shown in Fig. 9c. (Later we will see how to solve the problem if the noise is
correlated between successive observations.)

The minimization (3.6) can be shown (see Deutsch, 1965) to be a maximum
likelihood estimate under the assumption that the noise statistics are Gaussian.
That is, it finds those values a, b which make the observations the most likely to
have been observed. But if there was reason to believe that (for example), the noise
statistics had an exponential probability density, then we would be better advised
to minimize

$$\delta_1 = \sum_{i=1}^{N} |y_i - (\mathbf{A}\mathbf{q})_i| \tag{3.13}$$

(see Arthnari and Dodge, 1981). An advantage of this latter minimization is its
insensitivity to noise outliers ('blunder points'); it is a so-called robust estimator.
The solution obtained for the data of Fig. 9a is shown in Fig. 9d and is different

from the least squares one. Another useful possibility is the so-called min/max estimate; find

$$\min : \max \left| y_i - \sum_j A_{ij} q_j \right| \tag{3.14}$$

It should now be clear that the warm, intuitive feeling that somehow least-squares over-determined problems render unique answers is unjustified. The uniqueness is the direct result of the (possibly unwitting) statistical assumptions made about the observational noise. This point is belabored because the slightly more complex solution for an underdetermined problem (taken up in Section 3.2) has been misunderstood by many as somehow much more uncertain than the supposedly overdetermined one. We will return to this important issue later.

We now have an estimate of all $N + 2$ unknowns of the original problem; many investigators stop at this point — but that is a mistake. Two things at least, remain to be done. We need to understand how well we have estimated $\mathbf{q}^T = (a, b)$, and we should really ask whether there is any evidence that the model we used, the straight line, had anything to do with the data we fit to it. Eq. (3.10) will usually always give some non-zero estimate for a, b, no matter what the data look like. Merely finding values is the beginning of least squares estimation, not the end.

Consider some examples depicted in Fig. 10. In 10a, the observations are so noisy that one might well question whether the best straight line has any meaning at all. In 10b, one's eye suggests that perhaps the data are not fit by a straight line but by some other relationship, i.e. some other model. The situation in 10b is particularly worrisome; indeed the data were generated using the rule

$$\mathbf{y}(t) = \theta(t) + n(t) = a + ct^2 + n(t)$$

But we fit a straightline to it as shown — and the result is clearly erroneous in the sense that our model is 'wrong'. While 10b is extreme, one might be bothered a bit more by 10c, where the noise is sufficiently great that by eye, one cannot really tell that an incorrect model has been fit. Does that mean that least squares methods have failed? Regarding this situation as a failure of least squares would be silly — what has really happened is that we have failed to finish the job, using tools that least squares provides us with. We will show below that the statistical uncertainty of our solution can be estimated from the rule

$$\langle (\hat{\mathbf{q}} - \mathbf{q})(\hat{\mathbf{q}} - \mathbf{q})^T \rangle \equiv \mathbf{E} = (\mathbf{A}^T \mathbf{A})^{-1} \langle \mathbf{n}\mathbf{n}^T \rangle \tag{3.15}$$

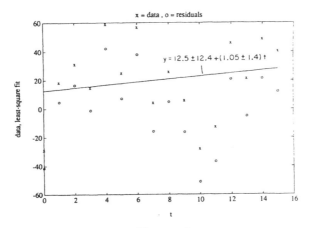

x = data , o = residuals

$$y = 12.5 \pm 12.4 + (1.05 \pm 1.4)\,t$$

data, least-square fit

t

Figure 10a
Linear fit to linear data of Fig. 9a, except the noise level has been raised
so far (variance 30^2; sample variance 26^2) that the parameters of the best-
straight line are indistinguishable from zero. These data are inadequate
either to confirm or to refute the hypothesis of a linear law; such a result
is not a failure of least-squares, but rather a failing of the data.

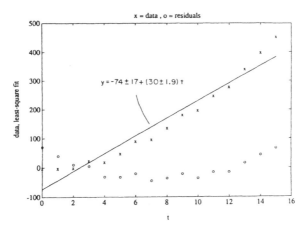

x = data , o = residuals

$$y = -74 \pm 17 + (30 \pm 1.9)\,t$$

data, least-square fit

t

Figure 10b
Linear fit to data $(\times)$ generated from a quadratic law $y = 1 + 2t^2 +$ noise,
where the noise variance was 10^2. Although the residuals left are reasonably small, they show systematic effects (a negative bias and a curvature)
suggesting that one should reject the linear explanation of the data.

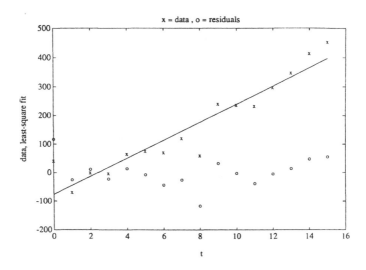

Figure 10c
Same as 10b, except that the noise level has been raised sufficiently (variance 40^2) that one can no longer tell definitively by eye that the linear fit should be rejected. More sophisticated statistical tests on the residuals would show that there is suspiciously much structure in the residuals.

Notice that the estimate depends upon the prior estimate of the noise in the data as well as the coefficient matrix $\mathbf{A}$. It does not depend upon the data values themselves. The diagonal elements of $\mathbf{E}$ are the standard errors of the estimates of q_i; the off-diagonal elements are the covariances of q_i and q_j.

If upon examining the result of the least squares estimates shown in Figs. 9 or 10, we find that the standard error shows that a, b are indistinguishable from zero (a common reference point) what have we learned? To a great extent, the answer depends upon what we were after. If the question was whether there was any linear relationship between $\mathbf{y}, t$ we may regard as very important the conclusion 'there is no evidence for such a connection'. If we badly need non-zero values of a, b to make a prediction of $\mathbf{y}$ at some future time, then we must conclude, perhaps regretfully, that the data are inadequate to permit any such useful prediction. If we are testing the notion that $a = a_0$, $b = b_0$, then there are further possibilities. If $|a - a_0| < \sigma_a$; $|b - b_0| < \sigma_b$ where σ_a, σ_b are the respective diagonal elements of $\mathbf{E}$, we can conclude that 'there is nothing in the data that conflicts with the hypothesis'.

If to the contrary, $|a - a_0| \gg \sigma_a$, then noisy as the data are, they may nonetheless permit us to reject the hypothesis $a = a_0$. We are still not done however. The values of $\langle nn^T \rangle$ appearing in (3.15) were based upon our prior estimate of the noise statistics in the measurement of mean 0, and variance σ_0^2. Suppose we calculate each noise value n_i, and their sample mean

$$\bar{n} = \frac{1}{N} \sum_{i=1}^{N} n_i$$

and sample standard deviation

$$s^2 = \frac{1}{N-1} \sum_{i=1}^{N} (n_i - \bar{n})^2$$

Now we do not expect to find $\bar{n} = 0$ or $s^2 = \sigma_0^2$ (the probability that they will be equal to the prior value is vanishingly small). Nonetheless, if the prior estimates of the noise mean and variance are close to correct, then $\bar{n}$ and s^2 should not deviate very far from our initial estimates. We can use standard statistical methods (e.g. Seber, 1977; Cramér, 1946) to decide how close we should come in having the sample averages equal the true averages. To take an extreme case, suppose we find the sample noise variance is 5 times the prior estimate, and that the probability of that occurring by chance is 1 in 1000. What is the intrepretation? The conclusion is that our entire picture of what we are doing is incorrect in some fundamental way; somehow the noise in our measurements appears to be 5 times larger than we thought it was, strongly suggesting that the physical basis of what we are doing is *wrong* in some way. At this stage, the entire solution including our estimate of a, b should be thrown out or at least regarded with grave doubts, and the problem re-examined *ab initio*. This point is belabored because the statistical estimates of the noise are as much a part of our model as is the straightline assumption, and we have just rejected this statistical component. If the statistical component is incorrect, it should make us have grave doubts about all the rest. The problem is not completed until this examination of the residuals is undertaken.

Residuals, although boring, are often the most informative part of the whole analysis.

If one fits a straight line to data generated from a rule

$$y = a + bt + ct^2 \tag{3.16}$$

(Fig. 10), then the least-squares procedure produces a definite value for a, b. That these values are incorrect is determined by examination of the residuals of the fit (Fig. 10b). One sees, by eye in this case, that there is a structure (a curvature) to the residuals, strongly suggesting that they deviate from our specific statistical hypothesis that the residuals should be completely unstructured ('white noise'). Formal statistical tests exist to test for such non-random residuals (e.g. tests applied to the autocorrelation of the residuals), but these are discussed in texts on regression and we will not pursue them further here. Suffice it to say that structured residuals are a strong warning that the model being used is insufficiently complex to explain the data.

Suppose that the data really did come from rule (3.16) but that our statistical tests fail to show a significant deviation of the residuals from white noise? Has the procedure failed? The answer is definitely 'no', but raises the question of how any result of statistical inference should be described: we would have shown that our model is wholly consistent with the data. We never claim that a model has been 'verified' — that would be a foolish claim — someone may come along tomorrow with a new observation which shows the model to fail. The conclusion to draw is however, a very powerful one: nothing in the data requires a more complex hypothesis than the linear relationship and the data will neither prove nor disprove the necessity of something more elaborate.

3.2 Underdetermined Systems

What does one do when the number of equations is less than the number of unknowns and no more observations are possible? The conventional response to a situation like this one is to find some way of reducing the number of unknowns, so that the system at least appears to be fully determined. Such a problem parameter reduction procedure may be sensible, but is fraught with pitfalls. In the problem described in section 3.1, one may have fewer observations than the order, m, of the polynomial believed to be necessary to capture the structure of the data. If the data came from a law

$$y = 1 + a_m t^m \qquad (3.17)$$

an attempt to fit

$$y = \sum_{j=0}^{m-1} a_j t^j \qquad (3.18)$$

may give a very good fit, but be totally incorrect. One is often better off (as will be shown) retaining the underdetermined system, and making inferences concerning the possible values of a_m rather than using the form (3.18) in which any possibility of learning something about a_m has been eliminated. In more general terms (discussed by Wunsch and Minster, 1982) parameter reduction can lead to model errors, i.e. bias errors, which can give wholly illusory results. A specific example is provided in Wunsch (1988a); there a two dimensional ocean model gave very specific values for the rates of apparent oxygen utilization (AOUR). But when the parameterization was made more realistic in scope (a three dimensional parameterization), it was shown that AOUR was essentially undeterminable to within limits which are of any interest. The conclusions drawn from the underparameterized model are erroneous; the virtue of the second model, even though no numerical value is found, is that it prevents one from drawing an erroneous conclusion.

A closely related problem is one in which one appears to have an overdetermined system as in (3.3), but the equations are not all independent in practice — meaning that the conventional inverse of $\mathbf{A}^T \mathbf{A}$ does not exist or is ill-conditioned. In large, complex systems, one may not know ahead of time whether the system is in fact overdetermined as it appears, or is ill-conditioned.

Among several available methods for solving inverse problems, one of the most popular, powerful and illuminating is based upon a remarkable factorization of the matrix $\mathbf{A}$, usually called the 'singular value decomposition' (SVD). Apart from its application to inverse problems, the SVD has the property of uniting in one simple piece of mathematical machinery the discussion of over, under, and just determined systems — further undermining any sense that there is a fundamental distinction between them.

The singular value decomposition asserts that an arbitrary $M \times N$ rectangular matrix $\mathbf{A}$ can be written as

$$\mathbf{A} = \mathbf{U}\Lambda\mathbf{V}^T \tag{3.19}$$

where

$$\Lambda = \begin{pmatrix} \lambda_1 & 0 & . & . & . & . & . & 0 \\ 0 & \lambda_2 & . & . & . & 0 & . & . \\ . & . & . & . & . & . & . & . \\ . & . & . & . & . & . & . & . \\ . & . & . & . & . & . & . & . \\ 0 & . & . & . & . & \lambda_k & . & 0 \\ . & . & . & . & . & . & . & . \\ 0 & 0 & . & . & . & 0 & . & 0 \end{pmatrix}$$

is $M \times N$ with diagonal elements called the 'singular values'. The columns of matrix $\mathbf{U}$ are formed from the orthonormal vectors $\mathbf{u}_i$ and the columns of $\mathbf{V}$ from the orthonormal $\mathbf{v}_i$. There are M of the $\mathbf{u}_i$ vectors of length M and N of the $\mathbf{v}_i$ vectors each of length N. Thus the two sets each forms a complete spanning basis in the two different dimensions of the $\mathbf{A}$ matrix.

Among the important properties of this system are the following:

$$\mathbf{A}\mathbf{v}_i = \lambda_i \mathbf{u}_i, \quad 1 \le i \le k \quad (3.20\text{a}) \qquad \mathbf{A}^T \mathbf{u}_i = \lambda_i \mathbf{v}_i \quad 1 \le i \le k \quad (3.20\text{b})$$

$$\mathbf{A}\mathbf{A}^T \mathbf{u}_i = \lambda_i^2 \mathbf{u}_i \qquad\qquad (3.20\text{c}) \qquad \mathbf{A}^T \mathbf{A} \mathbf{v}_i = \lambda_i^2 \mathbf{v}_i \qquad\qquad (3.20\text{d})$$

$$\mathbf{A}\mathbf{v}_i = 0, \quad k+1 \le i \le N, \quad (3.20\text{e}) \qquad \mathbf{A}^T \mathbf{u}_i = 0 \quad k+1 \le i \le M \quad (3.20\text{f})$$

$$k \le \min(M, N)$$

Lanczos (1961) has a particularly clear account of this system which we do not have the space to repeat. An important point however, is that it follows from equations (3.20a,b) that the maximum number of λ_i which can be non-zero is the minimum of M and N, and the actual number may be smaller than both. This number, k, is the system rank. The SVD permits discussion of all combinations of over and underdetermined systems, including underdetermined systems that are not of full rank. A system for which the rank is less than the number of equations, and less than N, is said to be 'rank-deficient'. Such a system is one in which the number of useable equations is less than the apparent number of equations. In problems involving data (as opposed to mathematical problems), considerable attention must be paid to singular values which are effectively, as opposed to mathematically, zero. We will take up this issue below. Notice that by (3.20b), the mapping from the observation space of the $\mathbf{u}_i$ vectors to the solution space of the $\mathbf{v}_i$ vectors is carried out by the matrix $\mathbf{A}^T$. We can anticipate, as will be seen, that the most natural inverse of $\mathbf{A}$ is intimately related to its transpose.

Suppose that $k < M$ and N. It follows immediately from (3.19), that the last $M - k$ columns of $\mathbf{U}$ and the last $N - k$ columns of $\mathbf{V}$ are multiplied by zero, permitting us to drop from the SVD (3.12) both these sets of columns. Λ is then $k \times k$ square and diagonal, $\mathbf{U}$ is $M \times k$ rectangular, and $\mathbf{V}$ is $N \times k$ rectangular. Unless otherwise stated, the use of the SVD (3.19) implies this collapsed form. Because of the orthonormality of the columns of $\mathbf{U}$, $\mathbf{V}$, we have

$$\mathbf{V}^T \mathbf{V} = \mathbf{I}, \quad \mathbf{U}^T \mathbf{U} = \mathbf{I}$$

but $\mathbf{VV}^T \neq \mathbf{I}$ and $\mathbf{UU}^T \neq \mathbf{I}$ unless $k = N$ or M respectively. As we will see however, the dropped columns nonetheless continue to play an extremely important role in the inverse method.

Proof of the SVD is easy (see Lanczos, 1961; Strang, 1986; Wunsch, 1978) and is not reproduced here. It follows immediately from the theorem that says real, symmetric or Hermitian matrices, have a complete set of real eigenvalues and a complete (spanning) set of orthonormal eigenvectors and is applied to the matrix,

$$\begin{pmatrix} 0 & \mathbf{A}^T \\ \mathbf{A} & 0 \end{pmatrix}$$

(This important theorem is apparently due to Eckart and Young (1939); it may represent Carl Eckart's most enduring contribution to oceanography.)

In practical use, it is important to recognize that the size of the eigenvector problem which must be solved is the smaller of (3.20c,d), the $\mathbf{u}_i$ vectors being determined from the $\mathbf{v}_i$ by (3.20a), or *mutatis mutandis,* the $\mathbf{u}_i$ being determined from the $\mathbf{v}_i$ by (3.20b), $i = 1$ to k, and for $i > k$ by a Gram-Schmidt process.

The power of the SVD lies in its capability to solve and describe completely arbitrary linear algebraic systems, with exhaustive display of the information content of the system both structural and statistical, including not only the adequacy of the model itself, but of the relative importance of each of the observations.

Consider now any set of algebraic equations

$$\mathbf{Ax} = \mathbf{y} \tag{3.21}$$

$\mathbf{A}$ being $M \times N$, $\mathbf{x}$ is $N \times 1$, $\mathbf{y}$ is $M \times 1$, and where the relative size of M, N and rank k are not specified. Using the SVD, we can write the solution to (3.21) in a number of equivalent ways. One way is to take advantage of the completeness of the $\mathbf{v}_i$ vectors, meaning we can express the correct solution as

$$\mathbf{x} = \sum_{i=1}^{N} \alpha_i \mathbf{v}_i \tag{3.22}$$

where the α_i must be determined. Because the $\mathbf{u}_i$ are complete too, we can write the observations $\mathbf{y}$ as

$$\mathbf{y} = \sum_{i=1}^{M} \beta_i \mathbf{u}_i \tag{3.23}$$

where $\beta_i = \mathbf{u}_i^T \mathbf{y} = \mathbf{u}_i \cdot \mathbf{y}$

It then follows immediately from substituting (3.22) into (3.20a) that

$$\mathbf{x} = \sum_{i=1}^{k} \frac{\mathbf{u}_i^T \mathbf{y}}{\lambda_i} \mathbf{v}_i + \sum_{i=k+1}^{N} \alpha_i \mathbf{v}_i \qquad (3.24)$$

(assuming for definiteness that the system is underdetermined).

The $\mathbf{v}_i$ vectors in the second term on the right of (3.24) are called the null-space of $\mathbf{A}$, and satisfy (3.20e), corresponding to the singular values which vanish. No information is available, in the problem as stated, about the coefficients of these null-space vector parts of the solution. Eq. (3.20e) says that the structures corresponding to these vectors have actually been annihilated in the process of observation (the reader should compare this statement to the discussion of the indeterminate constant of the Neumann problem discussed in section 1). If a problem is underdetermined, then there will be a null-space in $\mathbf{v}_i$ — the solution space. If the problem is truly over-determined, then the null space is part of the $\mathbf{u}_i$, or observation space and belongs to $\mathbf{A}^T$ as in (3.20f)

In a fully-determined square problem there is no null space of either solution or observation. In a rank-deficient overdetermined problem, there will be null spaces of both. Although we may not (yet) be able to say anything about the missing null-space vector coefficients, we can nonetheless calculate the solution nullspace vectors themselves — and thus describe those features of the full solution which we have not determined. Knowing what we have not been able to determine is obviously the next best thing to actually determining it.

Understanding of the null-space is critical to interpretation of the solution. Its structure is analogous to the indeterminate structures in the partial differential equations considered in section 1. One can also think of the first term on the right of (3.24) as an analogue of the particular solution of a differential equation, and the second term as analogous to the homogeneous solution, which must be determined using additional information such as boundary conditions or radiation conditions.

If the system is overdetermined, the only null space lies in the observation space. Eq. (3.23) remains exactly correct, but if $M > N$, and $k = N$, only the first N of the coefficients in (3.23) appear in the solution (3.24). What do the remaining $M - k$ represent? Substituting solution (3.24) back into the original equations we have

$$\mathbf{A}\mathbf{x} = \sum_{i=1}^{N} \frac{\mathbf{u}_i^T \mathbf{y}}{\lambda_i} \mathbf{A}\mathbf{v}_i = \sum_{i=1}^{N} (\mathbf{u}_i^T \mathbf{y}) \mathbf{u}_i \overset{?}{=} \sum_{i=1}^{M} (\mathbf{u}_i^T \mathbf{y}) \mathbf{u}_i = \mathbf{y}, \quad M > N$$

Clearly left and right-sides are equal, if and only if

$$\mathbf{u}_i^T \mathbf{y} \equiv 0, \quad i = M - N + 1 \text{ to } M \tag{3.25}$$

The demand (3.25) thus represents a consistency, or solvability, condition on the over-determined set. If (3.25) is not satisfied for all $i = k + 1$ to M, the solution leaves a residual in the governing equations, and is the noise left as residuals in section 2. In the rank deficient case, there will be both a solution null space, and either a solvability condition or (more realistically) a system residual.

Although we will not dwell on it here, all the discussion of the structure of the residuals in section 3.1 applies to the residuals left by these solutions, including the underdetermined systems. That is, the values of the residuals must be tested against the prior estimates of their magnitudes and statistical structure before the solution and the model could be deemed acceptable. Even grossly underdetermined models may thus be rejectable with available data if the residuals fail to behave properly.

From here we need to treat the solution in two pieces, corresponding to the two different sums on the right of 3.24. Set $\alpha_i = 0$, $i = k + 1, M$, and denote this solution with the nullspace contribution set to zero as $\tilde{\mathbf{x}}$. We will call it the 'SVD solution'.

3.3 Errors in the Observations

The effects of noise in the observations can be anticipated from (3.24). Notice that if $\mathbf{u}_i^T \mathbf{y}$ is not perfectly calculated, there will be an error in the corresponding coefficient α_i; if the expected size of this error is small compared to λ_i, then the solution is insensitive to the error. On the other hand, if the expected error is large compared to λ_i, the error can be greatly amplified. Size of singular values is measured against the noise level. To compute the expected error in the solution, we have

$$\tilde{\mathbf{E}} = \langle (\hat{\mathbf{x}} - \mathbf{x})(\hat{\mathbf{x}} - \mathbf{x})^T \rangle = \sum_i \sum_j \frac{\mathbf{u}_i^T \langle \mathbf{nn}^T \rangle \mathbf{u}_j}{\lambda_i \lambda_j} \mathbf{v}_i \mathbf{v}_j^T$$

$$= \sum_i \sum_j \frac{\sigma^2 \delta_{ij}}{\lambda_i \lambda_j} \mathbf{v}_i \mathbf{v}_j^T = \sum_i \frac{\sigma^2}{\lambda_i^2} \mathbf{v}_i \mathbf{v}_i^T \tag{3.26}$$

and the simplifying assumption that $\langle \mathbf{nn}^T \rangle = \sigma^2 \mathbf{I}$ was made (noise uncorrelated from one equation to another). Wiggins (1972) discusses the more general (and realistic) situation. Plugging the SVD into (3.15) reduces that error estimate to (3.26).

The strong uncertainty introduced by small λ_i is now clearly apparent. One strategy for reducing the variance is to truncate the sum (reduce the system rank) at a value k beyond which the solution variance grows too large to be acceptable. Singular values can then be effectively zero, if they are smaller than an appropriate measure of the noise. Another possibility is to taper down the contributions to the solution of the highly uncertain terms dependent upon small λ_i. Wiggins (1972) and Wunsch (1978) discuss both these strategies; the second strategy is a type of 'ridge-regression'.

3.4 Resolution

In any over, or underdetermined system one is not finished with the analysis until the error estimates owing to the noise are attached to the solution as described in Eq. (3.26). This error estimate is the one discussed in all textbooks on linear regression, or least-squares estimation (although it is not usually written in this form) and represents an uncertainty of the solution.

The solution nullspace represents elements of the solution about which we lack information. In the absence of any further information about these structures (and if we had information about them we should have included it in the model to begin with) the best estimate we can make about the missing α_i is zero. This choice can be defended on several independent grounds; the most compelling is simply to say that if our model is unbiassed, the expected value of any of the coefficients is zero, and making this choice leads to the simplest solution consistent with the observations. The idea of 'simplest' in turn derives from the empirical observation that as the λ_i diminish, the corresponding $\mathbf{v}_i$ often tend to get more 'wiggly'; and thus one refrains from adding structures not actually demanded by the data (Occam's razor). On the other hand, there are simple counterexamples: the indeterminate constant of the Neumann problem represents a null-space which is a constant and thus has no structure at all. We can still argue that a solution to the Neumann problem with the unknown constant set to zero is simpler than any other choice.

Let us agree to accept momentarily the solution with zeroed solution nullspace. One might then wonder whether the structure of the nullspace is such that some of the elements of $\tilde{\mathbf{x}}$ were nonetheless fully determined (except for the noise uncertainty discussed above), while some were completely indeterminate, and some partly determined. The quest to find a measure of degree of determination of the individual elements of $\mathbf{x}$ (as opposed to the discussion in terms of the vectors $\mathbf{v}_i$), leads to the concept of 'resolution', which in the context of the finite dimensional methods being used here, is probably most clearly discussed by Wiggins (1972).

One way to motivate the final result is to pose the following question: suppose the true solution was unity in element j_0 of $\mathbf{x}$, and zero everywhere else; could we tell? If we can tell with hundred percent confidence, then we would say element j_0 is 'fully resolved'. Otherwise it is partially, or unresolved. Suppose that the effective rank of the system has been detemined to be k. The correct representation of this true solution would be

$$\mathbf{x} = \delta_{j_0} = \sum_{i=1}^{N} \gamma_i \mathbf{v}_i$$

where δ_{j_0} is defined as the vector with unity in element j_0 and 0 everywhere else, and

$$\gamma_i = \delta_{j_0}^T \mathbf{v}_i = V_{ij_0} \quad (i.e. \text{ element } j_0 \text{ of vector } \mathbf{v}_i) \tag{3.27}$$

so that

$$\delta_{j_0} = \sum_{i=1}^{N} V_{ij_0} \mathbf{v}_i \tag{3.28}$$

But with an effective rank of k, the last $k + 1$ to N coefficients in (3.28) are indeterminate and the best we could do is

$$\delta_{j_0} \sim \sum_{i=1}^{k} V_{ij_0} \mathbf{v}_i = \text{row } j_0 \text{ of } \mathbf{V}\mathbf{V}^T \tag{3.29}$$

The deviation of (3.29) from a true Kronecker delta at element j_0 measures the deviation from full resolution.

A little algebra shows that the SVD solution, $\tilde{\mathbf{x}}$, is a filtered version of the true solution $\mathbf{x}$ in the form

$$\tilde{\mathbf{x}} = \mathbf{V}\mathbf{V}^T \mathbf{x} \tag{3.30}$$

The matrix $\mathbf{V}\mathbf{V}^T$ is called, following Wiggins (1972), the 'resolution matrix'. If $\mathbf{V}\mathbf{V}^T$ is the identity matrix, $\mathbf{I}_N$, then the solution is fully determined, and all solution parameters are fully resolved. If $k < N$, $\mathbf{V}\mathbf{V}^T \neq \mathbf{I}$ and the SVD solution has determined the element $\mathbf{x}_{j_0}$ in linear combination with some or all of the remaining elements. Study of the resolution matrix for small systems is rewarding. If the elements of $\mathbf{x}$ represent physically adjacent variables, and if $\mathbf{V}\mathbf{V}^T$ is diagonally dominant, then following Wiggins, we have 'compact resolution', and the interpretation is particularly simple — we determine not individual $\mathbf{x}_{j_0}$, but a local average of $\mathbf{x}_{j_0}$. Unfortunately, in any given problem, there is no guarantee that resolution will be compact, and in the examples described by Wunsch (1978) and Wunsch and Grant (1982), resolution is not normally so nice (but Rintoul, 1988, found compact resolution in his solutions). These examples tell one that the information available does determine certain linear combinations of $\mathbf{x}_i$; the precise combinations are determined by the geography of the water mass structures of the system.

If an *a priori* statement about expected solution magnitude is available, then one can convert the resolution into an equivalent error. Suppose $\langle \mathbf{x}_{j_0}^2 \rangle = D$; and the mean square of the SVD solution is a fraction

$$\tilde{\mathbf{x}}_{j_0}^2 = \Delta, \quad D \geq \Delta$$

Then the fractional error owing to the failure to resolve is approximately

$$\frac{D - \Delta}{D} = 1 - \frac{\Delta}{D} \tag{3.31}$$

The total estimation error can then be written as the sum of the error (3.31) and the diagonal element of (3.26); the mapping error (2.10) derived from the Gauss-Markov theorem is automatically the sum of these two errors. The Gauss-Markov estimate is often easier to use; however, the resolution analysis contains a great deal of information concerning the solution, and how it might be improved with further observations, which is lost in the expression (2.10), albeit the computational load is much higher.

In an overdetermined system, $M > N$, the resolution discussion applies to the observations. In this situation the measurements can be written as in (3.23) where only the first $k \leq N$ terms contribute to the solution, the remaining terms being the residuals. In complete analogy to the discussion of $\mathbf{V}\mathbf{V}^T$, if any diagonal element of $\mathbf{U}\mathbf{U}^T = 1$, then that equation (observation), is fully used in determining the solution. At the other extreme, if a diagonal element of $\mathbf{U}\mathbf{U}^T$ is zero, that equation or observation contributes nothing to the solution. More generally, the rows of $\mathbf{U}\mathbf{U}^T$ express the linear combinations of observations which have contributed to the solution.

SVD procedures permit an exhaustive exploration of the properties of both observation and solution. In the other methods discussed below, one sacrifices much of the information content of the SVD for other desireable properties, but the SVD is in many ways the most interesting and useful of all the finite dimensional methods and is one way of understanding other methods. (The real issue is computational feasibility: one must solve an eigenvector problem of the size of the smaller dimension of $\mathbf{A}$. Such solutions are reasonable for systems of $O(1000)$ rows or columns, but beyond that one is unlikely to have adequate computing resources. Furthermore, the size of the resolution and variance matrices is square of the largest dimension of the problem, and it becomes difficult to display, much less assimilate, the information that is being made available to one about the problem at hand. Nonetheless, if the problem is important enough, the full resolution and variance matrices can be used.)

3.5 Row and Column Scaling

Overdetermined least-squares systems solved conventionally assume that the expected error in each equation has the same magnitude. Assuring that this is true is usually accomplished through 'rowscaling', i.e. by left-multiplying the system (3.21) by a matrix, which we call $\mathbf{S}^{-\frac{1}{2}}$:

$$\mathbf{S}^{-\frac{1}{2}}\mathbf{A}\mathbf{x} = \mathbf{S}^{-\frac{1}{2}}\mathbf{y} \qquad (3.32)$$

The diagonal elements of $\mathbf{S}^{\frac{1}{2}}\mathbf{S}^{\frac{1}{2}} = \mathbf{S}$ are the variances of the noise in each equation. If the noise in the equations are correlated, then the multiplication (3.32) can be demonstrated (see Wiggins, 1972) to be a coordinate rotation into a new observation space in which the errors are both of equal expected value and also uncorrelated. The solution then proceeds in this new space with $\mathbf{A}$ replaced by $\mathbf{A}_1 = \mathbf{S}^{-\frac{1}{2}}\mathbf{A}$. (The definition of the matrix square root is discussed by Wiggins and by Lawson and Hanson, 1974).

An equivalent scaling, often known as 'column scaling' is normally applied to the solution space. Let $\mathbf{W}$ be the solution covariance; then one column scales $\mathbf{A}$ in the following way:

$$\mathbf{A}\mathbf{W}^{\frac{1}{2}}\mathbf{W}^{-\frac{1}{2}}\mathbf{x} = \mathbf{A}'\mathbf{x}' = \mathbf{y} \qquad (3.33)$$

and proceeds to solve the system for $\mathbf{x}'$, and $\mathbf{x} = \mathbf{W}^{\frac{1}{2}}\mathbf{x}'$ (see for example, Wunsch, 1978). If $\mathbf{W}$ is non-diagonal, one is working in a new solution space of uncorrelated solution parameters.

We are now in a position to solve the formally overdetermined problem in its formally underdetermined form (3.10):

$$[\mathbf{A}, \mathbf{I}] \left[\begin{pmatrix} \mathbf{x} \\ \mathbf{n} \end{pmatrix} \right] = \mathbf{y}$$

One merely introduces an arbitrarily small solution variance $\mathbf{W}$ for $\mathbf{n}$ corresponding to the demand that the noise be as small as possible, relative to the expected parameter variance and proceeds to the SVD solution. The equations are of course then satisfied exactly ($k = M$). Readers should convince themselves that this solution gives exactly the same answer as (3.9). One can conveniently regulate noise size relative to parameter size, should that be desireable. The main disadvantage to this procedure is that it increases the solution space size from the original dimension of N, to a new dimension of $N + M$, which renders it inefficient compared to the convention (which we have been following) of leaving the noise in the observation space. But the possibility of formally treating the noise components as part of the solution is a powerful reminder of the necessity to examine the noise solution as carefully as everything else.

3.6 Generalized Inverses

Although much more can be said about the SVD than we have space for here, some parting further demonstration of its unification powers is helpful. Consider the conventional least-squares solution (3.9), rewritten in terms of the SVD:

$$\hat{\mathbf{x}} = \left(\mathbf{A}^T \mathbf{A} \right)^{-1} \mathbf{A}^T \mathbf{y} = \left(\mathbf{V} \Lambda \mathbf{U}^T \mathbf{U} \Lambda \mathbf{V}^T \right)^{-1} \mathbf{V} \Lambda \mathbf{U}^T \mathbf{y} = \left(\mathbf{V} \Lambda^2 \mathbf{V}^T \right)^{-1} \mathbf{V} \Lambda \mathbf{U}^T \mathbf{y}$$

The inverse of $\mathbf{V} \Lambda^2 \mathbf{V}^T$ can be found by inspection, using the column orthonormality of $\mathbf{V}$ to be, $\mathbf{V} \Lambda^{-2} \mathbf{V}^T$ and we have

$$\hat{\mathbf{x}} = \left(\mathbf{V} \Lambda^{-2} \mathbf{V}^T \right) \mathbf{V} \Lambda \mathbf{U}^T \mathbf{y} = \mathbf{V} \Lambda^{-1} \mathbf{U}^T \mathbf{y} = \mathbf{A}_1^+ \mathbf{y} \qquad (3.34)$$

as the usual least-squares solution, where the operator $\mathbf{A}_1^+ = \mathbf{V} \Lambda^{-1} \mathbf{U}^T$. The reader should confirm that (3.34) is the same as (3.24) without the null space contribution. We have employed the condition $\mathbf{V} \mathbf{V}^T = \mathbf{I}$ for $k = N$ in the calculation. $k = N$ is the condition for $(\mathbf{A}^T \mathbf{A})^{-1}$ to exist.

The usual approach to the underdetermined problem is to minimize $\mathbf{x}^T\mathbf{x}$ subject to (3.21) introduced through a vector of Lagrange multipliers. The canonical result is

$$\hat{\mathbf{x}} = \mathbf{A}^T\left(\mathbf{A}\mathbf{A}^T\right)^{-1} = \mathbf{A}_2^+\mathbf{y} \tag{3.35}$$

Re-writing (3.35) in the SVD, we have

$$\hat{\mathbf{x}} = \left(\mathbf{V}\Lambda\mathbf{U}^T\right)\left(\mathbf{U}\Lambda\mathbf{V}^T\mathbf{V}\Lambda\mathbf{U}^T\right)^{-1}\mathbf{y} = \mathbf{V}\Lambda\mathbf{U}^T\left(\mathbf{U}\Lambda^2\mathbf{U}^T\right)^{-1}\mathbf{y}$$
$$= \mathbf{V}\Lambda^{-1}\mathbf{U}^T\mathbf{y}$$
$$= \mathbf{A}_2^+\mathbf{y}$$

But we see that both $\mathbf{A}_1^+$ and $\mathbf{A}_2^+$ can be written as $\mathbf{A}^+ = \mathbf{V}\Lambda^{-1}\mathbf{U}^T$, thus giving us a single analytical form for the 'left-inverse' of $\mathbf{A}$ in simultaneous equations of any dimension M, N. The matrix $\mathbf{A}^+$ satisfies what are sometimes known as the Penrose conditions, and it is an example of a 'Moore-Penrose inverse'. See Nashed (1976).

3.7 Other Estimation Procedures

Two approaches to handling arbitrarily determined systems have already been described, that based upon the Gauss-Markov theorem (notice that the mapping of a field is a solution estimation procedure), and the SVD. In fact, these procedures are really different aspects of general optimal estimation methods using an ℓ_2-norm and can be shown (see Wunsch, 1978) to be the same procedure in certain limiting cases.

Briefly, what one does in the Gauss-Markov procedure is to derive an estimate for $\mathbf{x}$, using the constraint equations (3.15) subject to the *a priori* estimates of the field covariance $\mathbf{R} = \langle x_i x_j \rangle$ and the covariance of the noise in $\mathbf{y}$, $\mathbf{N} = \langle n_i n_j \rangle$. The references already cited in section 2 show that the best (minimum expected square error) solution is

$$\mathbf{x} = \mathbf{R}\mathbf{A}^T(\mathbf{A}\mathbf{R}\mathbf{A}^T + \mathbf{N})^{-1}\mathbf{y} \tag{3.36}$$

with estimated error of

$$\mathbf{E} = \langle(\hat{\mathbf{x}} - \mathbf{x})(\hat{\mathbf{x}} - \mathbf{x})^T\rangle = \mathbf{R} - \mathbf{R}\mathbf{A}^T(\mathbf{A}\mathbf{R}\mathbf{A}^T + \mathbf{N})^{-1}\mathbf{A}\mathbf{R}^T \tag{3.37}$$

(compare these to expressions 2.9, 2.11, 2.12 for the mapped fields). The error which is being minimized in (3.37) is the summation of the statistical error (3.26) plus the error owing to a failure to resolve (3.31).

The two methods differ mainly in their handling of the null-space. The SVD solution sets the null-space components to zero; the Gauss-Markov solution (3.36) activates the solution null-space by using it to produce a solution as consistent as possible with both the *a priori* estimates of solution covariance and the linear constraints. It is reasonably obvious that having constructed the SVD solution, we could then manipulate the null-space to produce best fitting statistics of the solution without affecting the system residuals (Roemmich and Wunsch, 1982 provide some simple examples); the Gauss-Markov method is 'automatic' in this regard.

The SVD solution can be regarded as the simplest; the Gauss-Markov procedure is best regarded as a maximum likelihood solution, if the field and noise statistics are Gaussian (the Gaussian assumption is not necessary to its derivation, however) and in that sense is readily interpreted as the most likely solution.

The SVD solution can be recovered from the Gauss-Markov solution by letting both **R** and **N** be diagonal of constant value, and writing **A** in its SVD form (cf. Lawson and Hanson, 1974, Ch. 25, Section 4).

For many purposes, particularly in problems that are grossly underdetermined (the circulation inversions discussed in Wunsch, 1978, are grossly underdetermined), one may not particularly care to see any individual solution, but want only to find out how uncertain some particularly interesting parameters might be. For example, Wunsch (1984) used geostrophic inversions to understand the range of uncertainty of the meridional flux of heat in the North Atlantic. Schlitzer (1987) estimated the possible upwelling range in the eastern Atlantic abyss, and Schlitzer (1988) the uncertainty of the meridional fluxes of nutrients, etc. Thus one may have a different goal in mind than determining 'the' solution (see the discussion by Parker, 1972, of bounding strategies).

The procedure most often used for finding bounding solutions is a form of linear programming, but the reader should be aware of methods based upon Monte Carlo simulation — e.g. Press (1968). Any system of noisy linear equations of the form (3.10) can be manipulated into the following canonical form

$$\mathbf{y} - \mathbf{e}_1 \leq \mathbf{Ax} \leq \mathbf{y} + \mathbf{e}_2 \tag{3.38a}$$

$$\mathbf{x}_{\min} \leq \mathbf{x} \leq \mathbf{x}_{\max} \tag{3.38b}$$

$$\min : H = \mathbf{h}^T\mathbf{x} \qquad (3.38c)$$

(or max)

where H represents an 'objective function' to be maximized or minimized. H takes the place of the objective functions in the ℓ_2-norm procedures, which were, e.g. to minimize $(\mathbf{Ax} - \mathbf{y})^T(\mathbf{Ax} - \mathbf{y})$, in the overdetermined least-squares case.

The difference between the set (3.38) and the ℓ_2-norm solutions is not as great as might first appear: set (3.21) is solved with residuals which are bounded statistically (i.e. any residual which is overly large would lead to rejection of the solution) — these are 'soft' inequalities — whereas the bounds in (3.38a) are 'hard' — equation residuals must not lie outside these bounds at all. The range constraints (3.38) (see Wunsch and Minster, 1982; Wagner, 1969), including non-negativity constraints, are a natural part of the linear programming framework. They become extremely useful for example, in demanding positivity of tracer concentrations and of eddy coefficients. While they can be incorporated into the ℓ_2-norm system (see discussions by Lawson and Hanson, 1974; Fu, 1981; Tziperman, 1988) the procedures are cumbersome and inelegant compared to those in linear programming.

Linear programming problems are normally solved by a form of efficient search algorithm called the Simplex method (owing to G.B. Dantzig) and which is described in detail in many textbooks (e.g. Luenberger, 1973; Noble, 1969). Because linear programming has found widespread application in such diverse fields as industrial planning and anti-submarine warfare (it emerged from World War II, having been used to determine optimal strategies for destroyer searches for submarines — operations research), a very large effort has gone into developing efficient codes (see Strang, 1986 for discussion of the new Karmackar algorithm which has attracted enormous attention recently), textbooks, and commercial solution packages, and I will not attempt to repeat any of this material here.

The fundamental result however, is that systems such as (3.38) have solution sets which are (1) non-existent (the system is contradictory), (2) unique, (3) infinitely many, or (4) unbounded. Set (3) will give rise to a unique extreme of H, but there will be infinitely many solutions producing that extreme. The infinity of solutions can be understood from the null space of $\mathbf{A}$ augmented with h.

Simplex methods thus permit one to rapidly and easily explore bounds on interesting solution properties. The equivalent to the ℓ_2-norm discussions of error and resolution analysis is the system sensitivity determination, usually most easily discussed through the so-called 'dual problem', and which is often produced as a by-product of the Simplex method ('dual Simplex method'). The dual corresponds to the Lagrange multipliers, λ, imposing the constraints (3.38a,b) on the objective function.

Without much difficulty, one can show that for an equation or bound, i, lying at the inequality limits, one has $\lambda_i = \partial H / \partial(\text{bound})$, where λ_i is the corresponding element of the dual, i.e. the dual is a measure of the sensitivity of the objective function to differential variations in the bounds. (These λ_i should not be confused with the singular values.) Constraints not at their bounds of course, do not differentially affect the value of H and the corresponding $\lambda_i = 0$. Use of Lagrange multipliers has the great virtue of unifying a number of optimization procedures (e.g. see Scales, 1985).

Examples of linear programming methods applied to geostrophic and tracer inversions can be found in Wunsch and Minster (1982); Wunsch (1984); Schlitzer (1987, 1988).

Both ℓ_1- and ℓ_2-norm inverse methods have generalizations to non-linear problems; the ℓ_1-norm extensions can generally be found under the category of 'mathematical programming'; the ℓ_2-norm version involving 'total least squares' (see Golub and van Loan, 1980) or the Bayesian methods of Tarantola (1987), employed by Mercier (1986).

Many techniques have been developed over the years for systematically comparing observations with data, sometimes accounting for ill-conditioning, *a priori* statistics, positivity, etc. Much reinvention plagues this subject, leading to opaque vocabularies disguising an underlying unity. In Table 1, I have made an incomplete and inhomogeneous listing of methods, ideas and algorithms having some bearing on inverse problems and inverse methods.

Chief amongst the most desireable characteristics of an inverse method I would list the following:

(1) Ability to cope with underdetermined, noisy, contradictory constraints.

(2) Provision of complete statements of solution variance and resolution for testing against *a priori* hypotheses.

(3) Rank ordering of the data by importance to the solution.

(4) Ability to invoke diverse *a priori* beliefs, especially statistical ones.

Table 1
Inhomogeneous and incomplete listing of algorithms, methods, and mathematical topics with a relationship to inverse methods. The general problem is that of systematically comparing a model to data, often modifying certain model parameters in the process. Many of these methods and algorithms are intimately related.

Related Methods and Algorithms

Backus-Gilbert Procedure

Singular Value Decomposition

Objective Mapping

Objective Analysis

A priori filter theory (e.g. Kalman form)

Control Theory

Optimal Estimation Theory

Linear Programming

Quadratic Programming

Dynamic Programming

Rank deficient regression

Ridge Regression

Assimilation

Regularization

Least-squares (many variations)

Gauss-Markov estimators

Generalized inverses

Adjoint methods

Duality theory

Constrained estimation

Krigging

Collocation (in geodesy)

Different choices of procedures are made, sometimes sacrificing different characteristics for special needs, e.g. the imposition of positivity constraints, or reducing computational cost. No method will be the best choice under all practical conditions.

3.8 Inverse Methods and Inverse Problems

We have come far enough into the inverse problem formalism to now make certain useful distinctions. It is necessary to carefully distinguish between forward and inverse *problems*, and forward and inverse *methods*.

Without attempting scholastic nit-picking, many practical problems can reasonably easily be divided into forward, inverse, or mixed types. The conventional Laplace's equation of section 1.2 provides a good prototype: the forward problem is Dirichlet's; an inverse problem could consist of specification of φ, and its boundary conditions and one seeks to estimate the source-term ρ; a mixed problem might consist of specification of φ in some parts of the domain, ρ in others, and one seeks an estimate of the boundary data φ_0.

We can solve the inverse problem by forward methods: an obvious (and often used) approach is brute force iteration. Guess φ, calculate ρ, and compare to the data; on the basis of the differences modify the initial guess and repeat. The advantage of such an approach is the existence, efficiency and familiarity of forward methods. A principle disadvantage is the loss of the sensitivity and resolution products of inverse methods.

Similarly, one can solve forward problems by methods best regarded as inverse ones. Wunsch (1987) solved the fully posed form of tracer advection/diffusion equation (1.7) both by linear programming and by use of the SVD. Examples are given in this same reference of solutions to the mixed problem by inverse methods.

Inverse methods are probably best characterized by their contrast with forward methods: they are directed primarily at problems which in some way are ill-posed and which need to be regularized. As part of the regularization, the more effective methods define explicitly the parts of the solutions which are fully determined and which not, and yield quantitative statements of solution uncertainty owing to noise, and a rank-ordering of data importance.

In summary, many inverse problems are stable and well-posed (a good example is the Abel inverse problem discussed by Aki and Richards, 1980) which can be solved by purely conventional 'forward' methods. Similarly, forward problems can be ill-posed in many different ways, and thus best-solved by an inverse method. Generally speaking, any problem which attempts to directly relate real observations to a model will be ill-posed.

4. USING STEADY TRACERS

4.1 The Background

Return now to the problem in the title of these lectures, assuming that we know how to map, form means etc. with known errors for any given tracer. How might one use tracer properties, such as those displayed in Figs. 1–4 to make some sense of the ocean circulation?

It is simplest to start with a discussion of steady tracers, there being several major additional practical issues which arise when we come to transients. From the various collections of flow/tracer linkages we have written down, as in equations (1.3), (1.6), it should be reasonably clear that we are faced in the steady tracer problem with a set of quasi-linear (i.e. ignoring the errors implicit in the coefficient matrices A) set of noisy, underdetermined relationships from which we wish to extract whatever inferences we can about the ocean.

I will not dwell on this steady problem here: other lecturers will deal with particular applications at great length during this school, and there is now more than a decade of published explicit applications of the methods outlined above. An incomplete list of applications to the geostrophic problem, with detailed working out of examples are: Wunsch (1977, 1978), Roemmich (1980, 1981), Fu (1981, 1986), Wunsch, Hu and Grant (1983), Wunsch and Grant (1982), Roemmich and McCallister (1988), Rintoul (1988), Pollard (1983), Schlitzer (1987, 1988), Bolin *et al.* (1987).

But some comments on the context of the steady tracer problem are worthwhile. Recall that the application of inverse methods to the geostrophic problem was provoked by the forceful assertions of Worthington (1976, and verbally for 20+ years preceding) that there were no geostrophic flows capable of balancing the observed water masses of the North Atlantic. It is not widely appreciated that the now familiar, much reproduced, circulation diagrams of Worthington (1976) represent flow fields which are not in geostrophic balance, probably because the only place he informs the reader of that is in the footnote on p. 36. I have put a little note on this 'case of whisky problem' into an Appendix to these lectures. One of the major

accomplishments of inverse methods was their use to show decisively, and quantitatively, that determination of the circulation through the conventional dynamic
method was intrinsically underdetermined. The production of specific examples of
possible circulation schemes was a secondary consideration.

Over the years, the dynamic method, coupled with assumptions of 'levels of
no motion' became so ingrained that it seems to have been forgotten that no one
had ever produced a dynamical reason why there should be a depth $z(\mathbf{r})$ where
the velocity vanished. Despite the fact that different authors assumed that $z(\mathbf{r})$
corresponded with a constant pressure, constant depth, constant isopycnal, constant
temperature etc., there was no reason at all, other than convenience, why z should
be a simple function of position. Difficulties in using the assumption of a simple
level of no motion reached their culmination in the work of Worthington (1976)
who ended by rejecting geostrophy as the appropriate dynamical balance.

At the present time, theory provides no general framework for determining
what the velocity should be at any given $z(\mathbf{r})$. Thus the most conservative approach
is to regard it as a vector of unknowns, each element of the vector being the velocity
at $z(\mathbf{r})$ between each pair of stations.

Another seemingly different method, that of Schott and Stommel (1978), is to
attempt a higher order dynamical balance than encompassed in pure geostrophy.
Their method, known as the beta spiral (see also Schott and Zantopp, 1979), has
been used with some success (Olbers, *et al.*, 1985) and has spawned a series of
theoretical offshoots (e.g. Killworth, 1983, Olbers and Willebrand, 1984). In general
terms, one can say this approach sets out to find a dynamical balance which is
satisfied only by the correct absolute velocity. Davis (1978) showed that the method
of Schott and Stommel (1978) could be reduced to that of Wunsch (1978) with the
addition of *a priori* scale assumptions; it is thus a version of the Gauss-Markov
procedure, where the prior scales are derived from higher order dynamical balances.

4.2 Where Steady Tracer Inverse Methods Are Going

As already mentioned, perhaps the greatest accomplishment of the application of
inverse methods to the geostrophically based dynamic method was the explicit
demonstration of how greatly underdetermined that problem really was. Examples extant in the literature show formally that qualitatively different flows are
all compatible with the constraints set out by Worthington: geostrophic balance,
plus near-conservation of various water mass types. It is now clear (I hope) that

the methods developed to handle this particular problem are in no way restricted to Worthington's special demands. The technique subsequently has developed in a number of different directions. The geostrophic restriction was removed almost immediately with the introduction of Ekman fluxes and mixing (e.g. see Wunsch, Hu and Grant, 1983); the beta-spiral explicitly introduces a quasi-geostrophic vorticity balance. Inequality constraints have been imposed as suggested in Wunsch (1978), by Fu (1981), Wunsch (1984), Tziperman (1988) and others. A variety of constraints has been added to the original temperature, salinity and mass balances, including budgets for nutrients and oxygen (Wunsch, Hu and Grant, 1982; Rintoul, 1988; Schlitzer, 1987). The use of inequalities permits semi-qualitative (or 'Wüstian') constraints such as 'the flow goes that way but I don't know how fast'; see the discussion in Wunsch (1984).

In general, one sees the continuing search for additional equations to add to the system; Pollard (1983) constrained an inversion to be consistent with current meters and many of the inequality constraints used in published solutions derive from float movements, etc.

Reducing the very large underdeterminism is a challenging task. One prefers, for the reasons described, to add information to a model believed to carry adequate degrees of freedom, rather than to reduce the degrees of freedom. But each additional observation demands quantitative analysis before it can be used (inverse methods force one to be quantitative — it is their great power — but drives one toward quantifying beliefs one often would prefer to leave comfortably vague). For example, the combination of current meter or float data with hydrography leads to awkward questions about Eulerian versus Lagrangian compatibility and of consistency of averaging intervals. The introduction of biologically active tracers (e.g. oxygen) in turn raises questions as to whether the information apparent in the distributions is greater than the uncertainty contained in the source/sink terms; an example using oxygen as a tracer is described by Wunsch (1988a).

4.3 Eclectic Modelling

Because for the foreseeable future, it appears that the oceanic data base of pure hydrography and tracers will remain inadequate for removing the remaining large-scale underdeterminancy, Wunsch (1984) was led to the concept of 'eclectic modelling'. The basic suggestion is that one should recognize that a great deal of the ocean circulation is known — but in fragmentary form: we know some things statistically (deep eddy kinetic energies), some things regionally (the mass flux through the Florida Straits), some things qualitatively (that the heat flux should be poleward in winter), etc. The argument is that the time has come to synthesize these different things into models, so that this fragmentary information may be tested against itself, and the sum lead to a more complete synthetic picture of the circulation.

The time has probably passed where it is useful to produce (say) a model of the oxygen balance of the North Atlantic that ignores the question as to whether the model properly balances salt, or heat, or has the known transport properties of the Gulf Stream. Eclectic modelling is directed at finding mathematical methods which permit the quantitative testing of extremely non-uniform bits of knowledge against each other and synthesizing the independent, self-consistent parts of each. Further discussion of these issues has to be left to the published papers.

4.4 Remaining Steady Tracer Issues

A number of outstanding problems remain. The primary one continues to be the data base: the overall paucity of mesoscale resolving, top to bottom, hydrographic plus tracer observations of the type which started to become available in the early 1980's.

In addition, it is increasingly clear that the hope biologically active tracers such as oxygen and nitrates would help to strongly constrain the general circulation will not be realized in the near-future. What would be needed is much better purely biochemical understanding of the source/sinks than is now possible. The current debates over the importance of dissolved organic concentrations, and the disconcerting uncertainty of rates of primary production suggest the magnitude of this problem. It would appear that for the foreseeable future, knowledge of the circulation deduced from other data will be used to make estimates of the source/sink rates of these tracers, rather than vice-versa.

Finally, the restriction of the flow models to be effectively steady will eventually have to be relaxed. None of the existing inverse calculations based on steady flow models has yet led convincingly to contradictions which would indicate the need to build time-dependent flows explicitly into the models. However, the increasing size of the data base in some regions, especially the North Atlantic, and what is hoped for in the World Ocean Circulation Experiment, indicates that such contradictions will soon appear, and we will have to deal with models which at the very least, have realistic annual convection cycles in them.

5. TIME DEPENDENT PROBLEMS

5.1 Observational Realities and Boundary Controls

We distinguish time dependent tracers in steady flows, from steady or time dependent tracers in the time dependent flows, just alluded to. Here we discuss only the problem of transient tracers in nominally steady circulations.

When turning from steady to transient tracers a number of practical problems arise. The finite differences, Eq. (1.8), for a transient tracer, or any equivalent form, show that mathematically there is no fundamental difference between inverting or regularizing a transient and a steady tracer. What one must however recognize is that the increase in dimensionality, from the three space dimensions of the steady problem to the added dimension of time in the transient one, introduces observational realities which must be treated from the outset — otherwise, one develops procedures which are hopelessly impractical.

Consider one simple observational issue. Suppose we do an inversion with a steady tracer and are convinced that more data are required to improve the result (increase the resolution, decrease the variance). Then in principle, we can go to sea and acquire the missing data, assuming the tracer is really steady. But if, with a transient tracer, the missing data correspond to a survey at sea at an earlier time, then no known physics permits the acquisition of that data.

Furthermore, modern oceanography has reached the stage of coming close to the ability to produce the equivalent of nearly synoptic three-dimensional coverage of an entire ocean basin, for a full suite of tracers, as now exists for the North Atlantic. Such a survey fills the observation space of the steady models we have discussed above. But to fill the observation space of a model covering three space dimensions and time over decadal scales, remains beyond us. Thus the presence of time in the model not only introduces a coordinate which can be examined at sea in one direction only, but nearly automatically makes the ratio of observations to unknown parameters very much smaller than in the steady case.

The practical use of transient tracers thus forces us to deal with the realities of observation from the beginning rather than as an afterthought. Following this idea (see Wunsch, 1987, 1988a,b) let us caricature the transient tracer problem as follows, thinking of ^{3}H (tritium) as prototypical.

We know that at $t = 0$ the concentration of tracer $C(\mathbf{r}, t = 0)$ in the system vanished. Some years later, at time $t = t_f$, we do a survey of tracer $C(\mathbf{r}, t_f)$ and map it in three-dimensions by the techniques of section 2.2. What can we do with this data to make inferences about the ocean?

One thing we know we can do, if we have been smart enough to have been measuring the atmospheric concentration, C_a, during this entire time, is to compute the *net uptake* of tracer by the ocean, through air/sea transfer processes. Given the complexities of air/sea exchange, such estimates are extremely useful, and are a non-trivial application of transient tracers. The development of observational strategies and estimates of uncertainties of the results can be based upon the procedures of section 2.

But if we wish to apply the procedures to determine flow and mixing parameters of the ocean circulation, we are faced with grappling with systems of equations analogous to (1.8). In view of the prior discussion of steady tracers, we can see that obtaining estimates of u, v or K from (1.8) demands that we specify $C_{ij}^{t+1} - C_{ij}^t$. But I have ruled out that possibility by the realistic supposition that the tracer survey exists at $t = t_f$ only. Being optimistic I might think I could do another survey and plug the concentration differences into (1.8), thus constraining u, v, etc. But how big a time difference is permitted? Eqs. (1.8) are a system of finite differences and carry with them demands on the time step Δt for stability and accuracy. Plausible estimates from real data and models suggest that if solved in the forward direction the time steps must be $O(1$ month) or less. Examination of the inverse problem suggests that a similar time step is required there. The possibility of basin-scale three dimensional surveys of any tracer at monthly time intervals is implausible anytime soon.

Suppose that we instead have such a survey a year or two apart. Can we do anything with this data? The answer is 'yes', but is best understood if the problem as originally posed is considered first in a slightly modified form:

It is supposed that by some means we have made a first estimate of the ocean circulation, so that we have a specific flow/mixing model. The fully-posed *forward* problem for C is then:

> Solve (1.8), or (1.9) subject to initial conditions $C(\mathbf{r}, 0)$, and boundary conditions $C = C_B(t)$ on ∂D.

We solve the forward problem using our best estimates of initial conditions (which of course, need not be zero) and boundary conditions, computing out to $t = t_f$. We then compare our forward solution to the tracer survey. Usually, there will be some discrepancy between what was calculated and what was observed (if there were no such discrepancy we would be in the unfortunate position of having done a tracer survey that provided no information about the ocean not already contained in the model — surely a disappointing outcome for so much effort. See Fig. 11). Now the problem gets interesting. What should we change in the model to improve agreement? The number of possible choices is embarrassing. The calculation leading to the circulation is dependent upon the interior specification of the model flows, topography, mixing rates, vertical and horizontal resolution, etc. It is also dependent upon estimates of boundary conditions including wind-stress curls and air/sea thermodynamic transfers. Changes in any of these, either alone, or in combinations, regional or global, with the others, will all lead to changes in the flow field and hence in the final tracer distribution. If these were all that were involved, the problem of what to change would be difficult.

But the introduction of time into the system has introduced a whole new set of system unknowns. To see this, let us stipulate that the circulation model is perfect, giving exactly the right flow field and mixing everywhere. Let the tracer boundary condition data $C_B(t)$ be slightly in error. In particular, suppose that over some region at the surface (see Fig. 12), one has overestimated the surface concentration by about 10% for part of the duration of the transient. It is quite clear then what will happen: the model will drive too-large a concentration into the interior, and give rise to model concentrations of tracer which are larger than were observed. If the flow is strong, the calculated model values at t_f will be incorrect far from the place of initial error. The modeller may conclude that his circulation is wrong, when the source of the difficulty lies wholly in the tracer field boundary specifications. Similar effects are possible if the interior source/sink terms are systematically incorrect over large regions. How does one know what to modify?

For steady tracers there are of course equivalent boundary condition difficulties: if the boundary concentrations are mis-specified, the whole interior distribution may be gotten wrong, and the error incorrectly ascribed to the circulation. With steady tracers, however, if in doubt about the boundary conditions, one can go back and look. But if our question in 1988 is 'what was the surface concentration of ^{3}H in the Greenland Sea, or at the northern boundary of our model at depth, in 1976?', we have no recourse.

These and related difficulties have led me (Wunsch, 1988a,b) to propose that transient tracers must necessarily be handled in a two stage process:

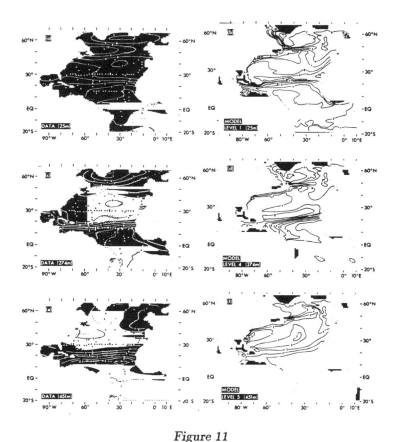

Figure 11

Comparison, by Sarmiento (1983), of the observed tritium distribution at different depths in 1972 (these maps were constructed by the methods of section 2.2), with calculations from a general circulation model. If the differences between the two are deemed significant, the number of circulation model parameters that could be changed to improve the fit, is very large. We argue that before changing the dynamical model or its dynamical boundary conditions, one needs to confirm that small systematic errors in the tracer boundary conditions do not account for the discrepancies.

Recognizing that boundary conditions are always subject to uncertainty we first ask whether the boundary conditions imposed can be modified within acceptable limits to bring model and observations together within the error bars. If this first stage is successful, we stop: there is no conflict between initial model and the

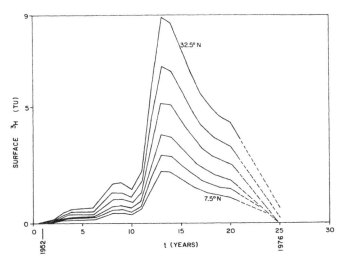

Figure 12

Example of complex time/space structure of the boundary conditions for a
transient tracer. Figure depicts estimated surface concentration of tritium
at several latitudes in the eastern North Atlantic as constructed by Wunsch
(1988a) from calculations of Dreisigacker and Roether (1978). Close examination of the original observations, and the complex calculations which
must be done to interpolate over the world ocean, strongly suggests the
possibility of systematic over or under-estimates of actual concentrations
for many years in different locations. Such systematic errors can accumulate in the model in regions remote from those of the original errors. The
tritium problem is even more difficult if one seeks to use fluxes rather than
concentrations; in principle, the problem is simpler for fluorocarbons. In
this latter case however, interior distribution of fluorocarbons calculated
in a model would be extremely sensitive to slight systematic modelling
errors in high-latitude convective regions. Unless the model is extremely
accurate there, the entire subtropical distribution of tracer may be incorrectly computed. One may be forced to consider the true tracer boundary
as lying equatorward of the convective regions, raising the same problems
discussed in the text concerning the paucity of early observations.

observations, except that we have modified our initial boundary condition estimates
and made them better ones.

If stage 1 is unsuccessful, e.g. that the modifications to the boundary conditions
are unacceptable, then we move to stage 2: modify the model. We need therefore,
to explore how one might go about solving stage 1.

Control theory encompasses a wide class of problems, many of them involving
direct connections between system evolution equations and noisy observations: a

concise description of the transient tracer problem. The specific problem we have posed is that described as 'distributed system boundary control with a terminal constraint'; the terminology 'distributed system' refers to the fact that we are faced with a partial differential equation rather than an ordinary one, and the 'terminal constraint' refers to the demand for the system to end at the observed tracer distribution, within error bars. Control by interior sources and sinks differs from boundary control, but the procedures are essentially the same.

A problem analogous to the tracer one, closer to a classical control problem, may be helpful. Suppose an automobile is to be driven from Paris to Rome. The automobile is in Paris on day t=0, and it must arrive in Rome on day $t = t_f \pm 0.5$ days. In going between the two cities, several routes are possible, and we seek that route which minimizes fuel consumption (or travel time, or maximum required acceleration, or any other plausible objective). A classical control problem would be to find the best route (or best speed and acceleration versus time, if the route is fixed) given a mathematical description of the automobile (fuel consumption against speed, acceleration and braking characteristics, etc.).

The analogue to the tracer problem puts it slightly differently. We observe that the automobile was in Paris at $t = 0$, and we observe that it arrived in Rome at $t = t_f \pm 0.5$ days. We ask whether there was any route, or strategy, which satisfies the two observed constraints given the physical characteristics of the automobile. If the observed travel time is shorter than the minimum possible driving distance divided by the maximum possible travel speed, we would be led to reject either the observations, or the physical model of the vehicle. One may elaborate the restrictions, by adding observations that the vehicle was seen in Milan on a certain date, etc. thereby further constraining the possible controls.

Thus we first ask whether there is any set of boundary conditions, which satisfies all our physical understanding (for example, negative tracer concentrations are not permitted; if intermediate observations are available, the calculated boundary conditions and interior concentration must pass through them within error bars). If no acceptable boundary conditions are determinable, the model is rejected, and we go on to stage 2.

The solution to the stage 1 problem is very interesting. We can only sketch it; more details can be found in any of the very large numbers of textbooks on control methods (e.g. Luenberger, 1979; Brogan, 1982; Stengel, 1986). Let $\mathbf{C}_d(t_f)$ be the observed tracer concentration from the survey. We minimize the difference between

$C(t_f)$ and $C_d(t_f)$ while simultaneously minimizing some function of the control variable (q in Eq. 1.9), e.g. its meansquare over the evolution period. The strategy is to constrain the evolution to obey the equations (1.9) by introducing them into the objective function through a set of vector Lagrange multipliers λ thus seeking to find stationary values of

$$J = \left[(C(t_f) - C_d(t_f))^T \, G \, (C(t_f) - C_d(t_f)) \right]$$
$$+ f(\mathbf{q}) + \lambda^T(t) \left[C(t+1) - A_1 C(t) - B\mathbf{q}(t) \right]$$

where G is a matrix best interpreted as the inverse of the relative error covariance of the terminal constraint $C_d(t_f)$ and f is a function of $\mathbf{q}$ whose minimum we seek. (The $\lambda(t)$ should not be confused with the singular values.) The above mentioned textbooks show how to find $\mathbf{q}$. One is led to solving the following coupled problems (the Pontryagin minimum principle): Equation (1.9) plus

$$C(0) = C_0(\mathbf{r}, 0)$$
$$\mathbf{q}(t) = -\tfrac{1}{2} B^T \lambda(t)$$
$$\lambda(t) = A_1^T \lambda(t+1)$$
$$\lambda(t_f) = 2G \, (C(t_f) - C_d(t_f)) \tag{5.1}$$

Notice that the Lagrange multpliers satisfy an evolution equation involving the transpose of the matrix A_1^T and that time runs backwards from $t = t_f$ in this equation. λ are sometimes known as the 'adjoint solution'. The 'adjoint method', which proved so fashionable at this School, is derived through minor variations of the problem we have stated. For matrix/vector systems, adjoints are matrix transposes and emerge naturally as the connection between given observations and model parameters. (Recall Eq. 3.20b connecting the $\mathbf{u}_i$ vectors of the observation space and the $\mathbf{v}_i$ vectors of the solution space in the SVD involves the matrix A^T, which also appears intrinsically in the generalized inverse (3.34).) Green's functions, which are often thought of as inverse to differential operators, normally also satisfy the adjoint of the given system. A grand unification of a lot of seemingly disparate approaches is possible, but is not pursued here for want of time and space. Finding the solution to system (1.9) plus (5.1) is non-trivial because of the different time directions for the forward and adjoint systems, but there are practical procedures for doing so. Fig. 13 is taken from Wunsch (1988b) where the system was explicitly solved for a simple example. Wunsch (1988a) solved a much more complex real system by parameterizing the adjoint solution using Green's functions. The method used in Wunsch (1987) avoided the adjoint system altogether, but at the non-trivial price of a system which was much larger in size than (1.9) plus (5.1).

As discussed in Wunsch (1988b), and as hinted above, one can add demands that the system pass through an intermediate survey at $t = t_1 < t_f$, thus accommodating the two survey problems we described above. Much work remains to be done to fully understand how to use these methods most effectively; the subject is in its infancy. The questions about it range from computational efficiency, to the best strategy for imposing, if necessary, positivity constraints upon the values of $\mathbf{C}$.

5.2 Modifying the Model

If no acceptable solution is found, one is then faced with systematically changing the model, presumably in some minimizing fashion. This subject has not, to my knowledge, been addressed except in principle, because there has not yet been an example of an acceptably realistic model being rejected because of its failure to reproduce the terminal constraints. But control methods (see particularly the compilation of Stavroulakis, 1983) obviously lend themselves directly to this problem as well. The problem is intrinsically non-linear, but many methods are available for addressing the requisite model parameter changes. The future of this subject, including oceanographic data of all kinds, almost surely lies in the pursuit of these methods to gradually bring models into accord with observation.

ACKNOWLEDGEMENTS

My studies of the application of inverse methods to tracers and the general circulation have been consistently supported in part by the National Science Foundation for a decade, support for which I am grateful. The most recent grant goes under the label OCE 8521685. Much of what I have learned about inverse methods over the years has been the result of discussions with a large number of students and post-docs, among whom I would include D. Roemmich, L. Fu, B. Cornuelle, E. Tziperman, S. Rintoul, R. Schlitzer, L. Memery, J.-F. Minster, H. Mercier.

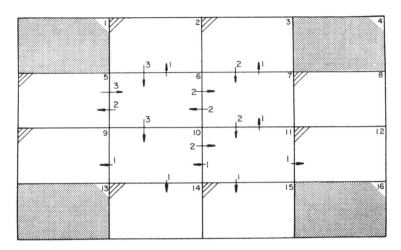

Figure 13a

A simple box model (Wunsch, 1988b) used to illustrate the use of control methods for a transient tracer. The estimated exchanges J_{ij} between boxes are supposed given a priori as shown. Boundary boxes are partially shaded (e.g. numbers 2, 3, 5, ...) and the tracer concentrations are supposed to be known there in the forward problem for all time (no budgeting of these boundary boxes is carried out).

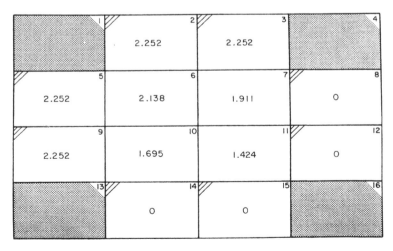

Figure 13b

'Terminal distribution' supposed observed (with some error) in the model of Figure 13a.

68

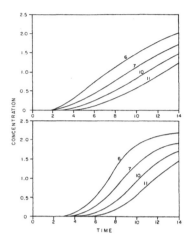

Figure 13c

Lower panel depicts the 'correct' time history in the interior boxes which led to the terminal distribution of Fig. 13b. Upper panel depicts a different time history, calculated from the control solution of Eqs. (1.9) and (5.1), leading to the same terminal state (within the observational errors).

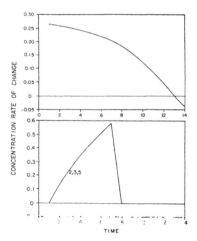

Figure 13d

Lower panel depicts the time rate of change of the 'correct' concentrations in boundary boxes 2, 3, 5 of Fig. 13a, and the upper panel the equivalent values from the control solution. That two such different histories lead to the same terminal state, and with the different time histories of Fig. 13c (which are not strikingly different) is strongly suggestive of the great freedom one has to satisfy the observed terminal constraint with a fixed model.

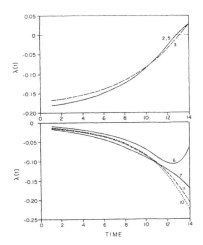

Figure 13e
Adjoint solution (Lagrange multipliers) of the control problem depicted in
Figs. 13a–d.

APPENDIX

Some Notes on the History of Inverse Methods in Ocean Circulation Problems

Long ago, Hidaka (1940) recognized that temperature, mass, and salt could all be used as tracers, and to a first approximation as conserved, passive ones. He defined a volume as in Fig. 14 with four station pairs and wrote conservation equations for the volumes depicted so that he had more equations than unknown b_j reference level velocities. Unfortunately, Defant (1961) demolished Hidaka's method by demonstrating that very slight changes in the data specification gave rise to very large changes in the solution, concluding (rightly) that Hidaka's equations were ill-conditioned. Defant's demonstration seems to have killed any approach to using the conservation equations to calculate reference level velocities until the mid-1970's, by which time it was obvious that the ill-conditioning was real, but not fatal.

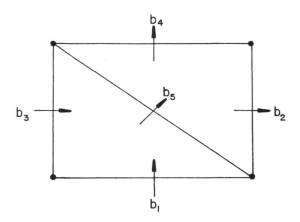

Figure 14
Geometry like that used by Hidaka (1940) to determine unknown reference level velocities involved with four station pairs. One writes conservation of mass and salt for the volumes depicted formally overdetermining the solution. But as Defant (1961) discusses, the system as used by Hidaka was ill-conditioned, and he did not know how to control the solution through the SVD or otherwise.

Riley's (1951) paper was a very interesting attempt to combine geostrophy with mass, oxygen and nutrient balance equations in a scheme very like those now being employed. He clearly knew what had to be done, but lacked the mathematical machinery and the data base for carrying out the calculations. But Riley deserves a lot of credit for having perceived the direction in which the subject had to go — although it took another 25 years before real progress could be made.

The use of inverse methods for determining the oceanic circulation as inroduced in Wunsch (1977, 1978) has been controversial, bordering on the polemical. It is important to recall the situation at the time the first of these papers was written. The prime motivation came from L.V. Worthington of Woods Hole Oceanographic Institution, whose book (Worthington, 1976) was about to appear. Val (as he is known) gave a seminar at WHOI (I think sometime in 1975) in which he summarized the content of the book, and it was only at that point that I realized in a specific way what he had been saying informally for years.

As I understood him, he claimed that one could not calculate the flow of water in the North Atlantic while simultaneously obeying the geostrophic relationship and conserving water types. As a result of grappling with this problem for many years, Val had at last concluded that geostrophy should be abandoned, and his book therefore presented a complete North Atlantic circulation scheme in which geostrophy was violated over large regions of the ocean.

Now the abandonment of geostrophy is no small step; all scale analyses of the large-scale ocean circulation show that the dominant balance of forces is between the Coriolis force and the pressure force, known in short-hand as 'geostrophic balance'. If one wishes to move water around the ocean on the large-scale, but cannot find pressure gradients to support the Coriolis forces, then there are only two other possibilities: a gross violation of Newton's laws of motion, or the presence of some other balancing force. The difficulty with the first alternative is clear; unfortunately, no one has ever succeeded in finding any other reasonable way to balance the Coriolis forces.

As already noted, apart from the one cryptic footnote, Worthington essentially ignored the problem in his book. At the time of his lecture in WHOI, it was widely known that Worthington years before had bet a case of whisky that no one could balance the North Atlantic geostrophically. I was stimulated by the lecture to tackle what by then was known as the 'case of whisky problem'. (Later, Fiadeiro and Veronis (1982) made a claim to the case of whisky on the basis of a circulation pattern for the Tasman Sea. To my knowledge however, Worthington never said that there were any problems with geostrophic balance there).

Listening to Worthington, it seemed to me that this was a nearly ideal opportunity to try out inverse methods, which had come to be widely used in geophysics following the pioneering efforts of Backus and Gilbert (1967).

The introduction of inverse methods into the classical reference level problem (Wunsch, 1977, 1978) elicited a considerable critical published response which has generated a lot of heat. Their use in other areas of oceanography (e.g. tomography — see Munk and Wunsch, 1979) has not been anywhere near as controversial. That the geostrophic problem stimulated a confused response is, I suspect, owing to its classical nature — it is a problem that all practicing oceanographers grew up with; the nature of the fundamental problem is theoretically somewhat trivial (a missing integration constant), and it is one that touches immediately on 40 years of theoretical work on the general circulation. Thus a number of theoreticians, with

72

little experience in working with real data, proceeded to misunderstand what was being said.

Of course with the benefit of over 10 years' experience, one comes to a better understanding of the problems, and (one can hope) better and simpler ways of explaining what one is doing than is possible when one is still groping for complete understanding of a problem.

REFERENCES

Aki, K. and P.G. Richards, 1980: *Quantitative Seismology,* W.H. Freeman, San Francisco, 2 vols., 932 pp.

Arthnari, T.S. and Y. Dodge, 1981: *Mathematical Programming in Statistics,* Wiley, New York, 413 pp.

Backus, G.E. and J.F. Gilbert, 1967: Numerical applications of a formalism for geophysical inverse theory. *Geophys. J. Roy. Astr. Soc.,* **13,** 247–276.

Bolin, B., A. Björkström, K. Holmén and B. Moore, 1987: *On inverse methods for combining chemical and physical oceanographic data: A steady state analysis of the Atlantic Ocean.* Tech. Rep., Meteorologiska Institutionen (MISU), Stockholm Universitet, Arrheniuslaboratoriet, 134 pp.

Bracewell, R.N., 1978: *The Fourier Transform and Its Applications,* McGraw-Hill, New York, 444 pp.

Bretherton, F.P., R.E. Davis and C. Fandry, 1976: A technique for objective analysis and design of oceanographic instruments applied to MODE-73. *Deep-Sea Res.,* **23,** 559–582.

Brogan, W.L., 1982: *Modern Control Theory,* Prentice-Hall/Quantum, Englewood Cliffs, N.J., 393 pp.

Buffham, B.A., 1985: Residence-time distributions in regions of steady-flow systems. *Nature,* **314,** 606–608.

Claerbout, J.F., 1976: *Fundamentals of Geophysical Data Processing. With Applications to Petroleum Prospecting,* McGraw-Hill, New York, 274 pp.

Cramér, H., 1946: *Mathematical Methods of Statistics,* Princeton U. Press, 574 pp.

Dantzig, G.B., 1963: *Linear Programming and Extensions,* Princeton U. Press, Princeton.

Davis, R.E., 1978: Estimating velocity from hydrographic data. *J. Geophys. Res.* **83,** 5507–5509.

Defant, A., 1961: *Physical Oceanography*, Vol. 1. Pergamon, N.Y., 598 pp.

Deutsch, R., 1965: *Estimation Theory*, Prentice-Hall, Englewood Cliffs, N.J., 269 pp.

Dreisigacker, E. and W. Roether, 1978: Tritium and 90-Sr in North Atlantic Surface Water. *Earth and Planet. Sci. Lett.*, **38**, 301–312.

Eckart, C. and G. Young, 1939: A principal axis transformation for non-Hermitian matrices. *Bull. Amer. Math. Soc.*, **45**, 118–121.

Fiadeiro, M.E. and G. Veronis, 1982: On the determination of absolute velocities in the ocean. *J. Mar. Res.*, **40**(supplement), 159–182.

Fiadeiro, M.E. and G. Veronis, 1984: Obtaining velocities from tracer distributions. *J. Phys. Oc.*, **14**, 1734–1746.

Freeman, H., 1965: *Discrete-Time Systems. An Introduction to the Theory*, John Wiley, New York, 241 pp.

Fu, L., 1981: The general circulation and meridional heat transport of the subtropical South Atlantic determined by inverse methods. *J. Phys. Oc.*, **11**, 1171–1193.

Fu, L., 1986: Mass, heat and freshwater fluxes in the South Indian Ocean. *J. Phys. Oc.*, **16**, 1683–1693.

Gelb, A. (Ed.), 1974: *Applied Optimal Estimation*, The MIT Press, 374 pp.

Golub, G.H. and C.F. Van Loan, 1980: An Analysis of the Total Least Squares Problem. *SIAM J. Num. Anal.*, **17**, 883–893.

Golub, G.H. and C.F. Van Loan, 1983: *Matrix Computation*, Johns Hopkins U. Press, Baltimore, 384 pp.

Hidaka, K., 1940: Absolute evaluation of ocean currents in dynamic calculations. *Proc. Imp. Acad.* Tokyo, **16**, 391–393.

Kawase, M. and J.L. Sarmiento, 1985: Nutrients in the Atlantic thermocline. *J. Geophys. Res.*, **90**, 8961–8979.

Killworth, P.D., 1983: Absolute velocity calculations from single hydrographic sections. *Deep-Sea Res.*, **30**, 513–542.

Lanczos, C., 1961: *Linear Differential Operators*, Van Nostrand, Princeton, 564 pp.

Landau, H.J. and H.O. Pollak, 1962: Prolate spheroidal wave functions, Fourier analysis and uncertainty — III: The dimensions of the space of essentially time and bandlimited signals. *Bell System Tech.*, *J.* **41**, 1295–1336.

Lawson, C.L. and R.J. Hanson, 1974: *Solving Least Squares Problems*, Prentice-Hall, Englewood Cliffs, N.J., 340 pp.

Leith, C.E., 1973: The standard error of time-averaged estimates of climatic means, *J. Appl. Meteor.,* **12,** 1066–1069.

Liebelt, P.B., 1967: *An Introduction to Optimal Estimation,* Addison-Wesley, Reading, Mass., 273 pp.

Luenberger, D.G., 1969: *Optimization by Vector Space Methods,* John Wiley and Sons, New York, 326 pp.

Luenberger, D.G., 1973: *Introduction to Linear and Non-Linear Programming,* Addison-Wesley, Reading, MA, 356 pp.

Luenberger, D.G., 1979: *Introduction to Dynamic Systems. Theory, Models and Applications,* John Wiley, New York, 446 pp.

Mercier, H., 1986: Determining the general circulation of the ocean: a non-linear inverse problem. *J. Geophys. Res.,* **91,** 5103–5109.

Munk, W. and C. Wunsch, 1979: Ocean acoustic tomography: A scheme for large scale monitoring. *Deep-Sea Res.,* **26A,** 439–464.

Nauman, E.B. and B.A. Buffham, 1983: *Mixing in Continuous Flow Systems,* Wiley-Interscience, New York, 271 pp.

Nashed, M.Z. (ed.), 1976: *Generalized Inverses and Applications.* Proceedings of an Advanced Seminar Sponsored by the Mathematics Research Center, The U. of Wisconsin-Madison, October 8–10, 1973, 1054 pp.

Nayfeh, A.H., 1973: *Perturbation Methods,* John Wiley, New York, 425 pp.

Noble, B., 1969: *Applied Linear Algebra,* Prentice-Hall, Englewood Cliffs, N.J., 523 pp.

Olbers, D.J. and J. Willebrand, 1984: The level of no motion in an ideal fluid. *J. Phys. Oc.,* **14,** 203–212.

Olbers, D.J., M. Wenzel and J. Willebrand, 1985: The inference of North Atlantic circulation patterns from climatological hydrographic data. *Revs. Geophys.* **23,** 313–356.

Östlund, H.G. and R.A. Fine, 1979: Oceanic distribution and transport of tritium. *Behavior of Tritium in the Environment,* International Atomic Energy Agency, Vienna, 303–314.

Parker, R.L., 1972: Inverse theory with grossly inadequate data. *Geophys. J. Roy. Astr. Soc.,* **29,** 123–138.

Petersen, D.P. and D. Middleton, 1962: Sampling and reconstruction of wave-number-limited functions in N-dimensional Euclidean space, *Inform. and Control,* **5**, 279–323.

Pollard, R.T., 1983: Mesoscale (50–100 km) circulations revealed by inverse and classical analysis of the JASIN hydrographic data. *J. Phys. Oc.* **13**, 377–394.

Press, F., 1968: Earth models obtained by Monte Carlo inversion, *J. Geophys. Res.,* **73**, 5223–5234.

Riley, G.A., 1951: Oxygen, phosphate and nitrate in the Atlantic Ocean. *Bulletin of the Bingham Oceanographic Collection,* **13: 1**, 126 pp.

Rintoul, S., 1988: *Mass, heat and nutrient fluxes in the Atlantic Ocean determined by inverse methods.* PhD Thesis, Massachusetts Institute of Technology/Woods Hole Oceanographic Institution, 275 pp.

Roache, P.J., 1976: *Computational Fluid Dynamics,* Hermosa, Albuquerque, NM, 446 pp.

Roemmich, D., 1980: Estimation of meridional heat flux in the North Atlantic by inverse methods. *J. Phys. Oc.,* **10**, 1972–1983.

Roemmich, D., 1981: Circulation of the Caribbean Sea: a well-resolved inverse problem. *J. Geophys. Res.,* **86**, 7993–8005.

Roemmich, D., 1983: Optimal estimation of hydrographic station data and derived fields. *J. Phys. Oc.,* **13**, 1544–1549.

Roemmich, D. and C. Wunsch, 1982: On combining satellite altimetry with hydrographic data. *J. Mar. Res.,* **40**(supplement), 605–619.

Roemmich, D. and C. Wunsch, 1985: Two transatlantic sections: Meridional circulation and heat flux in the subtropical North Atlantic Ocean. *Deep-Sea Res.,* **32**, 619–664.

Roemmich, D. and T. McCallister, 1988: *Large scale circulation of the North Pacific Ocean.* Unpublished manuscript.

Scales, L.E., 1985: *Introduction to Non-Linear Optimization,* Springer-Verlag, New York, 243 pp.

Schlitzer, R., 1987: Renewal rates of East Atlantic deep water estimated by inversion of 14C data. *J. Geophys. Res.,* **92**, 2953–2961.

Schlitzer, R., 1988: Modeling the nutrient and carbon cycles of the North Atlantic. Part 1: Circulation, mixing coefficients, and heat fluxes. *J.Geophys.Res.,* **93**,10699–10723.

Schott, F. and H. Stommel, 1978: Beta-spirals and absolute velocities in different oceans. *Deep-Sea Res.*, **25**, 961–1010.

Schott, F. and R. Zantopp, 1979: Calculation of absolute velocities from different parameters in the western North Atlantic. *J. Geophys. Res.* **84**, 6990–6994.

Schröter, J. and C. Wunsch, 1986: Solution of non-linear finite difference ocean models by optimization methods with sensitivity and observational strategy analysis. *J. Phys. Oc.*, **16**, 1855–1874.

Seber, G.A.F., 1977: *Linear Regression Analysis,* John Wiley and Sons, New York, 465 pp.

Sheppard, C.W., 1962: *Basic Principles of the Tracer Method,* John Wiley, New York, 282 pp.

Stavroulakis, P., ed., 1983: *Distributed Parameter System Theory, Part 1, Control; Part 2, Estimation,* Hutchinson Ross, Stroudsburg, Penn., 396 pp. and 391 pp.

Stengel, R.F., 1986: *Stochastic Optimal Control,* Wiley-Interscience, N.Y., 638 pp.

Strang, G., 1986: *Introduction to Applied Mathematics,* Wellesley-Cambridge Press, Wellesley, Mass., 758 pp.

Tarantola, A., 1987: *Inverse Problem Theory. Methods for Data Fitting and Model Parameter Estimation,* Elsevier, Amsterdam, 613 pp.

Tziperman, E., 1988: Calculating the time-mean oceanic general circulation and mixing coefficients from hydrographic data, *J. Phys. Oc.,* **18** 519–525

Wagner, H.M., 1969: *Principles of Operation Research. With Applications to Managerial Decisions,* Prentice-Hall, Englewood Cliffs, N.J., 937 pp., plus appendices.

Weiss, R.F., J.L. Bullister, R.H. Gammon and M.J. Warner, 1985: Atmospheric chlorofluoromethanes in the deep equatorial Atlantic. *Nature* **314**, 608–610.

Wiggins, R.A., 1972: The general linear inverse problem: Implication of surface waves and free oscillations for earth structure. *Revs. Geophys. and Space Phys.,* **10**, 251–285.

Worthington, L.V., 1976: *On the North Atlantic Circulation,* Johns Hopkins U. Press, Baltimore, 110 pp.

Wunsch, C., 1977: Determining the general circulation of the oceans: A preliminary discussion. *Science,* **196**, 871–875.

Wunsch, C., 1978: The North Atlantic general circulation west of 50°W determined by inverse methods. *Revs. Geophys. and Space Phys.,* **16**, 583–620.

Wunsch, C., 1984: An eclectic Atlantic Ocean circulation model. Part 1: The meridional flux of heat. *J. Phys. Oc.,* **14,** 1712–1733.

Wunsch, C., 1985: Can a tracer field be inverted for velocity?, *J. Phys. Oc.,* **11,** 1521–1531.

Wunsch, C., 1987: Using transient tracers: The regularization problem. *Tellus,* **39**B, 477–492.

Wunsch, C., 1988: Transient tracers as a problem in control theory. *J. Geophys. Res.,* **93,** 8099–8110

Wunsch, C., 1988: Eclectic modelling of the North Atlantic, Part 2: Transient tracers and the ventilation of the eastern basin thermocline. *Phil. Trans. Roy. Soc. A,* **325,** 201–236

Wunsch, C. and B. Grant, 1982: Towards the general circulation of the North Atlantic Ocean. *Prog. in Oceanogr.,* **11,** 1–59.

Wunsch, C., D.-X. Hu, and B. Grant, 1983: Mass, heat, salt and nutrient fluxes in the South Pacific Ocean. *J. Phys. Oc.,* **13,** 725–753.

Wunsch, C. and J.-F. Minster, 1982: Methods for box models and ocean circulation tracers: Mathematical programming and non-linear inverse theory. *J. Geophys. Res.,* **87,** 5647–5662.

Wüst, G., 1935: Schichtung und Zirkulation des Atlantischen Ozeans. Die Stratosphäre. Wissenschaftliche Ergebnisse der Deutschen Atlantischen Expedition auf dem Forschungs-und Vermessungsschiff 'Meteor' 1925–1927. English translation by W.J. Emery, 1978, of Vol. VI, Section 1, *The Stratosphere of the Atlantic Ocean,* Amerind Publishing Co., New Delhi, 112 pp.

A GEOMETRICAL INTERPRETATION OF INVERSE PROBLEMS

by
Dirk Olbers,
Alfred-Wegener-Institut
für Polar- und Meeresforschung, Bremerhaven

Given the answer, what was the question. (N. FOFONOFF)

Most data interpretation problems in geophysics have no unique solutions since the information collected about a set of unknown parameters is frequently insufficient or contradictory or even both. Consider for simplicity a linear problem

$$\sum_{k=1}^{K} D_{lk}\, p_k = b_l \qquad (l = 1, \ldots, L) \tag{1}$$

with unknowns p_k, $k = 1, \ldots, K$. This has no solution at all if $L > K$ (overdetermined case) and an infinity of solutions if $L < K$ (underdetermined case). We assume here for the moment that (1) is not further degenerated, i.e. that D_{lk} has the full rank (either L or K). Unique solutions for the underdetermined case can be only given if some additional criteria of selection on the infinity of solutions is assumed. One can of course also consider the complete $K - L$-dimensional hyperplane defined by (1) as "the solution" of the problem. Solutions for the overdetermined case can only be meaningfully defined if $L - K$ equations are discarded. We will outfit here the possible solutions to these problems with geometrical interpretations. The content of this paper is a supplement to WUNSCH's outline of inverse problems in this book.

1   THE OVERDETERMINED CASE

The general procedure with (1) is simply demonstrated by some schematic pictures. Consider first the overdetermined case. Think of the b_l, $l = 1, \ldots, L$ as your data and the p_k, $k = 1, \ldots, K$ as your parameters and the $\sum_{1}^{K} D_{lk}\, p_k$ as your model you are trying to fit to the data. The data $(b_l) = \mathbf{b}$ is an L-dim vector in the data space D_L. The model space defined by all vectors

$$\sum_{k=1}^{K} p_k\, \mathbf{a}_k = \sum_{k=1}^{K} p_k\, (D_{1k}, \ldots, D_{Lk}) \tag{2}$$

D. L. T. Anderson and J. Willebrand (eds.), Oceanic Circulation Models: Combining Data and Dynamics, 79–93.

formed for all p_k is a K-dim subspace M_K in D_L, i.e. $M_K \subset D_L$, as indicated in figure 1. In this situation you will never be able to satisfy (1) with any set of parameters. Only if $\mathbf{b} \in M_K$, you will succeed.

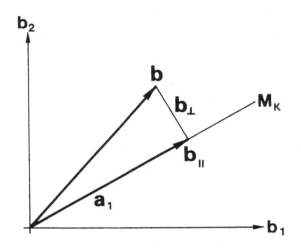

Figure 1: A schematic view of the data vector $\mathbf{b}$ and the K-dimensional model subspace M_K in the data space

Now, what can one do if $\mathbf{b} \notin M_K$? Apparently, the best one could achieve is to come in model space as close to $\mathbf{b}$ as possible, i.e. to define $\mathbf{b}_{\parallel} \in M_K$ as "the solution" of the problem (1) where

$$\left\| \mathbf{b} - \mathbf{b}_{\parallel} \right\| = \|\mathbf{b}_{\perp}\| = \min \tag{3}$$

with some meaningful norm in the data space. If W_{lj} is the metric in D_L, i.e. a positive definite and symmetric matrix, then (3) may be written in the form

$$\varepsilon^2 = \left\| \mathbf{b} - \mathbf{b}_{\parallel} \right\| = \left\| \mathbf{b} - \sum_{k=1}^{K} p_k\, \mathbf{a}_k \right\|$$

$$= \sum_{lj} \left(b_l - \sum_k D_{lk}\, p_k \right) W_{lj} \left(b_j - \sum_{k'} D_{jk'}\, p_{k'} \right) = \min \tag{4}$$

with the yet to be determined parameter set p_k corresponding to the vector $\mathbf{b}_{\parallel}$. From $\partial \varepsilon^2 / \partial p_k = 0$ one arrives at

$$\sum_{lj} D_{lk}\, W_{lj} \left(b_j - \sum_{k'} D_{jk'}\, p_{k'} \right) = 0$$

or in matrix notation

$$\mathcal{D}^{\dagger}\, \mathcal{W}\, \mathcal{D}\, \mathbf{p} = \mathcal{D}^{\dagger}\, \mathcal{W}\, \mathbf{b} \tag{5}$$

where $\mathcal{D} = (D_{lk})$ is a matrix with L rows and K columns, $\mathcal{W}$ is the square matrix (W_{lj}) and $\mathbf{p}$ and $\mathbf{b}$ are the parameter and data vector, respectively. Furthermore $\mathcal{D}^{\dagger} = (D_{kl})$ is the transpose of $\mathcal{D}$.

The remaining task obviously is to invert the $K \times K$-matrix $\mathcal{D}^{\dagger}\, \mathcal{W}\, \mathcal{D}$. If it exists then our solution of the overdetermined case is found to be

$$\mathbf{p} = \left(\mathcal{D}^{\dagger}\, \mathcal{W}\, \mathcal{D} \right)^{-1} \mathcal{D}^{\dagger}\, \mathcal{W}\, \mathbf{b}. \tag{6}$$

Unfortunatly in many problems occurring in geophysics a regular inverse does not exist, i.e. the $\mathcal{D}^{\dagger}\, \mathcal{W}\, \mathcal{D}$ is almost singular. If then the values of some of the D_{lk} are only slightly changed (if they come from measurements with errors) the solution (6) will change drastically and hence has no real value. The case which is most frequently encountered in this respect is one in which it is attempted to determine parameters which are not constrained enough by the model. As an example consider a matrix D_{lk} with $D_{l1} \ll 1$ for all $l = 1, \ldots, L$ and all other $D_{lk} = O(1)$. Obviously the parameter p_1 is not very meaningful in such a model and trying to determine it will spoil the entire solution $\mathbf{p}$ (and not only yield a "wrong" p_1) because of the near-singularity of the matrix $\mathcal{D}$. To avoid such behaviour one must study the condition of $\mathcal{D}^{\dagger}\, \mathcal{W}\, \mathcal{D}$ carefully and, in case of singularity, come up with a unified procedure which still gives a meaningful solution. This, of course, is possible. Just some bad values in D_{lk} do not make the entire information worthless! A procedure which extracts the meaningful information and points out the bad parameters is based on the singular value decomposition (SVD) explained below in section 3.

2   THE UNDERDETERMINED CASE

Let us now consider the underdetermined case and give (1) a unique solution when $K > L$. For this case view $\mathbf{p} = (p_1, \ldots, p_K)$ as a vector in the parameter space P_K of K dimensions as in figure 2. Equation (1) defines a $K - L$-dimensinal hyperplane

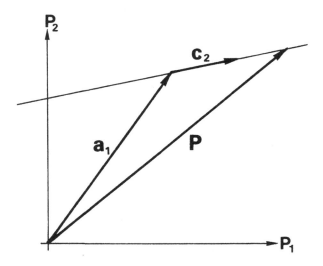

Figure 2: A schematic view of the $K - L$-dimensional subspace of admissible solutions $\mathbf{p}$ of the underdetermined case

in P_K. We may characterize this surface by any solution of (1), say $\mathbf{a}$, and a set of $K - L$ vectors $\{\mathbf{c}_l, l = L + 1, \ldots, K\}$ so that any point in the hyperplane can be represented as

$$\mathbf{p} = \mathbf{a} + \sum_{l=L+1}^{K} \alpha_l \, \mathbf{c}_l \tag{7}$$

by varying the set of coefficients α_l. The task, in principle, would be, to determine $\mathbf{a}$ and the $\mathbf{c}_l$ from

$$\begin{aligned} \mathcal{D}\,\mathbf{a} &= \mathbf{b} \\ \mathcal{D}\,\mathbf{c}_l &= \mathbf{0} \end{aligned} \tag{8}$$

Equation (7) gives all solutions to our problem. If one is not satisfied with this infinity and rather likes to pick out a unique one, there must be an additional constraint. Now, in general an underdetermined problem as (1) arises in an iterative procedure. A first guess $\mathbf{p}^0$ of the parameter vector is known and the true value $\mathbf{p}$

must satisfy some set of relations

$$g_l(\mathbf{p}) = 0 \qquad (l = 1, \ldots, L). \tag{9}$$

Expanding this about $\mathbf{p}^0$

$$g_l(\mathbf{p}^0) + \sum_{k=1}^{K} \left.\frac{\partial g_l}{\partial p_k}\right|_{\mathbf{p}^0} \left(p_k - p_k^0\right) + \cdots = 0 \tag{10}$$

one arrives at (1) with

$$
\begin{aligned}
D_{lk} &= \left.\frac{\partial g_l}{\partial p_k}\right|_{\mathbf{p}^0} \\
b_l &= -g_l(\mathbf{p}^0) + D_{lk}\, p_k^0
\end{aligned}
\tag{11}
$$

In this case, because of termination of the Taylor series (10), one would look for small improvements $\mathbf{p} - \mathbf{p}^0$ of the first guess and then eventually repeat the expansion about the improved parameter value $\mathbf{p}$.

A meaningful procedure is thus to choose the vector $\hat{\mathbf{p}}$ on the hyperplane given by (1), which has the smallest distance to $\mathbf{p}^0$ as indicated in figure 3. If $M_{kk'}$ is the metric in the parameter space the solution of (1) will thus be made unique by requiring

$$\varepsilon^2(\hat{\mathbf{p}}) = \sum_{k,k'} \left(\hat{p}_k - p_k^0\right) M_{kk'} \left(\hat{p}_{k'} - p_{k'}^0\right) = \min \tag{12}$$

The solution of (12) and (1) is found by considering the variational problem

$$V(\hat{\mathbf{p}}, \boldsymbol{\gamma}) = \varepsilon^2(\hat{p}) + \sum_{l=1}^{L} \gamma_l \left(\sum_k D_{lk}\, \hat{p}_k - b_l\right) = \min \tag{13}$$

where $\boldsymbol{\gamma} = (\gamma_l)$ is an L-dim vector of Lagrange multipliers. Variation of V with respect to $\boldsymbol{\gamma}$ recovers (1) and variation with respect to $\hat{p}$ yields the K equations

$$2 \sum_{k'} M_{kk'} \left(\hat{p}_{k'} - p_k^0\right) + \sum_l \gamma_l D_{lk} = 0. \tag{14}$$

In matrix notation (1) and (14) are

$$
\begin{aligned}
\mathcal{D}\,\hat{\mathbf{p}} &= \mathbf{b} \\
2\,\mathcal{M}\,(\hat{\mathbf{p}} - \mathbf{p}^0) &= \mathcal{D}^\dagger\,\boldsymbol{\gamma}
\end{aligned}
\tag{15}
$$

Since $\mathcal{M}$ can be inverted this yields an equation for $\boldsymbol{\gamma}$

$$\mathcal{D}\,\mathcal{M}^{-1}\,\mathcal{D}^\dagger\,\boldsymbol{\gamma} = 2\,\mathcal{D}\left(\hat{\mathbf{p}} - \mathbf{p}^0\right) = 2\left(\mathbf{b} - \mathcal{D}\,\mathbf{p}^0\right). \tag{16}$$

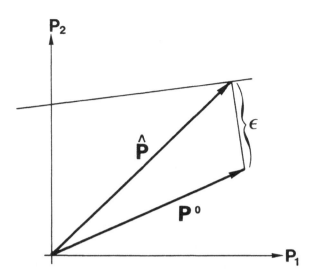

Figure 3: Choosing the unique solution $\hat{p}$ with minimum distance to a first guess P_0

Here $\mathcal{D}\,\mathcal{M}^{-1}\,\mathcal{D}^\dagger$ is a $L \times L$ matrix. If it can be inverted we have

$$\gamma = 2\left(\mathcal{D}\,\mathcal{M}^{-1}\,\mathcal{D}^\dagger\right)^{-1}(\mathbf{b} - \mathcal{D}\,\mathbf{p}^0) \tag{17}$$

and from (15) the final solution is written in the form

$$\begin{aligned}
\hat{\mathbf{p}} &= \mathbf{p}^0 + \mathcal{M}^{-1}\,\mathcal{D}^\dagger\left(\mathcal{D}\,\mathcal{M}^{-1}\,\mathcal{D}^\dagger\right)^{-1}\left(\mathbf{b} - \mathcal{D}\,\mathbf{p}^0\right) \\
&= \left[1 - \mathcal{M}^{-1}\,\mathcal{D}^\dagger\left(\mathcal{D}\,\mathcal{M}^{-1}\,\mathcal{D}^\dagger\right)^{-1}\mathcal{D}\right]\mathbf{p}^0 + \\
&\quad + \mathcal{M}^{-1}\,\mathcal{D}^\dagger\left(\mathcal{D}\,\mathcal{M}^{-1}\,\mathcal{D}^\dagger\right)^{-1}\mathbf{b}.
\end{aligned} \tag{18}$$

This is a rather complicated expression but the two contributions

$$\begin{aligned}
\mathbf{c} &= \left[1 - \mathcal{M}^{-1}\,\mathcal{D}^\dagger\left(\mathcal{D}\,\mathcal{M}^{-1}\,\mathcal{D}^\dagger\right)^{-1}\mathcal{D}\right]\mathbf{p}^0 \\
\mathbf{d} &= \mathcal{M}^{-1}\,\mathcal{D}^\dagger\left(\mathcal{D}\,\mathcal{M}^{-1}\,\mathcal{D}^\dagger\right)^{-1}\mathbf{b}
\end{aligned} \tag{19}$$

which make up the solution $\hat{p}$ have a simple interpretation. It is easy to show the following properties by simple matrix operations

1. $\mathbf{c}^\dagger \mathcal{M} \mathbf{d} = 0$

   i.e. $\mathbf{c}$ and $\mathbf{d}$ are orthogonal with respect to the metric $\mathcal{M}$ in the parameter space

2. $\mathcal{D} \mathbf{d} = \mathbf{b}$

   i.e. $\mathbf{d}$ is a solution of (1) (but not the one with minimum distance to $\mathbf{p}^0$, it has minimum distance to the origin, see figure 4)

3. $\mathcal{D} \mathbf{c} = 0$

   i.e. $\mathbf{c}$ is contained in the vector space $\{\mathbf{c}_l, l = L+1, \ldots, K\}$ introduced above in (8).

Hence (18) has just the form (7) with $\mathbf{c}$ equal to a linear combination of the $\mathbf{c}_l$, i.e.

$$\mathbf{c} = \sum_{l=L+1}^{K} \alpha_l \, \mathbf{c}_l \tag{20}$$

and

$$\mathbf{d} = \mathbf{a}. \tag{21}$$

The values for the coefficients α_l and the vectors $\mathbf{c}_l$ will become evident in chapter 3.

The vector $\mathbf{c}$ is constructed from $\mathbf{p}^0$ by projecting off the part which is not in $K - L$-dim space $\{\mathbf{c}_l\}$ and $1 - \mathcal{M}^{-1} \mathcal{D}^\dagger \left(\mathcal{D} \mathcal{M}^{-1} \mathcal{D}^\dagger \right)^{-1} \mathcal{D}$ is just the projection operator to acheive this. Hence

$$\mathbf{p}^0 = \mathbf{p}^0_\| + \mathbf{p}^0_\perp \tag{22}$$

as shown in figure 4, where

$$
\begin{aligned}
\mathbf{p}^0_\| &= \mathbf{c} &= \left[1 - \mathcal{M}^{-1} \mathcal{D}^\dagger \left(\mathcal{D} \mathcal{M}^{-1} \mathcal{D}^\dagger \right)^{-1} \mathcal{D} \right] \mathbf{p}^0 \\
\mathbf{p}^0_\perp &= \mathbf{p}^0 - \mathbf{c} &= \mathcal{M}^{-1} \mathcal{D}^\dagger \left(\mathcal{D} \mathcal{M}^{-1} \mathcal{D}^\dagger \right)^{-1} \mathcal{D} \mathbf{p}^0
\end{aligned}
\tag{23}
$$

Notice that by varying $\mathbf{p}^0$ the entire hyperplane of solutions can be obtained.

The $\mathbf{c}$-part is the trivial part of the solution. It does not depend on the data $\mathbf{b}$ and only comes about because of improper representation. If we had taken the improvements $\mathbf{p} - \mathbf{p}^0 = \mathbf{p}'$ as the parameters to be determined in (10) the $\mathbf{c}$-part would disappear. In fact, from (10) in the form

$$\mathcal{D} \mathbf{p}' = -\mathbf{g} = \mathbf{b} - \mathcal{D} \mathbf{p}^0 \tag{24}$$

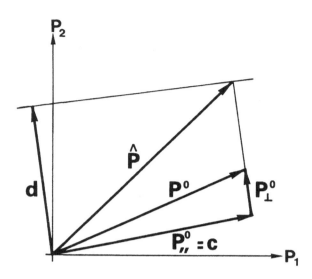

Figure 4: Separation of the minimum distance solution $\hat{\mathbf{p}}$ into the canonical parts $\mathbf{d}$ and $\mathbf{c}$

we find

$$\mathbf{p}' = \hat{\mathbf{p}} - \mathbf{p}^0 = \mathcal{M}^{-1} \mathcal{D}^{\dagger} \left(\mathcal{D} \mathcal{M}^{-1} \mathcal{D}^{\dagger} \right)^{-1} (-\mathbf{g}) \qquad (25)$$

which of course is identical to (18). Let us continue with this simpler expression in the form which is the heart of our solution. Remember that (25) is the solution of (24) with minimum norm, $\|\mathbf{p}'\| = \min$.

There are some formal similarities between the solutions (6) of the overdetermined case and (25) for the underdetermined case. In the first case $(L > K)$ the oversupply of information in the L-dim data vector $\mathbf{b}$ is reduced to K dimensions by operating with $\mathcal{D}^{\dagger} \mathcal{W}$ and this reduced information is then distributed on the parameter set via the transformation $\left(\mathcal{D}^{\dagger} \mathcal{W} \mathcal{D} \right)^{-1}$, a $K \times K$ matrix. In the other case $(K > L)$ we have an undersupply of information in the L-dim data vector $\mathbf{b}$. This is first redistributed by the $L \times L$-transformation $\left(\mathcal{D} \mathcal{M}^{-1} \mathcal{D}^{\dagger} \right)^{-1}$ and then blown up to the higher dimensional parameter vector of K dimensions by operating with $\mathcal{M}^{-1} \mathcal{D}^\dagger$. The common task in both cases is the inversion of a matrix, in each case in the space of lesser dimensions.

3   THE SINGULAR VALUE DECOMPOSITION (SVD)

The overdetermined case boiled down to solving (5), i.e. inverting $\mathcal{D}^\dagger \mathcal{W} \mathcal{D}$, and we pointed out that a regular inverse may not exist. In this case one is advised to consider the spectrum of eigenvalues $(\lambda^{(k)})^2$, $k = 1, \ldots, K$ (they are positive since $\mathcal{D}^\dagger \mathcal{W} \mathcal{D}$ is by construction a positive definite matrix). For sake of clearness (to get rid of the weighting matrix $\mathcal{W}$ and step into WUNSCH's notation), define

$$
\begin{aligned}
\mathcal{A} &:= \mathcal{W}^{1/2} \mathcal{D} \\
\mathbf{y} &:= \mathcal{W}^{1/2} \mathbf{b} \\
\mathbf{x} &:= \mathbf{p}
\end{aligned}
\tag{26}
$$

so that the model equations

$$
\mathcal{A}\mathbf{x} = \mathbf{y}
\tag{27}
$$

require that

$$
\mathcal{A}^\dagger \mathcal{A}\mathbf{x} = \mathcal{A}^\dagger \mathbf{y}
\tag{28}
$$

has to be solved. The eigenvalue problem of $\mathcal{A}^\dagger \mathcal{A}$

$$
\mathcal{A}^\dagger \mathcal{A} \mathbf{V}^{(k)} = \left(\lambda^{(k)}\right)^2 \mathbf{V}^{(k)} \qquad (k = 1, \ldots, K)
\tag{29}
$$

can be solved by standard techniques and, since $\mathcal{A}^\dagger \mathcal{A}$ is symmetric and selfadjoint, the eigenvectors $\mathbf{V}^{(k)}$ are pairwise orthogonal if the eigenvalues differ and can be normalized so that

$$
\left(\mathbf{V}^{(k)}\right)^\dagger \cdot \mathbf{V}^{(k')} = \delta_{kk'} \qquad \text{if } \lambda^{(k)} \neq \lambda^{(k')}.
\tag{30}
$$

The matrix $\mathcal{A}^\dagger \mathcal{A}$ can be represented as

$$
\mathcal{A}^\dagger \mathcal{A} = \sum_{k=1}^{K} \left(\lambda^{(k)}\right)^2 \mathbf{V}^{(k)} \left(\mathbf{V}^{(k)}\right)^\dagger.
\tag{31}
$$

Assuming for the moment that all $\left(\lambda^{(k)}\right)^2$ are non-zero the inverse would be

$$
\left(\mathcal{A}^\dagger \mathcal{A}\right)^{-1} = \sum_{k=1}^{K} \frac{1}{(\lambda^{(k)})^2} \mathbf{V}^{(k)} \left(\mathbf{V}^{(k)}\right)^\dagger
\tag{32}
$$

and (28) is then solved in the form

$$\mathbf{x} = \sum_{k=1}^{K} \frac{1}{\left(\lambda^{(k)}\right)^2} \mathbf{V}^{(k)} \left(\mathbf{V}^{(k)}\right)^\dagger \cdot \mathcal{A}^\dagger \mathbf{y}. \tag{33}$$

The matrix $\mathcal{A}$ appears in this solution in the form of the L-vector $\left(\mathbf{V}^{(k)}\right)^\dagger \cdot \mathcal{A}^\dagger = \left(\mathcal{A}\mathbf{V}^{(k)}\right)^\dagger$ and it will turn out suitable to define an L-vector $\mathbf{U}^{(k)}$ by

$$\mathcal{A}\mathbf{V}^{(k)} = \lambda^{(k)}\mathbf{U}^{(k)} \tag{34}$$

which implies from (29) that

$$\mathcal{A}^\dagger \mathbf{U}^{(k)} = \lambda^{(k)}\mathbf{V}^{(k)} \tag{35}$$

and

$$\mathcal{A}\mathcal{A}^\dagger \mathbf{U}^{(k)} = \left(\lambda^{(k)}\right)^2 \mathbf{U}^{(k)}. \tag{36}$$

The $\mathbf{U}^{(k)}$ are thus the K non-trivial eigenvectors of the (big) $L \times L$-matrix $\mathcal{A}\mathcal{A}^\dagger$ which has the rank K and the same eigenvalues as the (small) $K \times K$-matrix $\mathcal{A}^\dagger\mathcal{A}$.

The $\mathbf{U}^{(k)}$ are orthogonal and are taken normalized, i.e.

$$\left(\mathbf{U}^{(k)}\right)^\dagger \mathbf{U}^{(k')} = \delta_{kk'} \qquad \text{if } \lambda^{(k)} \neq \lambda^{(k')}. \tag{37}$$

The solution of (28) may then be expressed in the form

$$\mathbf{x} = \sum_{k=1}^{K} \frac{1}{\lambda^{(k)}} \mathbf{V}^{(k)} \left(\mathbf{U}^{(k)}\right)^\dagger \cdot \mathbf{y} \tag{38}$$

This form has a very vivid interpretation since it makes explicit what part of the data $\mathbf{y}$ are discarded. It is just the part which lies not in the K-dim subspace spanned by the eigenvectors $\mathbf{U}^{(k)}$, $k = 1, \ldots, K$. Notice further that (38) is the expansion of the parameter set $\mathbf{x}$ in terms of the K-dim eigenvectors $\mathbf{V}^{(k)}$. To repeat the usefullness of these two sets of eigenvectors: the $\mathbf{V}^{(k)}$ form of complete orthogonal basis in the K-dim parameter space and the $\mathbf{U}^{(k)}$ determine the K-dim subspace of the L-dim data space, from which information is extracted to construct the solution (38). Going back to the figure 1 the $\mathbf{U}^{(k)}$ are an orthogonal basis of the subspace M_K containing the parallel component $\mathbf{b}_\parallel$.

We may even go one step further to make the picture even more translucent. Multiplying (38) by $\left(\mathbf{V}^{(k)}\right)^\dagger$ it is seen that the linear combination

$$\xi^{(k)} := \left(\mathbf{V}^{(k)}\right)^\dagger \cdot \mathbf{x} = \frac{1}{\lambda^{(k)}} \left(\mathbf{U}^{(k)}\right)^\dagger \cdot \mathbf{y} =: \frac{1}{\lambda^{(k)}} \eta^{(k)} \quad (k = 1, \ldots, K) \tag{39}$$

of parameters (i.e. the projection of $\mathbf{x}$ on the direction $\mathbf{V}^{(k)}$) gets all its information from the component $\eta^{(k)} = \left(\mathbf{U}^{(k)}\right)^{\dagger} \cdot \mathbf{y}$ of $\mathbf{y}$ in the direction of $\mathbf{U}^{(k)}$. Hence the $\xi^{(k)}$ and $\eta^{(k)}$ can be viewed as the canonical variables of this problem.

Evidently the SVD formalism may also be used to represent the solution (25) of the underdetermined case (the full power, however, will be laid out later in section 3.2). We just list the corresponding relations:

$$
\begin{aligned}
\mathcal{A} \quad &:= \quad \mathcal{D}\,\mathcal{M}^{-1/2} \\
\mathcal{A}^{\dagger} \quad &= \quad \mathcal{M}^{-1/2}\,\mathcal{D}^{\dagger} \\
\left(\mathcal{A}\,\mathcal{A}^{\dagger}\right)^{-1} \quad &= \quad \left(\mathcal{D}\,\mathcal{M}^{-1}\,\mathcal{D}^{\dagger}\right)^{-1} = \sum_{l=1}^{L} \frac{1}{(\lambda^{(l)})^{2}}\, \mathbf{U}^{(l)} \left(\mathbf{U}^{(l)}\right)^{\dagger}
\end{aligned} \tag{40}
$$

The parameters appear as $\mathbf{f} = \mathcal{M}^{1/2}\,\mathbf{p}'$ and the solution of $\mathcal{A}\mathbf{f} = -\mathbf{g}$ takes the form

$$
\mathbf{f} = \mathcal{M}^{1/2}\,\mathbf{p}' = \sum_{l=1}^{L} \frac{1}{\lambda^{(l)}}\, \mathbf{V}^{(l)} \left(\mathbf{U}^{(l)}\right)^{\dagger} \cdot (-\mathbf{g}) \tag{41}
$$

Here the components $\gamma^{(l)} = \left(\mathbf{U}^{(l)}\right)^{\dagger} \cdot (-\mathbf{g})$ contain all the available information. As before, one can find the canonical variables

$$
\sigma^{(l)} := \left(\mathbf{V}^{(l)}\right)^{\dagger} \mathbf{f} = -\frac{1}{\lambda^{(l)}} \left(\mathbf{U}^{(l)}\right)^{\dagger} \mathbf{g} = \frac{1}{\lambda^{(l)}}\,\gamma^{(l)} \qquad (l = 1,\ldots,L) \tag{42}
$$

but notice that here you have less (L) canonical parameters $\sigma^{(l)}$ than original parameters p_k. Only the L combinations $\sigma^{(l)}$ get independent information (in form of the $\gamma^{(l)}$) which then is blown up to the K parameters $\mathbf{p}$.

It is worth to rewrite the general solution (18) which still includes the non-zero first guess vector $\mathbf{p}^0$. We have identified the solution as the sum of the "first-guess part" $\mathbf{c}$ and "data-part" $\mathbf{d}$ given in (19). In the transformed variables these relations read

$$
\begin{aligned}
\hat{\mathbf{f}} \quad &= \quad \mathcal{M}^{1/2}\hat{\mathbf{p}} = \mathbf{c}^0 + \mathbf{d}^0 \\
\mathbf{c}^0 \quad &= \quad \mathcal{M}^{1/2}\mathbf{c} = \mathbf{f}^0 - \mathcal{A}^{\dagger} \left(\mathcal{A}\,\mathcal{A}^{\dagger}\right)^{-1} \mathcal{A}\,\mathbf{f}^0 \\
\mathbf{d}^0 \quad &= \quad \mathcal{M}^{1/2}\mathbf{d} = \mathcal{A}^{\dagger} \left(\mathcal{A}\,\mathcal{A}^{\dagger}\right)^{-1} \mathbf{b}
\end{aligned} \tag{43}
$$

with $\mathbf{f}^0 = \mathcal{M}^{1/2}\,\mathbf{p}^0$. Expressing the matrices in the SVD form we find

$$
\begin{aligned}
\mathbf{c}^0 \quad &= \quad \left\{ 1 - \sum_{l=1}^{L} \frac{1}{\lambda^{(l)}}\, \mathbf{V}^{(l)} \left(\mathbf{V}^{(l)}\right)^{\dagger} \right\} \cdot \mathbf{f}^0 \\
\mathbf{d}^0 \quad &= \quad \sum_{l=1}^{L} \frac{1}{\lambda^{(l)}}\, \mathbf{V}^{(l)} \left(\mathbf{U}^{(l)}\right)^{\dagger} \cdot \mathbf{b}
\end{aligned} \tag{44}
$$

The operator in the curly brackets projects the first guess vector $\mathbf{f}^0$ into the $K - L$-dim subspace of the parameter space which is orthogonal to the L-dim subspace spanned by L eigenvectors $\mathbf{V}^{(l)}$, $l = 1, \ldots, L$. If we supplement these vectors by $K - L$ additional $\mathbf{V}^{(l)}$ to form a complete orthogonal base $\mathbf{V}^{(l)}$, $l = 1, \ldots, K$ then

$$
\mathbf{c} = \mathcal{M}^{-1/2} \mathbf{c}^0 = \mathcal{M}^{-1/2} \sum_{l=L+1}^{K} \mathbf{V}^{(l)} \left(\mathbf{V}^{(l)}\right)^{\dagger} \cdot \mathbf{f}^0 = \sum_{l=L+1}^{K} \alpha_l \, \mathbf{c}_l \tag{45}
$$

and the connection to the representation (7) or (20) of the solution is immediately evident, i.e.

$$
\begin{aligned}
\mathbf{c}_l &= \mathcal{M}^{-1/2} \mathbf{V}^{(l)} \\
\alpha_l &= \left(\mathbf{V}^{(l)}\right)^{\dagger} \cdot \mathbf{f}^0
\end{aligned} \tag{46}
$$

3.1 The overdetermined-underconstrained case

How should one now proceed if one (or some) of the eigenvalues $\lambda^{(k)}$ vanish or are very small compared to the rest, or—speaking in mathematical terminology—if the condition of the matrix A measured in terms of

$$
C = \frac{\min\left\{\lambda^{(k)}\right\}}{\max\left\{\lambda^{(k)}\right\}} \tag{47}
$$

is small? As explained above this often results when the model does not sufficiently constrain the solution so that the problem though being overdetermined is in fact underconstrained.

Let us assume that the eigenvalue $\lambda^{(k')}$ vanishes (the following framework can easily be extended to a whole subset of vanishing eigenvalues). The eigenvector $\mathbf{V}^{(k')}$ is then annihilated by the matrix A, i.e.

$$
A \, \mathbf{V}^{(k')} = 0 \tag{48}
$$

it forms the null-space of A. Evidently, from (39), we cannot determine the parameter combination $\xi^{(k')}$, i.e. the projection of $\mathbf{x}$ onto the null-space of A, in any meaningful way. Indeed the model equation $\lambda^{(k')} \xi^{(k')} = \eta^{(k')}$ with $\lambda^{(k')} = 0$ has no solution unless incidently $\eta^{(k')} = 0$ in which case $\xi^{(k')}$ can have any value.

Now, in defining the solution for the overdetermined case we had to discard the part $\mathbf{b}_\perp$ outside M_K and required $\mathbf{b}_\perp$ to have minimum norm. Hence a straightforward recipe for the underconstrained case would be to continue to discard information and reduce now even $\mathbf{b}$ inside M_K. The recipe is then to discard $\eta^{(k')} =$

$\left(\mathbf{U}^{(k')}\right)^{\dagger} \cdot \mathbf{y}$ for vanishing $\lambda^{(k')}$ and define the solution as

$$\mathbf{x} = \sum_{\substack{k=1 \\ k \neq k'}}^{K} \frac{1}{\lambda^{(k)}} \mathbf{V}^{(k)} \left(\mathbf{U}^{(k)}\right)^{\dagger} \cdot \mathbf{y} =: \left(\mathcal{A}^{\dagger} \mathcal{A}\right)_{\text{SVD}}^{-1} \mathcal{A}^{\dagger} \mathbf{y} \qquad (49)$$

which defines the generalized inverse $\left(\mathcal{A}^{\dagger} \mathcal{A}\right)_{\text{SVD}}^{-1}$ in the SVD-framework. Notice that according to (49) $\xi^{(k')}$ gets the value zero. If from some other data one has a better estimate $\xi_0^{(k')}$ this may just be added as $\xi_0^{(k')} \mathbf{V}^{(k')}$.

There is one more property of the solution (49) which is worth to mention. Since the $\xi^{(k)}$ and x_k are connected by an orthogonal transformation one has

$$\sum_{k=1}^{K} x_k^2 = \sum_{k=1}^{K} \left(\xi^{(k)}\right)^2 = \sum_{\substack{k=1 \\ k \neq k'}}^{K} \frac{1}{\left(\lambda^{(k)}\right)^2} \left(\eta^{(k)}\right)^2 \qquad (50)$$

So the Euclidian norm of $\mathbf{x}$ cannot be reduced unless more information is discarded. In this sense the SVD-solution is the one with minimum norm of the parameters.

3.2 Again the underdetermined case

Now that we have learned to define a meaningful inverse of a singular matrix let us return to the underdetermined case and its solution (41). To repeat, (41) is solution of

$$\mathcal{D} \mathbf{p}' = -\mathbf{g} \qquad (51)$$

subject to

$$\mathbf{p}'^{\dagger} \mathcal{M} \mathbf{p}' = \min \qquad (52)$$

In terms of the matrix $\mathcal{A}$ defined in (40) and the vector $\mathbf{f}$ defined in (41) this is

$$\begin{aligned} \mathcal{A} \mathbf{f} &= -\mathbf{g} \\ \mathbf{f}^{\dagger} \mathbf{f} &= \min \end{aligned} \qquad (53)$$

with the solution

$$\begin{aligned} \mathbf{f} = \mathcal{M}^{1/2} \mathbf{p}' &= \mathcal{A}^{\dagger} \left(\mathcal{A} \mathcal{A}^{\dagger}\right)^{-1} (-\mathbf{g}) \\ &= \sum_{l=1}^{L} \frac{1}{\lambda^{(l)}} \mathbf{V}^{(l)} \left(\mathbf{U}^{(l)}\right)^{\dagger} \cdot (-\mathbf{g}) \end{aligned} \qquad (54)$$

This expression used the eigenvalue representation of the (small) $L \times L$-matrix $\mathcal{A} \mathcal{A}^{\dagger}$ given in (40). However we may likewise use the (big) $K \times K$-matrix $\mathcal{A}^{\dagger} \mathcal{A}$ to construct the solution. This has always rank L or less. In fact, it is easy to show that

$$\left(\mathcal{A}^\dagger \mathcal{A}\right)^{-1}_{\text{SVD}} \mathcal{A}^\dagger = \sum_{l=1}^{L} \frac{1}{\lambda^{(l)}} \mathbf{V}^{(l)} \left(\mathbf{U}^{(l)}\right)^\dagger \tag{55}$$

If any of L eigenvalues $\lambda^{(l)}$ should vanish (or be small), i.e. if $\mathcal{A}\mathcal{A}^\dagger$ is singular, these of course must be cut out in all these expressions.

Hence, comparing (55) with (54)

$$\mathbf{f} = \mathcal{M}^{1/2} \mathbf{p}' = - \left(\mathcal{A}^\dagger \mathcal{A}\right)^{-1}_{\text{SVD}} \mathcal{A}^\dagger \mathbf{g} \tag{56}$$

which is the SVD-solution of

$$\mathcal{A}^\dagger \mathcal{A} \mathbf{f} = -\mathcal{A}^\dagger \mathbf{g} \tag{57}$$

obtained by multiplying our original problem (51) or (53) with $\mathcal{A}^\dagger$, i.e. by blowing these L equations up to K equations. Comparing now (27), (28) and (49) with (53), (57) and (56) we notice that the distinction between overdetermination and underdetermination is lost in the SVD-framework.

3.3   The tapered cut-off solution

The SVD-solution of our original problem (1) or (27) may be written as

$$\mathbf{x} = \sum_{\{\lambda^{(k)} \neq 0\}} \frac{1}{\lambda^{(k)}} \mathbf{V}^{(k)} \left(\mathbf{U}^{(k)}\right)^\dagger \cdot \mathbf{y} \tag{58}$$

for the overdetermined case and in similar form for the underdetermined case. This solution—as explained above—disregards any information which corresponds to the zero eigenvalues. Now in practice, even with appropriate scaling of the parameters, the eigenvalues may span several decades and decay exponentially. If the set of coefficients D_{lk} in our problem (1) arise from observations, too, or are subject to errors for other reasons it is very difficult to judge the reliability of small eigenvalues. In this case the resolution of the parameters from the data relies on the data noise, i.e. the attempt to extract more and more information from the data by resolving more and more parameters leads to situations that more and more information is drawn from the noise structure of the data since the "good" information has already been exhausted. On the other hand, if we cut off more and more of terms with the small $\lambda^{(k)}$'s in (58) the parameters become increasingly dependent, i.e. the resolution decreases.

Apparently, there is a trade-off between resolution and reliability of the parameters which is controlled by the range of the eigenvalues in the sum (58). Replacing (58) by the "tapered cut-off" version

$$\mathbf{x} = \sum_{k=1}^{K} \frac{\lambda^{(k)}}{\left(\lambda^{(k)}\right)^2 + \lambda_c^2} \mathbf{V}^{(k)} \left(\mathbf{U}^{(k)}\right)^{\dagger} \cdot \mathbf{y} \tag{59}$$

allows thus a control of the parameter variance on account of the resolution by the value of λ_c. The choice of λ_c is a matter of subjective judgement of the parameter and variance structure of the obtained solution, i.e. λ_c a control variable in the hands of the analyst, coming as an subjective element from outside much in the same way as the metric W_{lj} in (4) and $M_{kk'}$ in (12).

DETERMINING DIFFUSIVITIES FROM HYDROGRAPHIC DATA BY INVERSE METHODS WITH APPLICATIONS TO THE CIRCUMPOLAR CURRENT

by
Dirk Olbers and Manfred Wenzel

Alfred-Wegener-Institut
für Polar- und Meeresforschung, Bremerhaven

1   INTRODUCTION ...
TO ILL-POSED AND NOISY PROBLEMS

Several inverse models have recently been put forward for obtaining estimates of the unknown reference velocities on the basis of the balances for momentum, vorticity, heat and salt of the large scale geostrophic circulation in the ocean. Notable forerunners are the inverse method proposed by Wunsch (1978) and the β-spiral method of Stommel and Schott (1977). These have first been applied in strictly adiabatic versions in the North Atlantic (see also Wunsch and Grant 1982, Schott and Stommel 1978). Both approaches can be extended to diabatic conditions. For the β-spiral model the consideration of the diffusion terms in the balances for tracers and vorticity are conceptionally simple (e.g. Olbers et al. 1985, Bigg 1985). Also Wunsch's method has been generalized in this respect (see e.g. Wunsch 1984). We should also mention Hogg's model (Hogg 1987) which is an inversion of the advective-diffusive balances on two isopycnals subject to the geostrophic constraint between the isopycnals. The model was applied in the central North Atlantic.

All these approaches are able to give estimates of the parameters which control the diffusion of heat and salt. The results appear reasonable in areas where intense mixing is expected to occur. In regions of low mixing, however, the coefficients often are only marginally significant (see e.g. Olbers et al. 1985). In this paper we review the basic physics and the various technical and conceptual problems of estimation procedures which attempt to determine mixing parametrisation from hydrographic data. The review is guided by applications of various different forms of the β-spiral method to the hydrography of the Southern Ocean and finally, it introduces a new inverse model of the Antarctic Circumpolar Current.

95

D. L. T. Anderson and J. Willebrand (eds.), Oceanic Circulation Models: Combining Data and Dynamics, 95–139.
© *1989 by Kluwer Academic Publishers.*

The recognition of mixing properties from hydrographic data meets with considerable difficulties. Some of these are fairly obvious, as e.g. the effects of errors in the data. Other difficulties are of a more subtle nature. Of the various assumptions which have to be made in the course of any estimation procedure some are hidden so deeply in the approximation of physical principles or the mathematical estimation techniques that they may even slip through the caution of experienced people.

Estimation procedures as considered here are based on a combination of the observations and the mathematical relations between the ideal observations and the unkown parameters (the physical model) in some kind of inversion formalism (the invers model) which extracts estimates of the parameters from the observed data. At present, the main body of information about the large scale circulation and mixing in the ocean has been drawn from observations of temperature, salinity and some passive tracers such as oxygen and ^{14}C. The spreading of these properties is not a mere diffusion process but predominantly affected by advection through the mean currents. The determination of diffusion coefficients is thus coupled to the classical level-of-no-motion problem—the determination of absolute velocities from a hydrographic section. The physical model then must incorporate the advective-diffusive balance equations for the tracers and the momentum balance, usually in some approximated form adequate for a large scale flow. In terms of the unknown parameters—the diffusion coefficients and the reference velocities—the mathematical relations of such a model are linear so that linear inversion schemes may be used. Still, if large amounts of data and relations are involved the problem may be rather complex since intrinsic interdependencies and inconsistencies are not immediately evident.

1.1 A singular example

One of the most frequent fallacies of an inverse solution arises from the attempt to extract information from the data which is not really contained there, i.e. parameters should be estimated which are not or not adequately constrained by the relations and data considered. At first, it seems that such a mistake can easily be avoided by simple inspection of the data and the relevant physical mechanisms which are responsible for the shape of the data. And indeed, in low-order problems inspection generally will sort out such failures. However, in case of problems with many data and unknowns the failure will only be revealed deep in the elements of the inversion tools. An example may help here to elucidate this behaviour.

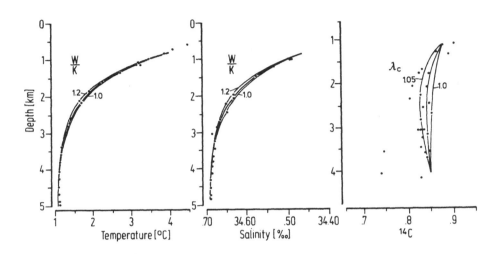

Figure 1

Potential temperature (a), salinity (b) and ^{14}C (c) from Fig. 3 and 5 of Munk (1966). Curves labeled w/K (in units km^{-1}) are based on (1.1) and curves labeled λ_c are based on (1.2).

A very early attempt to determine mixing parameters from tracer data can be found in Munk's abyssal recipes (Munk 1966). Assuming—not locally but for an entire ocean basin—a balance between vertical advection and vertical diffusion with a constant turbulent diffusivity K

$$w\,T_z = K\,T_{zz} \qquad (1.1)$$

for temperature and correspondingly for salinity, these properties vary in good agreement with observations (Fig. 1a and b) in an exponential depth dependence with a scale height K/w. Since temperature and salinity turn out to be highly correlated only one estimate K/w of .9 km can be obtained and Munk suggests to get further information about K and w from the ^{14}C-balance

$$K\,(^{14}C)_{zz} - w\,(^{14}C)_z = \mu\,(^{14}C) \qquad (1.2)$$

where $\mu = 3.93 \cdot 10^{-12} \, \text{sec}^{-1}$ is the known constant of radioactive decay. The exponential scale height of (1.2) is $(K/w)(1 \pm \alpha)$ where

$$\alpha^2 = 1 + 4\mu K/w^2 \tag{1.3}$$

Fig. 1c shows the fit to the ^{14}C data for $\alpha = 1.0$, 1.025 and 1.05. The first value yields an infinite w and (with $K/w = .9 \, \text{km}$) the latter yield $w = 2.8 \cdot 10^{-7} \, \text{m/sec}$, $K = 2.6 \cdot 10^{-4} \, \text{m}^2/\text{sec}$ and $w = 1.4 \cdot 10^{-7} \, \text{m/sec}$, $K = 1.3 \cdot 10^{-4} \, \text{m}^2/\text{sec}$, respectively. This is an uncomfortably large range for w and K in view of the 5% increase in the fitted parameter α. Obviously we are dealing here with a singular problem. This becomes apparent when expressing the two relations for K/w and w in a linear form

$$\begin{aligned} K/w &= .9 \, \text{km} \\ 4\mu K/w - (\alpha^2 - 1)\, w &= 0 \end{aligned} \tag{1.4}$$

The corresponding matrix with the two eigenvalues 1 and $1 - \alpha^2$ (i.e. 0.0, .051 and .10 for the above three estimates of α) clearly reveals the ill-posedness of the inverse problem. This shortcoming is not immediately evident from the physical model (1.1) and (1.2). Munk actually chose the largest of the three α's so we were taking a K of 1 cgs unit for the last two decades as canonical mixing constant of the ocean. With an α only a few percent less (which is still acceptable in view of the fit in Fig. 1c) we would have struggled with 3 cgs units or more!

Notice that the ill-posedness of the problem does not imply invalidity of the physical model, we are just asking too much if we want too many independent parameters from this model and the data. Notice further the potential fallacy in such a situation: the inversion always provides a set of parameters (unless the condition index is exactly zero) and only deeper insight into the inverse machinery and possibly a priori information about the expected solution prevents the acceptance of unreliable results. We will present more examples later in chapter 4.

1.2   Some noisy examples

We have mentioned the possible failure of parameter fitting from noisy observations. Hydrographic data contain errors from different sources. Instrumental noise may be important in great depths and high latitudes where temperature and salinity vary only a few thousands of a degree or per mille in the entire water column. Individual sections generally contain aliasing by internal waves and small-scale eddies due to

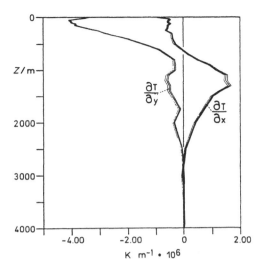

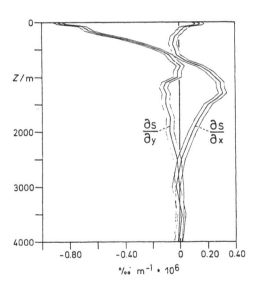

Figure 2
Gradients of temperature (a) and salinity (b) at 35° N, 25° W computed
from the Levitus atlas. Strips indicate the estimate of the standard deviation (from Olbers et al. 1985).

undersampling. The presence of data noise arising from such observational limitations apparently may lead to a masking of the mixing signatures so that mixing parameters become undeterminable. A more fundamental mistake may be introduced if the original data have been processed in some interpolation or objective analysis scheme which necessarily implies smoothing. The consequence may be a systematic bias of the mixing effects and the estimated mixing coefficients will be oversized.

If the noise statistics are known it is usually a simple task to study the effect on the estimated parameters by some sort of error propagation analysis. Most inverse techniques are formulated within a statistical framework and data noise is considered both in the weighting procedure of the model constraints as well as in the evaluation of second order statistics (variances and correlations) of the parameters (e.g. Wunsch this book). We will describe some mathematical aspects of the error analysis in the chapter 3.

It seems worth mentioning that the relevant quantities in our problem are not temperature and salinity or related scalars but their gradients since only these enter the physical model. Even though the scalars may be known accurately enough the gradients may be very small so that they are statistically not distinguishable from zero. Fig. 2 exemplifies this situation for Levitus' analysis (Levitus 1982) in the central North Atlantic. The estimation of the variance structure of this data set is discussed in detail in Olbers et al. (1985). At depths greater than 2500m horizontal variations of temperature and salinity become smaller than the estimated standard deviation and reliable gradients can not be given. This of course destroys any further attempt of parameter estimation in this depth range. The statement on erroneous deep gradients of temperature and salinity can however not be generalized as we will demonstrate later with the application of the Circumpolar Current.

Frequently the data situation is characterized by a considerable amount of ignorance about the possible structure of noise. One often does not even know how to get a reliable value for standard deviations. Hydrographic data are a vivid example in this respect since in many areas of the world ocean—particular high latitudes and greater depth—we have only very few observations and these are subject to a heavy seasonal bias. One hardly can construct a climatological mean not to speak of variances at all.

Finally we would like to address a problem which at first sight seems to be a mere manipulation of equations and not related to noise at all. Consider e.g. the physical model described in chapter 2, which consists of the advective-diffusive balance of a tracer supported by the geostrophic relations and the linearized vorticity balance. Suppose for simplicity that the profiles of the absolute horizontal velocities and the mixing parameters are already known. The remaining task is the determination of the profile of the vertical velocity w. In a straight forward way one would calculate w at each level writing the advective-diffusive tracer balance in the form

$$w = (D - u\tau_x - v\tau_y)/\tau_z \tag{1.5}$$

Alternatively the vorticity balance $fw_z = \beta v$ may be integrated with an unknown reference velocity w_0. This parameter can then be estimated by requiring that the imbalances $\varepsilon(z)$ at all levels z, given as

$$\varepsilon(z) = u\tau_x + v\tau_y + (w_0 + w')\tau_z - D \tag{1.6}$$

be a minimum in a least-square sense. At first it can be argued that (1.5) should give better results since the tracer balance—apart from the problem of parametrisation—appears to be more accurate than the linearized vorticity equation used in (1.6). The example in Fig. 3 taken from Olbers et al. (1985) shows that this need not be true. While the estimate of w from (1.6) looks rather smooth the direct calculation appears very noisy, in particular near the surface.

The relation of this surprising result to the noise structure is as follows. When (1.5) is used it is implicitly assumed that the noise originating from the tracer gradient data in this equation is small, in fact much smaller than any other term in this equation so that it can be entirely neglected. The method based on (1.6), however, allows for noise but denies any noise in the vorticity equation. Obviously, the noise structure in the data does not agree with the assumption leading to (1.5): the noise is not small and the estimate w collects all of it at every level. The better result of (1.6) is due to the consideration of noise—the tracer balance need not be exactly satisfied. Furthermore, the vorticity balance is used here in an integrated form which can be expected to reduce noise amplitudes.

2   THE PHYSICAL MODEL

In this chapter we introduce the basic physics of the large scale ocean circulation used in the inverse models mentioned above. Particular emphasis is put on the presentation of the various ideas and concepts which are behind the currently discussed parametrisations of the mixing terms in the balance of tracers.

102

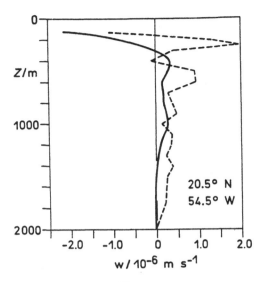

Figure 3
Profiles of the vertical velocity. Solid lines represent w computed from the
linear vorticity balance and the least square problem (1.6), and dashed
lines give w computed from the tracer balance (1.5) (from Olbers et al.
1985).

2.1 Equations of motion and the level-of-no-motion problem

The generally accepted dynamical and thermodynamical concepts of large scale
oceanic flow are the geostrophic and hydrostatic momentum balances

$$
\begin{aligned}
-\varrho f v &= -p_x \\
\varrho f u &= -p_y \\
g\varrho &= -p_z,
\end{aligned}
\tag{2.1}
$$

the mass conservation in form of the incompressibility constraint

$$u_x + v_y + w_z = 0, \tag{2.2}$$

and an advective-diffusive balance

$$u\tau_x + v\tau_y + w\tau_z = D[\tau] \tag{2.3}$$

where ϱ and p are the in-situ density and pressure and τ any—in adiabatic conditions conserved—functional $\tau(\Theta, S)$ of potential temperature Θ and salinity S

such as potential density or veronicity (for a definition of this tracer see Veronis 1972 or Olbers et al. 1985). The operator D represents diffusive effects. The set of equations may be augmented by the vorticity equation

$$\beta v + f\left(u_x + v_y\right) = F. \tag{2.4}$$

We have included here turbulent diffusion of vorticity by a term F since all terms in (2.4) are of higher order than the geostrophic momentum terms.

Integration of the hydrostatic balance in (2.1) yields the pressure field as the sum of two parts, one arising from the surface displacement ζ and a second part associated with the stratification of the mass field

$$p = g\varrho(0)\,\zeta + g\int_z^0 dz'\varrho(z') \tag{2.5}$$

Hydrographic data only allow to determine the second part—the baroclinic pressure. The surface topography ζ is essentially unobservable by standard oceanographic instrumentation (except for the future benefits from satellite altimetry). From a dynamical point of view the estimation of ζ is a non-local problem since the sea level depends on the entire state of the motion in the ocean basin. Without any knowledge about ζ we are thus faced with the classical problem of large-scale oceanography—the determination of the profile of the absolute velocity given the vertical shear by hydrography in form of the thermal wind relations

$$
\begin{aligned}
u_z &= \frac{g}{f}\,\varrho_y \\
v_z &= -\frac{g}{f}\,\varrho_x
\end{aligned} \tag{2.6}
$$

obtained by eliminating the pressure in (2.1), and the local version of the vortex stretching equation

$$w_z = (\beta v - F)/f \tag{2.7}$$

obtained from (2.2) and (2.4). Integration of these relations from a fixed reference level $z = z_0$

$$
\begin{aligned}
u &= u_o + \frac{g}{f}\int_{z_0}^z dz'\varrho_y = u_0 + u' \\
v &= v_o - \frac{g}{f}\int_{z_0}^z dz'\varrho_x = v_0 + v' \\
w &= w_0 + \frac{1}{f}\int_{z_0}^z dz'(\beta v - F) \\
  &= w_0 + \frac{\beta}{f}\,(z - z_0)\,v_0 + w'
\end{aligned} \tag{2.8}
$$

reveals the set of unknowns—the reference velocities u_0, v_0 and w_0—which any estimation scheme must determine in conjunction with the mixing parameters.

2.2 Parametrization of mixing

In our diagnostic approach ϱ and τ are given from the hydrographic observations and we consider here some problems concerning the estimation of a parameterization of $D[\tau]$ which by (2.3) is inherently coupled to the estimation of the velocities u, v and w as well. The relative role of advection and diffusion can be addressed by scaling the above set of equations (see e.g. Pedlosky 1979). Taking here for the purpose of demonstration

$$D[\tau] = K\,\tau_{zz}$$

as in (1.1) we obtain for motions of planetary scale the well-known ratio of diffusive to advective terms

$$\text{diffusion/advection} \approx \delta_d/\delta_a$$

where δ_a is the advective vertical scale (a = earth radius, $\delta\varrho$ = scale of ϱ)

$$\delta_a = \sqrt{fa^2W/g\delta\varrho}$$

and δ_d the diffusive vertical scale

$$\delta_d = K/W$$

as in Munk's global scale problem above. With a K of $10^{-4}\,\mathrm{m^2/sec}$ (or less as presumably appropriate for local conditions) and a vertical velocity scale W of $10^{-6}\,\mathrm{m/sec}$ the ratio of diffusive to advective terms becomes 0.1 or less and it appears that mixing is only a small contribution to the tracer balance. We must expect that determination of mixing parameters from such a balance might be a delicate problem unless applied in regions of very strong mixing.

But even in this case we are still faced with the severe problem of the correct or an adequate parametrization of the mixing effects. What is the appropriate form of the functional $D[\tau]$? A comprehensive theory in this field is still lacking (for a review refer to Gregg 1986, Holloway this book). We shall adopt here the traditional parametrization of the turbulent fluxes in terms of the mean gradients $\partial_i\tau$ ($i = 1, 2, 3$) of the tracer τ and write D in the general form

$$D[\tau] = \partial_i\,K_{ij}\,\partial_j\,\tau \tag{2.9}$$

where K_{ij} is the diffusion tensor. According to the traditional convention in modelling this tensor is considered diagonal in the Cartesian coordinate system

$$K_{ij} = K_h \, \delta_{ij} + (K_v - K_h) \, \delta_{i3} \, \delta_{j3} \qquad (2.10)$$

with diffusion coefficients K_h and K_v representing horizontal (i.e. on geopotential surfaces) and vertical diffusion, respectively.

In descriptive oceanography and particular in water mass analysis, however, it has long been recognized (e.g. Montgomery 1938, Solomon 1971) that spreading of tracers should preferrentially go along isopycnal surfaces which are surfaces of constant potential density. The word 'spreading' is chosen here as a deliberately undefined combination of advection and diffusion since these can hardly be destinguished by water mass properties alone. In a near-adiabatic flow regime both transport processes should be approximately oriented along the isopycnal surface but for different reasons. In an adiabatic flow advection conserves potential density exactly so that the velocity vector (u, v, w) must lie in the local isopycnal (see (2.3) with $D \equiv 0$). Based on a variety of arguments partly discussed below it is widely believed that mixing is favoured along isopycnals, too, so that the mixing tensor should be oriented with respect to these surfaes. The tensor then takes the form (Redi 1982, Olbers et al. 1985)

$$K_{ij} = A_l \, \delta_{ij} + (A_c - A_l) \, n_i \, n_j \qquad (2.11)$$

where n_i is the unit vector normal to the local isopycnal and A_l and A_c describe mixing along and across it, respectively.

A main motivation for the preference of isopycnals in mixing processes was sought in the argument that a displacement of a fluid parcel along this surface requires no work against the Archimedian force while a normal displacement would require work. However, this is strictly correct only at the reference level of the potential density and a thorough investigation leads to the replacement of isopycnals by the so-called neutral surfaces (McDougall 1987). This surface comprises the infinitesimal directions a particle can move without experiencing a buoyant restoring force. By this definition the neutral surface is locally well-defined. At every point it is tangent to the isopycnal if the potential density is referred to the local pressure. On larger scales the concept leads however to ambiguities (the neutral surface is not a real surface, see McDougall 1987) which are not our concern here. Taking the orientation with respect to the neutral surface the mixing tensor is of the same form as (2.11) but n_i is now the normal vector to this surface.

To be mathematically more specific and come to a further sensible parametrisation we write the equation of state as

$$\varrho = F(T, S, p) \tag{2.12}$$

where ϱ, T, S, and p are in-situ density, temperature, salinity and pressure, respectively. If the potential temperature $\Theta = \Theta(T, S, p; p_0)$ (referred to a pressure level p_0) is inverted for T the equation of state can be expressed as

$$\varrho = F\Big(T(\Theta, S, p; p_0), S, p\Big) = G(\Theta, S, p; p_0) \tag{2.13}$$

and the potential density referred to p_0 is defined by

$$\sigma_0 = G(\Theta, S, p_0; p_0) = F(\Theta, S, p_0) \tag{2.14}$$

Now consider a particle at the position $\mathbf{x}$ with in-situ density $\varrho(\mathbf{x})$ moving adiabatically (i.e. conserving Θ and S) to a neighbouring position $\mathbf{x} + d\mathbf{x}$ where the density is $\varrho(\mathbf{x} + d\mathbf{x})$. The particle will change its density by adiabatic compression and show a density excess

$$
\begin{aligned}
\delta\varrho &= \varrho(\mathbf{x})\,(1 + \gamma\,dp) - \varrho(\mathbf{x} + d\mathbf{x}) \\
&= \varrho(\mathbf{x})\,\{\gamma\,dp + (\alpha\,\mathrm{grad}\,\Theta - \beta\,\mathrm{grad}\,S - \gamma\,\mathrm{grad}\,p)\cdot d\mathbf{x}\} \\
&= \varrho(\mathbf{x})\,(\alpha\,\mathrm{grad}\,\Theta - \beta\,\mathrm{grad}\,S)\cdot d\mathbf{x}
\end{aligned}
\tag{2.15}
$$

where we have introduced the coefficients

$$
\begin{aligned}
\gamma &= (1/\varrho)\,(\partial G/\partial p) \\
\alpha &= -(1/\varrho)\,(\partial G/\partial \Theta) \\
\beta &= (1/\varrho)\,(\partial G/\partial S)
\end{aligned}
\tag{2.16}
$$

for compressibility as well as thermal and haline expansion, respectively. The derivatives in (2.16) are taken in the usual thermodynamic convention but also holding p_0 fixed.

The buoyancy force acting on the particle is

$$\mathbf{F} = \delta\varrho\,\mathrm{grad}\,\Phi \tag{2.17}$$

horizontal

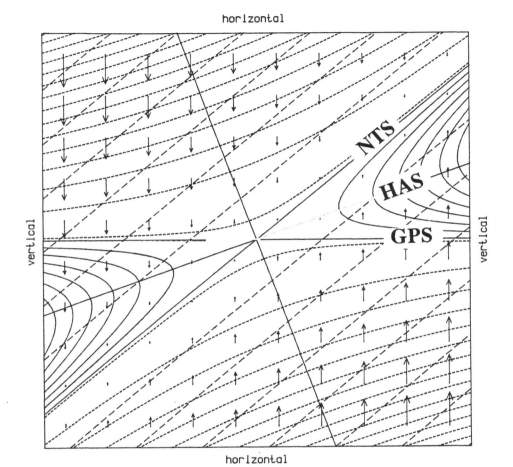

Figure 4
A schematic picture of the Archimedian force and work associated with the
displacement of parcels in a stratified fluid. GPS is the geopotential surface
and NTS the neutral surface. The arrows give direction and magnitude of
the Archimedian force (2.17) and the isolines represent the work dA done
by this force (given by (2.20)) on a particle shifted from origin to the point
of consideration. HAS is the half angle surface where dA is maximum.

where Φ is the geopotential. Displacement in a direction dx which places the particle in an environment where there is no excess of density—and hence no restoring buoyancy force—requires that

$$(\alpha \operatorname{grad} \Theta - \beta \operatorname{grad} S) \cdot d\mathbf{x} = 0 \qquad (2.18)$$

This is a local definition of the neutral surface (the vector in the brackets is the local normal). Notice that the isopycnal can be expressed in similar form (after some simple manipulations, see also McDougall 1987)

$$(\alpha_0 \operatorname{grad} \Theta - \beta_0 \operatorname{grad} S) \cdot d\mathbf{x} = 0 \qquad (2.19)$$

where α_0 and β_0 are evaluated at the reference pressure p_0, not the local pressure p as the α and β in (2.18). It is apparent from these two relations that neutral surfaces and isopycnals are tangent if potential density is referred to the local pressure p.

The vanishing of the Archimedian force hence seems to emphasize the neutral surface in the parametrisation of mixing. However, the work done by the Archimedian force during the displacement

$$\begin{aligned} dA &= \mathbf{F} \cdot d\mathbf{x} \\ &= \varrho(\mathbf{x})(\alpha \operatorname{grad} \Theta - \beta \operatorname{grad} S) \cdot d\mathbf{x} \, (\operatorname{grad} \Phi \cdot d\mathbf{x}) \end{aligned} \qquad (2.20)$$

vanishes on the neutral surface (where $\mathbf{F} = 0$) but of course also on the geopotential surface (since $\mathbf{F}$ and $d\mathbf{x}$ are normal here). So taking literally the energetics point of view it does not help up to choose between the isopycnal-neutral-surface mixing form (2.11) and the ordinary lateral mixing form (2.10)! But a closer inspection of 2.20 clarifies the energetics concept and points at a third possibility.

Eq. (2.20) describes a quadratic form of the displacement components dx_i. The work dA turns out to be constant on hyperbolic cylinders confined between the neutral and geopotential surfaces (Fig. 4). Work is needed ($dA < 0$) to move a particle from the center to a position outside the angle range between the neutral and the geopotential surfaces but work is done ($dA > 0$) (i.e. available potential energy is converted) if the particle is displaced into this range. The energy conversion is a maximum at half of the angle between the neutral and the geopotential surfaces. So based on the energy argument this should be the direction favoured by mixing. Mixing along this direction implies mixing both along and across neutral surfaces. A canonical parametrisation of mixing effects in this framework then appears in form of a diagonal tensor with three independent coefficients in the coordinate system given by the intersection of the neutral and the geopotential surfaces (where $dA = 0$), the orthogonal direction in the half angle plane (where dA is maximum, i.e. maximum energy conversion) and the normal direction (where dA is minimum, i.e. maximum input of energy is required). If the corresponding coefficients are A_c, A_0 and A_m (for components of the mixing tensor normal to the neutral surface, along the intersection line of this surface with the horizontal surface and in the direction of the maximum energy gain, respectively) the tensor takes the form

$$K_{ij} = A_c\, m_i\, m_j + A_0\, q_i\, q_j + A_m\, t_i\, t_j \qquad (2.21)$$

Here $\mathbf{m}$ is unit vector normal to the half angle plane

$$\mathbf{m} = 1/\sqrt{2\,(n_1^2 + n_2^2)} \cdot \left(n_1\sqrt{1 - n_3}, n_2\sqrt{1 - n_3}, \sqrt{n_1^2 + n_2^2}\sqrt{1 + n_3}\right) \qquad (2.22)$$

where $\mathbf{n}$ is normal to the neutral surface and $\mathbf{q}$ and $\mathbf{t}$ are the corresponding unit vectors orthogonal to $\mathbf{m}$

$$\begin{aligned}
\mathbf{n} &= (n_1, n_2, n_3) = \alpha\,\mathrm{grad}\,\Theta - \beta\,\mathrm{grad}\,S \\
\mathbf{q} &= (m_2, -m_1, 0)/\sqrt{m_1^2 + m_2^2} \\
\mathbf{t} &= (-m_1 m_3, -m_2 m_3, 1 - m_3^2)/\sqrt{m_1^2 + m_2^2}.
\end{aligned} \qquad (2.23)$$

3   THE INVERSE MODEL

As demonstrated with the example in the introduction we can set up the basic equations of the physical model in different ways when utilizing them for parameter estimation. Formally different methods can be viewed as originating from one unified inverse model which diversifies for different choices of weighting the equations of the physical model, i.e. for different statistical structures of the data. In practice, however, each method has developed in its own direction in the course of more or less successful applications.

In chapter 4 we will exemplify the possible fallacies and the power of inverse estimation procedures using an inverse model of the β-spiral method in the framework presented in Olbers et al. (1985). A short description of the outline of this model is given here.

3.1 The β-spiral method

Inserting the integrated thermal wind and vorticity equations (2.8) into the tracer balance (2.3) we find the β-spiral equation in the form

$$u_0 \tau_x + v_0 \left\{ \tau_y + \frac{\beta}{f} (z - z_0) \tau_z \right\} + w_0 \tau_z + M[K_{ij}, F] = b \qquad (3.1)$$

where M collects the terms arising from diffusion of the tracer τ and of vorticity. Considering a diffusive form for the vertical transport of relative vorticity $v_x - u_y$ the term F can entirely be expressed by density derivatives

$$
\begin{aligned}
F &= [A(v_x - u_y)_z]_z = \\
&= -\frac{g}{f} \left[A \{ \nabla^2 \varrho - \frac{\beta}{f} \varrho_y \} \right]_z
\end{aligned}
\qquad (3.2)
$$

where A is the eddy coefficient. Furthemore, b collects the baroclinic advection terms

$$b = -(u' \tau_x + v' \tau_y + w' \tau_z) \qquad (3.3)$$

which are known from the data. Abbreviating the unkown parameters u_0, v_0, w_0, K_{ij} and A by the vector p_k $(k = 1, \ldots, K)$ and applying (3.1) at several levels z_j $(j = 1, \ldots, J)$ we write this set of equations in the matrix form

$$D_{jk} p_k - b_j = 0 \qquad j = 1, \ldots, J \qquad (3.4)$$

with known coefficients D_{jk} and b_j. Physically meaningful are only solutions with positive diffusivities so we require additionally

$$B_{lk} p_k \geq 0 \qquad l = 1, \ldots, L \qquad (3.5)$$

with a correspondingly defined coefficient matrix B_{lk}.

In our case the system (3.4) is generally overdetermined since we have more levels than parameters. But frequently it is also underconstrained, i.e. one or more parameter combinations are not sufficiently constrained by the equations. We have pointed out such ill-posed situations in the introduction. A detailed discussion of solution techniques to such a problem can be found in text books on inverse theory such as Tarantola (1987). We have used the singular value decomposition (SVD) framework (see also Wunsch and Olbers this book) to construct a generalized inverse. The equations (3.4) were weighted using a diagonal approximation $\mathcal{W}$ of the noise covariance structure of these relations (for details we refer to Olbers et al. (1985)). The SVD-solution of

$$W^{1/2} \mathcal{D} p = W^{1/2} b \tag{3.6}$$

is then given by

$$p = \sum_{k=1}^{K} (1/\lambda_k) V_k U_k^\dagger \cdot b \tag{3.7}$$

where the V_k (K-dimensional) and U_k (J-dimensional) are eigenvectors of the matrices $\mathcal{D}^\dagger W \mathcal{D}$ ($K \times K$) and $W^{1/2} \mathcal{D} \mathcal{D}^\dagger W^{1/2}$ ($J \times J$), respectively, with simultaneous eigenvalues $(\lambda_k)^2$. The V-vectors form an orthogonal base of the K-dimensional parameter space and likewise the U-vetors form a base of the J-dimensional data space of the b's. Hence (3.7) shows that the parameter combination $V_k^\dagger \cdot p$ is determined by the data $U_k^\dagger \cdot b$ and a canonical way to write (3.7) is

$$V_k^\dagger \cdot p = U_k^\dagger b / \lambda_k \qquad (k = 1, \ldots, K) \tag{3.8}$$

where the eigenvectors are assumed normalized.

The above expressions refer to the well-posed case when $\mathcal{D}^\dagger W \mathcal{D}$ is regular and all eigenvalues non-zero. Otherwise only contributions with non-zero λ_k are included in (3.7). This in fact is implicit in the definition of the SVD inverse. It is obvious from (3.8) that this case leaves certain combinations of the parameter undetermined (one for each zero eigenvalue). The value of such a combination is set to zero by the SVD-inverse. If this is not intended another apriori value can of course be assigned. It should be noted that the information of the data $U_k^\dagger \cdot b$ corresponding to a zero λ_k is entirely discarded.

Finally we would like to discuss the problem of near-singularity of $\mathcal{D}^\dagger W \mathcal{D}$. In practice, even with appropriate scaling of the parameters, the eigenvalues may span several decades and decay more or less exponentially which make it very difficult to distinguish between very small and almost zero eigenvalues. This sounds as if we are addressing the computational error in computing machines but we rather mean a data problem. Eq. (3.8) shows that a small eigenvalue amplifies the effect of data structures in determining parameter values. This can be perfectly correct if the data structure in question and the structure of the matrix $\mathcal{D}$ responsible for the small eigenvalue in question are reliable. But $\mathcal{D}$ is—at least in our problem—determined from observations, too, and since data errors in $\mathcal{D}$ will mainly affect the low eigenvalues the problem arises to separate the range of unreliable low values from the eigenvalue spectrum.

There are several possibilities to repair the SVD-inverse once the cut-off in the eigenvalue range has been determined (or guessed). Just as for vanishing eigenvalues contributions in (3.7) with unreliable eigenvalues can be canceled. A softer cut-off is the replacement of $1/\lambda_k$ in (3.7) by a smoother function such as

$$\lambda_k^n/(\lambda_k^{n+1} + \lambda_c^{n+1})$$

This tapered cut-off includes the hard cut-off for an infinite n. In the applications reported later in this paper we have taken $n = 1$.

The covariance matrix of the parameters can be shown to have the approximate values

$$< \delta p \, \delta p^\dagger > \simeq \frac{< \varepsilon^2 >}{J - K} \sum_{k=1}^{K} (1/\lambda_k)^2 \, V_k \, V_k^\dagger \tag{3.9}$$

where ε^2 is the sum of squared residual of the model relations (3.4) and the cornered bracket denoted statistical averages. This expression is strictly valid if the weighting matrix is taken according to the maximum likelihood principle as the inverse of $< \delta b \, \delta b^\dagger >$ which in practice however is rarely known. Eq. (3.9) is still a good approximation if the number J of equations appearing in the denominator is replaced by an effective number counting the uncorrelated equations. The relation also reveals the statistical counterpart of the range of unreliable eigenvalues. The contributions from the low λ's may lead to an unacceptably large amplification of the data noise $< \varepsilon^2 >$ in the parameter variance. If this occurs the resolution of the model has been driven too high, i.e. the attempt to extract more and more information from the data by resolving more and more parameters leads to the situation that information is increasingly drawn from the data noise since the reliable information has already been exhausted. Hence, also from this point of view a cut-off procedure appears to be necessary and the same replacements as in (3.7) should be performed. We should however point out again that the variance control always means a reduction of the resolution since cutting off more and more terms with small λ's makes the parameters increasingly dependent.

3.2 Mass conservation

The estimate of the three-dimensional velocity field obtained by the β-spiral approach does generally not satisfy continuity. This is a consequence of a local estimation procedure which determines velocity profiles from the local gradients of density and tracers without taking care of the neighbouring velocities. One can of course design inverse methods which consider continuity from the beginning. However, with high resolution in a large region as e.g. the Southern Ocean the number of unknowns will be too large to obtain a manageable problem. Hence we decided to investigate in a separate step how severe the violation of continuity is and how much the β-spiral results must be changed in order to get a mass conserving velocity field. The procedure which we present here has been devised by Wenzel (1986) who applied it first to the β-spiral results of the North Atlantic (see Olbers et al. 1985).

Close to the surface we should include the Ekman part to define the total velocity vector by

$$\mathbf{u}^* = \mathbf{u}_G + \mathbf{u}_E \tag{3.10}$$

where $\mathbf{u}_G$ is the geostrophic part estimated from the β-spiral. The Ekman part $\mathbf{u}_E$ can be calculated from the windstress τ using the Ekman theory of a frictional wind-driven boundary layer. The simplest way would be to distribute the Ekman mass transport $(-\tau^y, \tau^x)/f$ over the (assumed) depth of the Ekman layer.

So consider the task to find a velocity field $\mathbf{u} = (u, v, w)$ which is free of divergence and as close as possible to a given field $\mathbf{u}^* = (u^*, v^*, w^*)$. Hence we have to solve the variational principle

$$\int d^3x \parallel \mathbf{u} - \mathbf{u}^* \parallel^2 = \min \tag{3.11}$$

subject to the constraints that $\operatorname{div} \mathbf{u} = 0$ or, in terms of a vector potential $\mathbf{A} = (A, B, C)$

$$\mathbf{u} = \operatorname{curl} \mathbf{A} \tag{3.12}$$

and the kinematic boundary conditions at vertical and lateral boundaries

$$
\begin{aligned}
w & = 0 && \text{at the surface} \quad z = 0 \\
w + \mathbf{u}\operatorname{grad} H & = 0 && \text{at the bottom} \quad z = -H \\
\mathbf{u}^\perp & = 0 && \text{at closed lateral boundaries} \\
\mathbf{u}^{\parallel} & = \mathbf{u}^{*\parallel} && \text{at open lateral boundaries}
\end{aligned}
\tag{3.13}
$$

where $\mathbf{u}^\perp$ and $\mathbf{u}^{\parallel}$ are the normal and tangential velocity vectors, respectively. The last condition appears best in view of our task to determine a closely neighbouring velocity field which is free of divergence and indeed, it is the canonical boundary condition for the variational principle. The norm in (3.11) is taken diagonal with a suitable weighting of the vertical part by a factor μ relative to the horizontal parts. Also the separate levels can be weighted differently. In view of increasing errors in the assumption of a linear vorticity balance without horizontal friction we chose smaller weights in the surface layers so that there w will essentially be determined by the continuity constraint.

Eq. (3.11) yields three coupled differential equations for the components A, B and C of the vector potential which we give here for simplicity in Cartesian coordinates (for the large region considered in the next chapter these equations have been taken in spherical coordinates)

$$
\begin{aligned}
A_{zz} + \mu\, A_{yy} &= v_{1z} - \mu\, w_{1y} \\
B_{zz} + \mu\, B_{xx} &= -u_{2z} + \mu\, w_{2x} \\
C_{xx} + C_{yy} &= u_{3y} - v_{3x}
\end{aligned}
\tag{3.14}
$$

with

$$
\begin{aligned}
v_1 &= v^* + C_x & w_1 &= w^* - B_x \\
u_2 &= u^* - C_y & w_2 &= w^* + A_y \\
u_3 &= u^* + B_z & v_3 &= v^* - A_z
\end{aligned}
\tag{3.15}
$$

These equation have been written in a form to suggest a possible iteration for determining the solution. Fast convergence is achieved when first C is computed with neglection of the vector potential terms on the rhs, then calculating B with $A = 0$ and C given, then calculating A with C and B given and iterating this loop (usually not more than two or three times). For further details such as weighting and numerical algorithms we refer to Wenzel (1986).

4 EXAMPLES ... CIRCULATION AND MIXING IN THE SOUTHERN OCEAN

In the last chapters we have described some of the techniques and problems of oceanographic inverse models and given some illustrations. Here we will continue with the β-spiral model and focus on the results obtained from a collection of hydrographic data from the Southern Ocean prepared by Gordon et al. (1982). The dynamic topography between various levels as well the geostrophic circulation at the surface relative to 1000 db is presented in Gordon et al. (1978). These maps which are based on the original section data still comprise the state of our knowledge on the circulation in this ocean. For our work the gridded version of this atlas was used for which the temperature and salinity fields are interpolated on a grid with dimension of 1° of latitude and 2° of longitude on 42 standard levels in the depth range 0 to 7000m. The objective analysis method is described in Gordon et al. (1982). The surface currents of this data set (relative to 2000m) are displayed in Fig. 5 (notice that here and on all the following maps only every second arrow of the current field is shown).

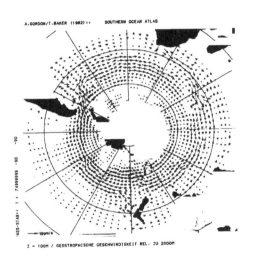

Figure 5
Horizontal geostrophic velocity at 100m relative to 2000m, computed from the atlas of Gordon et al.(1982).

4.1   A singular and some well-posed problems

As with the North Atlantic inverse model described in Olbers et al. (1985) and Wenzel (1986) we have followed the strategy of successive construction of the circumpolar model and started with the construction of purely adiabatic circulations using different depth ranges to estimate reference velocities. These investigations

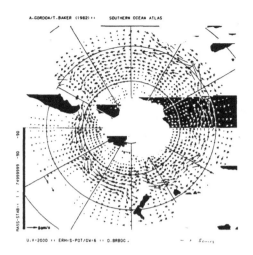

Figure 6a
The absolute horizontal velocity at 2000m depth, obtained from data 100m–2000m and an adiabatic β-spiral model.

Figure 6b
Condition index of the matrix $\mathcal{D}^\dagger \mathcal{W} \mathcal{D}$ for the model of Fig. 6a. The condition index is the ratio of the smallest to the largest eigenvalue. Shaded areas have values below 10^{-3}. The contours are spaced logarithmically with interval 0.5.

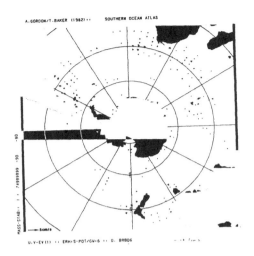

Figure 7a
Contribution of the first eigenvector to the horizontal reference velocities
for the model of Fig. 6a.

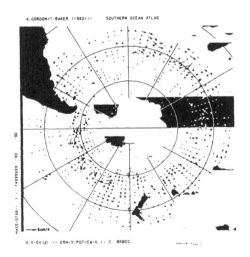

Figure 7b
Contribution of the second eigenvector to the horizontal reference velocities for the model of Fig. 6a.

have mainly been performed to test the ability of the β-spiral model in the circumpolar environment and get insight into the hydrographic data. Some of the results of these experiments may serve here as a lucid illustration of the power and pitfalls of inverse models.

The model we are considering now is given by equation (3.1) with vanishing M. As conserved tracer we used potential density referred to the surface. The reference level of the geostrophic velocities is taken at 2000m depth. The data have been vertically split into subsets for which separate fits have been obtained.

Adiabatic β-spiral from data between 100m to 2000m

Fig. 6a displays the absolute geostrophic circulation in 2000m depth obtained by fitting the model to the 100m–2000m part of the data. In view of the relative surface circulation (see Fig. 5) the level-of-no-motion in this solution must lie well above 2000m depth which entirely contradicts our expectation that the Antarctic Circumpolar Current (ACC) is an almost barotropic eastward current. The matrix condition of the solution with its very low values (below 10^{-3}) shown in Fig. 6b clearly points out that the problem should indeed be affected by some kind of ill-posedness as we have discussed above. This suggestion is confirmed when the separation of the total circulation of Fig. 6a into three components of the eigenvector basis is made according to (3.7). Figs. 7a and b demonstrate that the contributions from the two largest eigenvalues are small and mainly meridional whereas the contribution from the smallest eigenvalue is almost identical to the total solution (not shown for this reason). Also the solution for w (not shown here) has its peculiarity: the contribution from the largest eigenvalue is negligibly small and the second and third contribute by and large with opposing signs.

The reason for this behaviour must be sought in the zonality of the potential density field. Over most of the upper part of the water column the zonal gradient is much smaller than the meridional one so that the reference velocity u_0 is not sufficiently constrained by the β-spiral equation (3.1). This is the situation discussed above where unreliable data structure may be amplified as the consequence of near singular relations and spoil the solution. Cutting off the contribution from the smallest eigenvalue as suggested by the SVD approach, leaves us with the sum of the two contributions shown in the figures 7 and this is still an unsatisfactory circulation. The only way out appears explicitly to prescribe u_0 from some a pri-

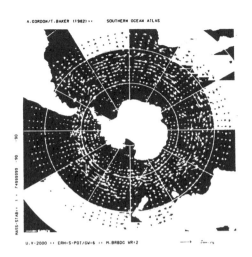

Figure 8a
The absolute horizontal velocity at 2000m depth, obtained from data
100m–2000m and an adiabatic β-spiral model with inclusion of the kinematic bottom boundary condition (4.2).

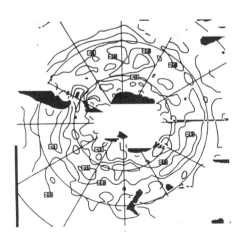

Figure 8b
Condition index of the matrix $\mathcal{D}^\dagger \mathcal{W} \mathcal{D}$ for the model of Fig. 8a.

ori knowledge but u_0 of course is the most essential part of the signal we actually would like to extract from the data. So, does the β-spiral not work in the Southern Ocean? We consider here two possible ways out of this dilemma: taking deeper data and/or expanding the model.

A topographically modified β-spiral

Let us first investigate a model expansion which from its structure should be able to overcome the zonality of the problem. Since the ACC has a strong barotropic component the current must be affected by the bottom topography. At the bottom it has to satisfy the kinematic boundary condition

$$w + \mathbf{u} \cdot \operatorname{grad} H = 0 \qquad \text{at the bottom } z = -H \qquad (4.1)$$

Strictly, this applies to the total current and not to the geostrophic part alone. If a frictional boundary layer is present each component of the total velocity vector must separately vanish at the bottom and the behaviour of the geostrophic velocity is a matter of the entire dynamical problem. Still, in case of weak frictional effects we expect that the geostrophic velocity adjusts to approximately satisfy (4.1) close to the bottom. Considering this in the form

$$\begin{aligned} w_0 - \frac{\beta}{f}\left(H + z_0\right) v_0 + \mathbf{u}_0 \cdot \operatorname{grad} H = \\ = -w' - \mathbf{u}' \cdot \operatorname{grad} H \qquad \text{at } z = -H \end{aligned} \qquad (4.2)$$

we see that in this equation u_0 is no longer multiplied by a small factor since the topography in the Southern Ocean shows almost everywhere strong non-zonal patterns. Adding this constraint to set of β-spiral equations we thus expect a better condition of the corresponding inverse problem.

The resulting reference velocities and matrix condition displayed in Fig. 8a and b by and large confirm this presumption. The condition index lies well above 10^{-3} almost everywhere and the reference velocities point eastward in the Atlantic and some parts of the Indian and Pacifc Oceans. Remember that this model still uses the identical upper ocean tracer relations as the one before. Deep data are only taken to calculate the relative geostrophic velocities u', v' and w' at the bottom to incorporate the relation (4.2) (in a suitably weighted form).

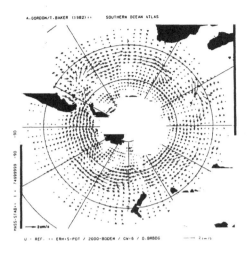

Figure 9a
The absolute horizontal velocity at 2000m depth, obtained from data
2000m to the bottom and an adiabatic β-spiral model without bottom
boundary conditon.

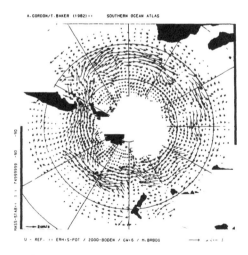

Figure 9b
The absolute horizontal velocity at 2000m depth, obtained from data
2000m to the bottom and an adiabatic β-spiral model with bottom boundary conditon.

Adiabatic models from data between 2000m and the bottom

Corresponding experiments have been made with data from other parts of the water column. For the results displayed in Fig. 9a and b only deep data below 2000m were used. The reference velocities for both cases—without (Fig. 9a) and with (Fig. 9b) bottom boundary condition—are in good agreement with our presumption of the circulation in the Southern Ocean. We recognize here a strongly barotropic current influenced considerably by topography even at the 2000m level. Also details of the subpolar gyres in the Weddell and Ross Seas are visible. The topographically modified β-spiral (Fig. 9b) generally yields a slightly enhanced circulation with more regional structures. We will come back to a more detailed description of the circulation in the next section.

4.2   The diabatic model

In this section we will present some features of our more elaborate diabatic inverse model of the Circumpolar Current. The emphasis will be put on the strategy and performance of the determination of mixing parameters for the circumpolar region. But in order to judge the usefulness of the results it is also necessary to describe the basic structure of the currents and related properties of the model.

Model strategy

From the experience with the adiabatic models we have to conclude that the data from the upper part of the water column are not useful to determine reference velocities. In the set of β-spiral relations for the different levels the upper levels should thus be weighted less than the lower levels. We decided to choose the extremum of such a weighting procedure and determined reference velocities exclusively from data below 2000m depth. Upper data were taken then to calculate the geostrophic velocity profiles and determine mixing parameters for this part of the water column. The results presented in this section were obtained according to a strategy which proved to be very effective in separating the entire estimation problem into manageable pieces:

1. Reference velocities (with consideration of diapycnal and vorticity diffusion) were determined from the balance of potential density (referred to 2000m) at

the levels between 2000m and the bottom. The bottom boundary condition was included.

2. A mass conserving circulation was determined as described above. The Ekman part was taken from the windstress data of Hellerman and Rosenstein (1983). This velocity field was then taken to get a final estimate of the diffusive part of the solution.

3. The balance equations of potential density and veronicity were used to determine values for the various diffusion coefficients of the parametrisations discussed in chapter 2.2. Three subsets of levels were used to determine the depth dependence of the coefficients which are assumed constant over the corresponding depth range. The ranges are 100m to 2000m, 800m to 2500m and 2000m to the bottom.

The mass conserving circulation

The mass conserving circulation is displayed in some examples in the following set of figures. The maps of the horizontal velocitites (Fig. 10) and the sections of the zonal velocity (Fig. 11) reveal a strong barotropic component of the circumpolar current. The velocities decrease from roughly 0.20 m/sec at the surface to about 0.01 m/sec near the bottom. The circulation in the upper ocean is thus in many aspects very similar to the one displayed in the surface topography and currents relative to 1000 db of Gordon et al. (1978). The reference level of our solution is however well below 1000m. If the current reverses it does so below 2000m.

The circulation in the upper ocean displays many of the well-known features of the Circumpolar Current. After the sudden northward shift by about 15 degrees behind the Drake Passage the current spirals slowly southward. In some areas there is a multiaxial shape of the current which however is lost below a few hundred meters (see Fig. 11). The influence of topography can clearly be seen in many areas. Even the upper level currents are affected as in some parts of the Indian Ocean and the western Pacific where guidance along the midocean ridge is most obvious. There is evidence of closed circulations in sub-basins such as the Weddell Sea. Sometimes topographic guidance becomes more evident in mid-depths than near the bottom. Examples can be seen in the 2000m map in the Crozet Basin (70° E) where a closed gyre appears in the basin, and in the southeastern Pacific at 150° W, where a closed

124

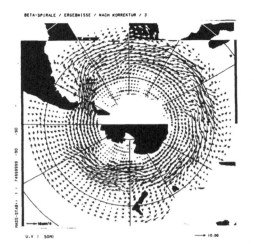

Figure 10a
Absolute mass conserving horizontal velocity field at 50m depth. This is obtained from data 2000m to the bottom, using the diabatic topographically modified β-spiral model in conjunction with the mass conservation step.

Figure 10b
Same as Fig. 10a but for 2000m depth. The map also shows the 500m depth contour.

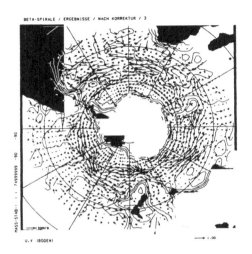

BETA-SPIRALE / ERGEBNISSE / NACH KORREKTUR / 3

U.V (BODEN)

Figure 10c
Same as Fig. 10a but for the bottom. This map shows contours of the
ocean depth in intervals of 500m.

gyre—the Ross Sea gyre—is partly on top of the ridge and partly over the USARP
fracture zone. Notice that the eastward current extends only rarely down to the
bottom but reverses by and large in the depth range 2500m to 3500m.

The vertical velocity in the deep ocean is strongly coupled to the topography
(this is not entirely due to the kinematic constraint at the bottom but already
appeared in solutions which do not use this constraint). Fig. 12 shows that w
generally changes its sign over the flanks of the ridges, upwelling of course occurs
upstream of the bottom current and downwelling downstream. Since the horizontal velocities are strongly barotropic the profile of the vertical velocity has also a
strong barotropic (i.e. linear) component, too, so that the bottom conditions can
be seen through the entire water column. Towards the surface the pattern of the
Ekman pumping velocity is approached, generally with upwelling south of 50° S and
downwelling northward.

The total mass transport of our solution compares favourably with the values
derived from ISOS data. Nowlin and Klinck (1986) report a value of 134 Sv (with

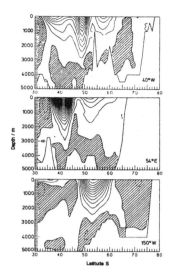

Figure 11
Meridional sections of the absolute zonal velocity along (a) 40° W,
(b) 54° E, and (c) 150° W. Eastward velocities have dashed contours,
the contour interval is 0.01 m/sec.

10% standard deviation and 20% time variability) for the transport though the
Drake passage. We get 125 Sv and should emphasize here again that it is not an
implication of the boundary conditions of the mass conservation scheme but a result
of the β-spiral fit.

Vertical flux of momentum and vorticity

The estimates for the coefficient A governing the vertical transport of horizontal
momentum and vorticity are displayed in Fig. 13. The values are as large as a few
m²/sec in the core of the current and drop down two orders of magnitude towards
the northward adjacent subtropical gyres and one order of magnitude towards the
subpolar gyres. In the deeper part of the water column (2000m to bottom, not
shown here) there is a general tendency of slight increase. These values are considerably larger than those assumed in most ocean circulation studies. Large scale
ocean circulation models use 10^{-4} to 10^{-3} m²/sec (see e.g. Cox and Bryan (1984)).
However, eddy resolving models (McWilliams et al. 1978, Wolff and Olbers 1988)

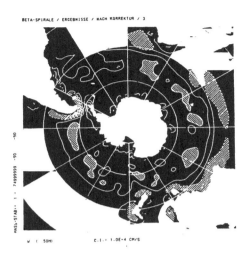

Figure 12a
The vertical velocity field of the mass conserving solution, at 50m. The contour interval is 10^{-6} m/sec. Negative contours are dashed. Areas with values above 10^{-6} m/sec and below -10^{-6} m/sec are shaded.

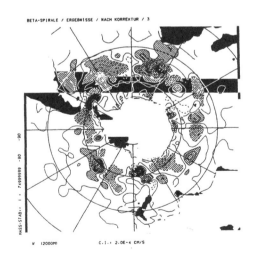

Figure 12b
The vertical velocity field of the mass conserving solution, at 2000m. The contour interval is $2 \cdot 10^{-6}$ m/sec, shading above $2 \cdot 10^{-6}$ m/sec and below $-2 \cdot 10^{-6}$ m/sec.

128

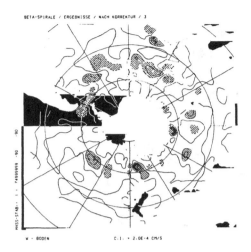

Figure 12c
The vertical velocity field of the mass conserving solution, at the bottom.
The contour interval is $2 \cdot 10^{-6}$ m/sec, shading above $2 \cdot 10^{-6}$ m/sec and
below $-2 \cdot 10^{-6}$ m/sec.

of the ACC strongly suggest that such values of A are far too small to allow for a realistic dynamical balance of the current. In these models there is a vigorous transport of momentum imparted by the wind at the surface through eddy form drag (corresponding to an effective vertical viscosity of a few m^2/sec) to the deep ocean where it leaves the ocean predominantly by bottom form drag. From a physical point of view such a balance is definitely more meaningful than the entirely diffusive balance of the ACC in coarse models: here—with the small vertical viscosity—the momentum is diffused by lateral friction with enormously high viscosities (a few 10^5 to 10^6 m^2/sec, depending on the model resolution) through the lateral boundaries (for further discussion see Wolff and Olbers 1988). In view of the large observed eddy activity in this oceanic area the eddy dominated balance seems a very likely representation of the real dynamical balance of the ACC. This balance is supported by the estimates obtained by our inverse model.

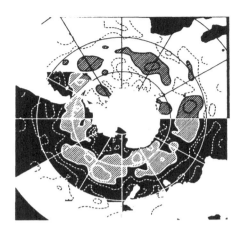

Figure 13
The vertical diffusivity for vorticity and horizontal momentum for the
depth interval 100m to 2000m. Contour are logarithmically spaced with
interval 0.5. Areas with values ≥ 1 m^2/sec are shaded.

Diffusion of heat and salt

Using the mass conserving velocity field described above the various parametriza-
tions (2.10), (2.11) and (2.21) of the mixing term $D[\tau]$ in the tracer balance (2.3)
were implemented in the inverse model (3.1) to determine the unknown parameters
of the diffusion tensor K_{ij}. We report here on three sets of experiments:

- the standard model which uses potential density to determine the diabatic
  coefficient (A_c in Eq. (2.11)) and veronicity to determine the isopycnal coefficient (A_l in (2.11), A_c is then taken fixed from the preceding experiment),

- models with either potential density or veronicity to determine the coefficients
  for horizontal and vertical mixing (K_h and K_v in (2.10)),

- and two experiments to search for differences between the isopycnal form
  (2.11) and the half-angle form (2.21) of the mixing tensor.

Figure 14a

Map of the diapycnal diffusivities for the depth range 100m–800m. Contour are logarithmically spaced with interval 0.5. Contours which are $\geq 10^{-4}$ m²/sec are full and $< 10^{-4}$ m²/sec dashed. Areas with values $\geq 10^{-3}$ m²/sec are shaded.

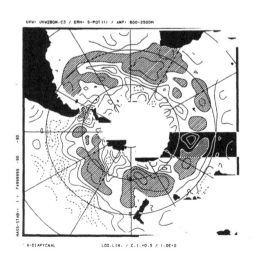

Figure 14b

Same as Fig. 14a but for the depth range 800m–2500m.

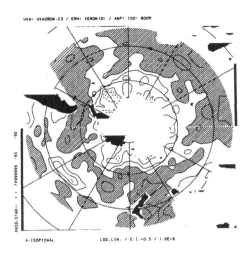

Figure 14c

Map of the isopycnal diffusivities for the depth range 100m–800m. Contour are logarithmically spaced with interval 0.5. Contours $\geq 10^2$ m^2/sec are full and $< 10^2$ m^2/sec dashed. Areas with values $\geq 10^3$ m^2/sec are shaded.

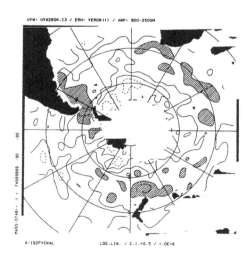

Figure 14d

Same as Fig. 14c but for the depth range 800m–2500m.

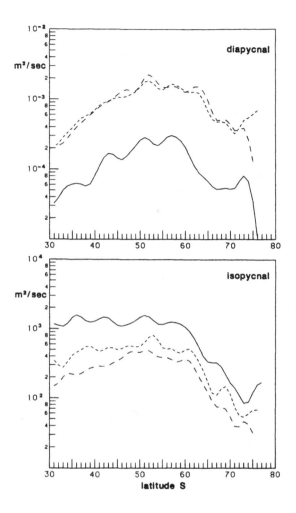

Figure 15
Zonal averages of (a) the diapycnal diffusivities and (b) the isopycnal diffusivity. Solid line corresponds to 100m–800m values, narrow dashed to 800m–2500m values and wide dashed to 2000m–bottom values.

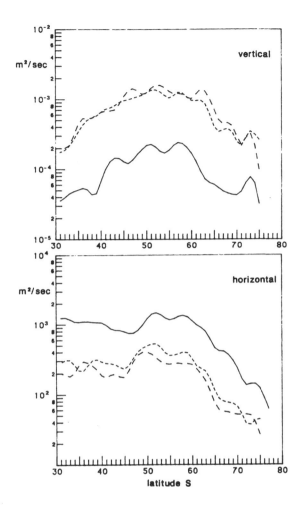

Figure 16
Same as Fig. 15, but for vertical and horizontal diffusivities. Solid lines
correspond to 100m–800m values, narrow dashed to 800m–2500m values
and wide dashed to 2000m–bottom values.

The differences between these models are quite small and the common pattern is displayed in Fig. 14 showing the regional distribution of the diapycnal and isopycnal diffusivities for the depth intervals 100–800m and 800–2500m. Fig. 15 gives the zonal averages of these quantities and the 2000m–bottom values in addition. The diapycnal diffusivity has a pronounced peak in the core of the ACC and increases significantly with depth. Along the current path there are some well defined maxima (with values of a few $10^{-4}\,\mathrm{m^2/sec}$ in the upper layer and exceeding $10^{-3}\,\mathrm{m^2/sec}$ in the deep ocean) which generally extend through the entire water column. The isopycnal coefficients reveal a similar horizontal pattern but a slight decrease from top to bottom with values of $10^3\,\mathrm{m^2/sec}$ in the upper layer dropping down to a few $10^2\,\mathrm{m^2/sec}$ at greater depth. It appears that the decrease away from the ACC to the northern subtropical gyres is less than towards the southern subpolar gyres (except for the eastern Pacific region).

The correlation of the pattern of the diffusivities with the current core is quite obvious. It is tempting to search for more connections to environmental conditions such as topography, current speed or eddy kinetic energy. With some intention one could easily find correspondences between areas of large diffusivities and the hot spots of eddy activity as visible in the maps of rms surface topography determined from Seasat altimetry (Cheney et al. 1983) or the maps of eddy kinetic energy determined from the FGGE drifters (Daniault and Menard 1985). We are reluctant to value such findings, partly because the current path itself is correlated with all these parameters and partly due to our inability to give sufficiently accurate estimates of the parameter variance. The objective scheme used by Gordon et al. (1982) did not provide the correlation structure of the analysed fields which is necessary to determine the effective number of data levels entering the covariance matrix (3.9) of the estimated parameters. Considering that the variances well may be underestimated by a factor of two to three any possibly found correlation of the diffusivities to environmental conditions may become questionable. We thus desist here from further hasty interpretations.

A comparison between the performance of the isopycnal and the Cartesian orientation of the mixing tensor can be drawn from Fig. 15 and Fig. 16, the latter being the result of fitting the Cartesian form (2.10) to potential density conservation. There are no striking differences between the corresponding coefficients, neither in the typical magnitudes nor the overall pattern. What fit should be judged superior, however? *Can we decide whether the ocean is mixed along isopycnal or geopoten-*

tial surfaces? A decent answer to this important question would require the careful statistical framework and an adequate knowledge of the data statistics which are necessary to formulate and test hypotheses about the models we are using. For an explanation of statistical testing on the basis of inverse model results the reader is referred to Wunsch's article in this book or e.g. to Müller et al. (1978). Again, the lack of a sensible model of the covariances of the analysed temperature and salinity fields of Gordon's atlas prevented us from going along this well-established path of inverse techniques.

As a substitute for statistical testing we consider here the skill achieved by the two competing models. The normalized sum of the squared imbalances of the model equations (3.6) in form of the expression

$$\text{skill} = 1 - \frac{(\mathcal{D}p - b)^\dagger \, \mathcal{W} \, (\mathcal{D}p - b)}{b^\dagger \, \mathcal{W} \, b} \qquad (4.3)$$

measures the performance of each of the diffusive models in comparison to an entirely advective model (remember that the parameters p are here only the diffusivities since according to our strategy the advection velocities have been determined in the first step). To make a meaningful decision between different models the number of parameters allowed in the fit must obviously be the same. This is not the case for the results of Fig. 15 and 16 since the isopycnal coefficient drops out of the balance of potential density. Fig. 17 shows the skills achieved with an isopycnal and a Cartesian diffusion tensor (i.e. (2.11) and (2.10), respectively) in the balance of veronicity. There is no apparent distinction between the models in the upper layer (100–800m). However, in the deep ocean south of 50° S the isopycnal diffusion seem to work slightly better than the Cartesian diffusion—the skill is up to 5% larger. The zonal averages shown here are in fact representative for the areal patterns. We have to conclude that diffusion along isopycnals is a slightly better parametrization of mixing in the Circumpolar Current than diffusion along geopotential surfaces. It may well be, however that these small differences become insignificant when the noise structure of the data is invoked to calculate variances of the skills considered here. In any case, what remains is the quite unexpected result that the isopycnal diffusion tensor is not drastically superior to the Cartesian tensor, even in an oceanic area as the Circumpolar Current where the isopycnal slopes are large.

The last set of model comparisons considered here aimed at the distinction between the half-angle orientation of the mixing tensor given by (2.21) based on the energy conversion arguments and the forms discussed above. With these results

136

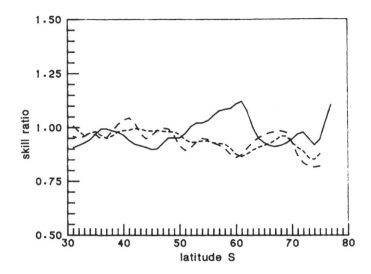

Figure 17
The skill of the two models with (a) Cartesian orientation and (b) isopycnal orientation of the mixing tensor. Solid lines correspond to 100m–800m values, narrow dashed to 800m–2500m values and wide dashed to 2000m–bottom values.

in mind one would of course not expect any great differences of the half-angle models to any of the other models. Indeed, corresponding isopycnal and half-angle models turned out almost indistinguishable. As an interesting side remark, however we like to point out that the tensor (2.21) is anisotropic in the neutral surface so that fair comparison with isopycnal or Cartesian diffusion requires to rewrite these mixing tensors in an anisotropic form, too. The result is exemplified in Fig. 18, showing the zonal means of the coefficients A_0 and A_m of the half angle tensor (describing mixing along the horizontal direction and along the direction of maximum energy gain in the neutral surface, respectively). There is clear evidence for anisotropy in the upper layer, the horizontal coefficient A_0 exceeds the maximum gain coefficient by a factor of two. With the orientation of the neutral surfaces in the area of the ACC this means larger diffusivity along the current.

Acknowledgements

These studies were partly supported by BMFT contract 07KF221/1. This is contribution number 149 of the Alfred-Wegener-Institut für Polar- und Meeresforschung.

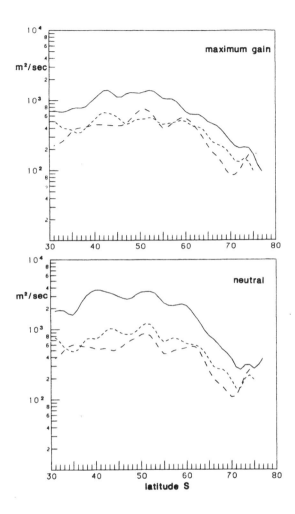

Figure 18
Zonal averages of the diffusivities for an anisotropic model in the half-angle
plane, (a) for direction of maximum energy gain and (b) for the neutral
direction. Solid lines correspond to 100m–800m values, narrow dashed to
800m–2500m values and wide dashed to 2000m–bottom values.

References

Bigg, G.R., 1985, 'The beta spiral method', *Deep Sea Res.*, **32**, 465–484

Cheney, R.E., Marsh, J.G. and B.D. Beckley, 1983, 'Global mesoscale variability from collinear tracks of Seasat altimeter data', *J. Geophys. Res.*, **88**, 4343–4354

Cox, M.D. and K. Bryan, 1984, 'A numerical model of the ventilated thermocline', *J. Phys. Oceanogr.*, **14**, 674–687

Daniault, N. and Y. Menard, 1985, 'Eddy kinetic energy distribution in the Southern Ocean from altimetry and FGGE drifting buoys', *J. Geophys. Res.*, **90**, 11877–11889

Gordon, A.L., Molinelli E. and T. Baker, 1978, 'Large-Scale Relative Dynamic Topography of the Southern Ocean', *Journal of Geophysical Res.*, **83**, 3023–3031

Gordon, A.L., Molinelli E. and T. Barker, 1982, 'Southern Ocean atlas', *Columbia University Press.*

Gregg, M.C., 1987, 'Diapycnal Mixing in the Thermocline: A Review', *J. Geophys. Res.*, **92**, 5249–5286

Hellermann, S. and M. Rosenstein, 1983, 'Normal monthly wind stress over the world ocean with error estimates', *J. Phys. Oceanogr.*, **13**, 1093–1104

Hogg, N.G., 1987, 'At least squares fit of advective-diffusive equations to Levitus Atlas data', *J. Mar.. Res.*, **45**, 347–375

Holloway, G., 1989, 'Parametrisation of small-scale processes', *this book*

Levitus, S., 1982, 'Climatological atlas of the world ocean', *NOAA Tech. Pap.*, **3**, 173

McDougall, T.J., 1987, 'Neutral Surfaces', *J. Phys. Oceanogr.*, **17**, 1950–1963

McWilliams, J.C., Holland, W.R. and J.H.S. Chow, 1978, 'A description of numerical Antarctic Circumpolar Currents', *Dyn. Atmos. Oceans*, **2**, 213–291

Montgomery, R.B., 1938, 'Circulation in upper layers of southern North Atlantic deducted with use of isentropic analysis', *Pap. Phys. Oceanogr. Meteorol.*, **2**, 55

Müller, P., Olbers, D. and J. Willebrand, 1978, 'The IWEX spectrum', *J. Geophys. Res.*, **83**, 479–500

Munk, W.H., 1966, 'Abyssal recipes', *Deep Sea Res.*, **13**, 707–730

Nowlin jr., W.D. and J.M. Klinck, 1986, 'The Physics of the Antarctic Circumpolar Current', *Reviews of Geophysics*, **24**, 469–491

Olbers, D.J., Wenzel, M. and J. Willebrand, 1985, 'The inference of North Atlantic circulation patterns from climatological hydrographic data', *Review of Geophysics*, **23**, 313–356

Olbers, D., 1989, 'A geometrical interpretation of inverse problems', *this book*

Pedlosky, J., 1979, 'Geophysical fluid dynamics', *Springer Verlag*, 624pp.

Schott, F. and H. Stommel, 1978, 'Beta spirals and absolute velocities from different oceans', *Dep. Sea Res.*, **25**, 961–1010

Solomon, H., 1971, 'On the representation of isentropic mixing in ocean circulation models', *J. Phys. Oceogr.*, **1**, 233–234

Stommel, H. and F. Schott, 1977, 'The beta spiral and the determination of the absolute velocity field from hydrographic station data', *Deep Sea Res.*, **24**, 325–329

Tarantola, A., 1986, 'Inverse problem theory', *Elsevier*, 335pp.

Veronis, G., 1972, 'On properties of seawater defined by temperature, salinity, and pressure', *J. Mar. Res.*, **30**, 227–255

Wenzel, M., 1986, 'Die mittlere Zirkulation des Nordatlantik auf der Grundlage klimatologischer hydrographischer Daten', *Berichte Institut für Meereskunde Kiel*, **157**, 109pp

Wolff, J.-O. and D.J. Olbers, 1988, 'The Dynamical Balance of the Antarctic Circumpolar Current Studied with an Eddy Resolving Quasigeostrophic Model', *submitted for publication*

Wunsch, C., 1978, 'The general circulation of the North Atlantic west of 50° W determined from inverse methods', *Rev. Geophys.*, **16**, 583–620

Wunsch, C., 1984, 'An eclectic Atlantic Ocean circulation model, I, The meridional flux of heat', *J. Phys. Oceanogr.*, **14**, 1712–1733

Wunsch, C., 1989, 'Tracer inverse problems', *this book*

Wunsch, C. and B. Grant, 1982, 'Towards the general circulation of the North Atlantic Ocean', *Prog. Oceanogr.*, **11**, 1–59

OCEAN ACOUSTIC TOMOGRAPHY: A PRIMER

Robert A. Knox
Scripps Institution of Oceanography A-030
University of California, San Diego
La Jolla, CA 92093

1. INTRODUCTION

The purpose of this paper is to survey the ocean observing technique known as ocean acoustic tomography, the term 'tomography' having been chosen to highlight a functional analogy with computer-assisted tomography (CAT) in medicine. The method was proposed by Munk and Wunsch (1979), and its practitioners, results and ramifications have since proliferated; it seems certain to be an active means of obtaining large-scale information about ocean circulation for some time to come.

The survey will explain how tomography works, give some basic feel for its strengths and limitations as a tool for observing the oceans, give some sense of its successes and failures to date, and conclude with a report on work in progress and on likely future uses of tomography, insofar as I know them. My two sources are a rather small group of colleagues with whom it has been my privilege to work and who are responsible for most of the results and ideas discussed below, and the published literature, almost all of which stems from the same set of individuals. But now tomographers are sprouting up in a number of new locations, which is a good thing; it certainly will mean that the next person who undertakes such a review will have to depend on more than his own direct knowledge of colleagues and their work.

D. L. T. Anderson and J. Willebrand (eds.), Oceanic Circulation Models: Combining Data and Dynamics, 141–188.
© 1989 by Kluwer Academic Publishers.

Despite the fact that it will not contain much discussion of modelling the ocean general circulation *per se*, the survey is still an appropriate element of this Advanced Study Institute for two reasons. First, tomography is a novel and significant weapon in the arsenal we can bring to bear on the problem of observing the ocean and obtaining the data needed to model the circulation in other than a fanciful way. It is important to recognize candidly how terribly data-poor oceanography is, for example by comparison with meteorology. We can ill-afford not to understand and use any sensible means of rectifying this situation; tomography is one such means. Secondly, tomography yields data — long path integrals of acoustic travel time — that are unfamiliar to most oceanographers and are not nearly as amenable to intuitive perusal or qualitative analysis (for instance, by drawing a map) as are more conventional point observations of temperature, salinity, or velocity. One inevitably must bring some sophisticated mathematics (inverse methods) to bear on tomographic data in order to extract oceanographically interpretable results. So there is a commonality of analytical machinery with other kinds of ocean modelling using inverse methods. Other chapters in this volume, especially that by Wunsch, explain this machinery in considerable detail, so here I give only a restricted account, to show how the problem is set up in the tomographic situation, and to exhibit some results of actual experiments. The point is that inverse methods are how we deal with tomographic data alone, and are also how one approach to ocean modelling deals with data of many kinds.

The paper begins with a brief review of some elementary hydrodynamics, obtaining the acoustic wave equation in the oceanic context, where the propagation speed (speed of sound) is a function of position in the medium, primarily of depth. Next I consider long range acoustic propagation in the ocean by means of ray theory, the mathematical representation in terms of which most, but not all, tomographic work has been done to date. Sound propagates for very great distances in the ocean in a waveguide formed by the vertical variation of the speed of sound; one must understand the features of this propagation and some of the limits posed by the waveguide which nature has handed us. I then describe the basic tomographic problem — making measurements of travel time for acoustic pulses along several different rays through an ocean region, repeating these measurements in time as the ocean changes, causing perturbations in the measured travel times, and inverting the travel time perturbations to make inferences about the oceanic variations which caused them. I discuss some sources of noise in the measurements and some limitations of hardware. Both lead to errors which set bounds on how well the travel time perturbations can be measured and thus on how well the method can 'see' the oceanic signals of interest. I conclude by showing results from several completed experiments and outlining plans for experiments that will be carried out in the foreseeable future.

2. ELEMENTARY HYDRODYNAMICS AND ACOUSTICS

We are concerned with sound waves in water. That is to say, we want to concentrate on the mechanics (hydrodynamics) of vibrations in the range of tens of Hz or higher frequencies, vibrations which cause only small departures from what is otherwise a static state in the fluid. This acoustic problem separates nicely from all the rest of complex ocean hydrodynamics. Because these vibrations are so fast, we ignore heat transfers between fluid elements; we also ignore viscosity since the fluid motions are small. In formal terms, then, consider an inviscid fluid without heat conduction. There are two equations to deal with, one expressing conservation of mass (the continuity equation) and one expressing Newton's second law for fluid parcels. Derivations are given in many textbooks. The results are:

$$\frac{\partial \rho}{\partial t} + \nabla . \rho \mathbf{u} = 0 \tag{1}$$

and

$$\rho(\frac{\partial \mathbf{u}}{\partial t} + \mathbf{u}.\nabla \mathbf{u}) = \rho \frac{D\mathbf{u}}{Dt} = -\nabla p - \rho g \hat{\mathbf{k}} \tag{2}$$

where ρ, $\mathbf{u}$ and p are the fluid density, velocity and pressure respectively in Eulerian coordinates (x, y, z, t), g is the acceleration of gravity, $\hat{\mathbf{k}}$ is a unit vector in the z (vertical) direction and the material or substantial derivative notation is as used in (2).

Equations (1) and (2) are the basic governing equations. But we are concerned with small vibrations, so we are going to linearize them. The vibrations are departures from a state of rest. Denote variables in the resting state with subscripts 0 and departures with subscripts 1. Since $\mathbf{u}_0 = 0$ by definition, eq. (1) becomes

$$\frac{\partial \rho_1}{\partial t} + \frac{\partial \rho_0}{\partial t} + \rho_0 \nabla . \mathbf{u}_1 + \rho_1 \nabla . \mathbf{u}_1 + \mathbf{u}_1 . \nabla \rho_0 + \mathbf{u}_1 . \nabla \rho_1 = 0 \tag{3}$$

$$\quad\ \text{A} \qquad \text{B} \qquad \text{C} \qquad\quad \text{D} \qquad\quad \text{E} \qquad\quad \text{F}$$

Term B vanishes because the rest state is time-independent, term D is dropped in comparison to term C because it is second order in small quantities, and similarly term F is dropped relative to E. Now compare terms C and E. C involves spatial variations of the vibration field, so its space scale is the acoustic wavelength λ, whereas term E involves spatial variation of the ocean density field in the resting state, with the scale of the ocean depth H. The ratio E/C is

$$\frac{|\mathbf{u}_1|\delta\rho_0}{H}\left(\frac{\rho_0|\mathbf{u}_1|}{\lambda}\right)^{-1} = \frac{\delta\rho}{\rho_0}\cdot\frac{\lambda}{H} \tag{4}$$

The first factor (fractional density change) is typically 10^{-3} in the ocean, and the second (ratio of acoustic wavelength to length scale of ocean density variation) is typically 10^{-2} at even the longest acoustic wavelengths we shall consider, say 15m. So E can be dropped, and we have

$$\frac{\partial\rho_1}{\partial t} + \rho_0(\mathbf{r})\nabla.\mathbf{u}_1 = 0 \tag{5}$$

as our linearized continuity equation, where we have now included possible dependence of the resting-state or mean density on position, to the degree that this variation (slight) arises in the real ocean.

We expand the dynamical equation (2) similarly:

$$(\rho_0 + \rho_1)(\underbrace{\frac{\partial\mathbf{u}_1}{\partial t}}_{A} + \underbrace{\mathbf{u}_1.\nabla\mathbf{u}_1}_{B}) = \underbrace{-\nabla p_0 - \nabla p_1}_{C} \underbrace{-\rho_0 g\hat{\mathbf{k}} - \rho_1 g\hat{\mathbf{k}}}_{D} \tag{6}$$

Terms A and B are dropped by the small vibration assumption, while terms C and D sum to zero by the definition of the resting state; with no motion the full dynamical equation reduces to this hydrostatic relation between p_0 and ρ_0. Thus

$$\rho_0\frac{\partial\mathbf{u}_1}{\partial t} = -\nabla p_1 - \rho_1 g\hat{\mathbf{k}} \tag{7}$$

is our linearized form. We need one more relation, because (5) and (7) are only two equations in three variables. The third equation is an equation of state connecting pressure and density in seawater. Without trying to explore the physics of it, we simply assume that the pressure can be expanded in a Taylor series for small departures of the density from the resting state:

$$p = p(\rho_0) + \left.\frac{\partial p}{\partial\rho}\right|_{\rho=\rho_0}(\rho - \rho_0) + ...$$

$$p_1 = p - p_0 = p - p(\rho_0) = \left.\frac{\partial p}{\partial \rho}\right|_{\rho=\rho_0} (\rho - \rho_0) = \rho_1 \left.\frac{\partial p}{\partial \rho}\right|_{\rho=\rho_0} = \rho_1 c^2 \qquad (8)$$

where $c^2 = \left.\frac{\partial p}{\partial \rho}\right|_{\rho=\rho_0}$ is defined as the indicated derivative, taken under adiabatic (no heat transfer) conditions. For water $c \approx 1.5$ km s^{-1}. Eq. (7) then is

$$\rho_0 \frac{\partial \mathbf{u}_1}{\partial t} = -\nabla p_1 - p_1 g c^{-2} \hat{\mathbf{k}} \qquad (9)$$

$$\text{A} \qquad \text{B}$$

But term A, in which the gradient involves the acoustic wavelength scale, is much larger than term B (15,000:1 for 100 Hz or 15m wavelength sound), so we drop B to obtain

$$\rho_0 \frac{\partial \mathbf{u}_1}{\partial t} = -\nabla p_1 \qquad (10)$$

We now use (8) and (10) in (5) to get one equation in one variable

$$\frac{\partial p_1}{\partial t} + \rho_0 c^2 \nabla . \mathbf{u}_1 = 0$$

$$\frac{\partial^2 p_1}{\partial t^2} + \rho_0 c^2 \nabla . \left(-\frac{\nabla p_1}{\rho_0}\right) = 0 \qquad (11)$$

$$c^{-2} \frac{\partial^2 p_1}{\partial t^2} - \nabla^2 p_1 + \frac{\nabla p_1 . \nabla \rho_0}{\rho_0} = 0$$

By the same arguments that simplified (5) we can drop the final term to obtain at last our acoustic wave equation in the perturbation or vibration pressure field, with c now emerging as the speed of sound in the medium, a function of position:

$$\frac{\partial^2 p_1}{\partial t^2} - c^2(\mathbf{r}) \nabla^2 p_1 = 0 = \frac{\partial^2 p}{\partial t^2} - c^2 \nabla^2 p \qquad (12)$$

The same equation governs ρ_1 or $\mathbf{u}_1$, to the same approximations used in obtaining (12). All the machinery to deal with wave propagation in inhomogeneous media can now be brought to bear on the ocean acoustics problem. Henceforth we will drop the subscripts 1 on the vibration or acoustic variables, as on the right side of (12).

3. OCEAN SOUND SPEED DISTRIBUTION

To afford a practical sense of how the speed of sound varies in the ocean we show some examples in Figs. 1–4. The overall fractional variation is slight. At depth, where the water is all roughly isothermal, c increases due to pressure; near the surface there is an increase due to warmer temperatures. In places where there is no warm surface water c increases monotonically with depth (Fig. 4). We can also see a limited region of this behavior in the Atlantic profile (Fig. 1) from late winter, where a thick mixed layer at the surface leads to a layer of increasing c. Pressure and temperature dominate salinity in causing variations of c; over oceanic ranges (0–20°C, 33–37‰, 0–4000m depth) one finds approximate ranges of c of 92, 5 and 64 ms^{-1} respectively. In the ocean T and S tend to be correlated over large regions (TS relation), say $S = S_0 + mT$. A typical gross value for m is 0.1‰(deg C)$^{-1}$, so for the relative magnitudes of sound speed changes due to ocean perturbations with such correlated T and S changes, we find

$$\frac{\delta c}{c} = \alpha \delta T + \beta \delta S = \alpha \delta T (1 + \beta m \alpha^{-1}); \quad \beta m \alpha^{-1} \approx 0.03$$

and the sound speed variations are primarily due to temperature variations. Depending upon the oceanographic goal, this sensitivity to temperature, or insensitivity to salinity, is either an advantage or a drawback of using acoustic propagation to sense ocean properties.

The formal variation of sound speed with position in (12) disguises the fact that most of the variation is in the vertical. Fig. 5 illustrates this; there is about 50 msec^{-1} change at shallow depths proceeding from 30°S to 50°S in the South Atlantic, a distance of 2000 km; a similar change over 3–4 km of depth is evident at many places. We will find it convenient to examine sound propagation first under the assumption that horizontal variations are negligible (range-independent), and then to mention how gradual range-dependent effects can be included to improve the analysis.

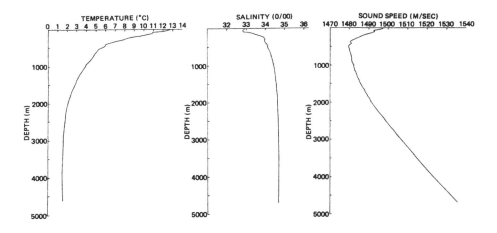

Figure 1
Midlatitude North Atlantic sound speed profile in early spring. Winter cooling and mixing have produced a 400m isothermal surface layer in which c increases approximately linearly with depth. This forms a surface duct or waveguide. From Boyles (1984).

4. RAYS AND MODES

As is well known from treatments of the wave equation (12) in many branches of physics, different approaches exist and are useful and informative in different situations and geometries. In simple geometries with suitably arranged coordinates and boundaries one can separate the equation, satisfy boundary conditions in three dimensions and express the general solution as a sum of these normal modes. Alternatively one can assume at the outset that the solution is nearly a plane wave, with only slow variation of amplitude and of sound speed on the scale of a wavelength, and arrive at the principles of geometric acoustics and ray tracing; rays are (gentle) curves normal to the (almost) plane wavefronts. We will sketch both approaches

148

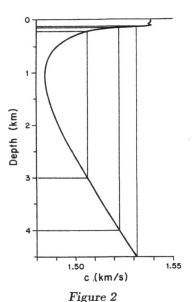

Figure 2

Profiles of temperature, salinity and sound speed for an April station in the midlatitude North Pacific. Note the shallower depth of the minimum (sound channel axis), compared to the North Atlantic profile of Fig. 1. Rays with lower turning points below about 2500m strike the sea surface at their upper extremes. From Boyles (1984).

but concentrate on the ray method, which has been the most useful for tomographic applications to date (but see Munk and Wunsch, 1983 and Romm, 1987).

We begin by discussing mode solutions, and we do so in the context of a sound channel or waveguide, such as occurs in Fig. 2, for example. An idealized case is drawn in Fig. 6. Sound speed increases above and below the channel axis at depth z_0, and we assume $c = c(z)$ only, i.e. a range-independent system. Since we are usually interested in problems involving point sources and receivers, we choose cylindrical coordinates with the z-axis vertical; a point source can be synthesized by adding up modes so that along this axis the result is a delta function at the required depth. Time-harmonic solutions of (12) are

$$p = R(x)P(z)e^{-i\omega t} \tag{13}$$

Substitution in (12) yields the radial (horizontal) equation

$$\frac{1}{x}\frac{d}{dx}\left(x\frac{dR}{dx}\right) + k_H^2 R = 0; \quad R = H_0^1(k_H x) \underset{x \to \infty}{=} \left(\frac{2}{\pi k_H x}\right)^{\frac{1}{2}} e^{i(k_H x - \pi/4)} \tag{14}$$

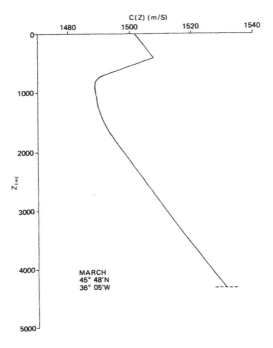

Figure 3
Sound speed profile on the Pacific equator at 150° W. Here the warm sur-
face layer leads to turning depths such that rays usually intersect the ocean
bottom while their upper turning points are still beneath the surface layer.
This profile tends to focus rays strongly, in the sense that distinct ray paths
have virtually the same travel times, posing difficulties for tomography in
such an environment.

where we have retained only the Hankel function solution H_0^1 corresponding to waves propagating outward from the radial origin. The separation constant is k_H. The vertical equation is

$$\frac{d^2 P}{dz^2} + (\omega^2 c^{-2} - k_H^2)P = 0 \tag{15}$$

Thus there is a local vertical wavenumber $k_v = (\omega^2 c^{-2} - k_H^2)^{\frac{1}{2}}$ which varies with depth, and which vanishes for $z = \hat{z}$ such that

$$c^2(\hat{z}) = \frac{\omega^2}{k_H^2} \tag{16}$$

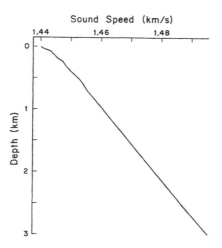

Figure 4a
Greenland Sea sound speed profile, from Romm (1987). It is well approximated by an adiabatic profile, which in turn is nearly a straight line.

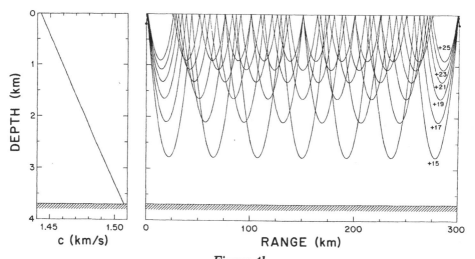

Figure 4b
Adiabatic profile and ray trace as an approximation to the Greenland Sea
profile of Fig. 4a, from Romm (1987). The profile is exponential in depth,
but nearly a straight line on this scale. Resultant rays are nearly arcs of
circles, as discussed in text (section 6).

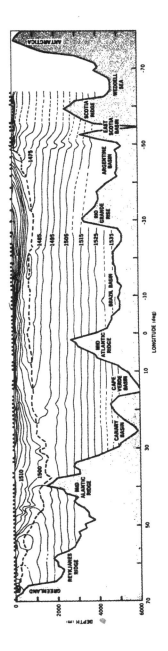

Figure 5
Section of sound speed in the Atlantic. Long dashed line is the sound channel axis; short dashed segments
indicate missing data. From Clay and Medwin (1977).

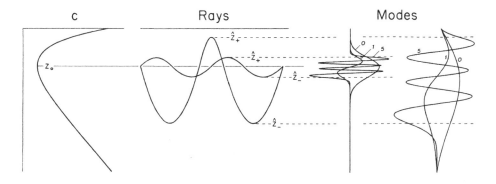

Figure 6
Schematic sound channel illustrating mode and ray structures and turning depths. Reference sound speed profile is shown at left. Two rays with different turning depths between a source and receiver separated by 96 km are shown in center. Modes 0, 1 and 5 for two different frequencies are shown at the right, with the individual mode frequencies chosen so that one set has the same turning depth as the flat ray, the other the same turning depth as the steep ray. From Munk and Wunsch (1983).

These depths are the turning points of the mode, where the vertical wavefunction changes from oscillatory near the channel axis to exponentially decaying far from it. For modes of practical interest, with turning points well-removed from the sea surface and bottom, we can obtain a good approximation by ignoring these boundaries and thus retaining only the solutions which decay exponentially beyond the turning points, not those which grow. Eq. (15) can be solved numerically in general, has solutions in terms of special functions for particular choices of $c(z)$, and can be solved approximately by WKBJ methods. Tolstoy and Clay (1987, sec. 2.9) give a good account of the WKBJ approximation. Pertinent results are i) in the region away from the turning points the WKBJ approximation is valid if the vertical variation of c is small on the scale of a local vertical wavelength, and ii) the approximation fails at turning points even if c varies slowly there. An improved approximate solution which is valid at the turning points and reduces to the WKBJ solution away from the turning points is discussed by Munk and Wunsch (1983; their Eq. (23)).

If one now superposes these modes to fit the condition of a point source on the

z-axis and examines the resulting field at some distant point, as a receiver would do, one finds that the interference of the modes gives rise to a discrete set of times at which the amplitude is large. These are the arrival times; if the source emits a single pulse, the receiver hears a series of pulses separated in time. Munk and Wunsch (1983) give the relevant mathematics, and show that these arrival times calculated using mode superposition are just the arrival times one obtains by the alternative analysis in terms of rays. We now turn to that representation, which is conceptually simpler in many ways.

Ray analysis begins with the idea that far from a source its field can be represented as a plane wave of slowly varying properties. That is, we assume a solution of the form

$$p = A(\mathbf{r})e^{i(k_0 L(\mathbf{r}) - \omega t)} \tag{17}$$

Substituting this into the wave equation and doing the indicated algebra we arrive at

$$|\nabla L|^2 - \left(\frac{\omega}{k_0 c}\right)^2 = \frac{\nabla^2 A}{k_0^2 A} \tag{18}$$

Wavefronts or surfaces of constant phase are given by $k_0 L - \omega t = $ const. Rays are curves which are everywhere normal to wavefronts. The variable L is referred to as the eikonal, and if the right side of (18) is suitably small, we have the eikonal equation governing L

$$|\nabla L|^2 = \left(\frac{\omega}{k_0 c}\right)^2 = \left(\frac{c_0}{c}\right)^2 = n^2 \tag{19}$$

where n is the index of refraction. Lindsay (1960, ch. 1, sec. 12) discusses the conditions under which the approximation (19) is a good one: ray direction $\nabla L/|\nabla L|$ must not change significantly over one wavelength, the change in wavelength over one wavelength must be small, and the change in amplitude over one wavelength must be small. Let ds be an increment of arc length along a ray, and let $\mathbf{r}$ be the position of a point on the ray. Then $d\mathbf{r}/ds$ is a unit vector in the direction of the ray, but so is $\nabla L/|\nabla L|$, so equating these and using (19) we have the equations for the geometry of the ray in terms of the eikonal

$$\frac{\omega}{k_0 c}\frac{dx}{ds} = \frac{\partial L}{\partial x}; \quad \frac{\omega}{k_0 c}\frac{dy}{ds} = \frac{\partial L}{\partial y}; \quad \frac{\omega}{k_0 c}\frac{dz}{ds} = \frac{\partial L}{\partial z} \tag{20}$$

A little further algebra gives

$$\frac{d}{ds}\left(\frac{\omega}{k_0 c}\frac{dx}{ds}\right) = \frac{d}{ds}\left(\frac{\partial L}{\partial x}\right) = \frac{\partial}{\partial x}\left(\frac{\partial L}{\partial x}\right)\frac{dx}{ds} + \frac{\partial}{\partial y}\left(\frac{\partial L}{\partial x}\right)\frac{dy}{ds} + \frac{\partial}{\partial z}\left(\frac{\partial L}{\partial x}\right)\frac{dz}{ds}$$

$$= \frac{\partial}{\partial x}\left(\frac{\partial L}{\partial x}\frac{dx}{ds} + \frac{\partial L}{\partial y}\frac{dy}{ds} + \frac{\partial L}{\partial z}\frac{dz}{ds}\right)$$

$$= \frac{\partial}{\partial x}\left(\frac{\omega}{k_0 c}\left[\left(\frac{dx}{ds}\right)^2 + \left(\frac{dy}{ds}\right)^2 + \left(\frac{dz}{ds}\right)^2\right]\right) = \frac{\partial}{\partial x}\left(\frac{\omega}{k_0 c}\right)$$

and similarly

$$\frac{d}{ds}\left(\frac{\omega}{k_0 c}\frac{dy}{ds}\right) = \frac{\partial}{\partial y}\left(\frac{\omega}{k_0 c}\right); \quad \frac{d}{ds}\left(\frac{\omega}{k_0 c}\frac{dz}{ds}\right) = \frac{\partial}{\partial z}\left(\frac{\omega}{k_0 c}\right) \tag{21}$$

so that we now have differential equations for the ray path in space, given the field of sound speed.

It is worth noting that the rays correspond to Fermat paths or paths of minimum time of travel. That is, for two fixed end points, the ray between them is such that any infinitesimal change of the path geometry yields a greater travel time. Tolstoy and Clay (1987, p. 53) demonstrate this by transforming the eikonal equation to a space in which the rays are straight lines and thus the shortest paths, while Lindsay (1960, p. 51) exploits the mathematical similarity between the eikonal equation and the Hamilton-Jacobi equation for the motion of a particle in a force field. The principle of least action then leads directly to the Fermat condition. This analogy with classical mechanics is pursued much further by Wunsch (1987).

The situation is simplified if c is a function of z only; we can consider just the x, z plane without loss of generality, and the two relevant parts of (21) become

$$\frac{d}{ds}\left(\frac{\omega}{k_0 c}\frac{dx}{ds}\right) = \frac{d}{ds}\left(\frac{\omega}{k_0 c}\cos\theta\right) = \frac{\partial}{\partial x}\left(\frac{\omega}{k_0 c}\right) = 0; \quad \frac{\omega}{k_0 c}\cos\theta = \text{const.}$$

and

$$\frac{d}{ds}\left(\frac{\omega}{k_0 c}\frac{dz}{ds}\right) = \frac{d}{dz}\left(\frac{\omega}{k_0 c}\right) \tag{22}$$

where θ is the angle the ray makes with the horizontal. The first of these equations is just Snell's Law in a medium with continuous variation of the index of refraction in one dimension. When $\theta = 0, \cos\theta = 1, z = \hat{z}$, so the constant is $\omega/k_0 c(\hat{z})$. We write the two equations as

$$\frac{\cos\theta}{c} \equiv K = \frac{1}{c(\hat{z})}$$

$$\frac{d\theta}{ds} = -K\frac{dc}{dz} \tag{23}$$

We can now write expressions for the horizontal and vertical positions of a point on the ray, and for the time of travel to that point, given an initial point and the ray parameter K or constant in Snell's law:

$$x_1 - x_0 = \int_{z_0}^{z_1} \frac{Kc}{(1 - K^2c^2)^{\frac{1}{2}}} dz$$

$$t_1 - t_0 = \int_{z_0}^{z_1} \frac{1}{c(1 - K^2c^2)^{\frac{1}{2}}} dz = \int_{s_0}^{s_1} \frac{1}{c} ds = \int_{z_0}^{z_1} \frac{1}{c \sin\theta} dz$$

$$s_1 - s_0 = \int_{z_0}^{z_1} \frac{1}{(1 - K^2c^2)^{\frac{1}{2}}} dz \tag{24}$$

Our ray concept of a source-receiver transmission then is to think of a source at some given position which emits a pulse along all possible rays (omnidirectional). These can be traced through space using (24). But only a discrete subset of these rays will also intersect a receiver at some other given point. The travel times along these particular rays can be computed as in (24), and when this is done one obtains again exactly the travel times found by the analysis of superposing modes. Again we refer to Munk and Wunsch (1983) for details.

For a sound channel such as that of Fig. 6 we again encounter the notion of a turning point. Rays with an inclination to the horizontal at the channel axis are refracted back toward the axis, becoming horizontal at depths $\hat{z}$ such that $c(\hat{z}) = c(z_0)/\cos\theta(z_0)$ by Snell's law. In fact, as shown by Munk and Wunsch (1983), a ray arrival is just the constructive interference of modes whose turning depths are the same as the ray turning depth. In the range-independent situation we have assumed the turning points are always at the same depths $\hat{z}$, and the rays are therefore periodic in x. In a range-dependent environment one can utilize the fact that the x-variation of c is much slower than the z-variation. The ray paths are almost periodic, and the classical mechanics methods for treating almost-periodic orbits can be used to obtain very satisfactory approximations (Wunsch, 1987).

156

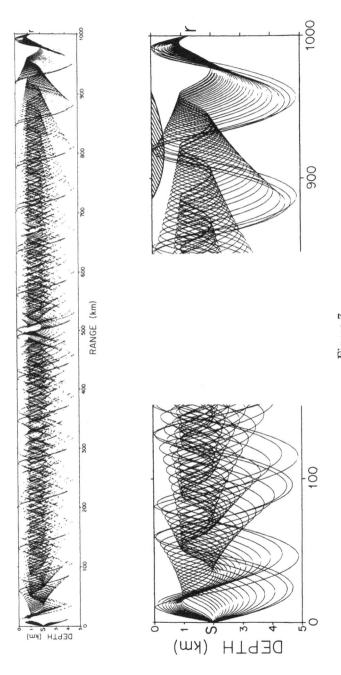

Figure 7

Ray trace for 1000 km range Eleuthera – Bermuda, with only those rays aimed downward at the receiver drawn in. From this it is easy to visualize qualitatively how a localized shallow anomaly or perturbation (shaded region) would affect travel time along some rays but not others, since some rays reach near the surface and bottom, while others are confined near the axis. This different effect on different rays is the basis of the tomographic method. From Munk and Wunsch (1979).

5. CAPSULE DESCRIPTION OF THE TOMOGRAPHIC METHOD

The basic idea of acoustic tomography can now be simply stated. Consider a sound channel such as that of Fig. 6. If a source within the channel sends a pulse to a receiver within the channel, the pulse will travel along each of several different rays that are continuously refracted within the channel (for example as in Fig. 7). If we know the sound speed field we can calculate the ray geometry and travel times as noted above. In fact, we wish to turn the problem around (invert). We measure the travel times, and wish to deduce the sound speed field. More exactly, we usually begin with an initial estimate of the sound speed field, say from historical data, and thus with an estimate of the ray pattern and arrival times. We then repeatedly observe the actual arrival times, obtaining a time series of these arrival time perturbations. Variations in the travel time perturbations over time, say from day to day, are due to ocean fluctuations, i.e. variations in c. For small perturbations we can set up a linear inverse problem to calculate the c-variations using the travel time perturbations as data. One can see intuitively from Fig. 7 that a shallow warm anomaly (shaded region), for example, would increase the sound speed over a portion of the ray paths for some rays (steep ones) but not others (flat or near-axial ones), and that the travel times would therefore change for some of the arrivals but not others. One ought to be able to work backward and use this information to map the warm anomaly, and that is what the inverse calculation accomplishes.

Obviously the method extends to three dimensions. With several sources and receivers in a region one obtains a network of sections across the region, each source-receiver section containing a set of multipaths as in Fig. 7. Fig. 8 shows the horizontal placement of sources and receivers in the first three-dimensional test of the method in 1981. We can see the reasons for the term 'tomography': instead of the medical procedure which transmits X-rays through a patient along a variety of paths and analyzes the received intensities by means of inverse calculations to produce a map of internal tissue structure, we transmit sound pulses through a volume of ocean along a variety of paths and analyze the received travel times by means of inverse calculations to produce a map of the internal sound speed field.

A number of advantages and disadvantages already can be seen or inferred. Advantages include:

1. Since in an arrangement like Fig. 8 each source can transmit to each receiver, the number of paths grows like the product of the number of sources and the

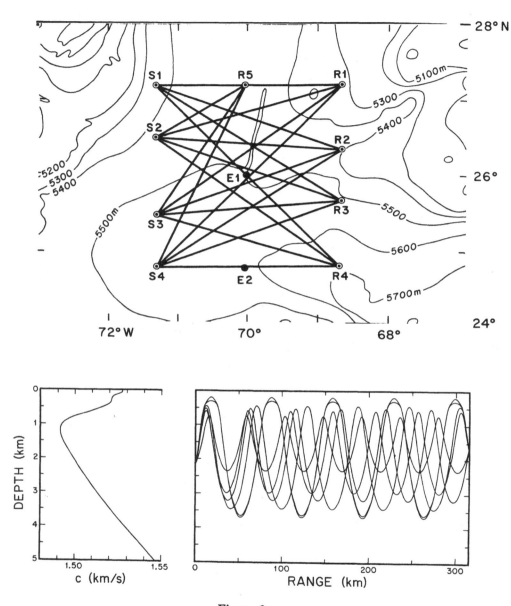

Figure 8

Geometry of the 1981 tomography experiment, with sound speed profile and paths of selected rays for one instrument pair. The plan view locates the sources (S) and receivers (R) over the bathymetry of the region.

number of receivers, so there are economies of scale relative to the costs of making point measurements.

2. The moorings which carry the sources and receivers are around the periphery of the region of interest, so one can avoid placing instruments in awkward locations such as high currents.

3. Low frequency sound along purely refracted rays is very gradually attenuated in seawater, so long transmission paths are possible and have been achieved.

4. Time, the basic observable quantity, is something that can be measured with great precision.

5. Sound speed is large, so the measurements in a region like Fig. 8 can be made truly synoptically, relative to time scales of mesoscale or large scale phenomena in the ocean. It is a bit like having a $c=3000$ knot ship.

A few of the disadvantages are:

1. A tomographic array also loses information like the source-receiver product if an instrument fails.

2. Not all ocean regions have a sound speed profile that is conducive to tomography. In particular, the equatorial profile of Fig. 3 tends to be strongly focussing, i.e. the different ray arrivals all occur at nearly the same time and cannot easily be separated from one another.

3. In much of the tropical ocean the warm surface water means that even the steepest rays which turn very near the bottom also turn well beneath the surface layer, so the shallow ocean is poorly sampled.

4. The method is primarily sensitive to temperature, as previously noted, but there are problems and regions where the field of interest (density) is more strongly dependent on salinity.

Finally, a caveat — success depends upon being able to identify rays and to associate a given acoustic arrival with a particular ray. If the sound channel is nearly symmetric above and below the axis, one often finds a degeneracy between rays that are sloped upward at the receiver and rays that are sloped downward, in the sense that pulses along the two rays will have traversed 'mirror image' paths and thus have the same travel time. It has been found very useful to expand the receivers with short vertical arrays of a few hydrophones, so that they can discriminate the incoming signal in terms of arrival angle as well as time. This solves the up-down degeneracy problem. Other kinds of degeneracies in the arrival pattern can arise; one of the first steps in any tomographic experiment is to do some exploratory ray tracing using historical data to find out if the regional sound speed profile yields a sufficient number of resolved arrivals corresponding to a sufficient depth range of turning points (see next section) for the method to be effective in the intended study.

6. A SIMPLE RAY EXAMPLE

It is useful to work one ray tracing example with real numbers, to get a feel for the refraction phenomenon and its scales. A simple textbook case is that of the constant gradient of c, viz. $c = c_s + \gamma z$. This is a very good approximation for nearly adiabatic regions such as the winter Greenland Sea (Fig. 4). In this case the ray paths are just arcs of circles with radius R; for the source and z-origin at the surface, we have from (23), (with the z-axis positive downwards)

$$\frac{d\theta}{ds} = K\frac{dc}{dz} = K\gamma, \quad R = \frac{1}{K\gamma}, \quad \text{and}$$

$$H \equiv R\cos\theta_s = \frac{c}{\gamma\cos\theta} \cdot \cos\theta_s = \frac{c_s}{\gamma\cos\theta_s} \cdot \cos\theta_s = \frac{c_s}{\gamma}, \quad \text{allrays} \quad (25)$$

Fig. 9 shows three different rays and the geometry. The dashed line is parallel to the sea surface and at height H above it. Any circle with center on this line and passing through the origin (source) is a ray. Only the first loops of each example ray are drawn; these loops repeat periodically. Only selected rays will intersect a receiver at a given range, i.e. will have an integral number of loop distances nD equal to the range (for a receiver at the surface), resulting in a countable set of multipaths analogous to Fig. 7. Taking $c_s = 1.45$ km s^{-1} and $\gamma = 0.02$ s^{-1}, typical of the Greenland Sea situation, we find

$\theta_s(°)$	h(m)	D(km)
0	0	0.0
5	217	12.7
10	1118	25.6
15	2557	38.8
20	4653	52.8

For shallow angles the maximum ray depth h varies approximately as θ_s^2, while the horizontal range of the loop varies linearly with θ_s. The numbers re-emphasize the

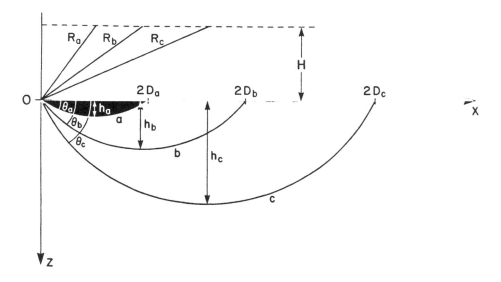

Figure 9
Sketch of ray paths in adiabatic ocean with source at surface and at origin.
For each of three rays the launch angle θ_s of Eq. (25) is shown as $\theta_a, \theta_b, \theta_c$,
the range to the surface reflection as $2D_a, 2D_b, 2D_c$, and the turning depth
as h_a, h_b, h_c. Rays are arcs of circles with centers on a line at height H
above sea surface and radii R_a, R_b, R_c.

flatness of the rays; even ones which turn near the bottom have modest maximum
inclinations (at the surface).

One important fact of ray geometry, obvious from considering Fig. 9, is that
the total travel time along a ray is sensitive primarily to the sound speed in the
depth range of the turning points. The sound speed really does not change very
much from its mean value (which is why the ray curvature is gentle), so in any layer
of fixed thickness the time of travel depends mostly on the ray length in the layer,
which is clearly greater for a layer next to a turning point than for a layer of equal
thickness elsewhere. This is why tomographic experiments in two-sided channels (c
increasing above and below the channel axis) usually place sources and receivers at
or near the axis, so that the suite of rays will include a variety of turning depths.
The farther from the axis one places the source, the fewer near-axial rays exist.
Phrased another way, the only way to obtain an axial ray is with an axial source
and receiver, because to touch a non-axial source or receiver a ray must cross the

axis at an angle, and thus the ray path must oscillate around the axis.

7. NOTES ON HARDWARE LIMITATIONS
AND OBSERVATIONAL ERRORS

Various errors place limits on the ability of a tomographic system to determine the desired oceanographic fields. One class of errors derives from limitations of practical hardware. As has been mentioned, sources and receivers usually are deployed on deep-sea moorings, in order to position them near the sound channel axis so that rays with different turning levels can be used. But no mooring is rigid, and mooring excursions change the travel time just as variations of c do. Typical excursions (100 m or 67 msec of travel time) are not insignificant. One must either measure the mooring motion, with a bottom mounted array of transponders, and remove this error, or one must include the mooring position changes in the set of unknowns to be solved for in the inverse computation, thereby using up information to determine mooring positions. Timekeeping is another problem. Absolute time is kept in each instrument separately, so that when travel times are later computed by differences between source transmission time and reception time in the receiver, any clock drifts introduce errors. The transmitted signal and its processing introduce further errors. Ideally one would transmit an infinitely sharp pulse, but sufficient energy cannot be coupled into the water in this way without highly nonlinear effects, and in addition we do not have any precisely timed transducers to do the job. Most acoustic transducers tend toward the other extreme, being optimized at a single frequency. But the time of travel determination in the receiver is done by cross-correlating a stored version of the expected signal against the actual incoming signal, and estimating the arrival time by the cross-correlation peak. For a sharp peak one wants a broadband signal, not a narrowband one.

The signals actually utilized are the subject of volumes in themselves; we touch on the basic elements. The transmitted waveform consists of a few cycles of a basic carrier sinusoid, at about 200 Hz in our current equipment, then an abrupt phase change at the time of a zero crossing, then a few more cycles, etc. The phase changes are of only two values, and one can think of the sequence of phase changes as a sequence of binary numbers, -1 and 1. Certain patterns, called linear maximal sequences, have attractive properties (see Metzger, 1983) which have proved useful in governing the modulation (sequence of phase changes) of the signals. These sequences have L terms, of which $(L+1)/2$ are one binary number and the rest are the other. $L = 2^n - 1$; n is called the degree. The most relevant feature is that the digital cyclic autocorrelation of a linear maximal sequence over one sequence length (cyclic means that the sequence is extended indefinitely by repeating it periodically) has value L at zero lag, value -1 at all nonzero lags. One can see in this last property the basis for forming the sharp cross-correlation in a digital receiver. A simple linear maximal sequence of length 7 is 1, -1, -1, 1, -1, 1, 1, and the autocorrelation property can be seen graphically in the following form. The correlation is the sum of products of pairs of elements from the same column, extending between the two columns with asterisks. One member of the pair is in row A, the second in row A, B C or D for lags of 0, 1, 2, or 3 respectively. It is easy to check that the other nonzero lags also give correlation $= -1$.

	2	3	4	5	6	7*	1	2	3	4	5	6	7*	1	2	3	4	5	corr.
A	-1	-1	1	-1	1	1	1	-1	-1	1	-1	1	1	1	-1	-1	1	-1	7
B	1	-1	-1	1	-1	1	1	1	-1	-1	1	-1	1	1	1	-1	-1	1	-1
C	1	1	-1	-1	1	-1	1	1	1	-1	-1	1	-1	1	1	1	-1	-1	-1
D	1	1	1	-1	-1	1	-1	1	1	1	-1	-1	1	-1	1	1	1	-1	-1

The optimum phase angles used to modulate the waveform are set by $\tan^2\phi = L$; this gives the best possible signal-to-noise ratio improvement, as discussed at length by Metzger. Some considerations and tradeoffs in the arrangement of such a signal for tomographic work are as follows.

1. The longer the sequence (L) the better, because one gets a bigger contrast between the zero and nonzero lag correlation values, L and -1 respectively.

2. But one cannot increase L indefinitely, because the total time for a transmission must not last longer than a typical decorrelation time due to internal wave activity along the ray path. These decorrelation times can be estimated given a model of internal waves (Flatté et al., 1979).

3. For a fixed total transmission time one can increase the sequence length L by cutting down on the time (number of carrier cycles) taken up for one 'digit' of the sequence. But there are electronic and mechanical limits to how rapidly a transducer can cycle from one digit to another in linear fashion.

4. Although from the point of view of processing gain in the ideal case one can use one long sequence or repeat a shorter sequence to make up the same total length, there are practical considerations of end effects and data storage that usually make it easier to adopt the second alternative. The processing gain, however, just depends on the product of the number of digits in the basic sequence and the number of sequence repetitions in the total transmission time.

When all is said and done one can think of the signal as being just a single pulse of the digit length repeated periodically and the reception summed in the receiver to increase the signal-to-noise ratio. Then the limit to the sharpness of the receiver crosscorrelation is set by the digit length, and the precision of travel time estimation depends on this digit length and on the gain in signal-to-noise achieved by the processing scheme. The standard formula is

$$\sigma_\tau = \left(2\pi\delta f\sqrt{2(S/N)}\right)^{-1}$$

where δf is the signal bandwidth (related to and of the same order as the reciprocal of the digit length) and $2(S/N)$ is the signal-to-noise ratio. Increasing this precision (reducing σ_τ) has been one of the major improvements of tomographic instrumentation in recent years. In early experiments, equipment limitations made the digit length and hence σ_τ so large that nearby ray arrivals could not be fully resolved. With better precision such rays can be resolved and the travel time data can be assigned usefully narrow error bars.

A further source of travel time error is due to the ocean itself. The precision σ_τ noted above pertains to a single resolved arrival along a deterministic ray path. But in the real ocean internal wave fluctuations split any single path that one would compute from the mean field of c into a set of so-called micro-multipaths, and this smears out the arrival peak in time, in an intrinsically random way. Statistics of this internal wave broadening can be estimated by means of the theory of sound propagation through a medium with random fluctuations (Flatté et al., 1979), given a spectral model of the internal wave field, and these estimates have proved quantitatively useful in the design of experiments and in assigning appropriate error bars to observed travel times. One assumes that the travel time variance due to internal waves is independent of the variance σ_τ^2 due to signal processing and noise, and so adds them to obtain total error variance. Typical values are $\sigma_\tau^2 = 0.09$ msec^2 and internal wave travel time variance $= 16.0$ msec^2 for 300 km range. Further error reduction is obtained by averaging; typically one forms daily average travel times by averaging hourly or bi-hourly measurements, thus dividing the total error variance by 24 or 12. Cornuelle *et al.* (1985) give such an error budget and a discussion of its terms.

8. INVERSIONS

We now sketch how tomographic data usually are inverted to obtain measures of the sound speed field. As in any inverse problem, the first step is to be clear about the forward problem, i.e. given a model, calculate the observations. We have now developed the machinery to do this by means of rays. In general, we suppose that there is a basic, steady sound speed field $c_0(z)$ and small perturbations to it, $\delta c(\mathbf{r}, t) \ll c_0$. We take the basic field to be a function of z only, for simplicity of illustration. In consequence of this perturbation any particular unperturbed ray path Γ_i will be perturbed to a nearby path Γ_i'. But because Γ_i is a Fermat path, these small perturbations of the path are second-order effects, and the first-order change of travel time along the ray is due to the perturbed sound speed field along the unperturbed ray. That is

$$\delta T_i = \int_{\Gamma_i'} \frac{1}{c_0 + \delta c} ds - \int_{\Gamma_i} \frac{1}{c_0} ds \approx \int_{\Gamma_i} \frac{-\delta c}{c_0^2} ds \qquad (26)$$

The travel time perturbations thus are linearly related to the sound speed perturbations. From this point one can proceed with various models of the ocean structure. In Munk and Wunsch (1979) the model was to cut the ocean up into boxes or cells; then it is a simple matter to calculate δT_i using (26) once the cell boundaries and ray paths are known. Most work to date has utilized models of ocean structure in terms of a truncated sum of modes, perhaps empirical functions computed from historical data or dynamically derived modes. Whatever the rationale for the modes chosen, we can write an expansion such as

$$\delta c = \sum_{l=1}^{N_l} A_l(t) f_l(\mathbf{r}) \sum_{n=1}^{N_n} Z_n(z) \qquad (27)$$

where the coefficients $A_l(t)$ are the model parameters to be estimated. For example, the Z_n might be the usual hydrostatic vertical modes of a stratified ocean and the

$f_l(\mathbf{r})$ might be Rossby waves. We can isolate the time-dependent coefficients by treating each combination of spatial functions as a separate expansion function

$$\delta c = \sum_{k=1}^{N_l N_n} p_k(t)\psi_k(\mathbf{r}, z) \tag{28}$$

Substituting this into the expression for the travel time perturbations gives a result which reduces to a matrix equation

$$\delta T_i \equiv d_i = \int_{\Gamma_i} \frac{-1}{c_0^2} \sum_{k=1}^{N_l N_n} p_k \psi_k ds = \sum_{k=1}^{N_l N_n} p_k \int_{\Gamma_i} \frac{\psi_k}{c_0^2 \cos\theta} dx = \sum_{k=1}^{N_l N_n} G_{ik} p_k; \quad \mathbf{d} = \mathbf{Gp} \tag{29}$$

that is, the vector of observations $\mathbf{d}$ can be written as the product of a matrix $\mathbf{G}$ that depends on assumed model functions and geometry and a vector $\mathbf{p}$ of unknown amplitudes or model parameters. In fact, we allow for some error or noise and write

$$\mathbf{d} = \mathbf{Gp} + \mathbf{e} \tag{30}$$

The next step in the 'stochastic inverse' which has been most utilized in tomography is to recall the Gauss-Markov procedure of minimum-mean-square estimation. The discussion here is very truncated; interested readers would do well to refer to Cornuelle (1983, especially chapter 5) which is still one of the most accessible expositions of the relevant mathematics, or to the chapter by Wunsch in this volume. More compact statements have been given by Howe (1986) and by Howe *et al.* (1987). The general problem is to estimate some unknown quantity as a linear sum of some given data. In our particular case the unknown is one of the model parameters p at a particular time t_0, and the available data are observations of acoustic travel times made at time t_0. Our estimate is:

$$\hat{p}(t_0) = \sum_i a_i(t_0)(d_i - \bar{d}_i) + \bar{p}(t_0) \tag{31}$$

where the overbars denote some estimate of the mean value and the a_i are weights to be determined. The expected value of the square error is

$$\langle R^2 \rangle = \langle (p(t_0) - \hat{p}(t_0))^2 \rangle \tag{32}$$

and we wish to choose the weights a_i of (31) so that $\langle R^2 \rangle$ is minimized. That is, we set the partial derivatives with respect to the weights $= 0$, obtaining:

$$\sum_j a_j \langle d_i' d_j' \rangle = \langle p' d_i' \rangle; \quad p' \equiv p - \bar{p}, \quad d' \equiv d - \bar{d}$$

or

$$a_i = \sum_j \langle p' d_j' \rangle \mathbf{D}_{ji}^{-1}$$

$$\mathbf{a}^T = \langle p' \mathbf{d}'^T \rangle \mathbf{D}^{-1}; \quad \mathbf{D} = \left(\langle \mathbf{d}' \mathbf{d}'^T \rangle \right) \tag{34}$$

where we recognize $\mathbf{D}$ as the matrix of data-data covariance; it is square and symmetric. If we wish to estimate p at other times we just add rows to (34), so in the more general case we have

$$\mathbf{A} = \left(\langle \mathbf{p}' \mathbf{d}'^T \rangle \right) \mathbf{D}^{-1}; \quad \hat{\mathbf{p}}' = \mathbf{A} \mathbf{d}' \tag{35}$$

We now substitute the matrix form of the forward problem into (35). The matrix algebra involves a number of steps, all given by Cornuelle (1983). One finally obtains a weighted version of the forward problem:

$$\mathbf{E}^{-\frac{1}{2}} \mathbf{d}' = \left(\mathbf{E}^{-\frac{1}{2}} \mathbf{G} \mathbf{W}^{\frac{1}{2}} \right) \left(\mathbf{W}^{-\frac{1}{2}} \mathbf{p}' \right) + \mathbf{E}^{-\frac{1}{2}} \mathbf{e} \tag{36}$$

or

$$\tilde{\mathbf{d}}' = \tilde{\mathbf{G}} \tilde{\mathbf{p}}' + \tilde{\mathbf{e}} \tag{37}$$

where the weightings are

$$\tilde{\mathbf{d}}' = \mathbf{E}^{-\frac{1}{2}} \mathbf{d}', \quad \tilde{\mathbf{e}} = \mathbf{E}^{-\frac{1}{2}} \mathbf{e}, \quad \tilde{\mathbf{G}} = \mathbf{E}^{-\frac{1}{2}} \mathbf{G} \mathbf{W}^{\frac{1}{2}}, \quad \tilde{\mathbf{p}}' = \mathbf{W}^{-\frac{1}{2}} \mathbf{p}' \tag{38}$$

and the two important matrices used in weighting are

$$\mathbf{W} = \left(\langle \mathbf{p}' \mathbf{p}'^T \rangle \right), \quad \mathbf{E} = \left(\langle \mathbf{e} \mathbf{e}^T \rangle \right) \tag{39}$$

These are the covariance matrices for model parameters and for errors, respectively, and reasonable estimates of what these covariances are must be supplied from prior knowledge, etc.; they are not computable from observations. The square root notation for matrices is defined as in standard texts:

$$\mathbf{Q}^{\frac{1}{2}} \mathbf{Q}^{\frac{1}{2}} = \mathbf{Q}; \quad \mathbf{Q}^{\frac{1}{2}} \mathbf{Q}^{-\frac{1}{2}} = \mathbf{I} \tag{40}$$

and presents no difficulty, since the matrices $\mathbf{W}$ and $\mathbf{E}$ are covariance matrices and hence positive definite (see Bellman, 1960, chapter 6).

The matrix $\tilde{\mathbf{G}}$ is not square in general (N x M for N data and M model parameters) and so does not possess a standard inverse; Cornuelle discusses the singular value decomposition (SVD), which solves the coupled eigenvalue problems for $\tilde{\mathbf{G}}$ and $\tilde{\mathbf{G}}^T$

$$\tilde{\mathbf{G}}\mathbf{v}_i = \lambda_i \mathbf{u}_i; \quad \tilde{\mathbf{G}}^T\mathbf{u}_i = \lambda_i \mathbf{v}_i \tag{41}$$

and constructs a pseudoinverse (Lanczos, 1961) as

$$\left(\tilde{\mathbf{G}}\right)^{-1} = \mathbf{V}\left(\Lambda^T\right)^{-1}\mathbf{U}^T \tag{42}$$

A practical consideration arises; inclusion of small eigenvalues leads to unstable computations. One procedure is to exclude them (truncate); doing this and rewriting in terms of the original matrices leads to

$$\mathbf{W}^{-\frac{1}{2}}\hat{\mathbf{p}}' = \left(\mathbf{E}^{-\frac{1}{2}}\mathbf{G}\mathbf{W}^{\frac{1}{2}}\right)^T \cdot \left(\mathbf{E}^{-\frac{1}{2}}\mathbf{G}\mathbf{W}\mathbf{G}^T\mathbf{E}^{-\frac{1}{2}}\right)_r^{-1} \cdot \left(\mathbf{E}^{-\frac{1}{2}}\mathbf{d}'\right) = \tilde{\mathbf{G}}^T\left(\tilde{\mathbf{G}}\tilde{\mathbf{G}}^T\right)_r^{-1}\tilde{\mathbf{d}}' \tag{43}$$

as the weighted estimation, where r denotes the truncation. The main qualitative feature to keep in mind is the way in which the model and error covariance matrices enter as weights. If we can construct our forward problem matrix $\mathbf{G}$, and can also supply estimates of the two covariance matrices, we have the elements of the calculation of our estimated model parameters. We also obtain, though we have not shown the algebra here, estimates of the size of our minimum-square estimation error, one of the most useful results of the machinery.

9. SOME RESULTS AND FUTURE PLANS

With this background we now look at the results of some completed tomography experiments in the ocean. The first proper test of the method in three dimensions was conducted in 1981, near the MODE region in the Atlantic. The objective was to test the technique as a means of mapping the mesoscale fluctuations, just as envisaged by Munk and Wunsch (1979). The region was deliberately selected to be one in which a good deal was known already about the mesoscale, so that comparisons between acoustic results and conventional ones could be as straightforward as possible. The arrangement is shown in Fig. 8. The array was in place by early March; by the latter part of April instrument batteries, from a bad batch, began to fail, so the useful data series is about two months long. Fig. 10 shows a tomographic map for yearday 82, a conventional map drawn from a CTD survey done between days 66–85, and the difference between the two. It also shows the error map. The pattern of the conventional observations is mirrored in the acoustic map, but the gradients are smoothed. This is a consequence of rather large errors in travel time determination, which led to the non-resolvability of some groups of rays. As time progresses (Fig. 11) the acoustic data show that the eddy changes little for many days, then rather abruptly moves to the west after day 100. This in itself points to an advantage of the acoustic technique and to a difficulty in comparing acoustic and conventional maps. Any CTD survey of this scale takes a significant time to complete, whereas the acoustic maps are virtually synoptic. They can reveal sudden changes that would escape the CTD maps; they cannot be expected to match the CTD maps exactly. But at the end of the series in Fig. 11 the new CTD map is at least consistent with what has taken place in the acoustic map evolution. Some further confidence is gained by comparing time series near a point from the tomographic results with actual point measurements on conventional moorings (Fig. 12). Though the error bars are large, there is a general correspondence between the two.

Perhaps the most important lesson to emerge from this experiment was the need to reduce the travel time estimation errors in order to improve the inversions. Fig. 13 shows the dramatic mapping improvement that would have obtained in a field like the 1981 field if we had been able to reduce the travel time error from 5 msec to 2 msec, as was done in 1983. The difficulty was due to the narrow bandwidth of the sources available in 1981. Since then we have carried out experiments with sources of greater bandwidth and have improved the travel time errors by nearly an order of magnitude.

In 1983 a test of so-called reciprocal tomography to measure ocean currents was carried out. The results are reported by Howe (1986) and by Howe and Worcester (1987). The reciprocal method, pioneered by Worcester (1977), is an extension of the basic idea in which each mooring carries both a source and a receiver. Then in simplest terms mooring A transmits to and receives from mooring B, and vice versa. If there is no component of ocean current along the path between A and B the two transmissions follow exactly reciprocal paths and yield exactly the same travel time. If there is a current component one transmission is advanced, the other retarded, and so the difference in the travel times contains information on the current along the path. This information can be extracted from the data by

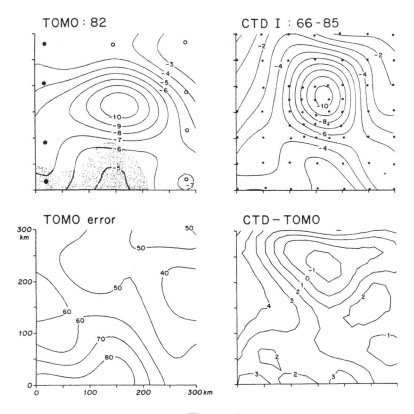

Figure 10
Comparison of tomographic and CTD maps at 700m in the 1981 experiment. Contours are sound speed anomaly in msec^{-1}. Both the tomographic map from yearday 82 and the approximately contemporaneous
CTD map (which required days 66–85 to make) show a cold eddy near
the center of the region. The expected error of the inverse calculation is
shown at the lower left; 70% corresponds to about 3msec^{-1}, which is the
boundary of the shaded portion of the upper left panel. The differences
between the two maps are contoured at the lower right. From Cornuelle
et al. (1985).

inverse calculations just as in the one-way case. The travel time perturbations are
very small, and timing precision requirements are severe (fraction of a millisecond)
but achievable.

Fig. 14 shows the experiment arrangement. The original plan was to have a
triangle of instruments, but due to an early failure only one leg was obtained. The

171

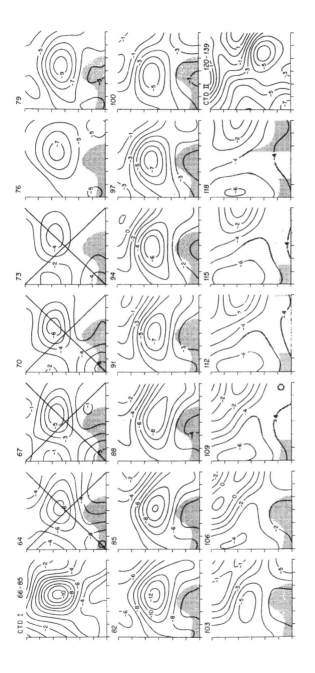

Figure 11

Time series of tomographic maps at 700m in the 1981 experiment, with CTD maps shown at beginning and end. The tomographic maps at three-day intervals show the evolution of the field. The cold eddy in the center of the region changes little until yearday 100, after which it moves rapidly to the west. Contour interval is 1 msec⁻¹ or about 0.2 deg C. Regions with expected error above 70% are shaded. Crosses denote maps degraded by large mooring motions. From Cornuelle et al. (1985).

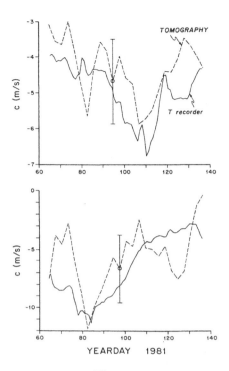

Figure 12
Comparison of time series estimated from acoustic data with moored time series in the 1981 experiment. Upper panel compares temperature at 876 m on mooring S3, lower panel compares temperature at 626m on mooring carrying conventional instruments within the region. Error bars for the acoustic estimates are indicated. A mean TS relation was used to convert temperature fluctuations to sound speed fluctuations, as plotted. From Cornuelle et al. (1985).

sound speed profile and ray trace are shown in Fig. 15, the arrival pattern in Fig. 16. Vertical ocean modes selected as expansion functions for the inversions are shown in Fig. 17. The time variation calculated by the inverse is shown in Fig. 18. The numbers are quite reasonable and in agreement with the rather sparse conventional data available for comparison (geostrophically derived baroclinic velocities from XBT data taken at the start of the experiment and from one AXBT survey about two weeks after the end of reciprocal transmissions).

A larger scale version of this experiment (Fig. 19) has recently been recovered

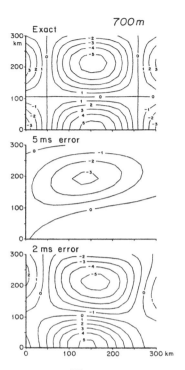

Figure 13
Results of computer simulations to assess the impact of travel time precision on tomographic mapping skill. Top panel shows an assumed sound speed anomaly pattern, of about the same scale and amplitude encountered in the 1981 experiment. Center panel shows map obtained if travel time precision is 5 msec, as in 1981, and bottom panel shows the markedly improved map obtained if travel time precision is 2 msec, as has been achieved in 1983. From Cornuelle et al. (1985).

from the North Pacific; it is too early to say much about results except that data processing is satisfactory thus far. The first-order objective, to make the reciprocal transmissions successfully over long ranges, thereby taking advantage of the intrinsic spatial averaging of the acoustic technique, has been realized. As in the case of the 1983 experiment the reason for the triangle was twofold: first and foremost it gives redundancy for the basic objective of obtaining at least one path, and second, if everything works, it affords a direct measure of relative vorticity in the region. Adding up the ocean currents on each side of the triangle one has the

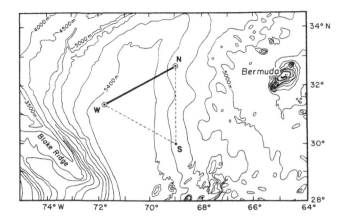

Figure 14
Geometry of the 1983 reciprocal transmission experiment. Of three moorings set only N and W returned useful data. From Howe (1986).

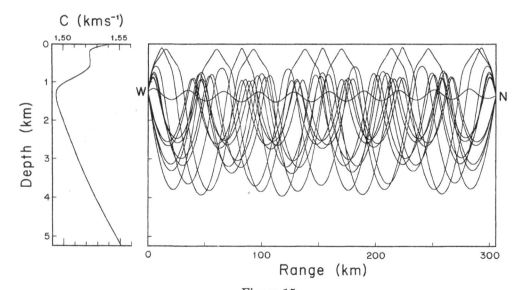

Figure 15
Average sound speed profile for the 1983 experiment, obtained by combining XBT data along the line between moorings N and W with historical deep data in the region (left), and traces of rays used in the inversions (right). From Howe (1986).

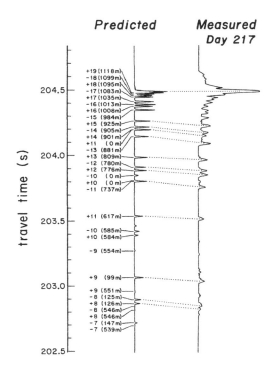

Predicted   Measured
                   Day 217

Figure 16
Predicted and measured arrival patterns in the 1983 experiment. Prediction is computed from the sound speed profile of Fig. 15, using the
WKBJ method of Brown (1982). Labeling is $n(z)$, where $+$ $(-)$ rays depart upward (downward) from the source, n is the total number of upper
and lower turning points and (z) is the upper turning depth in metres.
Dashed lines connect the observed and computed peaks for rays used in
the inversions. From Howe (1986).

circulation around the triangle and hence the area-average vorticity within it, by
Stokes' Theorem. This is a novel measurement that would be hard to obtain by
other methods.

For the sake of completeness I mention two experiments now in the planning
stages. The first is an array to be deployed for a year in the Greenland Sea,
beginning in summer 1988. A map is shown in Fig. 20. The objective is to utilize
the synoptic, rapidly-repeating features of tomography to study the time changes
of velocity and temperature in the interior, where we may expect to see an impact

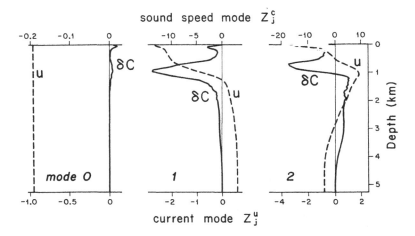

Figure 17
(Left to right) Quasigeostrophic vertical modes 0, 1, and 2 of sound speed
perturbation (solid lines) and horizontal current (dashed lines) used as
model functions to invert data from the 1983 experiment. From Howe
(1986).

of winter cooling and deep mixing (deep water formation) and also a spinup due to the stronger winds of winter. Using historical data from winter and summer seasons we have done preliminary inversions to suggest that seasonal changes in the sound speed profile can be determined to useful accuracies (Figs. 21–23). This is one of the situations referred to earlier in which the field best measured by tomography (sound speed and thus temperature, primarily) is not the only significant component in the oceanographic field of prime interest in the deep water problem (density or stability); salinity contributions are of about the same magnitude.

Finally, we have embarked on a study of how to use tomographic equipment from ships to make rapid and efficient surveys of large regions at high resolution (mesoscale mapping). Our original notion, which proved to be flawed, was that two ships steaming around the boundaries of a square could transmit to each other at very short intervals of time, thus produce numerous ray paths in a time short compared to mesoscale evolution, thus yield very well resolved maps. The notion is sketched in Fig. 24; there are two flaws in it. The first is that a set of transmission paths all in one direction, say the east-west lines of Fig. 24, gives information only about the north-south variation of the field, smearing out any east-west structure

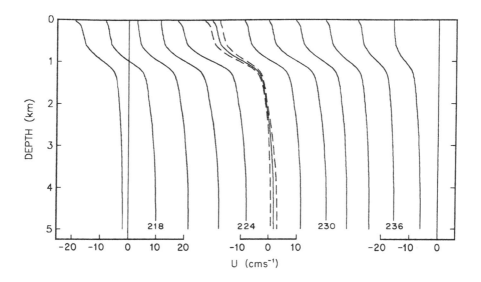

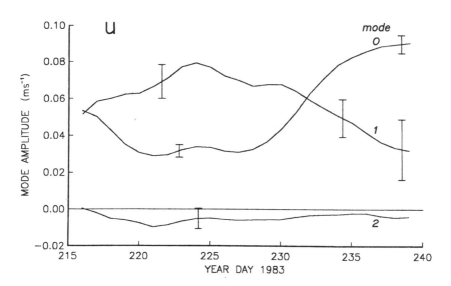

Figure 18
(Upper) Time series of range-averaged horizontal current profiles from inversion of the 1983 data, with error estimates (dashed lines). (Lower) Time series of the amplitude of the barotropic (0) and gravest two baroclinic (1, 2) modes of horizontal current. The vertical structure of these modes is shown in Fig. 17. From Howe (1986).

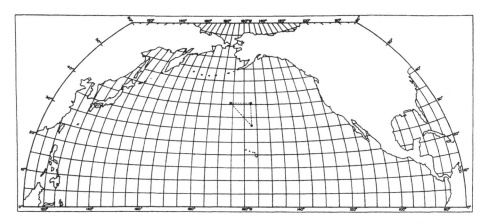

Figure 19
Location of the 1987 reciprocal transmission experiment in the North Pacific.

in the field. Phrased another way, the east-west lines alone yield 100% mapping error at all wavenumber vectors with magnitudes between one cycle per box edge length and one-half cycle per path separation (the Nyquist wavenumber), except for the purely north-south wavenumber vectors. The second flaw is that closely-spaced parallel paths see the same ocean and thus do not give independent data, so beyond a certain point (mesoscale spacing) there is no gain in making transmissions along parallel paths at small separation. What one really wants in high-resolution tomographic mapping, as is done in the medical application, is a network of paths such that any small interior volume has a number of paths through it at different angles; these will afford the independent data and resolution of wavenumber vectors. But maneuvering two realistic ships to accomplish this takes much longer than the scheme of Fig. 24, and by then the mesoscale has changed.

With a judicious combination of a few moorings and a ship, however, one can recapture the high-resolution objective. A possible geometry is shown in Fig. 25. Cornuelle, Munk and Worcester have been performing computer simulations with such geometries to quantify the pros, cons and error levels, while Worcester, Spindel, Howe, Birdsall and Metzger have begun to address the formidable technical problems of redesigned signal processing (to account for the substantial Doppler

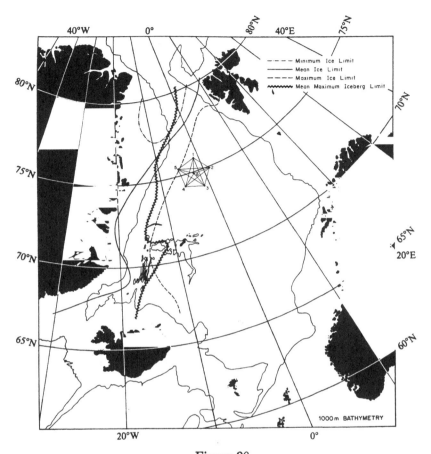

Figure 20
Arrangement of tomographic array to be set in the Greenland Sea in summer 1988.

shifts between fixed moorings and moving ships), precise ship positioning, etc. At the end of this work we hope to have a sensible method to make mesoscale-resolving maps of a 1000km × 1000km region in two weeks, using one ship and a few moorings, with a mapping error no worse than 10% at any point. To accomplish the same mapping accuracy in so large a region using conventional methods would require unacceptably large numbers of unattended instruments and/or unacceptably large numbers of ships or aircraft to deploy the necessary point sensors within so short a time window.

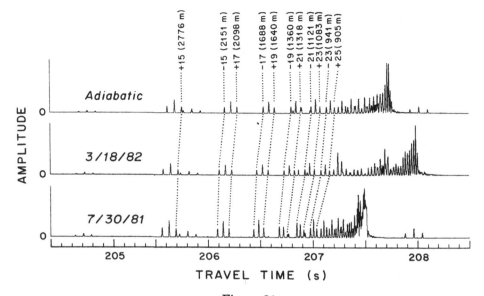

Figure 21

Predicted arrival patterns in the Greenland Sea experiment region of Fig. 20, for an adiabatic sound speed profile and for winter and summer profiles taken from individual stations on the indicated dates near 74° N, 1° W. Arrival labels are as in Fig. 15 except that lower turning depth is indicated; rays in this region are reflected at the sea surface. Dashed lines indicate arrivals used in trial inversions.

10. CONCLUSION

If this discussion has given non-specialists in ocean acoustics some sense of what ocean acoustic tomography is all about, how it works, and what kind of results it can produce, it will have served its purpose. The future of tomography now depends very much on whether other oceanographers beyond our small circle come to view it as a tool to be applied to oceanographic problems when it is advantageous to do so, like any other tool. The way is open for all sorts of good ideas about improving the method, applying it, making the hardware more economical, and

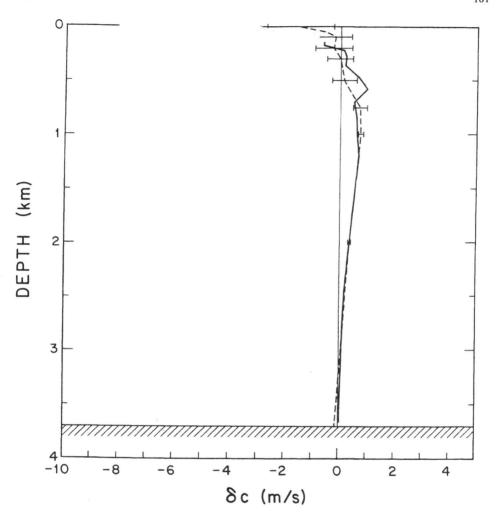

Figure 22
Measured (solid) and inverse solution (dashed) profiles of sound speed
perturbation for the winter data of Fig. 21, with estimated error bars.

182

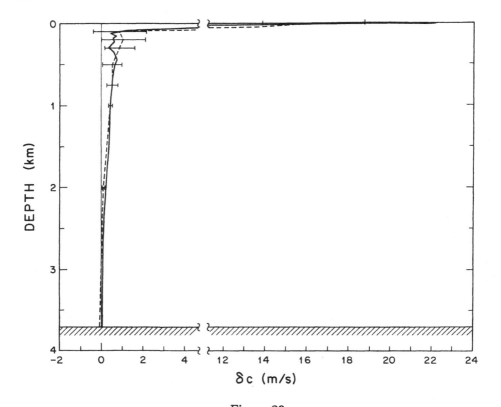

Figure 23
Same as Fig. 22, for the summer case. Note break in axis to accommodate
the large surface value.

combining tomographic data with other data or with numerical models to make maps, simulations and studies of various aspects of the ocean circulation. The two 'in the works' experiments noted above have barely scratched the surface of possibilities.

11. LITERATURE

The literature of tomography is becoming large. My purpose here is to point out a limited set of references that offer good entry points to the literature, so that the interested reader can get started on the major themes and then pursue refinements as interest dictates. For basic texts on acoustic propagation in the ocean, Tolstoy and Clay (1987) and Boyles (1984) are useful; most of the mathematics of rays, modes, etc. is contained in them. The paper by Munk and Wunsch (1983) gives a thorough discussion of rays and modes and the correspondence between the two representations in the specific context of the ocean sound channel. An elegant discussion of sound channel rays in terms of Hamiltonian mechanics is given by Wunsch (1987). Flatté et al. (1979) give an advanced account of sound transmission through a fluctuating ocean, showing how such quantities as travel time variance due to internal waves can be estimated, given a spectral model of internal waves.

Reading on tomography *per se* begins with the original paper of Munk and Wunsch (1979), still remarkable for the number of practical problems and features of the method that it recognized and addressed. Signals and their processing are the subject of the thesis by Metzger (1983), which is a good place to begin serious study of this topic. For the methods of inverse computations, Cornuelle's (1983) thesis is both readable and complete. The extensions to reciprocal tomography are well described in Howe's (1986) thesis and, more tersely, by Howe et al. (1987).

Results of the 1981 experiment are found in short report form in Ocean Tomography Group (1982) and in more extended form in Cornuelle et al. (1985). Howe (1986) and Howe et al. (1987) serve the same purpose for the 1983 reciprocal transmission experiment.

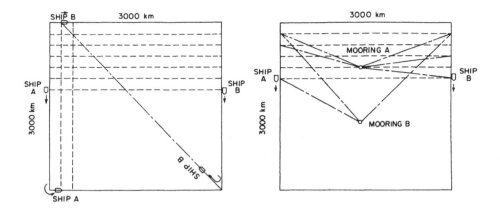

Figure 24
Possible schemes for ship-borne tomography. A purely ship-borne scheme is sketched at left, in which two ships transmit east-west across the region while steaming south, and then transmit north-south across the region while steaming east, after ship B has made a northwestward transit to reposition. This scheme yields insufficient mesoscale mapping resolution, as discussed in text. A hybrid ship/mooring scheme is sketched at right, and refinements of this idea are much more promising (Fig. 25). From Munk and Wunsch (1982).

ACKNOWLEDGEMENT

Most of the ideas and results discussed in this paper have been the work of others, and in nearly all cases it has been my good fortune to work directly with those individuals. Most of their names appear as co-authors in the report of the first full test of tomography at sea (Ocean Tomography Group, 1982). I am particularly indebted to my SIO colleagues B. Cornuelle, B. Howe (now at the University of Washington), W. Munk and P. Worcester for years of congenial collaboration and for enlightening and useful discussions too numerous to record.

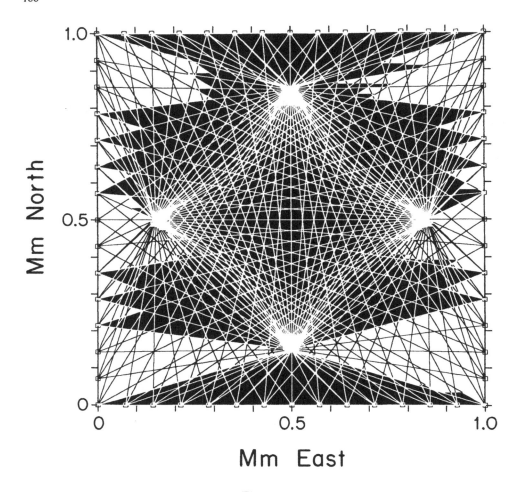

Figure 25
A hybrid ship/mooring scheme which should yield eddy-resolving maps
with no worse than 10% error over a 1000 km (1 Mm) square area. Ship
steams around the boundary, making measurements at the points indicated, and completing the pattern in about two weeks, a time less than
that of mesoscale evolution.

REFERENCES

Bellman, R. (1960). *Introduction to Matrix Analysis*. McGraw-Hill, 328 pp.

Boyles, C. A. (1984). *Acoustic Waveguides*. John Wiley and Sons, 322 pp.

Brown, M. C. (1982). Application of the WKBJ Green's function to acoustic propagation in horizontally stratified oceans. *J. Acoust. Soc. Amer.*, **71**, 1427–1432.

Clay, C. S. and H. Medwin (1977). *Acoustical Oceanography*. John Wiley and Sons, 544 pp.

Cornuelle, B. D. (1983). *Inverse methods and results from the 1981 ocean acoustic tomography experiment*. Ph.D. thesis, Massachusetts Institute of Technology/Woods Hole Oceanographic Institution, 359 pp.

Cornuelle, B. D., C. Wunsch, D. Behringer, T. Birdsall, M. Brown, R. Heinmiller, R. Knox, K. Metzger, W. Munk, J. Spiesberger, R. Spindel, D. Webb and P. Worcester (1985). Tomographic maps of the ocean mesoscale. Part 1: pure acoustics. *J. Phys. Oceanogr.*, **15**, 133–152.

Flatté, S. (Ed.), R. Dashen, W. Munk, K. Watson and F. Zachariasen (1979). *Sound Transmission through a Fluctuating Ocean*. Cambridge University Press, 299 pp.

Howe, B. M. (1986). *Ocean acoustic tomography: mesoscale velocity*. Ph.D. thesis, Scripps Institution of Oceanography, UCSD, 59 pp.

Howe, B. M., P. F. Worcester and R. C. Spindel (1987). Ocean acoustic tomography: mesoscale velocity. *J. Geophys. Res.*, **92**, 3785–3805.

Lanczos, C. (1961). *Linear Differential Operators*. Van Nostrand, 564 pp.

Lindsay, R. B. (1960). *Mechanical Radiation*. McGraw-Hill, 415 pp.

Metzger, K. (1983). *Signal processing and techniques for use in measuring ocean acoustic multipath structures*. Cooley Electronics Laboratory, Department of Electrical and Computer Engineering, University of Michigan, technical report 231, 316 pp.

Munk, W. and C. Wunsch (1979). Ocean acoustic tomography: a scheme for large scale monitoring. *Deep-Sea Res.*, **26A**, 123–161.

Munk, W. and C. Wunsch (1982). Observing the ocean in the 1990s. *Phil. Trans. R. Soc. Lond. A*, **307**, 439–464.

Munk, W. and C. Wunsch (1983). Ocean acoustic tomography: rays and modes. *Revs. Geophys. Space Phys.*, **21**, 777–793.

Ocean Tomography Group (1982). A demonstration of ocean acoustic tomography. *Nature*, **299**, 121–125.

Romm, J. J. (1987). *Applications of normal mode analysis to ocean acoustic tomography*. Ph. D. thesis, Massachusetts Institute of Technology, 114 pp.

Tolstoy, I. and C. S. Clay (1987). *Ocean Acoustics*. American Institute of Physics, for the Acoustical Society of America. 381 pp.

Worcester, P. F. (1977). Reciprocal acoustic transmission in a midocean environment. *J. Acoust. Soc. Amer.*, **62,** 895–905.

Wunsch, C. (1987). Acoustic tomography by Hamiltonian methods including the adiabatic approximation. *Revs. Geophys. Space Phys.*, **25,** 41–53.

THE CIRCULATION IN THE WESTERN NORTH ATLANTIC DETERMINED BY A NONLINEAR INVERSE METHOD

Herlé Mercier
Laboratoire de Dynamique des Circulations Planétaires dans l'Océan
Unité associée au Centre National de la Recherche Scientifique n°710
IFREMER centre de Brest, BP 70, 29263 PLOUZANE cedex, France

ABSTRACT

The circulation in the western North Atlantic is studied within the framework of a nonlinear finite difference geostrophic inverse model using hydrographic and currentmeter data. The nonlinear inverse formalism is first reviewed. A circulation scheme compatible with the data, the mass conservation and the potential density equation is then derived. The optimal circulation scheme presents variations in the Gulf Stream transport with longitude. The variations are explained by the interaction of the Gulf Stream with recirculating gyres.

1. INTRODUCTION

At mid-latitudes, the large scale circulation of the ocean is in near hydrostatic and geostrophic balance so that the vertical shear of the horizontal velocity can be computed from horizontal density gradients by using the thermal-wind relationship. Measurements of temperature, salinity and pressure consequently define the horizontal velocity relative to a reference level where the velocity is usually unknown. To reduce this indeterminacy, Wunsch (1977, 1978) proposed using mass and salt conservation to write a set of linear equations which could then be solved for the

189

D. L. T. Anderson and J. Willebrand (eds.), Oceanic Circulation Models: Combining Data and Dynamics, 189–201.

reference velocities. However, it appears that the problem is largely underdetermined, leading to a large range of circulation patterns compatible with geostrophic balance and the conservation equations (see Wunsch and Grant, 1982). Wunsch (1984) argued that the range of possible oceanic solutions could be narrowed by adding to the inversion information derived from other data sets (direct velocity or tracer measurements). Using an eclectic model, Wunsch (1984) succeeded in putting useful bounds on the meridional flux of heat in the Atlantic ocean.

Here, we use a similar approach to study the large scale time-averaged circulation in the western North Atlantic. In particular, we show that the circulation can be well determined by simultaneously inverting hydrographic and currentmeter data. Our approach has been to take into account explicitly the errors associated with each data set. Mercier (1986) showed that such an approach leads to a nonlinear inverse problem which will be solved, here, using the inverse formalism of Tarantola and Valette (1982a,b) (hereafter TV).

In the following section we review the nonlinear inverse formalism of TV. In section 3, the circulation in the western North Atlantic, north of 32°N, West of 50°W is discussed.

2. A NONLINEAR INVERSE FORMALISM

The general nonlinear inverse problem is examined by Tarantola and Valette (1982a,b). TV propose to formulate the inversion using probability density functions for data, unknowns and theoretical relationships. They show that in the particular case where the probability density functions are gaussian, the least squares formalism reviewed here, is obtained.

Let us denote as $\mathbf{x}$ the vector whose elements define the circulation. (In the geostrophic context, the elements of $\mathbf{x}$ are the reference level velocities and the densities). $\mathbf{x}_0$ is the *a priori* (before inversion) state of information about $\mathbf{x}$ ($\mathbf{x}_0$ is usually defined using measurements and the hypothesis of a deep level of no motion). $\mathbf{C}_0$ is the covariance matrix for $\mathbf{x}_0$. The constraints take the value $f(\mathbf{x})$ at point $\mathbf{x}$ and their error covariance matrix is denoted as $\mathbf{C}_T$. The least squares method looks for the best estimate $\mathbf{x}_*$ which minimizes

$$(\mathbf{x} - \mathbf{x}_0)^T \mathbf{C}_0^{-1}(\mathbf{x} - \mathbf{x}_0) + f^T(\mathbf{x})\mathbf{C}_T^{-1} f(\mathbf{x}) \tag{1}$$

Eq. (1) is made of two weighted sums. At a minimum, in the particular case where $\mathbf{C}_0$ and $\mathbf{C}_T$ are diagonal, the first term of Eq. (1) is a linear combination of the square distance between the *a priori* and *a posteriori* (best) estimates of the parameters. The second term is a linear combination of the square residuals of the constraints. When the constraints are linear, Jackson (1979) (see also TV) shows that x_* is given uniquely by

$$x_* = x_0 - C_0 F^T \left(F C_0 F^T + C_T\right)^{-1} f(x_0) \tag{2}$$

F is the matrix of the partial derivatives of f with respect to x:

$$F^{ik} = \frac{\partial f^i}{\partial x^k} \tag{3}$$

The error covariance matrix C_* of the solution is given by

$$C_* = C_0 - C_0 F^T \left(F C_0 F^T + C_T\right)^{-1} F C_0 \tag{4}$$

When the constraints are nonlinear TV propose to use an iterative approach. TV show that x_* satisfies the following implicit equation:

$$x_* = x_0 + C_0 F_{x_*}^T (F_{x_*} C_0 F_{x_*}^T + C_T)^{-1} (F_{x_*}(x_* - x_0) - f(x_*)) \tag{5}$$

where the derivatives F_{x_*} of the constraints are taken at point x_*. Writing $f(x_*)$ as a Taylor expansion around x leads to

$$f(x_*) = f(x) + F_x(x_* - x) + O\big((x_* - x)^2\big) \tag{6}$$

and

$$x_* = x_0 + C_0 F_x^T \left(F_x C_0 F_x^T + C_T\right)^{-1} \left(F_x(x - x_0) - f(x)\right) \tag{7}$$

TV define the following fixed point iterative procedure :

$$x_{k+1} = x_0 + C_0 F_{x_k}^T \left(F_{x_k} C_0 F_{x_k}^T + C_T\right)^{-1} (F_{x_k}(x_k - x_0) - f(x_k)) \tag{8}$$

where the derivatives are taken at the current point x_k. This algorithm is often referred to as the 'total inversion algorithm' (TI). Uniqueness of the solution which is guaranteed only if Eq. (1) is convex can be explored by running the algorithm with different starting points. (In the study presented below, the TI algorithm always finds a unique minimum). TV show that the *a posteriori* covariance matrix C_* derived for the linear problem also gives useful information for the nonlinear case. In the nonlinear problem the more the *a posteriori* variance of a parameter differs from the *a priori* one, the more we have improved our knowledge about this parameter. Using *a priori* information about the circulation is equivalent to choosing the rank in the Singular Value Decomposition approach (see Wunsch, 1978). The *a posteriori* covariance matrix C_* measures the importance of the null space. Assuming that we have m parameters (unknowns) and n constraints, x and x_0 are vectors of dimension m, $f(x)$ is a vector of dimension n, F is $n \times m$, C_T is $n \times n$, C_0 and C_* are $m \times m$.

More details can be found in Mercier (1986) where the nonlinear formalism is applied to the analytical model of Stommel and Veronis (1981). The TI algorithm has also been successfully applied by Wunsch and Minster (1982) in the context of box models. In the following section, we present another successful application of the nonlinear formalism.

3. THE CIRCULATION IN THE WESTERN NORTH ATLANTIC

3.1 Motivation

This study was motivated by the apparent differences the western North Atlantic between the circulation patterns derived from direct velocity measurements and those derived from the inversion of hydrographic data. Some discrepancies were already mentioned by Hogg (1983) who studied the time-averaged deep circulation. In addition note that the vertical structure of the circulation strongly depends on the data set which is analysed: Richardson's (1985) analysis of float trajectories and currentmeter data along 55°W shows a Gulf Stream which extends to the sea floor while Wunsch and Grant's (1982) inversions result in a Gulf Stream limited to the upper layers.

The discrepancies may reflect real changes in the ocean circulation (Wunsch and Grant, 1982, use the 1957–58 IGY hydrographic data; Hogg, 1983, and Richardson, 1985, use measurements carried out in the '70's and '80's). However, we know that the errors associated with the currentmeter time-averaged velocities (the typical record length is equal to one year) as well as the errors associated with the Wunsch and Grant's (1982) circulation schemes are large. Given these error bars, there may exist a circulation scheme compatible with the hydrographic and current-meter data as well as with geostrophic dynamics and the conservation equations. This paper examines that possibility.

3.2 The Data Base

393 hydrographic stations carried out during the IGY (see Fuglister, 1960) and Gulf Stream'60 (see Fuglister, 1963) experiments form the data set. The locations of the hydrographic stations are displayed in Fig. 1. To reduce the amount of data the densities were first projected onto the three most energetic density Empirical Orthogonal Functions (EOFs) computed from the vertical correlation matrix. (Three EOFs describe 91.2% of the total energy.) The EOF coefficients were then computed on a regular grid using an objective mapping technic (see Bretherton *et al.*, 1976). The spatial correlation functions of the field were taken to be gaussian with e-folding distances of 3° in latitude and 5° in longitude. The grid size is equal to 2° in latitude and 3.5° in longitude. The resulting *in situ* density field is depicted in Fig. 2 for 500 and 3750 m depths. Note the strong meridional gradient at the latitude of the Gulf Stream.

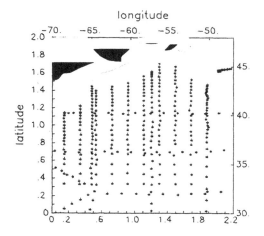

Figure 1
Location of the hydrographic stations. Labels of the bottom and left axes are in thousands of km.

Hogg (1983) and Hogg *et al.* (1986) have already given a complete description of the currentmeter data available in the region. The time-averaged velocities were

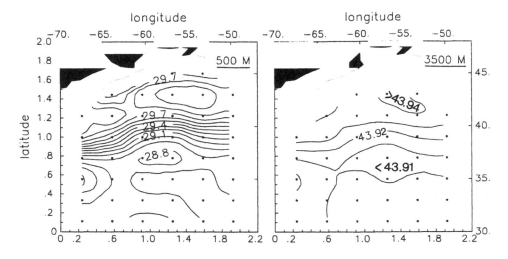

Figure 2
A priori in situ density field.

horizontally averaged within bins having the same size as the density grid. The statistical errors were then computed using an integral time scale of 10 days in water depth greater than 5000 m, and 5 days elsewhere. The resulting vectors are presented in Fig. 3 along with the error ellipses, the axes of which are three standard deviations long.

3.3 The Dynamical Model

Let us denote as x, y, z the East, North and vertical (positive upward) coordinates, g is gravity, f the Coriolis parameter, β its meridional gradient, ρ is the *in situ* density, ρ_0 the reference density, c the sound speed in sea water, h the bottom topography, H the ocean thickness, and u, v, w are respectively the East, North and vertical components of the velocity vector.

The flow is assumed to be steady, in hydrostatic and geostrophic balance. We assume that the planetary vorticity balance holds. Then the velocity components are

$$u(z) = u(z_0) + \frac{g}{f\rho_0} \int_{z_0}^{z} \rho_y dz' \tag{9}$$

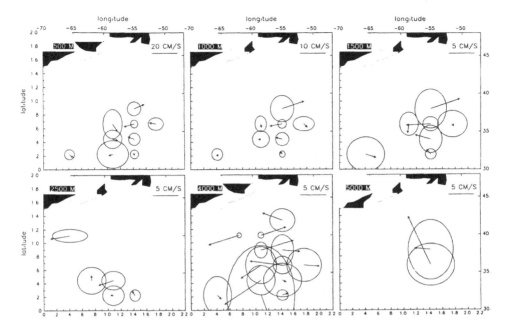

Figure 3
Currentmeter space- and time-averaged velocities and the error ellipses.

$$v(z) = v(z_0) - \frac{g}{f\rho_0} \int_{z_0}^{z} \rho_x dz' \tag{10}$$

$$w(z) = w(-H) + \int_{-H}^{z} \frac{\beta}{f} v dz' \tag{11}$$

where $u(z_0)$ and $v(z_0)$ are the horizontal components of the velocity at the reference level z_0. The bottom vertical velocity $w(-H)$ is given by the dynamical condition

$$w(-H) = -\big(u(-H)h_x + v(-H)h_y\big) \tag{12}$$

Continuity is ensured if

$$u_x(z_0) + v_y(z_0) + w_z(z_0) = 0 \tag{13}$$

while the conservation of potential density is approximated as

$$u\rho_x + v\rho_y + w\left(\rho_z + \frac{\rho_0 g}{c^2}\right) = 0 \tag{14}$$

3.4 A Priori Assumptions

The elements of $\mathbf{x}$ are the EOF coefficients and the reference level velocities. Given the above equations, the elements of $\mathbf{x}$ allow us to compute the three dimensional velocity field. The dynamical constraints $f(\mathbf{x})$ are mass and potential density conservation. We also add direct velocity measurement constraints written as

$$u(z) - u_{\text{obs}}(z) = 0 \tag{15}$$

$$v(z) - v_{\text{obs}}(z) = 0 \tag{16}$$

where $u(z)$, $v(z)$ are the geostrophic velocities computed from Eqs. (9) and (10) while $u_{\text{obs}}(z)$, $v_{\text{obs}}(z)$ are the space- and time-averaged currentmeter velocities. In the following, velocity components and constraints are computed using a centered finite difference scheme applied to the density grid.

The a priori vector $\mathbf{x}_0$ is built using the objectively analysed EOF coefficients and the assumption of a level of no motion at 1750 m. The a priori covariance matrix $\mathbf{C}_0$ is built taking 5 cm s^{-1} for the a priori standard deviation for $u(z_0)$, 3 cm s^{-1} for $v(z_0)$ and using the errors computed by objective analysis for the EOF coefficients. To force a large scale solution we assume correlated a priori errors for every kind of parameter. We took Gaussian correlation functions having the same e-folding lengths as those chosen for the objective analysis.

Mass conservation is applied with a negligible a priori standard deviation (This is the exact constraint limit of Mercier, 1986). For the direct velocity constraints, we assume uncorrelated errors and the diagonal terms of $\mathbf{C}_T$ are computed using the statistical variances of Fig. 3.

The inverse modeler has a lot of freedom for choosing the a priori information. The a priori choices have to be checked for consistency after inversion by examining the solution and the residuals of the constraints. An example of this procedure is given in Mercier (1986).

The potential density conservation is applied only in the deep ocean (below 1000 m) to avoid regions where the eddies are most active. Several choices for the associated a priori standard deviations are examined in the following subsection. We have 214 parameters, 94 currentmeter constraints and 117 dynamical constraints.

3.5 The Results

An integrated measure of the residuals in the density conservation constraint is given by w_c rms $= \langle w_c^2 \rangle^{\frac{1}{2}}$ where $\langle \cdot \rangle$ denotes an average over all grid points and w_c is the cross-isopycnal vertical velocity defined as

$$w_c = w - w_i \tag{17}$$

w_i is the isopycnal vertical velocity

$$w_i = -\frac{u\rho_x + v\rho_y}{\rho_z + (\rho_0 g/c^2)} \tag{18}$$

Table 1

$\langle w_c^2 \rangle^{\frac{1}{2}}$ (normalised such that the value for model M0 is 1) and number of unsatisfied currentmeter constraints for several choices of the a priori standard deviation for the potential density conservation constraint. For model M0, only the currentmeter constraints are applied.

Model	A priori STD for potential density conservation	$\langle w_c^2 \rangle^{\frac{1}{2}}$	No. of unsatisfied currentmeter constraints
M0	—	1	6
M7	10^{-7}	4.1×10^{-1}	8
M2	10^{-9}	1.4×10^{-2}	24
M5	10^{-11}	$.54 \times 10^{-2}$	31

The dependence of w_c rms upon the a priori standard deviation for the potential density conservation constraint is studied in Table 1 where we also indicate the number of unsatisfied currentmeter constraints. (A currentmeter constraint is said to be unsatisfied when, after inversion, the residual of the constraint is larger than three times the a priori standard deviation). Note that for model M0, the density conservation constraint is not applied.

As expected, a decrease in the a priori standard deviation for the potential density conservation constraint results in a decrease in the averaged a posteriori cross-isopycnal velocity. For model M0, w_c rms is $O(0.1 \text{ cm s}^{-1})$ which seems far too large to be credible (Hall, 1986, finds that the mean total vertical velocity in the GS

198

at 68°N is $O(5 \times 10^{-3} \mathrm{cm\,s}^{-1})$). Models M2 and M5 present plausible values for w_c rms. However, decreasing w_c rms results in an increase in the number of unsatisfied currentmeter constraints. Careful examination of the M0 solutions shows that, for this run, all incompatibilities were due to a strong vertical shear occurring in the currentmeter data between two adjacent vertical levels, the inverse model being unable to reproduce such a variation. The M0 solution (not shown) presents a complicated deep circulation, lacking in spatial coherence. (The behavior of the M0 solution is determined by the currentmeter constraints which are influenced both by large scale and small scale steady processes.) Decreasing the *a priori* standard deviation for the potential density conservation constraint (runs M2 and M5) moves the solution towards a larger scale circulation scheme. In the following we discuss the M2 run which presents a plausible value for w_c rms and satisfies most of the currentmeter constraints. The circulation is well determined, the *a posteriori* errors being on average equal to 0.4 cm s^{-1} for $u(z_0)$ and 0.2 cm s^{-1} for $v(z_0)$.

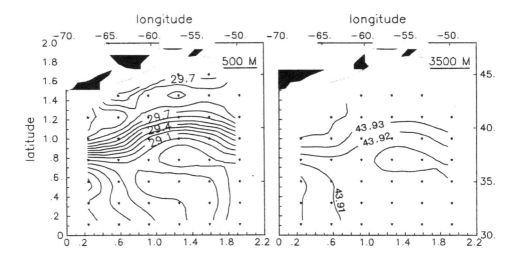

Figure 4
A posteriori in situ density field for model M2

The *a posteriori* density field is presented in Fig. 4 and should be compared with the *a priori* field of Fig. 2. (The density field was significantly adjusted by the inversion.) The vertical structure of the circulation can be apprehended from Fig. 5. The Gulf Stream extends to the sea floor and is bounded by westward flowing countercurrents. Note the southward and downward shift of the Gulf Stream axis. A similar shift was observed upstream by Joyce *et al.* (1986). A sketch of the vertically integrated geostrophic flow is given in Fig. 6. At 65.75°W the GS carries 92 Sv, then its transport increases downstream to reach 125 Sv at 58.75°W. This increase is due to the interaction with the southern countercurrent which, around 60°W, carries about 30 Sv of water into the GS. At 55°W the GS feeds the northern countercurrent with transport increasing from 32 Sv at 52°W to 56 Sv at 58.75°W. At 55°W part of the GS water also recirculates in the southern gyre. The northern and southern recirculations lead to a decrease in the GS transport downstream of 55°W and a GS transport of 90 Sv is measured at 51.75°W. More details are given in Mercier (1988).

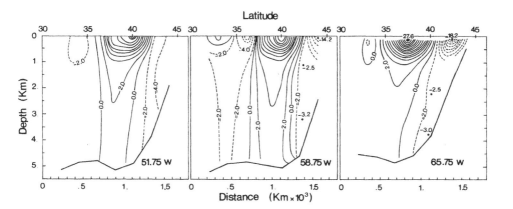

Figure 5
Zonal component of velocity for model M2. Units are cm s⁻¹

200

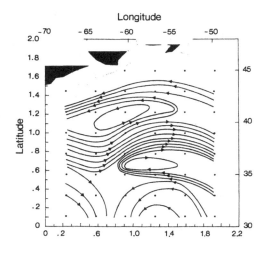

Figure 6
Sketch of the vertically integrated geostrophic transports for M2. Each
contour represents 10×10^6 m^3 s^{-1}.

ACKNOWLEDGMENTS

This work was begun when I was a visitor in the department of Earth, Atmospheric
and Planetary Sciences at the Massachusetts Institute of Technology, Cambridge,
USA, for whose hospitality I am grateful. It was completed at the laboratoire
de dynamique des circulations planétaires dans l'océan, IFREMER, Brest, France.
Many thanks go to Carl Wunsch for his advice during this work.

REFERENCES

Bretherton, F.P., R.E. Davis and C.B. Fandry (1976): A technique for objective
    analysis and design of oceanographic experiments applied to MODE-73, *Deep
    Sea Res.*, **23A**, 559–581.

Fuglister, F.C. (1960): Atlantic ocean atlas of temperature and salinity profiles
    and data from the International Geophysical Year of 1957–1958. *Woods Hole
    Oceanographic Institution Atlas Series*, 209 pp.

Fuglister, F.C. (1963): Gulf Stream '60, *Progr. Oceanogr.*, **1**, 265–373.

Hall, M.M. (1986): Horizontal and vertical structure of the Gulf Stream velocity field at 68°W, *J. Phys. Oceanogr*, **16**, 1814–1828.

Hogg, N.G. (1983): A note on the deep circulation of the western North Atlantic: its nature and causes, *Deep Sea Res.*, **30A**, 945–961.

Hogg, N.G., R.S. Pickart, R.M. Hendry and W.J. Smethie (1986): The northern recirculation gyre of the Gulf Stream, *Deep Sea Res.*, **33**, 1139–1165.

Jackson, D.D. (1979): The use of a priori data to resolve non-uniqueness in linear inversion, *Geophys. J. R. Astron. Soc.*, **57**, 137–157.

Joyce, T.M., C. Wunsch and S.D. Pierce (1986): Synoptic Gulf Stream velocity profiles through simultaneous inversion of hydrographic and acoustic doppler data, *J. Geophys. Res.*, **91**, 7573–7585.

Mercier, H. (1986): Determining the general circulation of the ocean: A nonlinear inverse problem, *J. Geophys. Res.*, **91**, 5103–5109.

Mercier, H. (1988): A study of the time-averaged circulation in the western North Atlantic by simultaneous nonlinear inversion of hydrographic and currentmeter data. Submitted to *Deep Sea Res.*

Richardson, P.L. (1985): Average velocity and transport of the Gulf Stream near 55°W, *J. Marine Res.*, **43**, 83–111.

Stommel, H. and G. Veronis (1981): Variational inverse method for study of the ocean circulation, *Deep Sea Res.*, **28**, 1147–1160.

Tarantola, A. and B. Valette (1982a): Inverse problems = quest for information, *J. Geophys.*, **50**, 159–170.

Tarantola, A. and B. Valette (1982b): Generalized nonlinear problems solved using the least squares criterion, *Rev. Geophys.*, **20**, 219–232.

Wunsch, C. (1977): Determining the general circulation of the oceans: a preliminary discussion. *Science*, **196**, No. 4292, 871–875.

Wunsch, C. (1978): 'The general circulation of the North Atlantic west of 50°W determined from inverse methods, *Rev. Geophys. Space Phys.*, **16**, 583–620.

Wunsch, C. and J.F. Minster (1982): Methods for box models and ocean circulation tracers: mathematical programming and nonlinear inverse theory, *J. Geophys. Res.*, **87**, 5647–5662.

ALTIMETER DATA ASSIMILATION INTO OCEAN CIRCULATION MODELS — SOME PRELIMINARY RESULTS

William R. Holland
National Center for Atmospheric Research
Boulder, Colorado 80307

1. INTRODUCTION

Assimilation of observations into ocean circulation models has become a worthwhile endeavor now that synoptic data from satellites are becoming available. In contrast to the atmospheric case, however, where fairly complete observations of the three dimensional structure of the atmosphere are available through an extensive observing system, the oceanic assimilation problem is much more difficult and will require quite novel techniques for blending observations and model predictions into a best possible, coherent picture of the four dimensional oceanic circulation. This is because ocean observations will be available mainly at the sea surface from the ocean satellite observing platforms. Ocean models will have to play a much more important role in that they must be able to successfully extrapolate surface observational information downward into the ocean interior if they are to achieve success in reconstructing a realistic ocean circulation.

There are three types of considerations we shall make in this paper. These are based upon a series of preliminary studies carried out recently to examine the prospects for using altimeter data to initialize an ocean model and to keep its forward integration in time on track by a continuous assimilation of sea surface height information. Firstly our main task in this paper will be to examine techniques for

203

D. L. T. Anderson and J. Willebrand (eds.), Oceanic Circulation Models: Combining Data and Dynamics, 203–231.
© 1989 by Kluwer Academic Publishers.

assimilating altimeter data in such a way that a dynamically adjusted, time dependent oceanic circulation, consistent with the data, can be constructed. The aim is to reconstruct the vertical structure of the time evolving currents in considerable detail. It is not clear *a priori* under what conditions, if any, this can be accomplished. Next, we shall examine the process by which such an assimilating ocean model can be initialized with sea surface height data, assumed to be available either on a grid at regular intervals or along satellite tracks in real time. Several issues regarding these approaches will be discussed. Finally, we shall describe a preliminary attempt to use actual altimeter data from GEOSAT to initialize and continuously assimilate data to give a dynamically adjusted, four dimensional description of the circulation in the Agulhas Current Retroflection Region south of South Africa.

2. A DYNAMIC INITIALIZATION SCHEME

In several exploratory studies, Holland and Verron (1988; hereafter HV), Verron and Holland (1988; hereafter VH) and Holland and Malanotte-Rizzoli (1988; hereafter HM) examined a simple technique for initializing a quasigeostrophic (QG) ocean model and continuously assimilating altimeter data into it. The basic idea is to add to the physics of the upper layer vorticity equation an additional term, proportional to the difference in vorticities in the model and the observations, a term that forces the assimilating model to be 'dynamically nudged' toward the data. The phrase 'dynamical initialization by nudging' was coined in the meteorological context by Anthes (1974) and was further studied by Hoke and Anthes (1976), who used a similar technique for an atmospheric assimilation study using nudging of the primitive equations. One important difference in using nudging of surface vorticity in the oceanic QG model context is that this process allows the deeper ocean circulation to be immediately updated, that is, the surface data are dynamically extrapolated downward. We shall discuss this point further below.

Note the similarity this nudging procedure has to the robust diagnostic approach of Sarmiento and Bryan (1982), who constrained their ocean temperatures and salinities to be near to observed climatological values. Here, however, we shall use sea surface height observations with high resolution space/time information to allow us to incorporate mesoscale eddy descriptions into an ongoing eddy-resolved ocean model calculation. The aim is to reconstruct the transient evolution of the turbulent ocean circulation. As we shall see, an important question arises as to the space and time structure of the observations, that is, the detailed way in which the observations are determined by an altimeter satellite coverage of a given portion of the World Ocean. Obviously, the nature of the errors in the observations and the efficacy of the assimilation process are and will continue to be important considerations for research in developing the best possible assimilation model for ocean circulation studies.

The quasigeostrophic model formulation with N arbitrary layers is a straightforward extension of the two-layer case described by Holland (1978). Here we shall present the semi-discrete form of the equations (in which the vertical discretization has already been done). The horizontal discretization and the form of the finite difference equations will not be discussed here.

The governing equations are the prognostic vorticity and interface height perturbation equations, and the diagnostic thermal wind relation:

$$\frac{\partial}{\partial t}\nabla^2\psi_k = J(f + \nabla^2\psi_k, \psi_k) + \frac{f_0}{H_k}\left(w_{k-\frac{1}{2}} - w_{k+\frac{1}{2}}\right) + F_k + T_k : k = 1 \text{ to } N \quad (1)$$

$$\frac{\partial}{\partial t}h_{k+\frac{1}{2}} = J\left(h_{k+\frac{1}{2}}, \psi_{k+\frac{1}{2}}\right) + w_{k+\frac{1}{2}} : k = 1 \text{ to } N-1 \quad (2)$$

$$h_{k+\frac{1}{2}} = \frac{f_0}{g'_{k+\frac{1}{2}}}\left(\psi_{k+1} - \psi_k\right) \quad (3)$$

Here whole number subscripts (k) denote the vertical layers $(k$ increasing downward) in which the quasigeostrophic streamfunction is defined (nominally at the center of each of the layers) while fractional subscripts $(k + 1/2)$ denote the interfaces between layers where vertical velocity and interface height perturbation are defined (Fig. 1). The variables are the quasigeostrophic streamfunction (ψ_k) with horizontal velocity components $(u = -\psi_y, v = \psi_x)$, the interface height perturbation $(h_{k+1/2})$, positive upward, and the vertical velocity $(w_{k+1/2})$, also positive upward. The horizontal coordinates are x (eastward) and y (northward), the Coriolis parameter is $f = f_0 + \beta y$, and the mean layer thicknesses are H_k. The values of f_0 and β are chosen to represent typical midlatitude gyre values. The basic background vertical stratification is written in terms of the reduced gravity $g' = g\Delta\rho_{k+1/2}/\rho_0$, where $\Delta\rho_{k+1/2}$ is the (positive) density difference between layers $k + 1$ and k. Frictional effects, written symbolically in Eq. (1) as F_k, are parameterized as lateral friction of the biharmonic kind (Holland, 1978), in which $F_k = -A_4\nabla^6\psi_k$. In addition, F_k

includes a bottom friction, $-\varepsilon\nabla^2\psi_N$, when $k = N$ (the bottom layer). Note that the effect of the wind forcing T_1, equal to $\nabla \times \tau/H_1$, produces an Ekman pumping/stretching tendency in the upper layer that is equivalent to a body force acting on the upper layer. The T_k for $k > 1$ are zero. At the sea surface, $w_{1/2} = 0$ and at a flat sea bottom $w_{N+1/2} = 0$. The advective velocities at the interfaces, needed in Eq. (2), are calculated from a weighted average of the velocities in the layers, i.e., $\psi_{k+1/2} = (\alpha_{k+1/2})\psi_k + (1 - \alpha_{k+1/2})\psi_{k+1}$, where $\alpha_{k+1/2} = H_k/(H_k + H_{k+1})$. The wind stress patterns used in this study are idealized double gyres of wind-forcing (Fig. 2). The eastward wind stress then is given by $\tau = -T_0\cos(2\pi y/L_y)$, where L_y is the north–south extent of the basin. The northward component of the wind stress is chosen to be zero.

For an assimilation version of this model, in which surface height observations are to be combined with the model streamfunction at each time step, we modify the upper layer vorticity equation by adding a Newtonian relaxation or nudging term:

$$\frac{\partial}{\partial t}\nabla^2\psi_1 = \text{`physics'} - R\left(\nabla^2\psi_1 - \nabla^2\psi^{\text{obs}}\right). \tag{4}$$

Here 'physics' represents the terms on the right-hand-side of (1). Also note that ψ^{obs}, equal to gh^{obs}/f_0, relates the upper layer streamfunction to sea surface height. R, an inverse time scale, is our nudging parameter, chosen by trial and error to achieve a best blending of model and observational estimates of the upper layer streamfunction. Haltner and Williams (1980) suggest that the time scale R^{-1} should be smaller than time scales in the changing observations. We have found that $R \sim (0.5\text{ day})^{-1}$ produces excellent results, and we have used this value in all the experiments here. Note that the special nature of nudging the vorticity is that, when the three-dimensional Helmoltz operator describing the new potential vorticity is inverted to find the new streamfunction (ψ^{n+1}), the entire streamfunction field is modified, the deep ocean as well as the surface.

Besides this 'nudging of vorticity' technique, used in the numerical experiments shown in this paper, HV investigated a further modification of the Eqs. (1)–(3) by introducing a nudging term also in the uppermost interface equation:

$$\frac{\partial}{\partial t}h_{\frac{3}{2}} = \frac{f_0}{g'_{\frac{3}{2}}}\frac{\partial}{\partial t}(\psi_2 - \psi_1) = \text{`physics'} + \frac{f_0}{g'_{\frac{3}{2}}}\hat{R}(\psi_1 - \psi^{\text{obs}}) \tag{5}$$

This is the only other place in the equation set (1)–(3) that ψ_1 occurs. The reason this is of interest is that, when both (4) and (5) are used and $\hat{R} = R$, the assimilation

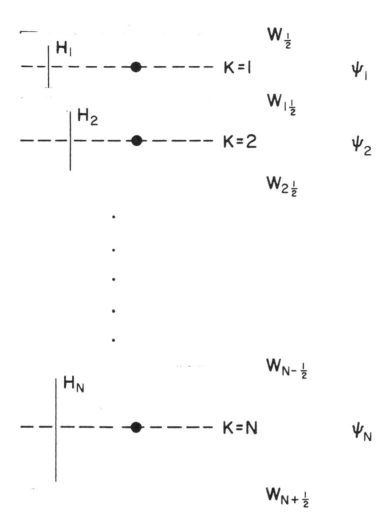

Figure 1
The vertical structure of the N layer QG model. The numerical experiments in this paper are carried out with a three layer version of the model.

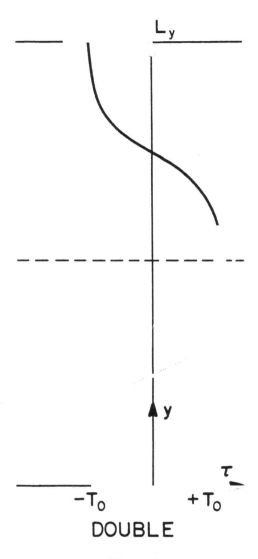

Figure 2
The meridional distribution of zonal windstress, creating a turbulent, two gyre oceanic circulation.

step is exactly equivalent to modifying the new upper layer streamfunction *after* performing a forward step of the pure physics model as follows:

$$\psi_1^{n+1} = \psi_1^{n+1}(\text{model}) + \alpha[\psi^{\text{obs}} - \psi_1^{n+1}(\text{model})] \tag{6}$$

where α is related to R multiplied by the time step of the model. The ψ_k for $k > 1$ (i.e., the deeper layers) are not modified at this point but can only feel the influence of surface observations at subsequent time steps through dynamical adjustments. Note that (6) represents a simple 'blending' of the model estimate and the observational estimate of the surface streamfunction.

Tests of the two techniques, which we might call nudging and blending, tended to show that nudging worked best in forcing the assimilation cases toward the control run, at least for the simple cases examined here. Blending allowed the deep ocean (the bottom two layers) in some cases to continue independently of the control run and not converge toward it. This might be due to the fact that, since the nudging technique introduces a modification in the deep circulation directly, the deep vorticity equations can immediately respond (at the next time step) by modifying the vorticity distributions there. The blending technique can result in modifying the deep vorticity distributions only through the vertical velocity (stretching) term linking the upper layer to the deeper ones.More work is needed to satisfactorily understand these results and to settle on a best strategy for data insertion.

Most of the comments above relate to the case where ψ^{obs} is available at every time step and every grid point of the model calculation. For the track assimilation experiments, using the pure nudging technique of (4), we found that simply setting R to zero at points where no data existed (between tracks) and to a constant (0.5 day)$^{-1}$ on the tracks worked quite well, even though the nudging term in (4) was not smooth spatially. Again this technique works because of the inherent smoothing that occurs when the modified potential vorticity distribution is solved for the new streamfunction. A vorticity point source (or a line source) at the sea surface introduces a smooth streamfunction correction as dictated by the vertical modes of the problem. Note also that we might eventually want to let R vary in both time and space, to take into account both where and when the data are available and to include a measure of expected errors in the data. This has the flavor of least squares techniques (optional estimation; Kalman filtering) for blending data into models.

The technique used by HV, VH and HM for studying the assimilation process has been a 'twin experiment' approach. Firstly, they ran numerical experiments without any assimilation ($R = 0$) to produce a control run that they considered a surrogate for the 'real world'. From this experiment they produced and saved for later use 'observations' of the surface layer streamfunction, data to be incorporated into the ocean in a separate set of assimilation experiments. These were usually experiments that had a very different initial condition or experiments that were based upon a somewhat different ocean model (for example with different basic parameters or wind forcing). Since the surrogate 'observations' come from a complete numerical experiment in its own right, there exists a complete description of this surrogate for the real world, including the deep ocean circulation. Thus we can examine carefully the convergence (or lack thereof) of the numerical experiments with various assimilation strategies or model characteristics toward this completely known substitute for the real world. This of course would be impossible for the real world itself, especially since the oceanic interior circulation is almost everywhere completely unknown.

There are a large number of issues to be understood in developing any initialization/assimilation scheme. Here we shall address only a few of these, but it is worth while to mention the others so as to make clear that we have a very long way to go before having a workable and well understood scheme. For a successful altimeter data assimilation scheme, ocean modellers must understand how best to combine models with observations; there are clearly a whole host of possible ways this blending process can be accomplished. A particular issue for using altimeter data is the fact that the oceanic geoid is relatively unknown so that the altimeter will give us mainly the transient component of the sea surface height. Other techniques will have to be used to determine the mean surface topography. We have ignored this issue in most of these preliminary experiments, but we shall come back to it when we consider the use of GEOSAT data in the Agulhas region. In addition to these considerations, modellers must understand how errors in the data and inadequacies in the models can contribute to the successes and failures of any assimilation scheme. The assimilation schemes themselves must be developed further to achieve a best strategy for combining data and model results together. The peculiarities of satellite sampling, in terms of a trade off between space and time resolution, is also an important consideration to be studied. The prospects for having a variety of kinds of satellite data from the sea surface in addition to altimetry

— temperature, scatterometer wind stresses, various components of the heat flux as well as more limited interior data of various kinds — will have to be examined in order to develop schemes that can most effectively accomplish the results needed by oceanographers: model improvement, model initialization for dynamical studies, dynamically interpolated observational descriptions of the circulation, and actual prediction of ocean currents and thermohaline structure.

Figs. 3, 4 and 5 show results from a basic experiment in altimeter data assimilation. In this case the data are considered to be perfectly known in space and time (but only at the sea surface) and the assimilation is carried out using a model with identical (i.e. perfect) physics. The only difference between the control experiment and this assimilation experiment, besides the assimilation procedure that operates upon the assimilation run, is in their respective initial conditions, shown in Figs. 3 and 4. Thus this case provides a baseline against which we can judge other assimilation cases. Here then we are examining the ability of the above assimilation scheme to reconstruct the structure and intensity of features existing in the control run, using only a knowledge of surface information from the control run. Fig. 3 shows the initial and final streamfunction fields (after 2160 days) for the control run and Fig. 4 shows the same fields for the baseline assimilation run. Note that the assimilation experiment has taken on all of the instantaneous characteristics of the control experiment. The assimilation, under these circumstances, has been highly successful. Fig. 5 shows the global root-mean-square differences between the two experiments as a function of time (for the first 1080 days of the experiment), indicating the nature of the convergence of the assimilation experiment toward 'reality' and showing that all three layers participate in this process. Moreover all layers have converged, for the most part, after only one year of assimilation. It is this last fact that suggests that the process of assimilating vorticity is indeed an effective one.

Why does this relaxation process (i.e. the inclusion of the assimilation term in Eq. 1) work as well as it does, even in the deep ocean? HV examined this issue by carrying out several simple examples, using linear models and barotropic and/or baroclinic control experiments, to show that it is the continuous insertion of space and time information over a barotropic or baroclinic time scale that results in a proper incorporation of the data. In essence, the space and time structure of the difference in the surface vorticities of the control and assimilation experiments leads to a resonance that allows the proper mode to grow while the other modes do not.

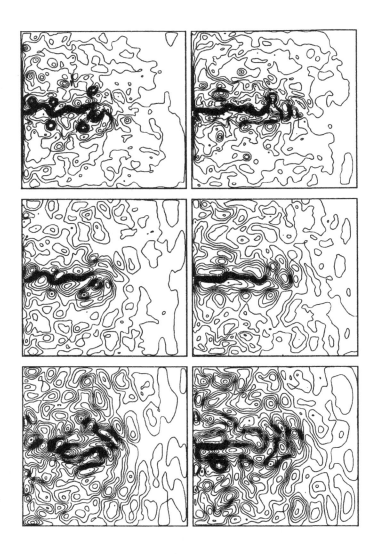

Figure 3
The streamfunction fields for the control experiment, which is a surrogate
for the real world. Top to bottom: surface, intermediate and bottom layers
respectively. Left: initial fields (at day 0). Right: fields at 2160 days.

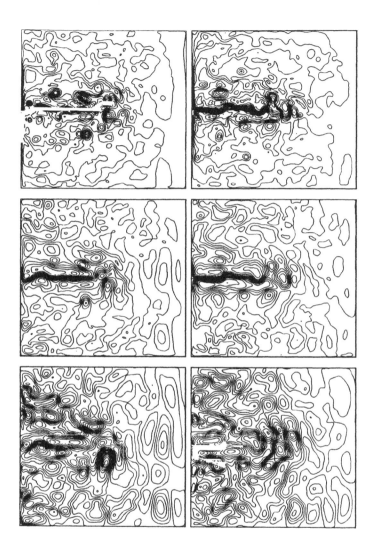

Figure 4
The streamfunction fields for a baseline assimilation experiment with perfect data. Top to bottom: surface, intermediate and bottom layers respectively. Left: initial fields (at day 0). Right: fields at 2160 days.

214

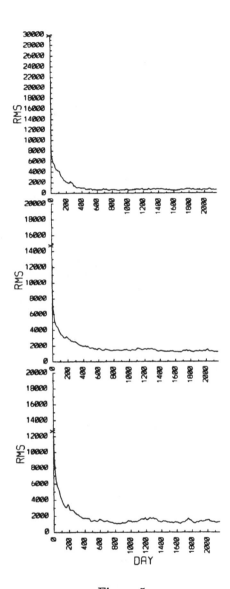

Figure 5
Time evolution of global RMS differences in streamfunction between the control and baseline assimilation experiments shown in Fig. 3 and 4. Top to bottom: surface, intermediate and bottom layers respectively.

Thus a barotropic mode control run will force a barotropic response in the model while a first baroclinic mode control run will force a first baroclinic mode response in the assimilation run. The model physics discriminates between the various possible responses and, as seen in the basic, fully nonlinear case shown in Figs. 3 and 4, the assimilation process can recreate the appropriate ratios of the modal amplitudes to give a nearly exact correspondence at all levels between the instantaneous currents in the control and in the assimilation experiment.

3. TIME AND SPACE DEPENDENCE OF THE OBSERVATIONS

Actual sea surface height observations from a single satellite will not only have errors in them due to a variety of sources, they will also be available only along certain tracks at certain times. Therefore there is a problem of how best to make use of such data. On the one hand, the data could be pre-processed by optimal interpolation in space and time to create a data set on a complete space-time grid. Such data would be 'smoothed' by the interpolation scheme and an estimate of the error field would be produced by the optimal estimation procedure. A second possibility would be to develop assimilation techniques that can incorporate the data into the assimilation model as the data becomes available, letting the model do the interpolation (as well as the downward extrapolation discussed earlier). The question of whether the assimilation is being carried out to provide a forward prediction, that is, only past data is available, or whether an existing data set is being used for initialization purposes or dynamical hindcast studies, is of course important in this consideration. Here we shall mainly be concerned with the uncertainites introduced by lack of complete data in time and space for the hindcast initialization problem.

The GEOSAT satellite covers the globe with a 17 day repeat cycle (that is the satellite will follow the same path every 17 days) and discussions are underway for TOPEX/POSEIDON to decide upon a repeat cycle between three and 20 days. There are different types of oceanic phenomena observable by an altimeter that are of interest to oceanographers, for example tides, mesoscale eddies, major midlatitude ocean currents such as the Gulf Stream, and transient equatorial currents. Thus there can be different arguments made for various sampling strategies. Here, following HM, we shall briefly examine how time and space structure in the observations might affect an eddying, midlatitude ocean circulation containing gyre-scale, as well as mesoscale, transient behavior.

Rather than first produce an assimilation model that assimilates data as it arrives along tracks, it is advantageous to examine the issues of time and space resolution in the data separately, in order to begin to understand whether it is lack of time resolution or lack of spatial resolution (or both equally) that is the major concern. HM carried out a number of twin experiments to examine this issue. Figs. 6–8 show some results from three assimilation experiments in which the observations are considered to be complete and accurate spatially but to be available only at discrete intervals in time. In order to accomplish the nudging type assimilation, the data N days apart are interpolated linearly between the discrete times at which they are available so that a continuous insertion is carried out. Note that this means that every N days the data are perfect but between times the data are only an approximation to the 'real' signal. The faster the time scales in the phenomena to be reconstructed, the more difficult that reconstruction will be, since the data will be badly aliased. Figs. 6, 7 and 8 show time series of the RMS differences between three assimilation experiments and the control run (right panels) and the final streamfunction fields at day 2160 (to be compared with Fig. 3, left panels). Note that all assimilation experiments start with the same initial condition, that shown in Fig. 4, left panels. The three assimilation experiments here are chosen such that maps of gridded data are made available at 2, 10 and 20 day intervals. When data are available every 2 days, the reconstruction of the streamfunction fields is excellent and the final RMS differences (between the control and this experiment) are small (Fig. 6). When the interval between maps is 10 days, the final RMS values are considerably bigger but the streamfunction maps are quite like the control (Fig. 7). The features are all in the right places but some lack of correspondence begins to be noticeable, in both structure and amplitude. This shows up mainly in the deep ocean eddy field. Finally, when the interval between observations is 20 days, the initialization/assimilation becomes less successful. The RMS curves show that the global errors undergo an oscillation in amplitude with a 20 day period as the assimilation procedure incorporates accurate data at the times the data is available (every 20 days) and inaccurate data in between. The streamfunction maps at day 2160 (Fig. 8, left panel) show that while the fields still have the 'flavor' of the control run streamfunctions (shown in Fig. 3, right panel), there are serious discrepancies in the location and strength of features in all layers, but particularly in the deep ocean.

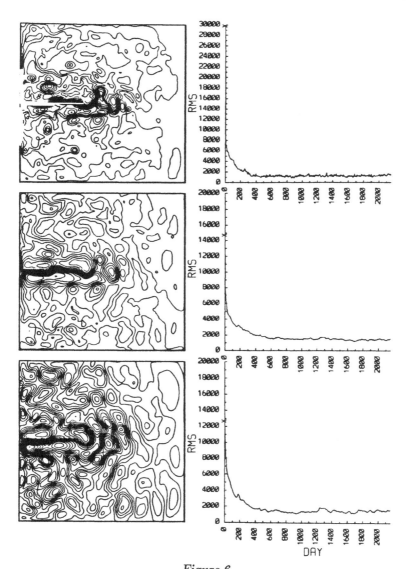

Figure 6
Time evolution of the global RMS differences in streamfunction from the
control run (right) and the streamfunction fields at 2160 days (left) for an
assimilation experiment in which surface vorticity is assimilated at every
grid point in space but with data available as a gridded map only every 2
days.

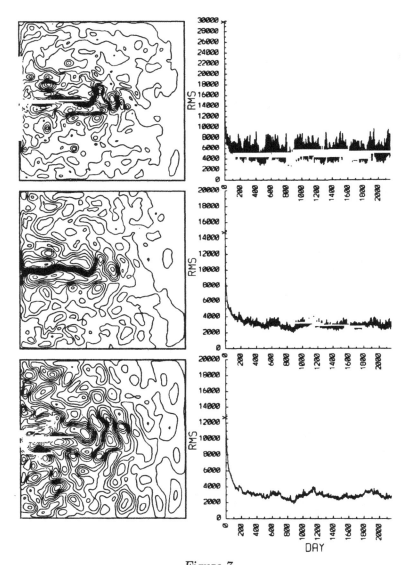

Figure 7
Time evolution of the global RMS differences in streamfunction (right) and
the streamfunction fields at 2160 days (left) for an assimilation experiment
in which surface vorticity is assimilated at every grid point in space but
with data available as a gridded map only every 10 days.

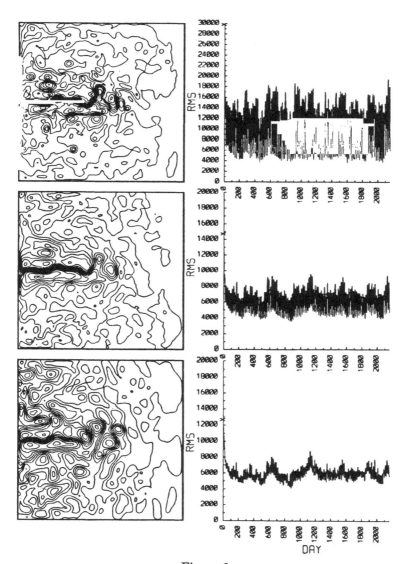

Figure 8
Time evolution of the global RMS differences in streamfunction (right) and
the streamfunction fields at 2160 days (left) for an assimilation experiment
in which surface vorticity is assimilated at every grid point in space but
with data available as a gridded map only every 20 days.

HM also examined the question of spatial sampling (but complete time information) in several assimilation experiments in which data was made available only along tracks. Fig. 9 (left panels) shows the coverage of three hypothetical satellite repeat tracks over a 4000 km square piece of ocean, adjusted to a spatial separation between ascending and descending tracks of 42, 99, and 198 km respectively (60, 140 and 280 km separation in the east–west direction). Note that the latter two separations are approximately the separations to be expected for a single satellite repeat cycle of 20 or 10 days. Fig. 9 (right panels) shows the same tracks crossing a portion of these basins (a 720 km square subdomain), as well as the grid points (at 20 km intervals) actually used in the numerical model. This shows the model grid points at which data is made available (points crossed by the tracks) and points at which data is not available (points between the tracks). Note that the tracks are adjusted to coincide with grid points in the model but that, while the 42 km case 'misses' only a few grid points in updating the model, the 198 km case misses a very large number. In all cases, no information is made available at the missed points. The nudging technique is implemented by choosing the nudging coefficient R to be constant and non-zero along the tracks and zero elsewhere. Here we choose to use the full vorticity from the observations, although only the component along the track is easily observable. This issue is currently being examined. Note that the observations are blended smoothly into the model, even though the data are sparse spatially, because the inversion process that determines streamfunction from potential vorticity is a smoothing process. Note also that this track assimilation procedure allows fine spatial scales of information along the track to be incorporated into the model. Figs. 10, 11 and 12 show results (global RMS errors; final streamfunction fields after 2160 days) from these assimilation experiments. For the 42 km track case (Fig. 10), the RMS values reach a level comparable to but a little larger than the 2 day time sampling case shown in Fig. 6. The fields faithfully reflect the control fields. When the spatial resolution of the observations is of the order of 99 km (Fig. 11), the RMS global errors are very much bigger (comparable to those in the 10 day sampling case of Fig. 7) and the streamfunction fields reflect this. The features are all present in the right places, but differences are noticeable. The discrepancies are as important in the surface layer as in the deeper layers. Finally, when the resolution is the order of 198 km or so, the model is not brought back to 'reality' very well at all (compare Fig. 3 and Fig. 12). The RMS values are very much larger even than those in the 20 day sampling experiment shown in Fig. 8 and the streamfunction fields are quite different from those in the control experiment.

Apparently the resonance phenomena that allows the nudging technique to work requires that the space-time structure of the observations be accurate enough not to generate spurious response in wrong modes of the system. The gyre situation is enlightening in that it is clear that the far fields can be sampled reasonably well by these space/time observing systems (which have reasonably realistic satellite characteristic coverage) but the Gulf Stream cannot. In addition, we should emphasize that, for a real satellite observing platform, both space and time interpolation will be needed and thus the problem will be harder. HR show some examples in which realistic satellite coverage is chosen.

We do not want to paint too black a picture concerning the prospect for single satellite coverage to be useful, even for a vigorous mesoscale eddy field. Firstly, the Gulf Stream time scales found in the present model may be faster than those in reality and there are certainly many places in the ocean where the phenomena are not so fast moving or such small scale. In fact, our final example, GEOSAT initialization for the Agulhas region, suggests that many regions of the World Ocean may have space/time scales of motion that are adequately sampled by a single satellite. In addition, even if such an initialization procedure recreates only the larger spatial and longer time scales of the phenomena, it may be possible to find ways to allow the model to develop its own fast time scale, small space scale characteristics with some modifications to the present assimilation procedure. The extent to which such phenomena, unresolved by the observing system, can be realistic is being investigated.

4. GEOSAT ASSIMILATION IN THE AGULHAS RETROFLECTION REGION

It is useful to attempt to carry out an actual case of data assimilation in order to begin to confront all of the problems at the same time. This is possible now using GEOSAT data, since there are at the time of this writing more than 18 months of continuous repeat cycle data available to the university community. Holland, Fu and Zlotnicki (1988; hereafter HFZ) have made a start at this problem, using the GEOSAT data from the period 18 December 1986 until 18 September 1987 in the Agulhas Retroflection Region south of South Africa. This particular region is

Figure 9
Hypothetical tracks along which data is made available for assessing spatial sampling questions. Right: the entire 4000 km square basin covered with tracks at 42 km (top), 99 km (middle), and 198 km (bottom) spacing diagonally (the distance between adjacent ascending or descending tracks). These represent a spacing of 60, 140 and 280 kms, respectively, in the east–west direction. Left: a 720 km square subregion showing the tracks and the grid points (20 km apart) of the model.

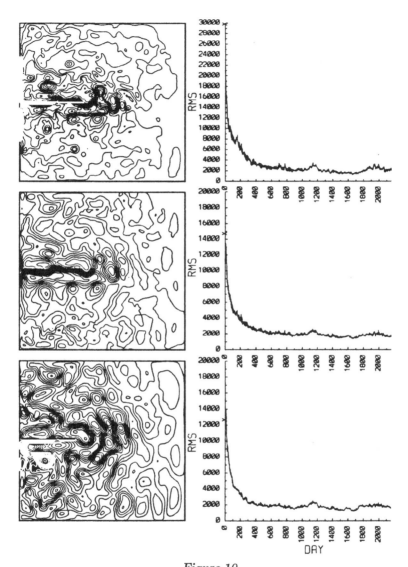

Figure 10
Time evolution of the global RMS differences in streamfunction (right) and
the streamfunction fields at 2160 days (left) for an assimilation experiment
in which surface vorticity is assimilated at every time step (complete time
information) but with data available only along tracks 42 km apart (60
km east–west). See the top panel of Fig. 9.

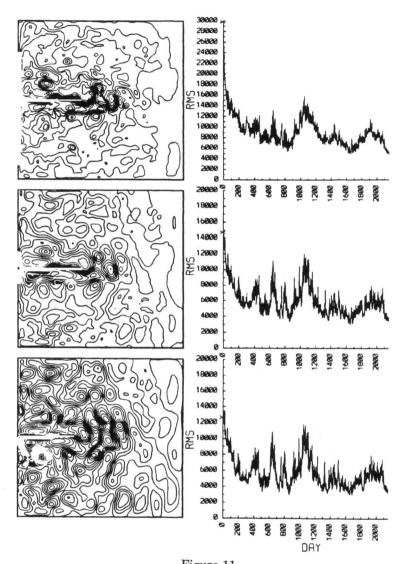

Figure 11
Time evolution of the global RMS differences in streamfunction (right) and
the streamfunction fields at 2160 days (left) for an assimilation experiment
in which surface vorticity is assimilated at every time step (complete time
information) but with data available only along tracks 99 km apart (140
km east–west). See the middle panel of Fig. 9.

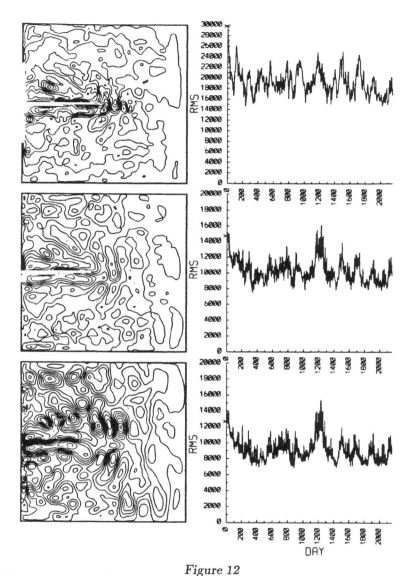

Figure 12
Time evolution of the global RMS differences in streamfunction (right) and
the streamfunction fields at 2160 days (left) for an assimilation experiment
in which surface vorticity is assimilated at every time step (complete time
information) but with data available only along tracks 198 km apart (280
km east–west). See the bottom panel of Fig. 9.

one of exciting transient oceanic behavior, a region of large amplitude rings that are spun off from the Agulhas Current as it rounds the southern tip of the African continent.

Fu and Zlotnicki (1988; hereafter FZ) carried out a space/time analysis to produce a gridded data set for sea surface height variability. The data is produced on a 1/4 degree grid every 10 days for the region 32°S–42°S and 10°E–35°E. Eighteen maps in all were available for ass imilation purposes. For our purposes here (and in HFZ), we shall take these data as given; FZ's analysis of the errors inherent in their analysis procedure can give us a handle on how to assess the resulting errors in the assimilation model by allowing us to run additional assimilation experiments with typical errors (from FZ's error maps) inserted into the 'observations'.

Holland (1986) had constructed an eddy resolving model of the Agulhas region earlier, for a considerably larger region (20°S–54°S latitude, 2.5°W–65°E longitude) but spanning the region dealt with by FZ. Those QG

give rise to periodic Ring production that is related to the retroflection process. The numerical experiments were basic studies in geophysical fluid dynamics rather than experiments in actual prediction. The way in which the interaction with the global circulation was handled, by way of 'pumps and baffles', was intended to deal with the larger scale circulation in which the Agulhas Current system is embedded in only an approximate way. Thus it became necessary in HFZ to impose a degree of realism on the region by creating a large scale mean sea surface topography that was in accord with the mean historical climatology for the region. FZ did this by calculating mean dynamic height using a level of no motion hypothesis. Then HFZ set out to assimililate data into the local subregion of the larger domain, using the analysed sea surface height variability from the GEOSAT altimeter and adding to it the mean sea surface height from these dynamic height calculations.

Fig. 13 shows the larger domain with a typical eddying circulation from Holland (1986). This three layer model was used for the assimilation experiment described here. The smaller domain outlined south of South Africa is the domain for which FZ made their analysis. Thus the 'observations' were available only for this sub-region and consisted of a mean dynamic topography and a set of sea surface height variability maps from FZ. Fig. 14 shows a sampling of the total sea surface topography maps gotten by adding these two components; these kind of observations served as the data for the assimilation experiment.

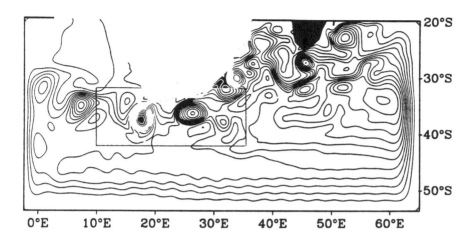

Figure 13
A regional model for the Agulhas Retroflection Region. The assimilation experiment shown in the next figures is carried out for the entire domain, but the altimeter data are available and the prediction of oceanic transience meaningful only in the subdomain. Fig. 14 and 15 show the subdomain only.

In order to carry out the assimilation experiment, two steps were necessary. This is because the ocean domain is a regional one with open boundaries and the boundary conditions had to be handled in a special way. We shall only outline the procedure here; see HFZ for details. First a mean assimilation experiment was carried out by extending the calculated mean sea surface topography (from dynamic height calculations) into the larger domain in a smooth fashion. Then an assimilation experiment was run in which the observations consisted of the mean alone. This allowed the subregion to have sensible and relatively realistic inflow and outflow conditions around its edges at the initial time. The initial condition itself was the mean circulation for the region. Finally, this mean topography, now extended into the larger domain, was added back to the transient observations of FZ (existant only in the sub-domain) to form a data set for assimilation purposes. Fig. 15 shows the sea surface topography in several snapshots from the model, at the same times as the 'observations' shown in Fig. 14; only the subregion is shown. The model has clearly taken in the observations and created a dynamically consistent, four dimensional 'prediction' of the currents in this region. The deeper layers also

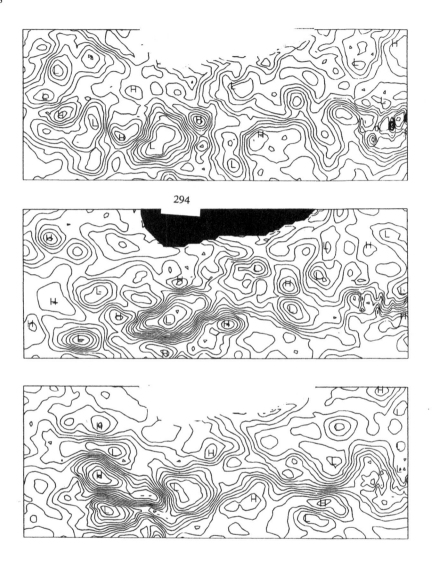

294

Figure 14
GEOSAT altimeter observations of sea surface height variability to which
has been added the mean sea surface topography based upon historical
hydrographic data for the region bounded by 32°S–42°S latitude, 10°E–
35°E longitude. Three gridded maps are shown at days 20, 90, and 160
relative to a day 0 that marks the beginning of the assimilation process.
See Fu and Zlotnicki (1988) for a description of the analysis of the data.

Figure 15
Model results for an assimilation experiment incorporating the above data
into a three layer QG model of the Agulhas Retroflection Region. The surface layer streamfunction is shown at days 20, 90, and 160, for comparison
with the data used in the assimilation procedure (Fig. 13).

have adjusted to some extent, but there are no independent observations available to us at the present time for checking the realism of these flows. As suggested in earlier work, another half year of data would probably be enough to largely complete the initialization task.

5. DISCUSSION

These preliminary studies suggest an exciting and fruitful future for the use of altimeter data in recreating a four dimensional description of the oceanic circulation in various regions of the World Ocean. The particular assimilation scheme used here is a simple but apparently effective one, although it has been used so far under the most favorable of circumstances. There are very likely other effective ways of combining observations at the sea surface with a four dimensional ocean model to get a description of the time evolving oceanic circulation, a description that is consistent with the observations and is (mostly) consistent with whatever dynamics one thinks is applicable. This gives the ocean dynamicist an opportunity to study problems that have a degree of realism heretofore unknown. It is likely that we will soon be operating a whole suite of models that have skill in actually making predictions. As demonstrated here, it is certainly possible to combine observations, even if available only at the sea surface, with ocean numerical models to provide both downward extrapolation and horizontal interpolations of synoptic information in a dynamically consistent framework. This will allow us to begin to study ocean circulation with new realism, to begin to make predictions, and to provide a much better description of the time evolving environment, for example for biological or geochemical models. The development of sophisticated numerical models of the ocean circulation, combined with the extraordinary ability of a new generation of ocean observing satellites to tell us what the ocean is actually doing synoptically, makes this a truly exciting time in ocean research.

REFERENCES

Anthes, R., (1974): Data assimilation and initialization of hurricane prediction models, *J. Atmos. Sci.*, **31**,702–718.

Fu, L. and V. Zlotnicki, (1988): [In preparation].

Haltner, G.J. and R.T. Williams, (1980): Numerical Prediction and Dynamic Meteorology, *John Wiley & Sons*, New York, 477 pp.

Hoke, J.E., and R.A. Anthes, (1976): The initialization of numerical models by a dynamic initialization technique. *Mon. Wea. Rev.*, **104**, 1551–1556.

Holland, W.R., (1978): The role of mesoscale eddies in the general circulation of the ocean. Numerical experiments using a wind-driven quasigeostrophic model, *J.Phys. Oceanogr.*, **8**, 363–392.

Holland, W.R., (1986): Quasigeostrophic modelling of eddy-resolved ocean circulation. In Advanced Physical Oceanographic Numerical Modelling, Editor J. J. O'Brien. *D. Reidel Pub.*, 608pp.

Holland, W.R. and P. Malanotte-Rizzoli, (1988): Along track assimilation of altimeter data into an ocean circulation model: space versus time resolution studies, *J. Phys. Oceanogr.*, submitted.

Holland, W.R., L. Fu and V. Zlotnicki, (1988): GEOSAT data assimilation into a model of the Agulhas Retroflection region using a dynamic initialization method. In preparation.

Holland, W.R. and J. Verron, (1988): Altimeter data assimilation into an eddy resolving ocean circulation model, *J. Geophys. Res.*, submitted.

Sarmiento, J.L. and K. Bryan, (1982): An ocean transport model for the North Atlantic, *J. Geophys. Res.*, **87**, 394–408.

Verron, J. and W.R. Holland, (1988): Impacts de donnes d'altimetrie satellitaire sur le simulations numeriques des circulations generales oceaniques aux latitudes moyennes, *Annales Geophysicae*, in press.

ASSIMILATION OF DATA INTO OCEAN MODELS

D. J. Webb

Institute of Oceanographic Sciences
Godalming, Surrey, GU8 5UB
U.K.

1. INTRODUCTION

Because of the size of the ocean we do not have, nor are we ever likely to have, good synoptic data on the ocean circulation. However, data from relatively isolated instruments within the ocean are usually available and these are now being joined by data from satellite mounted instruments which give routine surveillance of the ocean surface. The prime task facing us is to find methods for integrating this data to deduce the flow throughout the ocean.

This type of problem is not new but dates back at least to Gauss' (1809) work on the Moon's orbit. Expressed in terms of the orbital, elements the problem he faced was a non-linear one and had to be solved by first linearising the equations about an initial trial orbit. The method of least squares was then used to average out the errors in data collected over a period of time. This approach has been developed to give the Kalman type methods which are posed in terms of the error covariance matrices of the best estimate and the measurements.

As we shall see, one problem that arises in using these methods in oceanography results from the large number of degrees of freedom that the ocean possesses. The error covariance matrices are impracticably large and ways have to be found of either simplifying them or doing without them. It is also impossible to represent

233

D. L. T. Anderson and J. Willebrand (eds.), Oceanic Circulation Models: Combining Data and Dynamics, 233–256.
© 1989 by Kluwer Academic Publishers.

all the degrees of freedom of the ocean and so the models used with the methods will always incorrectly represent some aspect of the flow.

Another problem that arises, especially at midlatitudes, is the short time scale of non-linear effects in the ocean. A typical Rossby wave will have a period of 100 days and yet non-linear effects can be significant over periods of only 30 days. The amount of data available over such a period is never likely to be large and so schemes are needed which are efficient and robust in the presence of such strong effects.

To help in solving the problem we can draw on other peoples experience, especially from the fields of statistics, signal processing and control theory (Deutsch, 1965; Sorenson, 1966; Maybeck, 1979, 1982; Kwakernaak and Sivan, 1972). Meteorologists have also been very active in developing these methods for assimilating their own data (Bengtsson, 1975; Rodgers, 1976; Bengtsson, Gill and Kallen, 1981; Hollingsworth et al., 1985; Lewis and Derber, 1985; Lorenc, 1986; Talagrand and Courtier, 1987).

2. THEORY

Consider the ocean represented by the state vector $\mathbf{r}(t)$ and a model of the ocean with state vector $\mathbf{m}(t)$. In general the number of degrees of freedom of the ocean is essentially infinite whereas the model has only a finite number covering only the lowest wavenumbers. The model will also not follow exactly the same laws of motion, so if by chance the two vectors are ever identical they will soon drift apart.

A set of observations $\mathbf{z}(t_i)$ is made of the ocean. In general these are related to the state vector by a linear equation,

$$\mathbf{z} = \mathbf{H}^T \mathbf{r} + \mathbf{v}, \tag{2.1}$$

where $\mathbf{H}$ is a matrix describing the measurement and $\mathbf{v}$ is a random variable representing the error in the observation. The case of one observation, where $\mathbf{H}$ becomes a vector $(\mathbf{h})$, is shown in Fig. 1.

The first problem facing us is to use the data to fit the model $\mathbf{m}(t)$ as closely as possible to the real ocean $\mathbf{r}(t)$. In practice this means finding a model trajectory which gives the best fit to the observations and to any a priori information. A second problem is to determine the best measurement strategy given the resources available.

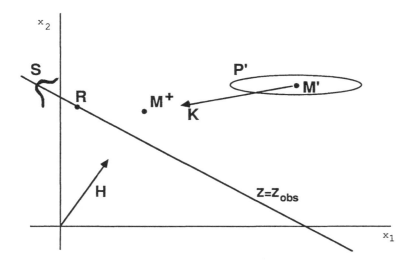

Figure 1
Second stage of Kalman filtering. If the measurement noise is zero the
model state $\mathbf{m}'$ is projected onto the surface $\mathbf{z} = \mathbf{z}_{\mathrm{obs}}$ defined by the data.
Non-zero noise gives an intermediate position $\mathbf{m}^+$.

2.1 The Kalman Filter

The scheme developed by Kalman (Kalman, 1960; Kalman and Bucy, 1961; Swerling, 1959; Battin, 1962; Carlton, 1962), was for a linear system whose time dependence satisfies the equation,

$$\mathbf{r}(t_k) = \mathbf{G}(t_k, t_{k-1})\mathbf{r}(t_{k-1}) + \mathbf{f}(t_k) + \mathbf{w}(t_k). \tag{2.2}$$

$\mathbf{G}$ is the Green's function (or resolvant) describing the evolution of the model from
time t_{k-1} to time t_k and $\mathbf{f}$ is the forcing during this period. $\mathbf{w}$ is an error term
representing the difference between the evolution of the model and the real ocean.
For the moment we assume that it is white noise and described by,

$$\langle \mathbf{w}(t_k) \rangle = 0, \qquad \langle \mathbf{w}(t_k)\mathbf{w}(t_l)^T \rangle = \mathbf{t}_k \delta_{kl}.$$

At some initial time t_0, an initial estimate of the state of the system and its
covariance is made,

$$\mathbf{m}(t_0) = \mathbf{m}_0, \qquad \langle (\mathbf{r} - \mathbf{m}_0)(\mathbf{r} - \mathbf{m}_0)^T \rangle = \mathbf{P}_0.$$

Then at each time t_k, the observations $\mathbf{z}_k$ are related to the true state of the ocean by,

$$\mathbf{z}_k = \mathbf{H}_k^T \mathbf{r}_k + \mathbf{v}_k. \tag{2.3}$$

$\mathbf{v}_k$ is the observational error with statistics,

$$\langle \mathbf{v}_k \rangle = 0, \qquad \langle \mathbf{v}_k \mathbf{v}_l^T \rangle = \mathbf{s}_k \delta_{kl}.$$

If $\mathbf{v}_k$ is zero then Eq. (2.3) defines a subspace in which $\mathbf{r}$ must lie. For a single observation, $\mathbf{H}$ becomes the vector normal to the subspace.

The Kalman problem is to obtain the 'best' estimate $\mathbf{m}_k$ of the state vector at time t_k, which is a linear combination of the measurements $\mathbf{z}_k$ and the previous best estimate $\mathbf{m}_{k-1}$. The solution depends on the definition of 'best', and for the Kalman filter the one used is that at each time t_k the expected value of the sum of squares of the error should be a minimum. Thus,

$$\langle (\mathbf{m}_k - \mathbf{r}_k)^T (\mathbf{m}_k - \mathbf{r}_k) \rangle = \text{ minimum.} \tag{2.4}$$

The solution of the problem is not difficult and is given in standard texts (Sorenson, 1966; Deutsch, 1965; Maybeck, 1979, 1982; see also Ghil $et\ al.$, 1981; Heemink, 1986).

If,

$$\mathbf{m}' = \mathbf{G}\mathbf{m}^- + \mathbf{f}, \tag{2.5}$$

then,

$$\mathbf{m}^+ = \mathbf{m}' - \mathbf{P}'\mathbf{H}(\mathbf{H}^T\mathbf{P}'\mathbf{H} + \mathbf{S})^{-1}(\mathbf{H}^T\mathbf{m}' - \mathbf{z}). \tag{2.6}$$

where $\mathbf{m}^+$ and $\mathbf{m}^-$ are the model states at times t_k and t_{k-1}. The corresponding covariance matrices $\mathbf{P}^+$ and $\mathbf{P}^-$ are related by the equations,

$$\mathbf{P}' = \mathbf{G}\mathbf{P}^-\mathbf{G}^T + \mathbf{T}, \tag{2.7}$$

and,

$$\mathbf{P}^+ = \mathbf{P}' - \mathbf{P}'\mathbf{H}(\mathbf{H}^T\mathbf{P}'\mathbf{H} + \mathbf{S})^{-1}\mathbf{H}^T\mathbf{P}'. \tag{2.8}$$

Given the initial estimate $\mathbf{m}^-$ and covariance $\mathbf{P}^-$, Eqs. (2.5) and (2.7) describe their evolution during the period t_{k-1} to t_k, and Eqs. (2.6) and (2.8) the corrections due to the new observations. Because Kalman theory was originally developed for signal processing, the terms describing the correction are often collected into the 'gain' matrix $\mathbf{K}$,

$$\mathbf{K} = \mathbf{P}'\mathbf{H}(\mathbf{H}^T\mathbf{P}'\mathbf{H} + \mathbf{S})^{-1}. \tag{2.9}$$

Although the above equations may initially seem daunting, their structure becomes more understandable if they are looked on as a projection (Fig. 1). First note that, as is the case with the Kalman scheme, if the only information we have about the state of the ocean is a best estimate $\mathbf{m}_0$ and the covariance $\mathbf{P}$, then the probability distribution for $\mathbf{m}$ is,

$$p(\mathbf{m}) = \left((2\pi)^{\frac{n}{2}}|\det\mathbf{P}|^{\frac{1}{2}}\right)^{-1} \exp\left(-0.5(\mathbf{m} - \mathbf{m}_0)^T\mathbf{P}^{-1}(\mathbf{m} - \mathbf{m}_0)\right). \tag{2.10}$$

This equation can be derived by maximising the entropy ($p\ln p$) using Lagrange multipliers for the constraints. The surfaces of the constant probability are ellipsoids centred on $\mathbf{m}_0$ and whose shape depends on the matrix $\mathbf{P}$. Thus their long axes are in directions for which $\mathbf{P}$ is large and correspond to directions so far unconstrained by the data.

If $\mathbf{P}'$ is isotropic, then, for a single measurement, $\mathbf{H}$ is a vector and the gain matrix describes a projection in the direction $\mathbf{H}$. If it is anisotropic then the projection is weighted towards the direction with largest error.

The term $(\mathbf{H}^T\mathbf{P}'\mathbf{H} + \mathbf{S})^{-1}$ is a normalising constant and determines the length of the projection vector. If the observations are error-free, $\mathbf{S}$ is zero and the projection is onto the surface defined by Eq. (2.3). If $\mathbf{S}$ is non-zero the size of the correction depends on the relative magnitudes of $\mathbf{S}$ and $\mathbf{P}'$.

It is pointed out by Deutsch (1965) that the Kalman and Bucy formalism gives the same result as earlier sequential estimators and that their main contribution is in the rigour and generality of their method. Similar results can be obtained using a Bayes estimator and by using classical least squares (see also Tzafestas, 1978; Jazwinski, 1970). In the latter case one chooses $\mathbf{m}$ to minimise,

$$J = (\mathbf{z} - \mathbf{H}^T\mathbf{m})^T\mathbf{W}(\mathbf{z} - \mathbf{H}^T\mathbf{m}). \tag{2.11}$$

If $\mathbf{W}^{-1}$ is taken to equal the error covariance matrix of the measurements and the initial state is unknown ($\mathbf{P}$ is infinite) then as shown by Sorenson (1966) the solution for $\mathbf{m}$ is the same as in Eq. (2.6). One can also show that the final result of applying Eq. (2.6) is the same whether the observations are projected concurrently or sequentially.

The connection between the Kalman and least-squares methods may be illustrated using probability functions. If just prior to assimilating data the Kalman method gives a best estimate $\mathbf{m}'$ and covariance $\mathbf{P}'$, then the probability distribution is,

$$p(\mathbf{m}) \propto \exp\left(-0.5(\mathbf{m} - \mathbf{m}')(\mathbf{P}')^{-1}(\mathbf{m} - \mathbf{m}')\right).$$

The observations also define a probability distribution for the state vector $\mathbf{m}$.

Consider a set of measurements $\mathbf{z}_{\mathrm{obs}}$ for which the error covariance matrix is $\mathbf{S}$. Then if $\mathbf{m}_0$ is any state of the system such that,

$$\mathbf{z}_{\mathrm{obs}} = \mathbf{H}^T \mathbf{m}_0,$$

then,

$$p(\mathbf{m}) \propto \exp\left(-0.5\left(\mathbf{H}^T(\mathbf{m} - \mathbf{m}_0)\right)^T \mathbf{S}^{-1}\left(\mathbf{H}^T(\mathbf{m} - \mathbf{m}_0)\right)\right),$$
$$= \exp\left(-0.5(\mathbf{m} - \mathbf{m}_0)^T(\mathbf{H}\mathbf{S}^{-1}\mathbf{H}^T)(\mathbf{m} - \mathbf{m}_0)\right).$$

The two sources of information are combined by forming the product of the probability functions. Thus,

$$p(\mathbf{m}) \propto \exp\left(-0.5(\mathbf{m} - \mathbf{m}')^T(\mathbf{P}')^{-1}(\mathbf{m} - \mathbf{m}')\right.$$
$$\left. -0.5(\mathbf{m} - \mathbf{m}_0)^T(\mathbf{H}\mathbf{S}^{-1}\mathbf{H}^T)(\mathbf{m} - \mathbf{m}_0)\right).$$

But this must equal the result obtained by the Kalman method,

$$p(\mathbf{m}) \propto \exp\left(-0.5(\mathbf{m} - \mathbf{m}^+)(\mathbf{P}^+)^{-1}(\mathbf{m} - \mathbf{m}^+)\right). \tag{2.12}$$

Note that the exponents in the above equations are all quadratic. As a result, if in the least squares method, the risk functions associated with the initial state and the measurements are taken to be,

$$(\mathbf{m} - \mathbf{m}')^T(\mathbf{P}')^{-1}(\mathbf{m} - \mathbf{m}'),$$

and,

$$(\mathbf{m} - \mathbf{m}_0)^T(\mathbf{H}\mathbf{S}^{-1}\mathbf{H}^T)(\mathbf{m} - \mathbf{m}_0),$$

then the joint risk function is,

$$(\mathbf{m} - \mathbf{m}')^T(\mathbf{P}')^{-1}(\mathbf{m} - \mathbf{m}') + (\mathbf{m} - \mathbf{m}_0)^T(\mathbf{H}\mathbf{S}^{-1}\mathbf{H}^T)(\mathbf{m} - \mathbf{m}_0)$$
$$= \mathrm{const} + (\mathbf{m} - \mathbf{m}^+)^T(\mathbf{P}^+)^{-1}(\mathbf{m} - \mathbf{m}^+). \tag{2.13}$$

The shape of the new risk function thus contains the same information as the vector $\mathbf{m}^+$ and matrix $\mathbf{P}^+$ of the Kalman method.

2.2 Non-Linear Systems

So far we have made the important assumption that the system is linear. Strictly speaking the methods can only be used with non-linear systems if a good first approximation is available which can be used as the basis of a linear perturbation expansion (Sorenson, 1966; Jazwinski, 1970; Maybeck, 1979, 1982; Anderson and Moore, 1979; Ghil *et al.*, 1981; Heemink, 1986).

Let m_0 be the initial state based on a *a priori* information and m_b the best estimate. Initially it will equal m_0, but later the calculation may be repeated with a revised value. For the linearised system the system variables are $(m_0 - m_b)$ and the Green's function in Eqs. (2.5) and (2.7) is that obtained by linearising about m_b.

The method first integrates m_b forwards in time using the full non-linear equations of motion. The linearised equations are then used to integrate forward the system variables and covariance matrices with new data being assimilated as before.

After this, two strategies are possible depending on how the assimilation is progressing. In many meteorological schemes where the synoptic data set is itself a good one and the errors in m are never very large, the corrections can be added to the base state and the process repeated. Thus the model is always integrated forward in time.

However when less data is available, as with the ocean, one needs to iterate the assimilation process using a forward-backward scheme (Morel *et al.*, 1971; Bengtsson, 1975). In this, after integrating forward in time, assimilating data, the linearised model is integrated backwards to give the updated value of m at the initial time. This is then used as the basis for the non-linear equation in the second iteration. Note that in order to make the correct use of the *a priori* information, the same initial state m_0 and covariance P_0 should be used at the start of each iteration and only the basis of the non-linear calculation changed (Rodgers, 1976).

Unfortunately there are at least four major problems in using such a scheme with ocean models. The first is the difficulty of finding a good basis state for the non-linear model. Experience has shown that the increase in the error vector due to a poor initial choice can soon swamp any improvement from assimilation of data.

A second problem arises with the highly damped short wavelength modes of the model. While integrating forward in time these are well behaved and any energy fed into them is soon dissipated. However while integrating backward, any small error grows rapidly and may swamp the solution. As a result the new basis state can be worse than the one available initially.

A third problem is that the solution to the full non-linear equation has to be stored while the model is being integrated forward in time for use later when integrating the perturbation equation backward in time. If this is by-passed by using the non-linear equations to integrate the updated solution backward in time, then the amplitude of the frictionally damped modes will grow even further.

Finally there is a problem in storing and updating the covariance matrix $\mathbf{P}$. With box models or other models with less than say one thousand degrees of freedom, it may be possible. However, with primitive equation models, which have typically over a million variables, it is impractical to construct or process such a matrix.

2.3 Alternative Approaches

Although the Kalman method has many advantages, the problems with highly damped waves and the impracticality of the large covariance matrices have stimulated other approaches. The simplest such scheme that might be considered is a shooting method in which an initial state $\mathbf{y}_0$ is chosen which is then integrated forward in time and used to calculate a cost function $J(\mathbf{y}_0)$. For a quadratic cost function,

$$J(\mathbf{y}_0) = \sum_i \tfrac{1}{2} \left(\mathbf{z}_i - \mathbf{H}_i^T \mathbf{m}(t_i) \right)^T \mathbf{W}_i \left(\mathbf{z}_i - \mathbf{H}_i^T \mathbf{m}(t_i) \right), \qquad (2.14)$$

where the sum is over all observations i. (A better alternative might be to use a modulus function because it does not give such weight to outliers). As before $\mathbf{W}^{-1}$ can be chosen to represent the error covariance matrix of the observations and a function dependent on the distance from the initial estimate can also be included. After the first pass through the data a new initial state is chosen and the scheme iterated, using a suitable descent algorithm, to minimise the value of J (Gill $et\ al.$, 1982; Hoffman, 1986). The scheme may be slow to converge because information on the gradient of the cost function is only built up slowly.

The scheme can be extended by using the methods of variational analysis with weak and strong constraints (Thompson, 1969; Sasaki, 1969, 1970; Sasaki $et\ al.$, 1986). However this is cumbersome for even simple problems (Lewis and Panetta, 1983) and with realistic problems drastic approximations are required (Bloom, 1983).

2.4 Adjoint System

If the system is linear then the model error at each observation can be written in terms of the initial state as,

$$
\begin{aligned}
\mathbf{z}_i - \mathbf{H}_i^T \mathbf{m}(t_i) &= \mathbf{z}_i - \mathbf{H}_i^T \mathbf{G}(t_i, t_0) \mathbf{y}_0, \\
&= \mathbf{z}_i - \left(\mathbf{G}(t_0, t_i)^T \mathbf{H}_i \right)^T \mathbf{y}_0, \\
&= \mathbf{z}_i - \mathbf{H}_i'^T \mathbf{y}_0.
\end{aligned}
\tag{2.15}
$$

Instead of the Green's function $\mathbf{G}$ transforming the initial state $\mathbf{y}_0$ into $\mathbf{m}$, its value at time t_1, the adjoint operator $\mathbf{G}^T$ is used to transform the measurement matrix into an equivalent matrix at time t_0 which keeps the product constant. If there is friction in the system, this requires that any reduction in amplitude produced by $\mathbf{G}$ in the forward direction must be matched by a similar reduction produced by $\mathbf{G}^T$ in the backward direction. If the system is oscillatory, then any increase in phase produced by $\mathbf{G}$ in the forward direction must be matched by $\mathbf{G}^T$ producing a decrease when going backward. Although this change in formulating the problem is a simple one, it opens up a number of new possibilities for both linear and non-linear systems. Methods based on the use of $\mathbf{G}^T$ instead of $\mathbf{G}$ are called adjoint methods.

Having transformed all the observations to a single time, almost any inverse method can be used to solve Eq. (2.15) (Tarantola, 1987; Wunsch, this volume). However it is usually solved by transforming it into a least squares problem.

The key operation with adjoint schemes is the transformation $\mathbf{G}^T \mathbf{H}_i$. This can be carried out using the adjoint of the model equations in the same way that the transformation $\mathbf{G} \mathbf{m}_0$ is carried out by integrating the normal linearised equations forward in time starting from the initial state $\mathbf{m}_0$. Methods for constructing the adjoint equations are given by Morse and Feshbach (1953). (See also Courant and Hilbert, 1952; Garrett and Greenberg, 1977; Talagrand and Courtier, 1987).

The normal form of the shallow water equations are,

$$
\begin{aligned}
\frac{\partial u}{\partial t} - fv &= -g \frac{\partial \zeta}{\partial x} - ku, \\
\frac{\partial v}{\partial t} + fu &= -g \frac{\partial \zeta}{\partial y} - kv, \\
\frac{\partial h}{\partial t} + \frac{\partial (hu)}{\partial x} + \frac{\partial (hv)}{\partial y} &= 0.
\end{aligned}
\tag{2.16}
$$

The adjoint variables and the adjoint equations depend on the way the dot product is defined. For the shallow water equations the best choice is (Garrett and Greenberg, 1977),

$$\mathbf{a}_1^T \mathbf{a}_2 = \int \left(\frac{\rho}{2}\right)(g\zeta_1\zeta_2 + hu_1u_2 + hv_1v_2)dxdy, \qquad (2.17)$$

where,

$$\mathbf{a}_i = \begin{pmatrix} \zeta_i \\ u_i \\ v_i \end{pmatrix}.$$

The corresponding adjoint equations are then,

$$\frac{\partial u}{\partial t} - fv = -g\frac{\partial \zeta}{\partial x} + ku,$$

$$\frac{\partial v}{\partial t} + fu = -g\frac{\partial \zeta}{\partial y} + kv,$$

$$\frac{\partial h}{\partial t} + \frac{\partial(hu)}{\partial x} + \frac{\partial(hv)}{\partial y} = 0. \qquad (2.18)$$

These are the same as the forward equations except for the change in sign of the friction term.

The advantage of the adjoint scheme is that whereas with the Kalman non-linear scheme the correction terms increased exponentially when integrated backwards in time, the corresponding adjoint variables, the components of $\mathbf{H}$, decrease exponentially. These still contain the information on how the initial state should be modified but they are not corrupted by the effect of amplified noise.

An iterative version of the scheme can be developed, just as for the Kalman equations, by linearising about an initial trial state. Again the full non-linear equations need to be integrated forward in time and the solution stored to enable the adjoint equations to be later integrated backwards in time.

For the main part of the calculation there are two possibilities. The first is to separately integrate each of the matrices backwards to time t_0, using the adjoint equations. The minimum of the cost function (Eq. 2.14) can then be calculated using standard techniques. The correction is added to the base state and the method iterated until the solution has converged.

Integrating the vector components of $\mathbf{H}_i$ backwards in time individually is computationally expensive and so an alternative method is usually used. Differentiating

Eq. (2.14) gives (Lewis and Derber, 1985; Talagrand and Courtier, 1987),

$$\nabla J = \sum_i \left(\mathbf{z}_i - \mathbf{H}_i^T \mathbf{m}(t_i) \right)^T \mathbf{W}_i \mathbf{G}(t_0, t_i)^T \mathbf{H}_i,$$
$$= \sum_i \mathbf{G}(t_0, t_i)^T \nabla J_i. \tag{2.19}$$

This is linear in the terms ∇J_i. The second method therefore starts from the last observation and integrates ∇J_i backward in time. As the other observations are reached each is added in turn until, at time t_0, the full value of ∇J is obtained. This is then used with a minimisation algorithm using gradients (Gill *et al.*, 1982; Press *et al.*, 1986; Powell, 1977) to improve the solution. Again the scheme is iterated.

The second method has the advantage that it requires less computation and storage during each iteration and for the first few iterations the reduction in the error variance will usually be large. However, as the vector in Eq. (2.19) gives only the local gradient of the cost function it needs more iterations than the first method. In particular if the system is linear the first method will converge in one iteration (given enough data), whereas the second will take a number of iterations equal to the number of degrees of freedom of the system. Both methods suffer, as does the Kalman scheme, from the problem of storing the solution of the forward integration in time of the full non-linear equations.

3. APPLICATIONS

3.1 Kalman Filters

Because of the problems of working with the large covariance matrices, applications of the Kalman method have concentrated on simplified systems. In the meteorological literature the scheme has been discussed by Jones (1965), Petersen (1968, 1973) and Ghil *et al.* (1981). Amongst oceanographers, Heemink (1986) applied the method to the one and two dimensional forms of the shallow water equations and used them to assimilate data in regions of the southern North Sea and the Dutch coast. For the 2-D case, the size of the calculation was reduced by using the asymptotic form of the gain matrix.

Applications to models of the deep ocean have been made by Miller (1986) and Bennett and Budgell (1987). Miller applied the Kalman filter to a one dimensional quasi-geostrophic model and tested the method using a number of sampling schemes. He also investigated the asymptopic form of the gain matrix and found, as did Ghil *et al.* (1981), that it rapidly converged to its final form.

Bennett and Budgell (1987) also studied the asymptotic form of the gain matrix using one and two-dimensional models. They worked primarily with the Kalman-Bucy form of the equations, in which measurements are made continuously, and investigated how accurately one needs to represent the gain matrix.

Other oceanographic related studies using a Kalman filter include Barbieri and Schopf (1982), Parrish and Cohn (1985), Budgell (1986) and Husain (1985).

3.2 Objective Analysis

Because of the problem of integrating the large covariance matrices forward in time, practical meteorological forecasting has used objective analysis (Gandin, 1965; Rutherford, 1972). This scheme sidesteps the time integration problem by calculating a gain matrix based on the long term statistics of the system. It should be equivalent to the limit of the Kalman scheme when regular observations are made over a long period of time.

In objective analysis, if $\mathbf{m}'$ is the latest model vector calculated as in Eq. (2.5), new data is assimilated using the equation,

$$\mathbf{m}_j = \mathbf{m}'_j + a_{ji}(\mathbf{m}_i^o - \mathbf{m}'_i), \qquad (3.1)$$

where m_i^o is the observed value at position i. The constants a_{ji} are calculated by minimising the mean square error of interpolation (Bengtsson, 1975), as in Eq. (2.4), but now using 'known' values of the error covariance of the fields m and the observations m^o.

In practice (Gustavsson, 1981), objective analysis schemes are usually simplified further, by assuming that the covariance matrix can be written as a product of two functions, one dependent on the height difference of the points and the other on their horizontal separation. For the horizontal function the most common choice is a Gaussian exponential.

An alternative scheme sometimes used, is a weighting function of the form,

$$a_{ji} = \frac{\mu(r_{ji})h(\rho_i)}{b + \sum_k \mu(r_{jk})h(\rho_k)},$$

where μ is a function of the distance r_{ji} from the observation to the model point, h is a function of ρ, the density of observation points and b is the weight given to the initial model field. However, as illustrated by Gustavsson (1981), all such approximations decrease the accuracy of the assimilation scheme.

A further problem with these and the Kalman schemes, and one which has produced a considerable literature (i.e. Bengtsson, 1975), arises from the way they project from the initial model point towards the surface defined by the data.

With a primitive equation model, such a projection will on average put as much energy into each of the two inertia-gravity wave modes at each wavenumber as it does into the Rossby wave modes. This extra gravity wave energy causes problems and has to be removed.

One method used to do this is to repeatedly assimilate the same data into the model over a period of time. The contributions from the high frequency gravity waves tend to cancel whereas the contribution from the low frequency Rossby waves is enhanced. Another approach is to filter the gravity waves using a bounded derivative scheme (Kreiss, 1979, 1980; Browning et al., 1980; Bube and Ghil, 1981), non-linear normal modes (Baer and Tribba, 1977; Machenhauer, 1977; Daley, 1980) or the 'slow' manifold (Leith, 1980). Examples of the use of these models with forecast models are given by McPherson et al. (1979), Lorenc (1981) and Hollingsworth et al. (1985).

3.3 Oceanographic Applications

Marshall (1985a) applied the method of objective analysis to the problem of determining the ocean circulation from satellite altimeter measurements in the presence of an imperfectly known geoid. Assume that at time t the altimeter measures an elevation z_0 with error estimate S_0. Let h'_g be the initial estimate of the geoid with horizontal covariance S'_g and h'_c the initial estimate of the ocean surface with covariance S'_c.

Marshall then introduces 'measured' values,

$$z_g = z_0 - Ah'_c, \qquad z_c = z_0 - Ah'_g, \qquad (3.2)$$

with variances,

$$S_{gz} = AS'_g A^T + S_0, \qquad S_{cz} = AS'_a A^T + S_0.$$

Minimising the expected mean square error, he shows that improved values are,

$$h_g = h'_g - K_g(Ah'_g - z_g), \qquad h_c = h'_c - K_c(Ah'_c - z_c),$$

$$S_g = (1 - K_g A)S'_g, \qquad S_c = (1 - K_c A)S'_c, \qquad (3.3)$$

where,

$$K_g = S'_g A^T (AS'_g A^T + S_{gz})^{-1}, \qquad K_c = S'_c A^T (AS'_c A^T + S_{cz})^{-1}.$$

In testing this scheme Marshall assumes that the errors in the geoid have a white noise spectrum, so S_g is a constant, and that $S_c(r)$ is proportional to $\exp(-r^2/2b^2)$ where b is the scale of the ocean features and is of order 100 km. A quasi-geostrophic numerical model is used to step the solution forward from one observation period to the next.

This scheme should be able to distinguish the time invariant geoid from time varying oceanic features with the same scale. However, the only way that the time invariant part of the flow can be determined is through its advection of and interaction with the time dependent part of the flow, and it is not obvious that the assimilation scheme will do this effectively. The method has also been used to study the efficiency of different altimeter sampling strategies (Marshall, 1985b).

An extensive discussion of methods for approximating and estimating the covariance matrix are given in Timchenko (1984). Unfortunately the translation is poor but the book also contains practical examples of the application of objective analysis to oceanographic problems in the Gulf of Lyons, the North-west Atlantic and the Polymode region. Objective analysis may also be used to analyse synoptic data without the use of models (Bretherton *et al.* , 1976; Freeland and Gould, 1976) and the data from different times combined at a later stage (Miyakoda and Talagrand, 1971; Kindle, 1986). Other examples of the use of objective analysis are given by Wunsch (this volume).

3.4 Projection Schemes

Because of the complexity of both the Kalman and objective analysis schemes, a number of authors have used schemes in which data is assimilated by making a simple correction to the model fields (Talagrand, 1981a, 1981b; Anderson and Moore, 1986; Hurlburt, 1986; Malanotte-Rizzoli and Holland, 1986; Thompson, 1986; Moore *et al.* , 1987; DeMey and Robinson, 1987; Holland, this volume). As with the previous assimilation schemes these methods can be represented by a projection, but now the direction of the projection is chosen *a priori*. They also have the advantage that it is simple to analyse their efficiency (Williamson and Dickinson, 1972; Talagrand, 1981c; Webb and Moore, 1986).

Let $\mathbf{q}$ be the projection vector corresponding to an observation z. The equation corresponding to (2.6) is then,

$$\mathbf{m}^{+} = \mathbf{m}' - \mathbf{q}(\mathbf{q}^{T}\mathbf{h})^{-1}(\mathbf{h}^{T}\mathbf{m}' - z).$$

If $\mathbf{m}'$ equals $\mathbf{G}\mathbf{m}^{-}$, where $\mathbf{m}^{-}$ is the model state at the end of the previous timestep, the change in the error $(\mathbf{r} - \mathbf{m})$ over one cycle of the assimilation scheme is given by,

$$\mathbf{E}^{+} = \left(1 - \mathbf{q}(\mathbf{q}^{T}\mathbf{h})^{-1}\mathbf{h}^{T}\right)\mathbf{G}\mathbf{E}^{-}.$$

The convergence properties of the scheme can be studied by calculating the eigenvalues of the operator on the right of the equation. For the scheme to converge the eigenvalues must have moduli less than one and the best scheme will be the one with the smallest maximum eigenvalue.

In this way, it can be shown that the scheme is unstable if, for any of the modes, $\mathbf{q}$ and $\mathbf{h}$ are of opposite sign. Thus the change in each mode must always be in the direction which reduces the error. It means for example, that in objective analysis, a gaussian weighting function, whose fourier transform is positive definite, is stable whereas a top hat function is not.

If two modes, otherwise indistinguishable, have similar frequencies w_i and w_j, the corresponding eigenvalue modulus is

$$|\lambda| = 1 - \text{const}(w_i - w_j)^2 dt^2,$$

where dt is the time interval between assimilating new data. From this it is straight-forward to show that the simple projection scheme is more efficient when dt is large than when it is small. This is because in the former case a large phase separation develops between the modes during each assimilation interval.

The above analysis can be extended to the corresponding forward-backward scheme. This is found to be unstable if there are two otherwise indistinguishable modes whose difference in decay constants is greater than their difference in angular velocity. If one was assimilating satellite altimeter data, which can distinguish different horizontal wavenumbers but not different vertical structure, one might find the scheme going unstable with the highly damped short wavelength modes.

It should be emphasised that these results are only for simple projection schemes in which there are no observational errors and no non-linear effects. If non-linearities are included the efficiency of the scheme should be reduced, because the vector r is always being moved away from the point to which the assimilation scheme is converging. However, Talagrand (1981c) working with a simple projection scheme and Ghil *et al.* (1981) who used the Kalman scheme, both reported improved convergence with non-linearities. In the latter case convergence was helped by the large scale flow advecting information obtained over the land out to sea.

One of the main results of the above analysis is the importance of the phase separation that develops between modes over each assimilation interval. Because of this tests of assimilation schemes using models with only the first two vertical modes, such as Hurlbert (1986) and Thompson (1986), where the barotropic and first baroclinic modes have well separated angular velocities, should be reasonably straightforward. A more stringent test is the use of models with three or more modes, because the higher baroclinic modes all have similar small angular velocities. Such tests have been made by Malanotte-Rizzoli and Holland (1986) and DeMey and Robinson (1987) and were succesful.

3.5 Adjoint Schemes

Adjoint methods appear to have first been introduced into meteorology by Peneko and Obraztsov (1976), but their extensive use with numerical models has only occurred recently with the work of Lewis and Derber (1985) and Talagrand and Courtier (1987). (See also LeDimet, 1980; LeDimet and Talgrand, 1986; and Lorenc, 1988).

Lewis and Derber considered both a simple one-dimensional advection equation problem and a two level quasi-geostrophic model. With the one-dimensional problem it was found that the cost function was reduced rapidly during the first few iterations but after that the improvement was very slow. Tests were made using upper air data measured from rawinsondes and by satellite during a two day period over the central USA and satisfactory results obtained.

Talagrand and Courtier (1987) carried out a test study on an atmospheric Haurowitz wave using a quasi-geostrophic model and a spherical harmonic expansion of the fields. As with the study of Lewis *et al.*, the conjugate gradient method was found to be more effective at minimising the cost function than steepest decent, reducing the cost function by about two orders of magnitude on each iteration. Courtier and Talagrand (1987) extended the work to assimilate all 500mb geopotential and wind observations from the northern hemisphere over a 24-hour period.

They found that during the first ten to fifteen iterations of the scheme a realistic flow pattern developed but that the main effect of further iterations was to introduce short wavelength noise which fitted the data in data intensive regions but produced spurious oscillations in data sparse regions. To overcome this they proposed that an extra cost function be introduced to reduce the development of energy in such modes.

4. CONCLUSIONS

As we have seen, the basic symmetry in many of the methods that have been discussed arises because finding the maximum of a gaussian probability function is equivalent to finding the minimum of a quadratic cost function. Methods based on estimation theory and the probability functions have the advantage that they can incorporate the effect of model errors but the disadvantage first that the covariance matrices are very large and almost impossible to handle and secondly that problems are encountered during the backward phase of iterative assimilation schemes.

Because of the latter problem, interest recently has focussed on adjoint methods which integrate backwards in time not the best estimate of the ocean state but the gradient of the cost function. The method is still computationally expensive and it has a serious problem in that in its present form it cannot allow for the errors in the numerical model used. It also needs substantial storage for the forward solution of the model which is used to integrate the adjoint equations backward in time. A final problem is that in the usual scheme used, the vector describing the gradient of the cost function does not point towards its absolute minimum, so in principal many iterations of the scheme may be needed.

Estimates of the efficiency of the different methods are for the moment mostly based on qualitative statements like the ones above. Some work has been carried out on the efficiency of the Kalman and simple projection schemes but more detailed analysis of both efficiency and robustness is required of all the methods discussed.

There are many possibilities for modifying the assimilation schemes to improve their efficiency and it may be found that the best schemes for oceanic data assimilation will differ from those that are best for the atmospheric models and data sets. In atmospheric data assimilation the best tests and stimulus for further development of the schemes arose from their routine use in forecast models. In the ocean we now need similar large scale tests.

REFERENCES

Anderson, B.D.O. and J.B. Moore (1979): *Objective Filtering*. Prentice-Hall, Englewood Cliffs, N.J.

Anderson, D.L.T. and A.M. Moore (1986): 'Data Assimilation'. Pp.437–464 in: *Advanced Physical Oceanographic Numerical Modelling*. (Editor: J.J. O'Brien). Dordrecht, D. Reidel, 608pp.

Baer, F. and J.J. Tribba (1977): On Complete Filtering of Gravity Wave Modes through Nonlinear Initialization. *Monthly Weather Review*, **105**, 1536–1539.

Barbieri, R.W., and P.S. Schopf (1982): *Oceanic applications of the Kalman filter*. NASA/Goddard Technical Memo TM83993,26pp.

Battin, R.H. (1962): 'A Statistical Optimizing Navigation Procedure for Space Flight'. *American Rocket Society Journal*, **32**, 1681–1696.

Bengtsson, L. (1975): *4-Dimensional assimilation of Meteorological Observations.* GARP Publication Series No. 15. World Meteorological Organisation and International Council of Scientific Unions, 75pp..

Bengtsson, L., M. Ghil and E. Kallen (1981): *Dynamic Meteorology: Data Assimilation Methods.* Springer-Verlag, New York, 330pp.

Bennett, A.F. and W.P. Budgell (1987): 'Ocean Data Assimilation and the Kalman Filter: Spatial Regularity'. *Journal of Physical Oceanography,* **17**(10), 1583–1601.

Bloom, S.C. (1983): 'The use of Dynamical Constraints in the use of Mesoscale Rawinsonde Data'. *Tellus,* **35A**, 363–378.

Bretherton, F.P., R.E. Davis and C.B. Fandry (1976): 'A Technique for Objective Analysis and Design of Oceanographic Experiments Applied to MODE-73'. *Deep-Sea Research,* **23**, 558–582.

Browning, G., A. Kasahara and H.O. Kreiss (1980): 'Initialisation of the Primitive Equations by the Bounded Derivative Method'. *Journal of Atmospheric Science,* **37**, 1424–1436.

Bube, K.P. and M. Ghil (1981): 'Assimilation of Asynoptic Data and the Initialisation Problem' Pp 111–138 in: *Dynamic Meteorology: Data Assimilation Methods.* (Editors: Bengtsson, L., M. Ghil and E. Kallen). Springer-Verlag, New York, 330pp.

Budgell, W.P. (1986): 'Nonlinear Data Assimilation for Shallow Water Equations in Branched Channels'. *Journal of Geophysical Research,* **91**(C9), 10633–10644.

Carlton, A.G. (1962): 'Linear Estimation in Stochastic Processes'. The John-Hopkins University, Applied Physics Laboratory, *Bumblebee Series Report 311,* Baltimore, Maryland.

Courant, R. and D. Hilbert (1962): *Methods of Mathematical Physics. Vols I and II.* Interscience. 830pp.

Courtier, P. and O. Talagrand (1987): 'Variational Assimilation of Meteorological Observations with the Adjoint Vorticity Equation. II: Numerical Results'. *Quarterly Journal of the Royal Meteorological Society,* **113**(478), 1329–1347.

Daley, R. (1981): 'Normal Mode Initialisation'. Pp 77–110 in: *Dynamic Meteorology: Data Assimilation Methods.* (Editors: Bengtsson, L., M. Ghil and E. Kallen). Springer-Verlag, New York, 330pp.

DeMey, P. and A.R. Robinson (1987): 'Assimilation of Altimeter Eddy Fields in a Limited-Area Quasi-Geostrophic Model'. *Journal of Physical Oceanography,* **17**(12), 2280–2293.

Deutsch, R. (1965): *Estimation Theory.* Prentice-Hall, Englewood Cliffs, N.J., 269pp.

Freeland, H.J. and W.J. Gould (1976): 'Objective Analysis of Meso-Scale Ocean Circulation Features'. *Deep-Sea Research,* **23**, 915–923.

Gandin, L.S. (1965): *Objective Analysis of Meteorological Fields.* Israel Program for Scientific Translations, Jerusalem, 242pp.

Garrett, C. and D. Greenberg (1977): 'Predicting Changes in the Tidal Regime: The Open Boundary Problem'. *Journal of Physical Oceanography,* **7**(2), 171–181.

Gauss, K.F. (1809): it Theory of the Motion of the Heavenly Bodies Moving about the Sun in Conic Sections. rm Dover, New York. [Republished 1963]

Ghil, M., S.Cohn, J.Tavantzis, K.Bube and E.Isaacson (1981): 'Applications of Estimation Theory to Numerical Weather Prediction'. Pp 139–224 in: *Dynamic Meteorology: Data Assimilation Methods.* (Editors: Bengtsson, L., M. Ghil and E. Kallen). Springer-Verlag, New York, 330pp.

Gill, P.E., W. Murray and M.H. Wright (1982): *Practical Optimisation.* Academic Press, London.

Gustavsson, N. (1981): 'A Review of Methods for Objective Analysis'. Pp 17–76 in: *Dynamic Meteorology: Data Assimilation Methods.* (Editors: Bengtsson, L., M. Ghil, and E. Kallen). Springer-Verlag, New York, 330pp.

Heemink, A.W. (1986): 'Storm Surge Prediction using Kalman Filtering'. *Rijkswaterstaat Communications,* **46**,194pp.

Hoffmann, R.N. (1986): 'A Four-Dimensional Analysis Exactly Satisfying the Equations of Motion'. *Monthly Weather Review,* **114**,388–397.

Hollingsworth, A., A.C. Lorenc, M.S. Tracton, K. Arpe, G. Cats, S. Uppala and P. Kallberg (1985): 'The Response of Numerical Weather Prediction Systems to FGGE Level IIb Data. Part I: Analyses. *Quarterly Journal of the Royal Meteorological Society,* **111**,1–66.

Hurlburt, H.E. (1986): 'Dynamic Transfer of Simulated Altimeter Data into Subsurface Information by a Numerical Ocean Model'. *Journal of Geophysical Research,* **91**(C2), 2372–2400.

Husain, T. (1985): 'Kalman Filter Estimation in Flood Forecasting'. *Advances in Water Resources,* **8**(1), 15–21.

Jazwinski, A.H. (1970): *Stochastic Processes and Filtering Theory.* Academic Press, New York.

Jones, R.H. (1965): 'Optimal Estimation of Initial Conditions for Numerical Prediction'. *Journal of Atmospheric Sciences,* **22**, 658–663.

Kalman, R.E. (1960): 'A New Approach to Linear Filtering and Prediction Problems'. *Journal of Basic Engineering,* Transactions ASME, **82D**, 33–45.

Kalman , R.E. and R.S. Bucy (1961): 'New Results in Linear Filtering and Prediction Theory'. *Journal of Basic Engineering,* Transactions ASME, **83D**, 95–108.

Kindle, J.C. (1986): 'Sampling Strategies and Model Assimilation of Altimeter Data for Ocean Monitoring and Prediction'. *Journal of Geophysical Research,* **91**(C2), 24181–2432.

Kreiss, H.O. (1978): 'Problems with Different Time Scales'. Pp 95–106 in: *Recent Advances in Numerical Analysis.* (Editors: C. de Boor and G.H. Golub). Academic Press, New York.

Kwakernaak, H. and R. Sivan (1972): *Linear Objective Control Systems.* Wiley-Interscience.

LeDimet, F.X. (1982): *A General Formalism of Variational Analysis.* CIMMS Report 22, Cooperative Institute for Mesoscale Meteorological Studies, 815 Jenkins Street, Norman, Oklahoma, USA, 34pp.

LeDimet, F.X., and O. Talagrand (1986): 'Variational Algorithms for Analysis and Assimilation of Meteorological Observations: Theoretical Aspects'. *Tellus,* **38A**, 97–110.

Leith, C.E. (1980): 'Nonlinear Normal Mode Initialisation and Quasi-Geostrophic Theory'. *Journal of Atmospheric Sciences,* **37**, 958–968.

Lewis, J.M. (1982): 'Adaption of P.D. Thompson's Scheme to the Constraint of Potential Vorticity Conservation'. *Monthly Weather Review,* **110**, 1618–1634.

Lewis, J.M. and L. Panetta (1983): 'The Extension of P.D. Thompson's Scheme to Multiple Time Levels'. *Journal of Climatology and Applied Meteorology,* **22**, 1649–1653.

Lewis, J.M. and J.C. Derber (1985): 'The Use of Adjoint Equations to Solve a Variational Adjustment Problem with Convective Constraints'. *Tellus,* **37A**, 309–322.

Lorenc, A. (1981): 'A Global Three-Dimensional Multivariate Statistical Interpolation Scheme'. *Monthly Weather Review,* **109,** 701–721.

Lorenc, A. (1986): 'Analysis Methods for Numerical Weather Prediction'. *Quarterly Journal of the Royal Meteorological Society,* **112**(474), 1177–1194.

Lorenc, A. (1988): 'Optimal Nonlinear Objective Analysis'. *Quarterly Journal of the Royal Meteorological Society,* **114**(479), 205–240.

Machenhauer, B. (1977): 'On the Dynamics of Gravity Oscillations in a Shallow Water Model with Application to Normal Mode Initialization'. *Beitrage zur Physik der Atmosphare,* **50,**253–271.

Malanotte-Rizzoli, P. and W.R. Holland (1986): 'Data Constraints Applied to Models of the Ocean General Circulation. Part I: The Steady Case'. *Journal of Physical Oceanography,* **16**(10), 1665–1682.

McPherson, R.D., K.H. Bergman, R.E. Kistler, G.E. Rasch and D.S. Gordon (1979): 'The NMC Operational Global Data Assimilation System'. *Monthly Weather Review,* **107,** 1445–1461.

Marshall, J.C. (1985a): 'Determining the Ocean Circulation and Improving the Geoid from Satellite Altimetry'. *Journal of Physical Oceanography,* **15**(3), 330–349.

Marshall, J.C. (1985b): 'Altimetric Observing Simulation Studies with an Eddy Resolving Ocean Circulation Model'. pp.73–76 in *The Use of Satellite Data in Climate Models,* compiled by J.J.Hunt. ESA Scientific and Technical Publications Branch, Noordwijk, The Netherlands, 191pp.

Maybeck, P.S. (1979): *Stochastic Processes, Estimation and Control, Vol I.* Academic Press, New York.

Maybeck, P.S. (1982): *Stochastic Processes, Estimation and Control, Vol II.* Academic Press, New York.

Miller, R.N. (1986): 'Toward the Application of the Kalman Filter to Regional Open Ocean Modelling'. *Journal of Physical Oceanography,* **16,** 72–86.

Miyakoda, K. and O. Talagrand (1971): 'The Assimilation of Past Data in Dynamical Analysis'. *Tellus,* **23,** 310–317.

Moore, A.M., N.S. Cooper and D.L.T. Anderson (1987): 'Initialization and Data Assimilation in Models of the Indian Ocean'. *Journal of Physical Oceanography,* **17**(11), 1965–1977.

Morel, P., G. Lefevre and G. Rabreau (1971): 'On Initialisation and Non-Synoptic Data Assimilation'. *Tellus*, **23**, 197–206.

Morse, P.M. and H. Feshbach (1953): *Methods of Theoretical Physics: Volume I.* McGraw-Hill, New York, 997pp.

Parrish, D.F. and S.E. Cohn (1985): *A Kalman Filter for a Two-Dimensional Shallow Water Model: Formulation and Preliminary Experiments.* Office Note 304, NOAA/NMC, 64pp.

Penenko, V.V. and N.N. Obraztsov (1976): 'A Variational Initialization Method for the Fields of the Meteorological Elements' (English Translation). *Soviet Meteorology and Hydrology*, **1976** (11), 1–11.

Petersen, D.P. (1968): 'On the Concept and Implementation of Sequential Analysis for Linear Random Fields' *Tellus*, **20**, 673–686.

Petersen, D.P. (1973): A Comparison of the Performance of Quasi-optimal and Conventional Objective Analysis Schemes. *Journal of Applied Meteorology*, **12**, 1093–1101.

Powell, M.J.D. (1977): 'Restart Procedures for the Conjugate Gradient Method'. *Mathematical Programming*, **12**, 241–254.

Press, W.H., B.F. Flannery, S.A. Teukolsky and W.T. Vettering (1986): *Numerical Recipes:* The Art of Scientific Computing. Cambridge University Press, Cambridge, 818pp.

Rodgers, C.D. (1976): 'Retrieval of Atmospheric Temperature and Composition from Remote Measurements of Thermal Radiation'. *Reviews of Geophysics and Space Physics*, **14**, 609–624.

Rutherford, I.D. (1972): 'Data Assimilation by Statistical Interpolation of Forecast Error Fields'. *Journal of Atmospheric Sciences*, **29**, *809–815.*

Sasaki, Y. (1969): 'Proposed Inclusion of Time Variational Terms, Observational and Theoretical, in Numerical Variational Objective Analysis'. *Journal of the Meteorological Society of Japan*, **47**, 115–124.

Sasaki, Y. (1970): 'Some Basic Formalisms in Numerical Variational Analysis'. *Monthly Weather Review*, **98**, 875–883.

Sasaki, Y.K., T. Gal-Chen, L.White, M.M. Zaman and C. Ziegler (1986): *Variational Methods in Geosciences.* Elsivier, New York.

Sorenson, H.W. (1966): 'Kalman Filtering Techniques'. *Advances in Control Systems, Vol 3.* (Editor: C.T. Leondes). Academic Press, New York, 346pp.

Swerling, P. (1959): 'First Order Error Propagation in a Stage-Wise Smoothing Procedure for Satellite Observations', *Journal of Astronautical Science*, **6**, 46–52.

Talagrand, O. (1981a): 'A Study of the Dynamics of Four-Dimensional Data Assimilation'. *Tellus*, **33**, 43–60.

Talagrand, O. (1981b): 'On the Mathematics of Data Assimilation'. *Tellus*, **33**, 321–339.

Talagrand, O. (1981c): 'Convergence of Assimilation Procedures'. Pp 225–262 in: *Dynamic Meteorology: Data Assimilation Methods.* (Editors: Bengtsson, L., M. Ghil and E. Kallen). Springer-Verlag, New York, 330pp.

Talagrand, O. and P. Courtier (1987): 'Variational Assimilation of Meteorological Observations with the Adjoint Vorticity Equation. I: Theory'. *Quarterly Journal of the Royal Meteorological Society*, **113**(478), 1311–1328.

Tarantola, A. (1987): *Inverse Problem Theory.* Elsevier, Amsterdam, 613pp.

Thompson, J.D. (1986): 'Altimeter Data and Geoid Error in Mesoscale Ocean Prediction: Some Results from a Primitive Equation Model'. *Journal of Geophysical Research*, **91**(C2), 2401–2417.

Thompson, P.D. (1969): 'Reduction of Analysis Error through Constraints of Dynamical Consistency'. *Journal of Applied Meteorology*, **8**(5), 738–742.

Timchenko, I.E. (1984): *Stochastic Modelling of Ocean Dynamics.* Harwood Academic Publishers, 311pp. [Translated by E.T. Premuzic.]

Tzafestas, S.G. (1978): 'Distributed Parameter State Estimation'. In: *Distributed parameter systems: Identification, Estimation and Control*, Eds: W.H. Ray and D.G. Lainiotis. Marcel Dekker, 135–208.

Webb, D.J. and A. Moore (1986): 'Assimilation of Altimeter Data into Ocean Models'. *Journal of Physical Oceanography*, **16**, 1901–1913.

Williamson, D. and R. Dickinson (1972): 'Periodic Updating of Meteorological Variables'. *Journal of Atmospheric Sciences*, **29**, 190–193.

DRIVING OF NON-LINEAR TIME-DEPENDENT OCEAN MODELS BY OBSERVATION OF TRANSIENT TRACERS—A PROBLEM OF CONSTRAINED OPTIMISATION.

by
Jens Schröter
Alfred-Wegener-Institut
für Polar- und Meeresforschung, Bremerhaven

Abstract

The problem of "driving" the phase space trajectory of an ocean model such that it comes "close" to observations distributed in space and time is discussed. The approach taken is one of optimization where a distance function I is minimized while the model equations act as hard (i.e. equality-) constraints. In this case model trajectories are subject only to model physics, initial- and boundary conditions. As a consequence changes in the trajectories can only be achieved via control variables u that describe the conditions mentioned like external forcing driving the circulation or that describe model physics like friction and diffusion parameterization. By changing only control variables and letting the model evolve according to its equations of motion the constrained problem in the space of model and control variables is changed into an unconstrained problem in the space of control variables. The optimization is performed iteratively: first an initial guess of u is taken, then the distance function I is evaluated by integrating the ocean model for a time interval in which tracer (or other) measurements are available. The gradient of I with respect to u is calculated with the aid of the adjoint model which is integrated backwards for the same time interval. Subsequently an unconstrained minimization algorithm is used to perform the optimization in u-space.

Examples are presented for a non-linear shallow water model which advects a passive tracer. The effect of measurement frequency in space and time, measurement kind (model variables included or tracer only) and measurement noise is discussed including questions of sensitivity and resolution.

1 INTRODUCTION

Data assimilation has been studied for a long time in meteorology where it is essential to make good weather forecasts. In oceanography assimilation of data into

257

D. L. T. Anderson and J. Willebrand (eds.), Oceanic Circulation Models: Combining Data and Dynamics, 257–285.

models is only at its beginning. The possible availability of satellite derived measurements that will cover the globe in space and time urges us to use this technique when we want to make profit of all the detailed information satellites can provide.

The analysis and interpretation of oceanographical and meteorological data is generally done in conjunction with a model. This model can for instance be statistical as in optimal interpolation where a linear regression model based on *a priori* statistical information is used to fill the space between measurements with data. Dynamical information is used e.g. when meteorologists allow sharp gradients in the vicinity of a front while the rest of the field is required to be smooth.

Data assimilation is a process which combines information of measurements with the corresponding field of a dynamical model to give better estimates of the measurements and simultaneously improve the model state. The model is then integrated forward for some time until new measurements are available and the process is repeated. After a number of assimilation cycles all model variables may have felt the influence of the measurements and even the part of the model state that is not observed can be estimated by this method. Time independent flow problems can also be dealt with: The same data are assimilated each cycle until an equilibrium solution has been reached and changes in the model state from cycle to cycle are very small. The weight given to measurements relative to that of model values in the assimilation determines whether the model is closer to observations or to steady state.

In oceanography the sparseness of data both in space and time until now has allowed to make estimates of global fields only for mean values and mean variances and in some cases for average seasonal cycles. Interannual variations or variations on short time scales are regarded as noise for these estimates and their information content is discarded. Optimal interpolation of temperature and salinity can lead to vertically unstable density profiles which of course cannot exist for average fields. To inhibit errors like this we have to use more than statistical information for the process of interpolation. In addition to the statistical requirements we can force the interpolated field to obey contraints. These constraints are either satisfied in a statistical sense (e.g. a least squares solution) or they are required to hold exactly as equalities. The first kind of constraint is denoted as "weak" or "soft" while the latter kind is called a "strong" or "hard" constraint.

In this paper I force the interpolated data to be a solution of a non-linear time dependent ocean model. The temporal evolution of the model variables $s(x, t)$—

the model trajectory in phase space—is required to be as close to measurements as possible in the least squares sense while the model equations are the hard constraints of the system.

The resulting problem is one of state estimation or optimal control. We have to determine the values of the model variables as a function of space and time. For large models and a long time of interest this can result in a very high number of unknowns. However, since the variables are interconnected through the dynamical model equations the problem can be simplified considerably. By defining initial and boundary conditions as well as adjustable model parameters the trajectory of the model is fully specified for the whole time interval of interest. Diffusive or turbulent models have a correlation time scale τ after which measured data depend no longer deterministically on the initial values. The integration time should not exceed τ.

Initial and boundary conditions plus adjustable model parameters can be treated as the independent control variables u from which all model variables s can be calculated. It is then our problem to find the appropriate values for the u. To illustrate the solution process I will discuss the classical question of retrieving flow field information by observation of tracers. A single time dependent tracer is used in this study.

The estimation of flow fields and mixing rates from measured tracer distributions is not a straightforward problem (Wunsch, 1988b). Simple models have been used for this purpose with either low resolution like box models (e.g. Wunsch and Minster, 1982) or with local dynamics (e.g. Olbers et al. 1985). Mercier (1986) uses an iterative "total inversion algorithm" to improve the estimated tracer field simultaneously with solving for the flow. At the solution the flow is in equilibrium with the improved (and not with the measured) tracer field.

Transient tracers are treated in a number of articles by Wunsch (1987, 1988a, 1988b). However all papers mentioned use stationary flow fields. In this work I attempt to use time dependent tracer observations to estimate a time dependent flow field, and give an improved estimate of the measured time dependent tracer field. The method applied here is often called "adjoint technique." It lends itself directly to infer mixing rates and provides the necessary information on the error covariance and the resolution of the solution. The observational analysis is simple and straightforward.

Adjoint equations have been used as a tool for sensitivity analysis in meteorology for some time, e.g. Marchuk (1975a, 1975b), Kontarev (1980), Cacuci (1981), Hall and Cacuci (1983) and Hall (1986). The adjoint technique, as presented here, has been used mainly in meteorology, e.g. Lewis and Derber (1985), LeDimet and Talagrand (1986), Courtier and Talagrand (1987), Talagrand and Courtier (1987), but also in oceanography, e.g. Thacker and Long (1988), Tziperman and Thacker (1988).

A major difference to conventional methods is that the model is interpolated to the data (and not vice versa). This is important when the measurement is a highly non-linear function of the model variables like radiation temperatures in meteorology or when we have e.g. one single measurement of some special kind. The adjoint technique is then capable of using sparse, asynoptic, redundant and even inconsistent data (Thacker, 1987) to estimate the state of the model. Every bit of information of maybe totally different origin can be exploited much in the sense that Wunsch (1984, 1988a) uses for his eclectic modelling. Information of the interior can then determine what happens at the boundaries and even estimates of the model state at times considerably earlier than the observational period can be given. Likewise the model fields in regions that are very hard to observe (e.g. under the shelf ice) can be infered.

The control problem that underlies the assimilation problem is described in the next part followed by a brief description on how to derive the adjoint equations. A simple but highly non-linear model and its adjoint are presented in section 4 and a number of examples are given in section 5. How to solve for adjustable model parameters is shown in part 6 followed by a discussion on how to compute resolution and error matrices. Conclusions are made in the last section.

2   A CONTROL PROBLEM

The problem of finding the optimal model trajectory can be written as

$$\text{minimize } I \text{ subject to } E = 0 \tag{1}$$

The model equations E act as so called "hard" constraints that is they are not only satisfied in the least-squares-sense but exactly. Problems as (1) where the constraints E consist of a number of equailities and/or inequalities have been solved by Schröter and Wunsch (1986), Navon and DeVilliers (1986) using augmented Lagrangian functions. It is my experience however that for non-linear problems the computation time becomes prohibitively large when the number of unknowns

N exceeds the order of 10^3. For time dependent problems which we want to solve here the total number of unknowns in space and time exceeds this threshold by far, even for the simple examples given in this paper N is larger than $2 \cdot 10^4$.

Another approach to solve (1) is the "reduction of the control variable" (LeDimet and Talagrand, 1986): The model trajectory in phase space is completely defined when model physics, initial- and boundary conditions (in their discretized form) are specified. A simple model integration in time yields a possible solution to (1) so that the constraints $E = 0$ are satisfied by definition. This solution has to be tested for optimality by calculation of the derivatives of I with respect to the control variables u which is the part of the model physics, initial- and boundary conditions that we are willing to change. If all derivatives of I vanish at least a local extremum of I has been found and (1) is solved. If on the other hand the gradient of the cost function with respect to the control variables $\nabla_u I$ is non-zero an iterative search algorithm can be applied to find the minimum of I in the space of control variables u only. Such an algorithm which changes u consistently until $\nabla_u I$ is equal to zero will need a number of evaluations of I as a function of u and in general this requires each time a full model integration in the respective time interval from t_1 to t_n.

This approach has been taken by Schröter and Oberhuber (1988) in an attempt to find a global model for the mixed layer of the upper ocean. In that paper a highly non-linear general circulation model with an isentropic coordinate in the vertical and an embedded mixed layer is used. The mixed layer model has a number of adjustable numerical parameters such as a coefficient for the efficiency of converting mean energy of the wind into turbulent energy of the ocean, the penetration depth of the sunlight, the decay scales for turbulence within the mixed layer etc. A total of seven parameters has been used as control variables while atmospheric- and initial conditions are left unchanged. The cost function to be minimized is the integral over space and time of the distance (L_2-norm) between model mixed layer depth and a mixed layer depth calculated from climatological measurements. The model covers the North Atlantic and is integrated for the seasonal cycle.

The gradient of this distance I with respect to the control variables is calculated by "brute force" i.e. by perturbation methods. The attribute "brute force" relates to the fact that for one model integration for a seasonal cycle more than 1 hour of CPU-time on a CYBER205 is needed. To calculate all seven components of the gradient eight integrations have to be performed with a cost of $\sim 10\,\mathrm{h}$ of CPU-time. Once the gradient has been computed by this expensive method an unconstrained, general minimization algorithm is applied to reduce the value of the cost function. However, as the problem is highly non-linear, the gradient has to be calculated successively at a number of different points in u-space until an iterative solution has been achieved. By our "brute force" method a total of $O(100)$ hours of CPU-time is spent to find optimal values for 7 adjustable parameters.

It is clear from this example that for complicated ocean models and high numbers of control variables the calculation of $\nabla_u I$ by perturbation methods is too expensive to be practical. A possible solution for this problem is given in the following part.

3 AN ELEGANT AND EFFICIENT WAY TO CALCULATE THE GRADIENT OF THE COST FUNCTION

The computation of $\nabla_u I$ can be simplified greatly by the use of the adjoint model equations. These equations which are adjoint to the tangent linear or linear perturbation equations of the model can be derived by a number of different methods, e.g. Marchuck (1975a), Luenberger (1979), LeDimet and Talagrand (1986), Sewell (1987), Thacker (1987). Here a derivation will be given that shows how the adjoint equations relate the gradient of I not only to the initial conditions (the case that has been studied most) but also to any other control variable one might consider. Furthermore there is no need to specify the cost function I for deriving the adjoint equations. This specification becomes necessary only when the adjoint equations are actually integrated in time where the dependence of I on s (e.g. model-data misfit, non-smoothness of s etc.) acts as a forcing on the adjoint model. It should be stressed again that the application of the adjoint equations only eases the calculation of the gradient $\nabla_u I$. An iterative, unconstrained minimization algorithm has to be used subsequently as has been mentioned above.

Let us rewrite Eq. (1) by defining a Lagrangian function L and determining its stationary points

$$L(u, s, \lambda) = I + \int_X \int_{t_1}^{t_n} \lambda(x, t) \cdot E(u, s, x, t) \, dx \, dt \qquad (2)$$

The integral is to be taken between t_1 and t_n and over the entire model domain X. A stationary point of L is reached when the following three conditions hold:

$$\frac{\partial L}{\partial \lambda} = 0 \qquad (3)$$

$$\frac{\partial L}{\partial s} = 0 \tag{4}$$

$$\frac{\partial L}{\partial u} = 0 \tag{5}$$

These are the Euler-Lagrange equations of problem (1). A direct solution of these equations can be very inefficient and we will solve (3)–(5) iteratively following LeDimet and Talagrand (1986).

Condition (3) recovers of course our model: $E = 0$. Thus if the model variables s evolve according to the model the equation (3) is verified by definition. Eq. (4) is the well known evolution equation for the Lagrange multipliers λ. Practical procedures to derive this equation—which we identify with the adjoint equation—by partial integration are given in many textbooks (e.g. Sewell, 1987) and will not be repeated here. Thacker (1987) describes a method for finding the adjoint of a numerical computer code. In general, for every subroutine in a computer program an adjoint subroutine can be defined. The adjoint model then consists of applying the adjoint subroutines in the reverse order as compared to the forward model. The partial integration also provides us with the appropriate boundary (terminal) conditions for λ at t_n (Luenberger, 1979) and by integration of the adjoint equations backwards in time from t_n to t_1 condition (4) can be satisfied. In general the integration of the adjoint model will not lead to a λ at t_1 that satisfies the corresponding boundary condition. This problem together with Eq. (5) is solved iteratively. A general unconstrained minimization algorithm is applied which iteratively finds the minimum of a function $I(u)$ when for all u the value of I and its gradient $\nabla_u I$ can be provided. In the minimum of I the gradient is zero and Eq. (5) is verified.

The solution of the adjoint equations can be used in an elegant and efficient way to calculate $\nabla_u I$: By construction the value of the Lagrange function $L(u)$ is always identical to that of the cost function $I(u)$. The integral in (2) does not contribute to L because we satisfy $E = 0$.

$$I(u) = L(u) \qquad \text{for } E(u, s) = 0 \tag{6}$$

Under this condition the gradient of I can be calculated as

$$\nabla_u I = \frac{dI}{du}$$

$$= \frac{dL}{du} \tag{7}$$

$$= \frac{\partial L}{\partial u} + \frac{\partial L}{\partial \lambda}\frac{d\lambda}{du} + \frac{\partial L}{\partial s}\frac{ds}{du}$$

The second and third term in this sum vanish because $\partial L/\partial \lambda = 0$ (Eq. 3) and $\partial L/\partial s = 0$ (Eq. 4). Only the first term contributes and inserting (2) yields

$$\nabla_u I = \frac{\partial I}{\partial u} + \int_X \int_{t_1}^{t_n} \lambda \frac{\partial E}{\partial u}\, dx\, dt \tag{8}$$

The gradient of the cost function I with respect to any possible control variable u is given by the partial derivative $\partial I/\partial u$ plus the integral of the adjoint variable λ weighted by $\partial E/\partial u$.

Among the most commonly used control variables are the initial conditions $s(t_1)$. It is obvious that in this case $\partial E/\partial u = \partial E/\partial s|_{t_1}$ is zero everywhere except at the initial time t_1. For this timestep Eq. (8) is equivalent to the evaluation of the adjoint equation. For discrete models the gradient is then given by the integration of the adjoint model for one timestep from t_1 to t_0.

$$\nabla_u I = \nabla_s I|_{t_1} = \lambda(t_0) \qquad \text{for } u = s(t_1) \tag{9}$$

Other control variables of course have other derivatives $\partial E/\partial u$ but the mechanism to calculate $\nabla_u I$ is the same as above. A number of examples will be giver in part 6. Once the trajectory $\lambda(x,t)$ of the adjoint model has been calculated it is straightforward to compute $\nabla_u I$ according to (8). Even if we are not prepared to change some of the control variables u during the optimizaion process (8) provides us with valuable information on likely candidates that could be used to diminish I and find a better fit between model and data. For example in a depth dependent ocean model the spacing of computational levels is not easily changed. However by applying (8) we will see immediately if the spacing is appropriate ($\nabla_u I$ is close to zero) or if major improvements could be made ($|\nabla_u I|$ is large).

In most cases the integration of the adjoint model is approximately computationally as expensive as an integration of the forward model. It is worth noting that the adjoint model is linear in λ. However, for non-linear forward models the adjoint equations depend on the state of the forward model as a function of time.

Thus the model trajectory (i.e. the time history) of the forward model has to be stored in order to be used for the integration of the adjoint model. For long time integrations of big models with many variables this can be a heavy burden.

4 THE OCEAN MODEL AND ITS ADJOINT

The ocean model used to illustrate the adjoint data assimilation technique describes non-linear shallow water waves in a one dimensional, cyclic domain.

$$\frac{\partial v}{\partial t} + v \frac{\partial v}{\partial x} - A_M \frac{\partial^2 v}{\partial x^2} + \frac{\partial \phi}{\partial x} = 0 \tag{10}$$

$$\frac{\partial \phi}{\partial t} + \frac{\partial (v\phi)}{\partial x} = 0 \tag{11}$$

Here v is the (depth average) velocity, A_M the viscosity and ϕ the potential (i.e. the time dependent water depth multiplied by the constant of gravity). The phase velocity is $v + \sqrt{\phi}$. Eq. (10) is the momentum equation in the absence of external forcing while Eq. (11) is the equation of continuity. The velocity field advects a passive tracer T according to the advection-diffusion equation (12):

$$\frac{\partial T}{\partial t} + v \frac{\partial T}{\partial x} - A_T \frac{\partial^2 T}{\partial x^2} + \varepsilon T = 0 \tag{12}$$

The tracer is subject to a decay with time constant ε and diffusion proportional to A_T. Sources are not included for the present. Equations (10)–(12) are made non-dimensional by scaling the variables with v_0, ϕ_0 and T_0 respectively. These values are taken as the average initial values of v, ϕ and T. The space variable x is scaled by the length of the cyclic domain L_x and the time is scaled by L_x/v_0 which yields

$$\frac{\partial v'}{\partial t'} + v' \frac{\partial v'}{\partial x'} - \frac{A_M}{L_x v_0} \frac{\partial^2 v'}{\partial x'^2} + \frac{\phi_0}{v_0^2} \frac{\partial \phi'}{\partial x'} = 0 \tag{13}$$

$$\frac{\partial \phi'}{\partial t'} + \frac{\partial (v'\phi')}{\partial x'} = 0 \tag{14}$$

$$\frac{\partial T'}{\partial t'} + v' \frac{\partial T'}{\partial x'} - \frac{A_T}{L_x v_0} \frac{\partial^2 T'}{\partial x'^2} + \frac{\varepsilon L_x}{v_0} T' = 0 \tag{15}$$

Non-dimensional diffusion and decay parameters can be defined as

$$A'_M = \frac{A_M}{L_x v_0}, \varepsilon' = \frac{\varepsilon L_x}{v_0} \text{ and } A'_T = \frac{A_T}{L_x v_0}$$

Primes denoting non-dimensional variables will be dropped hereafter for simplicity. The model equations (13)–(15) are discretized using a staggered grid and central differences in space and time. 40 gridpoints are used in space and a leap-frog integration scheme is used for an integration of 200 timesteps of $\Delta t = 3.125 \cdot 10^{-3}$. The time hereafter will be denoted by the number of the timestep starting from $t_1 = 1$ (initial conditions) to $t_n = 201$.

To derive the adjoint equations the cost function I is rewritten as an integral

$$I = \int_X \int_{t_1}^{t_n} J(x,t) \, dx \, dt \tag{16}$$

so that

$$L = \int_X \int_{t_1}^{t_n} F \, dx \, dt \tag{17}$$

with

$$
\begin{aligned}
F &= \lambda_v \left(\frac{\partial v}{\partial t} + v \frac{\partial v}{\partial x} - A_M \frac{\partial^2 v}{\partial x^2} + \frac{\phi_0}{v_0^2} \frac{\partial \phi}{\partial x} \right) \\
&+ \lambda_\phi \left(\frac{\partial \phi}{\partial t} + \frac{\partial (v\phi)}{\partial x} \right) \\
&+ \lambda_T \left(\frac{\partial T}{\partial t} + v \frac{\partial T}{\partial x} - A_T \frac{\partial^2 T}{\partial x^2} + \epsilon T \right) \\
&+ J(x,t)
\end{aligned}
\tag{18}
$$

In (18) the three model equations are multiplied by their respective adjoint variables λ_s where the subscript of λ denotes the type of variable to which λ_s corresponds. Then Eq. (4) (i.e. $\partial L / \partial s = 0$; $s \in (v, \phi, T)$) is partially integrated which yields

$$\frac{\partial F}{\partial s} - \frac{\partial}{\partial t} \left(\frac{\partial F}{\partial s_t} \right) - \frac{\partial}{\partial x} \left(\frac{\partial F}{\partial s_x} \right) + \frac{\partial^2}{\partial x^2} \left(\frac{\partial F}{\partial s_{xx}} \right) - \cdots = 0 \tag{19}$$

The subscript of s denotes its partial derivative and the dots stand for terms with higher order derivatives. The forward and the adjoint model equations have the same discretization in space and time. The adjoint of a leap-frog time integration scheme is again a leap-frog scheme. As the forward model is already non-dimensional no scaling is necessary[1] and by inserting (18) into (19) the adjoint equations are easily found:

$$-\frac{\partial \lambda_v}{\partial t} - v \frac{\partial \lambda_v}{\partial x} + \phi \frac{\partial \lambda_\phi}{\partial x} + \lambda_T \frac{\partial T}{\partial x} - A_M \frac{\partial^2 \lambda_v}{\partial x^2} = -\frac{\partial J}{\partial v} \tag{20}$$

[1]This is equivalent to using the "natural" inner product of two state vectors $s_i = (v_i, \phi_i, T_i)$: $\langle s_1, s_2 \rangle = \iint (v_1 v_2 + \phi_1 \phi_2 + T_1 T_2) \, dx \, dt$ for deriving the continuous form of the adjoint equations in the context of Hilbert spaces (see LeDimet and Talagrand, 1986)

$$-\frac{\partial \lambda_\phi}{\partial t} - v\frac{\partial \lambda_\phi}{\partial x} - \frac{\phi_0}{v_0^2}\frac{\partial \lambda_v}{\partial x} \;=\; -\frac{\partial J}{\partial \phi} \tag{21}$$

$$-\frac{\partial \lambda_T}{\partial t} - \frac{\partial(v\lambda_T)}{\partial x} - A_T\frac{\partial^2 \lambda_T}{\partial x^2} + \varepsilon\lambda_T \;=\; -\frac{\partial J}{\partial T} \tag{22}$$

These equations are linear in λ as has been mentioned before. The non-linearity of the forward model enters via the dependence of the adjoint model equations on variables of the forward model as a function of space and time. The trajectory of the forward model is stored and the values are used when Eq. (20)–(22) are integrated backwards in time from $t = 201$ to $t = 1$. Note that due to diffusion and decay the temporal evolution of the adjoint model is only stable for this direction of time.

The right hand side of (19) equals zero only when we can make the additional terms of the partial integration vanish. This is achieved by specifying appropriate boundary values of λ_s and of its derivatives at the spatial and temporal boundaries. This specification provides us with all the necessary initial and boundary conditions for the adjoint model, see e.g. Luenberger (1979) or Sewell (1987) for more detail. In our example the domain is cyclic in space, there are no spatial boundaries. The condition for λ_s at $t_n = 201$ that is at the terminal time of the forward model and the initial time of the adjoint model reads:

$$\lambda_s\big|_{t_n} \;=\; -\frac{\partial J}{\partial s}\bigg|_{t_n} \qquad s \in (v, \phi, T) \tag{23}$$

Finally the cost function I has to be specified. In the following we assume for simplicity that model variables are directly observed and not any functionals of model variables. The probability of observing a given realization of l measurements $s_{m,i}$ with $i = 1,\ldots,l$ (in vector notation $\mathbf{s}_m$) can be calculated from the l-dimensional multivariate probability density function $f(\mathbf{s}_m)$

$$f(\mathbf{s}_m) = \sqrt{\frac{\det(\mathcal{W})}{(2\pi)^l}}\;\exp\left[-\frac{1}{2}(\mathbf{s} - \mathbf{s}_m)^{\mathrm{T}}\,\mathcal{W}\,(\mathbf{s} - \mathbf{s}_m)\right] \tag{24}$$

where m stands for measured. $\mathbf{s}$ denotes the expected value of $\mathbf{s}_m$. For unbiased measurements $\mathbf{s}$ is the vector of the model counterpart of the observations. $f(\mathbf{s}_m)$ is maximized by minimizing the argument of the exponential function. This argument is used as the cost function I. It measures the square of the (weighted) distance between model and data. Note that the weighting matrix—or co-validity matrix—$\mathcal{W}$ is the inverse of the covariance matrix of the multivariate distribution given by (24). For uncorrellated observational errors $\mathcal{W}$ is a diagonal matrix. Here all diagonal elements of $\mathcal{W}$ are set equal ($w_{ii} = l^{-1}$ for $i = 1, \ldots, l$) which corresponds to the white noise used to simulate errors in the data.

The adjoint model Eq. (20)–(22) is driven by the gradient of the cost function with respect to the model variables. When I is taken to depend explicitly on derivatives of the model variables (e.g. when non-smoothness is penalized) additional driving terms arise for the adjoint model. Because of the special form of I used here we can identify the model-data misfit as the forcing term of the adjoint model. When this misfit vanishes, i.e. when we have a perfect fit between model and data, the forcing of the adjoint model becomes zero. As a consequence all adjoint variables are zero too. Eq. (8) shows that in this case $\nabla_u I = 0$ and the optimal solution has been found. In general however the model-data misfit cannot be made to vanish everywhere. In order to achieve $\nabla_u I = 0$ after Eq. (8) the driving of the adjoint model by data misfit in one place has to be compensated elsewhere in space and time by a misfit of the opposite sign.

5 RESULTS

To test the performance and practicability of the proposed method in the presence of measurement noise a large number of numerical experiments were carried out. The amplitude of the noise was increased from zero to $O(1)$ for finding out the highest acceptable noise level after which the solution starts to loose most of its information.

A variety of initial conditions and different values for diffusion and decay were used to study the performance of the method under many different strongly nonlinear situations. Out of these experiments only a small number of representative results are shown here owing to lack of space.

In this part only the initial values of the model are used as control variables while diffusion and decay coefficients remain unchanged. A simple example for retrieving their values simultaneously with solving for sources and initial values is given in the next part. For all experiments reported here the same initial values are used and only the temporal and spatial availability of measurements and their noise level are changed. The mean advection velocity v_0 and $\sqrt{\phi_0}$ are set equal. During the integration time the tracer is advected a little more than half way through the domain while the shallow water waves travel twice as far. The viscosity $A_M = 10^{-2}$

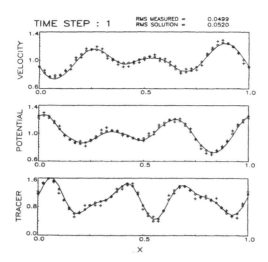

Figure 1a

'True' initial conditions (open circles), assimilated data (crosses) and retrieved solution for initial conditions (solid line) of the passive tracer/-shallow water model. Measurement error and residual error after assimilation are both approximately 0.05. Data are available for the whole space-time domain.

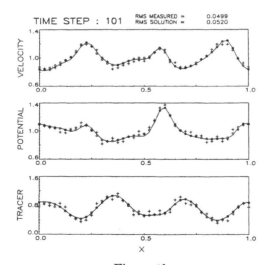

Figure 1b

State of the model after 100 time steps of integration beginning with initial conditions of Fig. 1a.

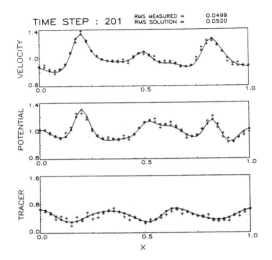

Figure 1c
Same as Fig. 1b, but after 200 time steps of integration.

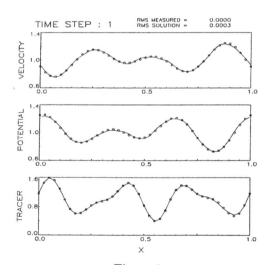

Figure 2
Same as Fig. 1a, but for assimilation of noise-free tracer measurements.
Data are given for timesteps 191 to 201 only.

is just about sufficient to inhibit wave breaking. $A_T = 2.5 \cdot 10^{-3}$ and $\varepsilon = 1.2$ are chosen such as to make their influence visible but not dominant.

Data to be assimilated into this model are taken from a model integration (identical twin) with the following initial conditions:

$$v(x, t = 1) = 1 + \sum_{i=1}^{2} v_i \, \sin(k_{v,i} x + \alpha_{v,i}) \tag{25}$$

$$\phi(x, t = 1) = 1 + \sum_{i=1}^{2} \phi_i \, \sin(k_{\phi,i} x + \alpha_{\phi,i}) \tag{26}$$

$$T(x, t = 1) = 1 + \sum_{i=1}^{2} T_i \, \sin(k_{T,i} x + \alpha_{T,i}) \tag{27}$$

Wave amplitudes, wave numbers and phase angles α are given in

Table 1

i	v_i	$k_{v,i}$	$\alpha_{v,i}$	ϕ_i	$k_{\phi,i}$	$\alpha_{\phi,i}$	T_i	$k_{T,i}$	$\alpha_{T,i}$
1	0.15	$3 \cdot 2\pi$	$\pi/2$	0.15	$3 \cdot 2\pi$	$7\pi/6$	0.4	$3 \cdot 2\pi$	$\pi/6$
2	0.12	$2 \cdot 2\pi$	$\pi/4$	0.12	$2 \cdot 2\pi$	$13\pi/12$	0.2	$5 \cdot 2\pi$	0

These initial conditions (Eq. (25)–(27)) are depicted in Figures 1 to 5 as open circles and referred to as the "true" solution. To allow steep gradients in ϕ to develop the chosen velocity field (25) and potential field (26) are not in dynamical equilibrium. Furthermore waves would break in the time interval of interest if the viscosity A_M were smaller by a factor of 2. To save space only some time steps of the trajectory are shown. In many cases the fields are depicted only for $t = 1$ where they describe the initial conditions, i.e. the control variables u that we solve for.

In Figures 1a to 1c the control or "true" trajectory is shown as open circles for the timesteps 1, 101 and 201. During the integration time the shallow water waves become steeper and the velocity and potential fields tend to equilibrate with the highest velocities at the waves crests and the lowest at the troughs. In the tracer field the wavenumber 5 wave is smeared out rapidly due to diffusion. The average value decays visibly.

Noisy data available for assimilation are shown as crosses in all figures. They are derived from the "true" data by the addition of a white Gaussian noise with a variance proportional to the variance of the three initial fields respectively. For the standard case shown here the ratio of variances was chosen as 0.25. When all possible data are available (i.e. $l = N$) this ratio can be increased to 1.0 without any loss of accuracy of the retrieved trajectory (depicted in all figures as the solid line). A further increase however leads to solutions that gradually become meaningless although they still fit the data to within measurement error.

In the next example the model state was retrieved by assimilation of tracer data only. Perfect tracer measurement were assimilated in this case, however data were given only from $t = 191$ to $t = 201$. The solution (Fig. 2) proves that we can go back for some time into the past and retrieve the correct state of the model if sufficient information is provided. This opens opportunities: When we have insufficient data to describe an oceanic phenomenon it may be possible to complete the picture by measurements to be made in future. The places and times where these measurements should be made can be calculated in advance by methods described in part 7 of this paper.

If the noise level is increased the solution process becomes unstable in the sense that many different initial states can explain the data within the measurement errors. Providing noisy tracer data for all time steps leads to the solution in Fig. 3. The fit for the tracer is almost perfect while for v and ϕ only some of the structure is correct.

When the model grid is finer than the observational grid it is possible to resolve structures like fronts or hydraulic jumps in the model that are not easily detectable in the measurements. A number of different data distributions were tried out to support this possibility. Regular and irregular spacing both in space and time were used. The ability to resolve fronts depended highly on the special observational pattern under consideration. The best results were achieved with high temporal resolution at some localities or with high spatial resolution (at least in some parts) for a few time steps. As a consequence it seems to be desirable and valuable to have both continuous time series of measurements at some places as well as quasi synoptic pictures (maybe satellite derived) at regular or irregular intervals.

The example in Fig. 4a is for continuous data in part of the domain. As the tracer is advected only about halfway through the cyclic domain not the whole field is observed at one time or another. The fit to measurements which are made in the region from $x = 0.25$ to $x = 0.5$ (shaded in Figures 4a and 4b) is reasonably well. At $t = 1$ the tracer field downstream of $x = 0.5$ differs strongly from the "true" solution. It never reaches the measurement region and is not compared to data at any time. As a consequence it is rather arbitrary. How much we can learn about that part of the tracer field is discussed in part 7 on sensitivity and resolution. The

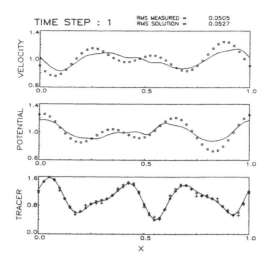

Figure 3
Same as Fig. 2, but with noise added to the tracer measurements. Data are given for all timesteps.

initial tracer field for $x < 0.25$ is retrieved quite well. This part reaches the region of observation during the model integration and can be compared to measurements at intermediate times. Fig. 4b shows the three fields at the terminal time $t = 201$. Now the well resolved tracer field is inside and leeward of the region of observation. Upstream of $x = 0.25$ the resolution is poor. The influence of measurement noise at $x = 0.25$ during the last few timesteps is rather high.

The shallow water waves on the other hand travel through the whole domain and feel the impact of data at earlier or later times. The fit to mesurements is always within the error limits. Outside the observational (shaded) region the initial fields of v and ϕ are quite different from the "true" fields (Fig. 4a). However they lead to a model trajectory that is close to observations at all times and also close to the "true" values towards the end of the integration time (Fig. 4b). Little change is found when in a final step only tracer data are assimilated and measurements of velocity and potential are withheld. The solution (Fig. 5) for tracer is almost unchanged. Of the velocity field only the mean value is of use, while the potential

274

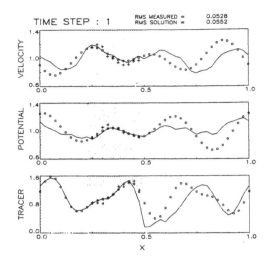

Figure 4a
Same as Fig. 1a, except data are given only between $x = 0.25$ and $x = 0.5$ (shaded region) and for all timesteps.

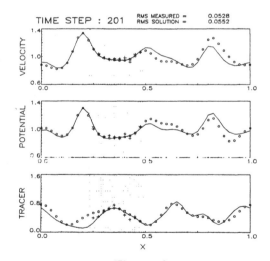

Figure 4b
Same as Fig. 1c, but for initial conditions of Fig. 4a.

is now of no importance in explaining the data. Its initial value is the almost unchanged first guess for ϕ.

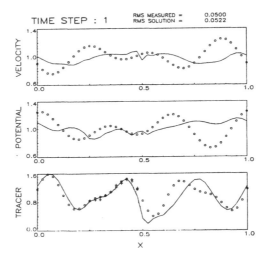

Figure 5
Same as Fig. 4a, but for assimilation of tracer data only.

All solutions were calculated with a conjugate gradient algorithm. Iterations were stopped when the value of the cost function I was reduced to 1.1 times the noise level. Further iterations will in general still decrease I but the solution becomes worse as more and more the measurement noise is modelled and not the large scale structure of interest. This is especially the case when most of the data are withheld and the distinction between random noise and retrievable structure becomes less obvious.

For a fast convergence of the minimization algorithm it is important to use a good initial guess for the u. In most cases presented here 7 to 10 iterations were needed to meet the convergence criterion. This number is independent of the particular initial guess for the tracer field and depends only little on the potential. As long as the first guess flow field has roughly the correct mean value of 1 its influence on the number of necessary interations is moderate, i.e. 3 to 5 additional iterations are needed. For cases with a totally different mean value of the velocity field and without velocity measurements the number of iterations increases by a factor of 2 to 5. For all cases the largest reduction of I is achieved during the first two iterations. Afterwards not only the absolute but also the relative improvement in the value of I per iteration is much smaller. Scaling of the u is equally important as a good first guess to achieve fast convergence. In many cases a linear transformation is sufficient. Instead of diffusion coefficients their logarithm can be used as control variable. For the cases shown here all necessary scaling was done by making the model non-dimensional.

To find out how much we can reduce the space needed to store the model trajectory $s(x,t)$ model variables were saved only every jth timestep. During the integration of the adjoint model a linear interpolation of s was used between storage times. As the adjoint model is now integrated along only an approximation of the model trajectory, the gradient computed is only an approximation of the true gradient. The minimization routine applied (E04DGF of the NAG-library) however was able to converge to the correct solution in a few iterations. Inaccurate values for the gradient are rejected in favor of information provided by the function values directly. For j smaller than 10 no significant change in the solution or in the rate of convergence was observed. When j is increased beyond 10 the number of iterations increases too until for j of the order of 50 no convergence within the first 100 iterations, or convergence to a different solution was found.

6   MODEL PARAMETERS AS CONTROL VARIABLES

In the preceeding examples only the initial conditions were uses as control variables. I will show in this part how adjustable model parameters can equally well be used for this purpose. We can e.g. solve simultaneously, for the initial state, friction and diffusion coefficients, unknown advection velocities, sources and sinks etc. The actual solution is found by the same iterative unconstrained minimization algorithm.

Let us consider tracer equation (12) again, where now the advection velocity v is prescribed and a source term Q added

$$\frac{\partial T}{\partial t} + v \frac{\partial T}{\partial x} - A_T \frac{\partial^2 T}{\partial x^2} + \varepsilon T + Q = 0 \tag{28}$$

The adjoint of this linear model is the same as before

$$-\frac{\partial \lambda_T}{\partial t} - \frac{\partial(v\lambda_T)}{\partial x} - A_T\frac{\partial^2 \lambda_T}{\partial x^2} + \varepsilon\lambda_T = -\frac{\partial J}{\partial T} \qquad (29)$$

This equation is formally identical to (22). The advection velocity $v(x,t)$ however is no longer a dynamical variable but an external parameter. Its value can either be prescribed or it is estimated by the assimilation process.

Suppose now that we want to adjust the diffusion coefficient A_T. With $u = A_T$ the gradient of the cost function I is derived after Eq. (8)

$$\begin{aligned}
\nabla_u I &= \frac{\partial I}{\partial u} + \int_X\int_{t_1}^{t_n} \lambda_T\frac{\partial E}{\partial u}\,dx\,dt \\
&= -\int_X\int_{t_1}^{t_n} \lambda_T\frac{\partial^2 T}{\partial x^2}\,dx\,dt \qquad \text{for } u = A_T
\end{aligned} \qquad (30)$$

This integral can easily be calculated during the integration of the adjoint model. The trajectory of λ_T need not be stored for this purpose. Likewise for the decay parameter ε we get

$$\nabla_u I = \int_X\int_{t_1}^{t_n} \lambda_T T\,dx\,dt \qquad \text{for } u = \varepsilon \qquad (31)$$

Next we solve for the advection velocity v. Let v be time independent then

$$\nabla_u I = \int_{t_1}^{t_n} \lambda_T\frac{\partial T}{\partial x}\,dt \qquad \text{for } u = v(x) \qquad (32)$$

and for a velocity that varies in time but not in space

$$\nabla_u I = \int_X \lambda_T\frac{\partial T}{\partial x}\,dx \qquad \text{for } u = v(t) \qquad (33)$$

The last two equations look very similar, however in (32) we have one control variable for every gridpoint in x while in (33) the control variables correspond to the time levels of the model. Solving for an advection velocity that depends both on space and time $u = v(x,t)$ with $\nabla_u I = \lambda_T\,\partial T/\partial x$ is only meaningful when there are considerably more data than unknowns, i.e. when the observational grid is finer in space and/or time than the computational grid of the model.

This also applies for a space and time dependent source $Q(x,t)$. Suppose now the source function can be separated into two parts $Q(x,t) = h(x)\cdot g(t)$ that depend on space or time only. With $h(x)$ given we solve for $g(t)$

$$\nabla_u I = \int_X \lambda_T\,h(x)\,dx \qquad \text{for } u = g(t) \qquad (34)$$

278

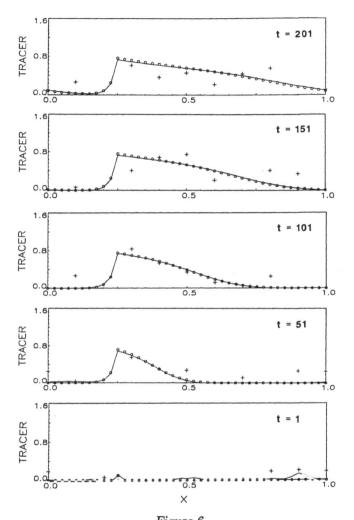

Figure 6
Lowest panel: 'true' initial conditions (open circles) and assimilated data
(crosses) of the advection-diffusion model. The source is located at $x =$
0.25. The solution is found simultaneously for initial conditions (solid
line), diffusion and decay coefficients and source strength as a function
of space. Data are available for all time steps and at every fourth grid-
point. Measurement error and residual error after assimilation are both
approximately 0.2.
Upper panels: successive model states every 50 timesteps apart until $t =$
201.

and for a known time history of the source we find

$$\nabla_u I = \int_{t_1}^{t_n} \lambda_T \, g(t) \, dt \qquad \text{for } u = h(x) \tag{35}$$

Let us take an example. We start from an initial field with no concentration $T(x, t = 1) = 0$ and inject tracer by a source $Q(x, t) = h(x) \cdot g(t)$ which is constant in time $g(t) = 1$. The spatial source function $h(x)$ is used to describe a point source at location $x = 0.25$. The advection is taken constant both in space and time $v(x, t) = v_0$. Parameters A_T, ε and v_0 are chosen in conjunction with the source strength Q_0 such that all play an important role in the temporal evolution of this model. Figure 6 shows the model trajectory every 50 timesteps from $t = 1$ to $t = 201$ as open circles. Data to be assimilated are taken from this experiment at every fourth gridpoint and every timestep. Subsequently a large amplitude noise is added. The data are plotted as crosses in Fig. 6.

The solid line depicts the solution after the assimilation where apart from the initial data $T(x, t = 1)$ the parameters A_T, ε, v_0 and $h(x)$ were used as control variables. They were estimated simultaneously. The unknown spatial source function $h(x)$ consists of 40 variables (one at each gridpoint). Their amplitude—or source strength—is calculated independently from each other. Although the data are very noisy the solution is almost perfect. Values for A_T, ε and v_0 were estimated correctly. Small noise in the initial T is soon smeared out due to the relatively strong decay and diffusion. The most important parameter here is the spatial dependence of the source. By adding the condition that $h(x)$ must be non-negative the correct structure could be retrieved: a source with strength Q_0 at $x = 0.25$ and zero strength elsewere.

To simulate strong non-linearities such as convective overturning in a general circulation model a source was introduced that is switched on or off depending on the model state

$$Q(x = 0.25) = \begin{cases} Q_0 & \text{for } T(x = 0.5) \leq T_{SW} \\ 0 & \text{for } T(x = 0.5) > T_{SW} \end{cases} \tag{36}$$

The value for T_{SW} is chosen such that a number of on-off cycles lie within the integration time. The source function is known to the adjoint model only indirectly through the trajectory $s(x, t)$ of the forward model. Experiments with this Q showed no complications compared to the other experiment made. When the model dependent source function was used in conjunction with a reduced storage of the

$s(x,t)$, convergence problems became severe when the cycle of storage and that of the source change were approximately equal.

7 SENSITIVITY

No inverse calculation is complete without the proper assessment of sensitivity and resolution. This is a well known fact in inverse theory. In practice however it is often overlooked because it involves further computations. Although the calculated solution is optimal it is not necessarily good at the same time. In fact it might be almost useless. For instance in some of the examples shown not all control variables were retrieved: In Fig. 5 the potential is practically without influence and is not resolved, of the velocity field only the mean value is found correctly. The part of the tracer field that does not travel through the observational region is arbitrary except that almost no information diffuses out into the observed part.

After the description of the solution we have to compute the corresponding error statistics: how stable is the solution with regard to noise, under what conditions can we distinguish between different sources located next to each other and how reliable is this distinction etc. In this chapter we present the necessary tools. The application, however, to our model has not yet been done.

7.1 Error covariance and resolution

To quantify the reliability of the solution the corresponding probability density function for the u is calculated

$$f(\mathbf{u}) = \sqrt{\frac{\det(\mathcal{E}^{-1})}{(2\pi)^m}} \exp\left[-\frac{1}{2}(\mathbf{u} - \overline{\mathbf{u}})^{\mathrm{T}} \mathcal{E}^{-1}(\mathbf{u} - \overline{\mathbf{u}})\right] \tag{37}$$

where $\mathbf{u}$ denotes the m-dimensional vector of the control variables and $\overline{\mathbf{u}}$ the solution. The covariance matrix of this distribution is $\mathcal{E}$. Note that (37) is similar to (24). When the model is linear the inverse of $\mathcal{E}$ is the Hessian matrix defined by

$$\mathcal{H} = \{h_{ij}\} = \frac{\partial^2 I}{\partial u_i \, \partial u_j} \tag{38}$$

For non-linear models $\mathcal{H}$ is only an approximation of $\mathcal{E}^{-1}$ in the vicinity of the solution $\overline{\mathbf{u}}$. When the difference $\mathbf{u} - \overline{\mathbf{u}}$ is large $f(\mathbf{u})$ will differ from (37).

The calculation of $\mathcal{H}$ can be simplified in the same way that was used to calculate $\nabla_u I$: combining perturbation methods with the adjoint technique (Tziperman and

Thacker, 1988) we get for the column vectors of $\mathcal{H}$

$$\mathbf{h}_i = \begin{pmatrix} h_{i1} \\ h_{i2} \\ \vdots \\ h_{im} \end{pmatrix} = \frac{\nabla_{u+\delta u_i} I - \nabla_u I}{\delta u_i} \tag{39}$$

When we solve for m control variables u we need m calculations of the gradient at perturbed u to build up the $m \times m$ matrix $\mathcal{H}$. This involves m integrations of the forward and the adjoint model and is in general considerably more expensive than the calculation of the solution alone. For serious studies however the price for calculating $\mathcal{H}$ has to be paid. One should therefore try to reduce the number of control variables as much as possible (e.g. by projection of the initial fields onto appropriate structures like EOF's, normal modes or into Fourier space...). The inverse of $\mathcal{H}$ is only meaningful when $\mathcal{H}$ is well conditioned.

By the singular value decomposition the symmetrical matrix $\mathcal{H}$ can be written as

$$\mathcal{H} = \mathcal{V} \lambda \mathcal{V}^{\mathrm{T}} \tag{40}$$

Where the elements of the diagonal matrix λ consist of the eigenvalues of $\mathcal{H}$ ordered by decreasing magnitude. $\mathcal{V}$ is the matrix of the eigenvectors. The error covariance matrix $\mathcal{E}$ is then given by

$$\mathcal{E} = \mathcal{H}^{-1} = \mathcal{V} \lambda^{-1} \mathcal{V}^{\mathrm{T}} \tag{41}$$

It is obvious from (41) how small eigenvalues λ_i amplify the error covariance matrix of the u.[2] $\mathcal{E}$ can become extremely noisy when we try to solve even for the u that are not determined by the data. The corresponding eigenvalues are zero to within machine precision.

To solve this problem one has to be content to retain only that part of the solution u that is well resolved and to reject the rest as totally unresolved. The truncated solution then is calculated as follows. First find the index k that is largest for which

$$\frac{\lambda_k}{\lambda_1} \geq \frac{\Delta I}{I} \tag{42}$$

ΔI is the expected error of I. The value for I is to be taken at its minimum. The truncated solution consists no longer of the original control variables u but of linear

---

[2]The meaning of λ has been changed from adjoint variable to eigenvalue to ease the comparison with other papers.

combinations thereof. These linear combinations are the first k eigenvectors of $\mathcal{H}$. Which part of the original u can be retrieved by the truncated solution is then given by the $m \times m$ resolution matrix $\mathcal{R}_k$ which describes the correllation between the original u.

$$\mathcal{R}_k = \mathcal{V}_k \mathcal{V}_k^{\mathrm{T}} \qquad (43)$$

where the $m \times k$ matrix $\mathcal{V}_k$ consists of the first k column vectors of $\mathcal{V}$. Note that for $k = m$ we have full resolution as $\mathcal{R}_m$ is the identity matrix. Likewise the corresponding (rank k) error covariance matrix $\mathcal{E}_k$ is given by

$$\mathcal{E}_k = \mathcal{V}_k \lambda_k^{-1} \mathcal{V}_k^{\mathrm{T}} \qquad (44)$$

The $k \times k$ matrix λ_k^{-1} is the inverse of λ_k, the matrix of eigenvalues truncated to the first k rows and columns. By this method we make optimal use of the information provided by the data. Poorly resolved linear combinations of control variables are rejected and cannot contaminate $\mathcal{E}_k$.

7.2   Observational analysis

For an observational analysis it is important to find the places in space and time where the observable part of the model is most sensitive to the individual control variables. By collecting data in these places only few measurements are necessary for the estimation of u. The $N \times m$ data sensitivity matrix $\mathcal{D}$ is defined as

$$\mathcal{D} = \{d_{ij}\} = \frac{\partial s_{m,i}}{\partial u_j} \qquad (45)$$

This matrix is enourmously large as N is the number of gridpoints times the number of time levels times the number of model variables. Even for the simple examples presented here $\mathcal{D}$ has more than $2 \cdot 10^6$ elements. In most cases we are interested only in those places in phase space where the sensitivity of the observable variables to changes in the control variables is high. It is then sufficient to store only the elements of $\mathcal{D}$ which have the largest modulus. This could be done separately for each control variable. The major part of $\mathcal{D}$ which describes small or vanishing sensitivity can then be discarded.

For non-linear models $\mathcal{D}$ has to be calculated at the solution as it depends on u. Linear models are simpler, $\mathcal{D}$ is independent of the solution and can be calculated in advance before any data are collected (the same holds for the Hessian matrix $\mathcal{H}$).

The evaluation of $\mathcal{D}$ is simple. Take the calculations that were used to set up the Hessian matrix and subtract the unperturbed trajectory from the perturbed one

$$d_{ij} = \frac{s_{m,i}(u + \delta u_j) - s_{m,i}(u)}{\delta u_j} \tag{46}$$

When model variables s are not measured but quantities derived from them the same operator which is used to simulate a model equivalent to the measured value from model variables has to be aplied to the $s_{m,i}$ in (46) to obtain the correct data sensitivity.

8 CONCLUSIONS

It has been shown that the adjoint method is a powerful and suitable sensitivity analysis (i.e. a gradient calculation) for complicated, time dependent non-linear models. The sensitivities can be used in an optimization procedure to assimilate data into the dynamical model and to derive its state. By this method asynoptic measurements are interpolated both in space and time consistent with the equations of motion to form a complete picture of the temporal evolution of the model. This evolution can be forecasted into the future or traced back into the past. Its reliability is given by the corresponding Hessian matrix. The method is suitable to estimate unobserved or even unobservable fields from model variables. Also uncertainties in the driving forces can be accounted for and a best estimate of the forcing conditions is given. All fields of interest are fully described including their temporal evolution which obeys the dynamical model equations.

The method lends itself directly to derive appropriate conditions for open boundaries by observation of the interior only. It can also be used to systematically improve the model representation of observed phenomena.

It is planned to apply the adjoint technique as described here to an eddy resolving, quasi-geostrophic ocean model. The most prominent problem to be studied is the impact of satellite altimetry and how surface information can be used to estimate the interior state of the ocean.

This is contribution number 152 of the Alfred-Wegener-Institut für Polar- und Meeresforschung.

References

Cacuci, D.G., 1981, 'Sensitivity theory for nonlinear systems. I. Nonlinear functional analysis approach', *J. Math. Phys.*, **22**, 2794-2812

Courtier, P. and O. Talagrand, 1987, 'Variational assimilation of meteorological observations with the adjoint vorticity equation. Part II, Numerical results', *Q. J. R. Meteorol. Soc.*, **113**, 1329-1347

Hall, M.C.G. and D.G. Cacuci, 1983, 'Physical interpretation of the adjoint functions for sensitivity analysis of atmospheric models', *J. Atmos. Sciences*, **40**, 2537-2546

Hall, M.C.G., 1986, 'Aplication of adjoint sensitivity theory to an atmospheric general circulation model', *J. Atmos. Sciences*, **43**, 2644-2651

Kontarev, G., 1980, 'The adjoint equation technique applied to meteorological problems', *Technical Report No. 21, European Center for medium Range Weather Forecasts*, 21pp.

LeDimet, F.X. and O. Talagrand, 1986, 'Variational algorithms for analysis and assimilation of meteorological observations: Theoretical aspects', *Tellus, Ser. A*, **38a**, 97-110

Lewis, J.M. and J.C. Derber, 1985, 'The use of adjoint equations to solve a variational adjustment problem with advective constraints', *Tellus, Ser. A*, **37a**, 309-322

Luenberger, D.G., 1979, 'Introduction to Dynamic Systems, Theory, Models and Applications', *John Wiley, New York*, 446pp.

Marchuk, G.I., 1975a, 'Formulation of the Theory of Perturbations for Complicated Models. Part I: The Estimation of Climate Change', *Geofisica International*, **15(2)**, 103-156

Marchuk, G.I., 1975b, 'Formulation of the Theory of Perturbations for Complicated Models. Part II: Weather Prediction', *Geofisica International*, **15(3)**, 169-182

Mercier, H., 1986, 'Determining the General Circulation of the Ocean: A Nonlinear Inverse Problem', *J. Geophys. Res.*, **91**, 5103-5109

Navon, I.M. and R. DeVilliers, 1986, 'Gustaf: A Quasi-Newton nonlinear ADI Fortran IV program for solving the Shallow-Water Equations with augmented Lagrangians', *Computers & Geosciences*, **12(2)**, 151-173

Olbers, D.J., M. Wenzel and J. Willebrand, 1985, 'The inference ofNorth Atlantic circulation patterns from climatological hydrographic data', *Rev. Geophys.*, **23**, 313-356

Schröter, J. and C. Wunsch, 1986, 'Solution of non-linear finite difference ocean models by optimization methods with sensitivity and observational strategy analysis', *J. Phys. Oceanogr.*, **16**, 1855–1874

Schröter, J. and J. Oberhuber, 1988, 'On the estimation of adjustable parameters of a mixed-layer model embedded in a general circulation model', *A manuscript*

Sewell, M.J., 1987, 'Maximum and minimum principles', *Cambridge University Press*, 468pp.

Talagrand, O. and P. Courtier, 1987, 'Variational assimilation of meteorological observations with the adjoint vorticity equation. Part I, Theory', *Q. J. R. Meteorol. Soc.*, **113** , 1311–1328

Thacker, W.C., 1987, 'Three lectures on fitting numerical models to observations', *A manuscript*

Thacker, W.C. and R.B. Long, 1988, 'Fitting dynamics to data', *J. Geophys. Res.*, **93**, 1227–1240

Tziperman, E. and W.C. Thacker, 1988, 'An optimal control/adjoint equations approach to studying the oceanic general circulation', *A manuscript submitted to J. Phys. Oceanogr.*

Wunsch, C. and J.-F. Minster, 1982, 'Methods for Box Models and Ocean Circulation Tracers: Mathematical Programming and Nonlinear Inverse Theory', *J. Geophys. Res.*, **87**, 5647–5662

Wunsch, C., 1984, 'An eclectic Atlantic Ocean circulation model, part 1, The meridional flux of heat', *J. Phys. Oceanogr.*, **14**, 1712–1733

Wunsch, C., 1987, 'Using transient tracers: The regularization problem', *Tellus, Ser. B*, **39b**, 477–492

Wunsch, C., 1988a, 'Eclectic modelling of the North Atlantic, part 2, Transient tracers and the ventilation of the eastern basin thermocline', *Philos. Trans. R. Soc., London, Ser. A, in press*

Wunsch, C., 1988b, 'Transient Tracers as a Problem in Control Theory', *J. Geophys. Res.*, **93**, 8099–8110

ASSIMILATION OF XBT DATA
USING A VARIATIONAL TECHNIQUE

J. Sheinbaum and D. L. T. Anderson
Hooke Institute for Atmospheric Research and
Department of Atmospheric, Oceanic and Planetary Physics
Clarendon Laboratory, Parks Rd.,
Oxford, OX1 3PU
U.K.

1. INTRODUCTION

For several years meteorologists have used different statistical techniques to combine model and observations to obtain a good analysis of the atmospheric state. One of the most successful and widely used methods in operational weather centres is optimal interpolation, denoted OI (Gandin, 1965; Lorenc, 1986; Hollingsworth, 1987), whereby one tries to minimize the mean square error between the truth state and the analysed state of the system at a given time. This is attempted by constructing an analysis which is a linear combination of *a priori* field plus the weighted sum of the observational deviations from this preliminary field. Operationally, a short range forecast constitutes the preliminary field. Constraints on the time evolution of the system are not imposed, since O.I. (as currently used) is an analysis at a single time.

Recently a new assimilation scheme based on the ideas of variational calculus has been considered, (Marchuk and Penenko, 1981; Le Dimet and Talagrand, 1986;

D. L. T. Anderson and J. Willebrand (eds.), Oceanic Circulation Models: Combining Data and Dynamics, 287–302.
© *1989 by Kluwer Academic Publishers.*

Talagrand and Courtier, 1987; Derber 1987; Thacker, 1988) which does try to include time constraints by finding the 'best' solution over a space-time trajectory. In this approach one tries to minimize the 'distance' between model and observation fields (represented by a cost function), while constraining the model variables to exactly satisfy the equations of motion (Sasaki's so called strong constraint). The variational method has been used only little in the past as the original formulation is computationally very expensive. The great simplification of the problem has come from the realisation that the initial conditions determine the space-time trajectory of the system and therefore the problem can be interpreted as that of finding the initial conditions that minimize a cost function or distance between model and observations over the time period considered. The constraint that the fields must satisfy the equations of motion is imposed using a set of generalized Lagrange multipliers which satisfy a new set of equations (which in fact is the adjoint of the linearised form of the model equations). The Lagrange multipliers contain sensitivity (gradient) information on how the cost function changes when the free parameters of the problem (in this case frequently, but not necessarily exclusively, the first guess initial conditions) are varied. When this gradient information is used in an iterative optimization procedure, the minimum of the cost function can be found and the problem solved. This powerful method incorporates the time dimension in the assimilation process quite naturally. Further, by incorporating the physics of the problem in the definition of the cost function and constraining dynamics it is very versatile and essentially imposes no limitation on the characteristics of the data to be assimilated, so long as data can be represented in terms of the model variables or functions or combinations thereof. The method can also be readily extended to adjust other physical quantities (including the forcing) to data, as long as there is sufficient data available to make this worthwhile.

We have used a linear reduced gravity model of the tropical Pacific to explore the potential of the variational adjoint method. The grid resolution is $1°$, the value of the Kelvin wave phase speed c is 2.8ms^{-1}. Laplacian friction is included with an eddy coefficient of $2 \times 10^{-4} \text{m}^2\text{s}^{-1}$. The model is driven by monthly mean wind data (Legler and O'Brien, 1984) which covers the period 1960–1983 and is converted to surface wind stress using the drag coefficient of Smith (1980), given by

$$C_d = 1 \times 10^{-3}(0.61 + 0.063u) \quad \text{if} \quad 6 \text{ ms}^{-1} < u < 22 \text{ms}^{-1}$$
$$C_d = 1.1 \times 10^{-3} \quad \text{if} \quad u < 6 \text{ ms}^{-1}$$

where u is the wind speed. The winds are also linearly interpolated to the model grid as well as in time. High frequency variability is absent in these winds and therefore in the ocean model as well.

Real XBT data from the period January – June 1980 are assimilated by identifying the depth of the 16°C isotherm depth with the model layer depth. Fig. 1 shows the data distribution for January and six regions where most of the data lies. The sparsity of the data is very noticeable, the number of measurements hardly exceeding 300 for any given month.

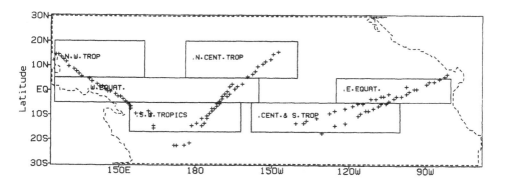

Figure 1
XBT stations for January 1980. Also shown are the model domain and the regions in which data are organised to facilitate the analysis.

2. IMPLEMENTATION OF VARIATIONAL ASSIMILATION USING LAGRANGE MULTIPLIERS

In variational data assimilation a solution of the problem is sought based on the minimization of a functional which represents the misfit between model and observations subject to the constraint that the solution satisfy the model equations.

Assuming we have a discretized model of the ocean (a finite difference model, for example), one form of this 'cost' functional would be:

$$C = \sum_{n=0}^{N} (\mathbf{m}_n - \mathbf{o}_n)^T \mathbf{W} (\mathbf{m}_n - \mathbf{o}_n), \tag{1}$$

where o_n represents the observation state vector and m_n represents the model counterpart of the observation vector. In general m_n is a function of the model state vector x_n. The matrix W is a weighting matrix which in principle should be given by the inverse of the observation error covariance matrix (Lorenc, 1986). The subscript n indicates time level and T indicates transpose. The natural way to impose the constraint that x must obey the model equations is by introducing a set of Lagrange multipliers. If the model equations are described by

$$x_n - A_{n-1}x_{n-1} - f_n = 0, \qquad n = 1, \ldots, N \tag{2}$$

with A a matrix which represents the model equations (with no forcing) and f the forcing, the new cost function (or Lagrange function) then takes the form

$$J = C + \sum_{n=1}^{N} \lambda_n^T \left(x_n - A_{n-1}x_{n-1} - f_n \right) \tag{3}$$

where λ_n is the vector of Lagrange multipliers and its dimension is the same as the dimension of x_n.

Note that the whole time history of x_n is determined by the initial conditions, i.e., the best fit space-time trajectory is found by establishing the correct initial state. The problem then becomes one of finding the best initial conditions, i.e., those in which the functional C has a minimum and J takes an extremum value. To find this extremum point of J, the gradient of J with respect to the initial conditions must be found. Obviously the dependence of x_n on the initial conditions is highly implicit, but the introduction of Lagrange multipliers actually allows us to take derivatives with respect to x_n as if they were independent. The condition for a minimum of C in terms of the extended functional J then becomes

$$\frac{\partial J}{\partial x_n} = 0, \quad \frac{\partial J}{\partial \lambda_n} = 0 \tag{4}$$

where the partial derivative with respect to a vector indicates a partial derivative with respect to each of the components of the vector. In Eq. (4), the derivative with respect to the components of the Lagrange multiplier vector is nothing more than the model equations, whereas the gradient with respect to the components of the vector x_n gives the equations for the Lagrange multipliers. In vector form these equations are:

$$\frac{\partial J}{\partial \mathbf{x}_n} = \lambda_n - \mathbf{A}_{n+1}^* \lambda_{n+1} + \frac{\partial C}{\partial \mathbf{x}_n} = 0, \quad n = 1, \ldots, N-1 \tag{5a}$$

$$\frac{\partial J}{\partial \mathbf{x}_N} = \qquad \lambda_N + \frac{\partial C}{\partial \mathbf{x}_N} = 0, \quad n = N \tag{5b}$$

$$\frac{\partial J}{\partial \mathbf{x}_0} = \qquad -\mathbf{A}_1^* \lambda_1 + \frac{\partial C}{\partial \mathbf{x}_0} = 0, \quad n = 0 \tag{5c}$$

$\mathbf{A}^*$ is the adjoint of matrix $\mathbf{A}$. Since the elements of $\mathbf{A}$ are real, $\mathbf{A}^*$ is the transpose of $\mathbf{A}$. Note that the elements of the first column of $\mathbf{A}$ are the coefficients that multiply the first element of the vector $\mathbf{x_n}$ in Eq. (5a) which is why $\mathbf{A}^*$ appears in the equations for the Lagrange multipliers rather than $\mathbf{A}$. Eq. (5c) is the gradient of J with respect to $\mathbf{x}_0$ which depends on observations and gradients at later times through λ_1.

Eqs. (2) and (5) consist of $2N + 1$ vector equations for the $2N + 1$ unknown vectors $(\lambda_1, \ldots, \lambda^N, \mathbf{x}_0, \ldots, \mathbf{x}_N)$. One way to solve this set would be to rewrite the equations in matrix form and use a Gaussian elimination procedure to find the solution. However, as is frequently the case to solve large matrix systems, an iterative method in which a first guess solution is modified until the right solution is reached (or a convergence criterion fulfilled), is used.

Suppose then that we start with first guess initial conditions introduced by appending to Eq. (2) the equation:

$$\mathbf{x}_0 - \mathbf{u} = 0$$

A new Lagrange multiplier λ^0 corresponding to this equation must be introduced in order to have as many equations as unknowns. Eq. (5c) now takes the form:

$$\frac{\partial J}{\partial \mathbf{x}_0} = \lambda_0 - \mathbf{A}^* \lambda_1 + \frac{\partial C}{\partial \mathbf{x}_0} = 0 + \Delta$$

where Δ is a residual which comes from $\mathbf{x}_0 = \mathbf{u}$ not being the right initial conditions. We want to reduce Δ iteratively to zero by modifying $\mathbf{u}$.

If we also add the dummy equation $\mathbf{x}_{N+1} = 0$ and its corresponding Lagrange multiplier $\lambda_{N+1} = 0$, which are identically zero, Eqs. (5a), (5b), (5c) can be rewritten in the concise form

$$\lambda_n - \mathbf{A}_{n+1}^* \lambda_{n+1} + \frac{\partial C}{\partial \mathbf{x}_n} = 0, \quad n = 0, \ldots, N \tag{5}$$

Note that x_{N+1} and λ_{N+1} are zero because there are no observations after time t, and λ_0 is equal to minus the gradient of J with respect to x_0 (see Eq. 5c). Although this gradient is not zero, its value contains information that can be used in an optimization algorithm like conjugate gradient or a Newton-type method which searches geometrically for the minimum[1]. Eqs. (5) are similar to the model equations but forced by the misfit between model and observations. They are integrated backwards in time, allowing the transmission of misfit information from the time of an observation to the initial time.

To summarize, we can formulate the method for solving Eqs. (2) and (5) as follows:

(a) Starting with first guess initial conditions integrate Eq. (2) forward and save the values of x at observation points to compute the forcing terms for the adjoint Eqs. (5);

(b) Integrate Eq. (5) backwards in time and find λ_0 which gives the value of minus the gradient of J with respect to the initial conditions;

(c) Use λ_0 in an optimization routine and modify the initial conditions in such a way that the value of C is reduced.

Repeat steps (a) (with the new initial conditions), (b) and (c) until the minimum is found or a convergence criterion is satisfied.

To apply the adjoint method, a cost function CF must be defined. There is no unique way of doing this and the first cost function we define, CF1, is essentially the total energy of the difference between model and observations and is given by

$$
\begin{aligned}
\mathrm{CF1} = \sum_{n=0}^{N} & (\mathbf{u}_n - \mathbf{u}_n^{obs})^T \mathbf{A}(\mathbf{u}_n - \mathbf{u}_n^{\mathrm{obs}}) \\
& + (\mathbf{v}_n - \mathbf{v}_n^{\mathrm{obs}})^T \mathbf{B}(\mathbf{v}_n - \mathbf{v}_n^{\mathrm{obs}}) \\
& + g'H(\mathbf{h}_n - \mathbf{h}_n^{\mathrm{obs}})^T \mathbf{C}(\mathbf{h}_n - \mathbf{h}_n^{\mathrm{obs}})
\end{aligned}
\tag{6}
$$

The weighting matrices $\mathbf{A}$ and $\mathbf{B}$ for the kinetic energy part are zero because there are no velocity observations and $\mathbf{C}$, the weighting matrix for the potential

---

[1] In this work a conjugate gradient method is used to solve Eqs. (2) and (5). This gives a good trade off between storage requirements and rapid convergence, though further investigation would be needed to determine an optimum strategy.

energy is the unity matrix. This means that the data has been considered to be essentially perfect and no attempt has been made in this work to introduce data and sampling errors via the inverse of the error covariance matrices although this should be done in a full implementation of the technique. The forcing for the adjoint equations, which is the difference between observations and model variables, was calculated by grouping the data into three periods of ten days, and the forcing terms calculated at days 5, 15, and 25 of each month. That is, we regarded the data of every period as if they were sampled synoptically at the above times.

3. RESULTS

Apart from the use of real data, another aspect that differentiates this study from other ones (Long and Thacker, 1988; Tziperman and Thacker, 1989; Schroter, 1989) is the fact that there are many fewer observations than model degrees of freedom. This means that we need a good first guess initial state not just to reduce the number of iterations required for convergence, but also, if possible, to have a realistic representation of what the field may be like in areas not constrained by the data (which we will call the solution null space). Fortunately the high degree of determinism of the system implies that by running the model with previous wind stress forcing, a moderately realistic initial state can be obtained. The first guess[2] (FG) initial conditions for most of the following experiments consists of the layer depth and velocity fields for the beginning of January 1980, obtained by spinning up the model from rest in January 1975 and forcing it with the appropriate winds from January 1975 to January 1980. However, to get a clearer picture of where and to what extent the data are determining the solution, results using other first guess fields will also be discussed.

In experiment CONT1, the model is run from January 1980 to June 1980 and no data is assimilated. It is considered as a control experiment and its fit to data will be compared to the results from experiments in which data is assimilated (see below). CF1A is a six-month (Jan. 1980–June 1980) assimilation experiment in which we try to minimize the cost function CF1 (Eq. 6).

---

[2] We use the convention that the First Guess (FG) is the initial state from which we start the assimilation and the Initial Conditions (IC) are the initial state obtained after assimilating the data.

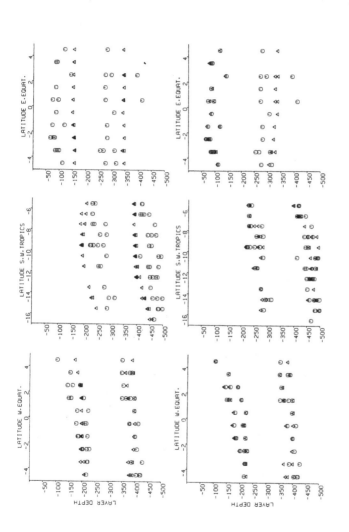

Figure 2

Latitude-depth plots of Observations vs. model layer depth for three of the six regions (Fig. 1). Panels (a), (b) and (c) show results for January and June for experiment CONT1 in which no data are used. The results for January are shown in the upper part of each panel whereas the results for June are shown at the bottom and are offset 200 m in depth. Panels (d), (e) and (f) show the results for experiment CF1A in which data is assimilated.

Several data are at the same latitude but they might not be at exactly the same longitude or might have been made at different times (days). Comparing panels (c) and (f) one can see that without data assimilation the model layer depth is 50–100 m too deep. This bias is corrected in January when data is assimilated but the fit is not maintained until June.

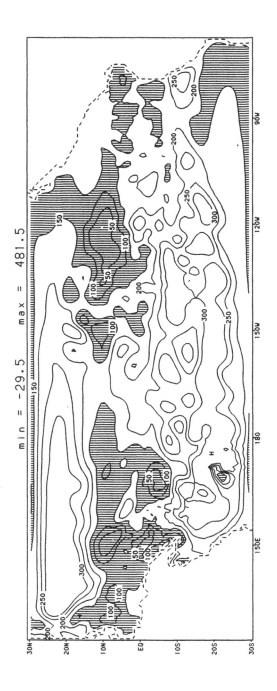

Figure 3
Initial conditions (thermocline depth) for experiment CF1A after ten iterations of the conjugate gradient algorithm.
The shading shows the regions where $h \leq 150$ m.

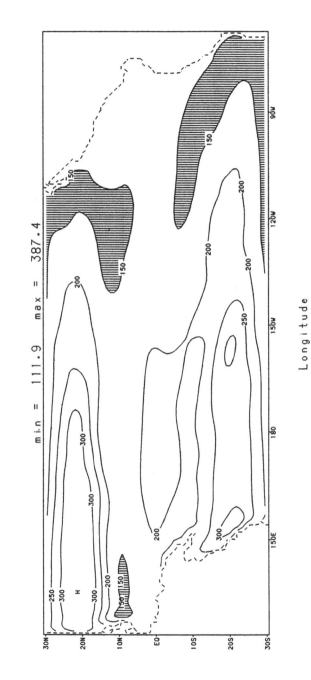

Figure 4

Thermocline depth for January 1980. This field was obtained by spinning up the model from rest for 5 years using the observed winds. The field serves as a First Guess for experiments CF1A and CF1SIM1. The shading shows the regions where h ≤ 150 m.

Fig. 2 panels (a), (b), (c) show the comparison of the model layer depths with the observations at the begining and end of the integration period *viz.* January and June for (a) the Western equatorial (b) the South West tropical and (c) the Eastern equatorial regions for the control run (CONT1) into which no data has been assimilated. The corresponding plots for the case when data is assimilated (experiment CF1A) are given in panels (d), (e) and (f).

In January, one can see that as a result of assimilating data, the fit in the SW tropics and the Eastern equatorial region has been substantially improved compared to the control run. For example, without data assimilation, the model layer depth is too deep in the eastern Pacific by 50 to 75 m whereas this bias has been corrected when data is assimilated. Encouraging changes also occur in the other regions of Fig. 1 (not shown). The fit was already quite good in the Western equatorial Pacific so no obvious improvement has taken place there as a result of the assimilation. However, looking at later times one can see that the improvement in the eastern equatorial Pacific is not maintained, and in fact the fit there is not much better in experiment CF1A than in CONT1.

In Fig. 3 the depth of the thermocline is contoured after ten iterations for the initial conditions of experiment CF1A. Compared with experiment CONT1 (Fig. 4) there is much more small-scale variability. Although the small scale corrections in Fig. 3 are very evident, larger scale corrections have also taken place e.g. in the South Pacific. The small scale variability characterized by large gradients in depth is the response of the model trying to fit the data exactly, data which are noisy and not necessarily consistent with the model.

Some of the changes seen in Fig. 3 such as the small scale features of 50–100 m depth in the western equatorial Pacific are produced by the model trying to fit the eastern equatorial data. On the other hand western equatorial data give rise to the 50–100 m feature centered at 10°N, 120°W. This highlights the nonlocal influence of the data on the IC in the variational assimilation procedure.

We return now to Fig. 2. As noted, the initial conditions for experiment CF1A fit the data very well, but the improvement observed in the initial month is not maintained. Indeed, by June, the fit in the eastern equatorial Pacific is now quite bad and only a little better than in the control run. The time evolution of the fit is shown in Fig. 5 where the mean residual in the western and eastern equatorial regions is plotted as a function of time for experiments CONT1 and CF1A. The vertical lines represent a measure of the scatter of the residuals. Comparing the results from the two experiments one can see more clearly the improvement of the fit in CF1A in the eastern Pacific as a result of assimilating data, and the loss of this fit in May and June. The results for the western Pacific show that there is little improvement as a result of assimilating data as the fit was already quite good there.

It is instructive to ask what has gone wrong? After all we set out to find the best trajectory through the data, all the data and not just a fit to the initial data. What we are seeing here is an inconsistency between the model, the forcing and the data. The first order balance in the equatorial region is between the zonal wind stress and the pressure gradient *viz.*

$$\rho g' H \frac{\partial h}{\partial x} \simeq \tau^z \tag{7}$$

The 'error' can not be partitioned unambiguously between the model and the forcing, but for the sake of argument, let us suppose the forcing is actually perfect and the problem is that the model parameters are incorrect[3]. Hence if the phase speed is too big, the thermocline slope will be weak in consequence. If we have data in both east and west then we will not be able to fit it all and satisfy Eq. (7) with a physically realistic state and so the cost function will never be zero. The initial conditions are, in principle, determined from data all through the time period, but the data may not all have equal influence. One might expect that data occurring at earlier times will have a greater influence than that at later times and hence it is possible that the data from the last two months do not really influence the determination of the Initial Conditions. In fact (not shown) this is not the case, and data from the last two months do have an influence on the initial conditions, although the response was not as large in the equatorial region as in off-equatorial regions of the Southern Hemisphere. This implies that the lack of fit to data in the equatorial region is indicative of a model or forcing deficiency.

This notion is supported by experiments using simulated data. The data are generated from a model with different parameters to that which is used to assimilate the data. Here we consider the case of different stratification between the two models, by using a value of c of $2ms^{-1}$ for the model used to generate the data

---

[3] An alternative hypothesis is that the model is correct but the forcing is in error.

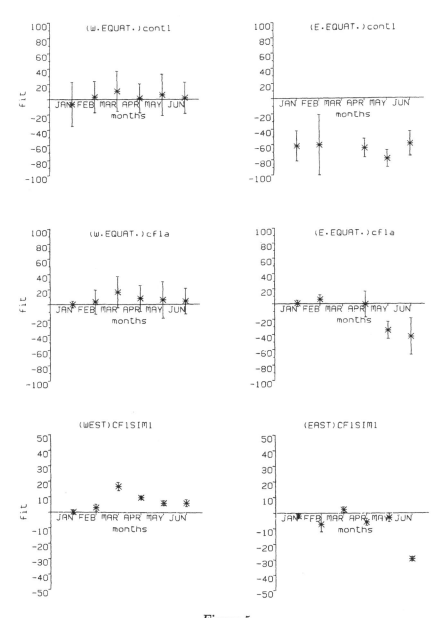

Figure 5
Mean of residuals vs. time plots in the western and eastern equatorial
regions for experiments CONT1, CF1A and CF1SIM1. The error bars are
standard deviations for each region and month.

as opposed to 2.8ms^{-1} in the model used to assimilate the data. The data are generated by first spinning up the model for five years (Jan. 1975 – Jan. 1980) starting from a flat field of $H = 200$m, $U = V = 0$ with $c = 2$ms^{-1} and using the FSU wind data. Then the model is integrated another six months (Jan. 1980 – June 1980) to generate the data.

In the first assimilation experiment (CF1SIM1) thermocline-depth data is provided once a month (day 15) along two sections equidistant from the eastern and western boundaries at 103.5°W and 157.5°E, from 30°S to 30°N. Ten iterations of the conjugate gradient algorithm are used to fit the data and obtain the IC. The FG is the model state for January from CONT1. The mean errors from experiment CF1SIM1 for the eastern and western equatorial regions are plotted in Fig. 5. The temporal evolution is very similar to that in experiment CF1A, except that the loss of fit in the east occurs one month earlier in CF1A than in the simulated case CF1SIM1. Further experiments show that the reason why the fit in the west is better than in the east (and not vice versa) is a consequence of dynamics rather than of data volume, although this can also contribute.

The above analysis shows how the time trajectory approach to fit the data can be used to highlight model-forcing data inconsistencies. These are important, for, if no further action is taken to correct this inconsistency, the impact of assimilating data will be quickly lost, and in fact in this experiment has even been lost within six months. If smoothing or other constraints are applied the loss of fit is even quicker. Similar inconsistencies were noted in Moore and Anderson (1989) when forecast experiments from the analysis were performed. Large amplitude Rossby waves were excited in that case.

The role of data assimilation in the tropical ocean where the system is deterministic, must be to highlight regions of inconsistency between the model, its forcing and observations, so that the model and forcing can be corrected. (This is very different from the atmospheric analogue where the fluid is essentially turbulent). The variational approach outlined here is particularly good at highlighting these inconsistencies.

REFERENCES

Courtier P. and O. Talagrand (1987): Variational assimilation of meteorological observations with the adjoint vorticity equation, Part II, Numerical Results, *Quart. J. Roy. Met. Soc.*, **113**, 1329–1347.

Derber J. C. (1987): Variational four dimensional analysis using quasi-geostrophic constraints, *Mon. W. Rev.*, **115**, 998–1008.

Gandin L. S. (1965): 'Objective analysis of Meteorological Fields', Translated from the Russian, *Israel Program for Scientific Translations*, Jerusalem, 242 pp.

Gill P. E., W. Murray and M. H. Wright (1981): 'Practical Optimization', Academic Press, Orlando Fl., 401 pp.

Hollingsworth A. (1987): 'Objective Analysis for Numerical Weather Prediction', *ECMWF Tech. Memo. No. 128*, available from ECMWF, Reading, U.K.

Le Dimet F. X. and O. Talagrand (1986): Variational algorithms for analysis and assimilation of meteorological observations: theoretical aspects, *Tellus*, **38A**, 97–110.

Long R. B. and W. C. Thacker (1988): Data Assimilation into a Numerical Equatorial Ocean Model, Part 2: Assimilation Experiments, to be published in *Dyn. Atmos. Oceans*.

Lorenc A. C. (1986): Analysis methods for numerical weather prediction, *Quart. J. Roy. Met Soc.*, **112**, 1177–1194.

Marchuk G. I. and V. V. Penenko (1981): 'Applications of Perturbation Theory to Problems of Simulation of Atmospheric Processes', In *Monsoon Dynamics*, Lighthill J. and Pearce R. P. eds., Cambridge University Press, 639–655.

Moore A. M. and D. L. T. Anderson (1988): The Assimilation of XBT Data into a Layer Model of the Tropical Pacific Ocean, to be published in *Dyn. Atmos. Oceans*.

Sasaki Y. (1970): Some basic Formalisms in Numerical Variational Analysis, *Mon. Wea. Rev.*, **98**, 875–883.

Smith S. D. (1980): Wind Stress and Heat Flux over the Ocean in Gale Force Winds, *J. Phys. Oceanogr.*, **10**, 706–726.

Schroter J. (1989): Driving of Non-linear Time Dependent Ocean Models by Observation of Transient Tracers: A Problem of Constrained Optimization, this volume.

Talagrand O. and P. Courtier (1987): Variational assimilation of meteorological observations with the adjoint vorticity equation, Part I: Theory, *Quart. J. Roy. Met. Soc.*, **113**, 1311–1328.

Thacker W. C., and R. B. Long (1988): Fitting dynamics to data, *J. Geophys. Res.*, **93**, 1227–1240.

Thacker W. C. (1988): The Role of the Hessian Matrix in Fitting Models to Measurements, submitted for publication.

Thacker W. C. (1988): Fitting to inadequate data by Enforcing Spatial and Temporal Smoothness, *J. Geophys. Res.*, **93**, 10655–10665.

Tziperman E. and W. C. Thacker (1989): An Optimal Control-Adjoint Equations Approach to Studying the Oceanic General Circulation, submitted for publication.

THE ROLE OF REAL-TIME
FOUR-DIMENSIONAL DATA ASSIMILATION IN THE
QUALITY CONTROL, INTERPRETATION, AND SYNTHESIS
OF CLIMATE DATA

A. Hollingsworth

ECMWF

Shinfield Park

Reading, RG2 9AX

U.K.

ABSTRACT

Over the next decade a vast quantity of remotely sensed data and *in situ* data will become available for climate studies. To gain maximum benefit from these large heterogeneous datasets, it is essential that they be quality-controlled, interpreted into directly measurable quantities, and then synthesised into a consistent description of the time-evolution of the atmosphere and ocean. Operational Numerical Weather Prediction (NWP) centres have developed considerable scientific insight and technical skill in problems of this kind. Experience in NWP has shown that the quality control and the interpretation (or inversion) procedures for remotely sensed data are considerably sharpened if all available *a priori* information, including current and earlier observations, are brought to bear on the interpretation of the new data. This idea is implemented in NWP centres in the form of four-dimensional assimilation systems.

Based on experience with FGGE and TOGA, it is clear that there will be a strong demand for the production of timely gridded III-a analyses of all the

D. L. T. Anderson and J. Willebrand (eds.), Oceanic Circulation Models: Combining Data and Dynamics, 303–342.

World Climate Research Programme (WCRP) data. The efficiency of the observing systems, and of the scientific work, will be considerably enhanced if the remotely-sensed WCRP data (preferably at level 1 or 1.5) is delivered to NWP centres in real time.

1. INTRODUCTION

Many new initiatives in remote sensing of the earth are planned for the next fifteen years as part of the 'Mission to Planet Earth', NASA/NOAA's 'Earth Observation System' (NASA, 1988a,b), or ICSU's International Geosphere-Biosphere Program. The acquisition and interpretation of this data are major concerns of the World Climate Research Programme (WCRP). The space agency study documents for these programmes tend to dwell at length on the hardware for the missions. On the whole, the agency planning documents are vague on methods to interpret (or 'retrieve') geophysical products from the radiometric data, and almost ignore entirely the problems involved in synthesising the heterogeneous data into a coherent whole.

The main WCRP effort is to describe, understand, and finally to predict the evolution of the climate of the atmosphere/ocean/biosphere system. Data assimilation systems can use and interpret data on processes for which they have a representation within the assimilating model. Operational data assimilation systems for medium range forecasts (3–10 days) have a rather complete representation of atmospheric processes, and represent sea, ice and soil conditions as slowly varying parameters. Assimilation systems for short term climate prediction consider atmosphere/ocean/ice/soil/vegetation/hydrology interactions more thoroughly.

It is not clear what possibilities there may be for short-term climate forecasting, at present. The rationale for weather prediction and for climate prediction breaks down if unpredictability of short lived phenomena has more effect on the prediction of large scale phenomena than do the slowly evolving resolved processes. In any event, both diagnostic and prognostic studies of the climate system will need the best possible description of the evolving climate system.

We argue in this paper that the best syntheses of WCRP data will be those which systematically exploit all our knowledge of the climate system and of the observing system. NWP Centres have much experience in these problems, and their formulations of four-dimensional data assimilation are already providing the research community with high-quality global atmospheric analyses and, increasingly in the future, oceanic analyses.

To gain maximum benefit from the data collection of the WCRP experiments, it is recommended that WCRP observational data be made available to operational centres in real time. It is desirable that the products delivered to the operational centres be calibrated earth-located radiance measurements. This will facilitate the application of comprehensive retrieval procedures to the data, using all available *a priori* information.

Several benefits will flow from this proposal. There will be rapid identification of problems in the observational data through on-line real-time quality control, and off-line long-term monitoring. Wide-spread study of the WCRP observational data will be accelerated if the III-a gridded analyses (which are made available to the community rather quickly) have the benefit of the WCRP observational data. Finally, if the community wishes to re-analyse the WCRP data at a later stage, the availability of the real time (III-a) quality control information will be of great value in the production of delayed-mode WCRP III-b analyses.

2. THE IMPORTANCE OF ACCURATE DATA ASSIMILATION FOR NWP

All prediction depends on an understanding of the system to be predicted, on a knowledge of the initial state, and on the formulation of a forecast model which gives quantitative expression to our theoretical and empirical understanding.

The resolution of conflicts between predictions and observations is at the heart of the scientific process. In using observations to identify reliably what is not yet known or understood, it is essential to take account of everything that is already known or explainable. Fruitful interpretation of observations of an evolving system requires that we bring to bear the most accurate expressions of our *a priori* information and understanding. In NWP the *a priori* information is summarised in

(i) Our understanding of the laws of the atmosphere, as expressed in the equations of the forecast model;

(ii) Our best *a priori* estimate of the current state of the atmosphere, which provides the background for the next analysis; and

(iii) The statistics of forecast and observation error.

The realisation of the importance of using all available *a priori* information to interpret new data is becoming a truism in many areas of science, from studies of computational visual processing to weather prediction. It is necessary that it be understood and acted upon also in the area of climate studies.

Accuracy of the initial state is vital for successful NWP. For that reason meteorologists have devoted considerable efforts to producing the most accurate analyses possible, using all available data. This is because the atmosphere is an unstable fluid system. Fig. 1, from Simmons and Hoskins (1979) illustrates how rapidly an isolated disturbance, on an otherwise undisturbed but unstable flow, can amplify and propagate downstream. Fig. 2, from Hollingsworth *et al.* (1985) is an illustration of the same phenomenon in a more realistic flow. In this figure we are looking at the amplification and downstream propagation of the difference between two objective analyses for the same time, where the differences occur only in the northeast Pacific. Within six days the modest initial differences amplify considerably and propagate to northwest Europe.

To predict the evolution of the atmosphere, one needs a good forecast model, and an accurate description of the initial state. The production of the initial state for NWP presents special difficulties because of the quantities of data which have to be handled. In NWP we deal every day with masses of data from land stations, ships, aircraft, and geostationary and polar orbiting satellites with multi-channel radiometers in the visible, infrared and microwave. Fig. 3, from the FGGE year, illustrates the data available within ±3 hours of 0000 UTC on a typical day during the FGGE second special observing period. (Bengtsson *et al.*, 1982). This large, heterogeneous (and noisy) database has many gaps and is typical of operational data coverage in 1989. Despite the great volumes of data involved, the observations at any single time are inadequate to determine the state of the atmosphere on the scales of interest. The observations at any given time have to be supplemented by knowledge of earlier observations in order to arrive at a more complete description of the system at any given instant.

The further information is provided through the process of data assimilation. In this process, information from an earlier time is projected forward in time to the current time using a sophisticated Numerical Weather Prediction (NWP) model. Fig. 4 shows the range of interactions and feedbacks in a typical NWP model. Most of the processes represented in the model are central concerns of the WCRP. Operational NWP models usually follow Parkinson's Law quite well — models

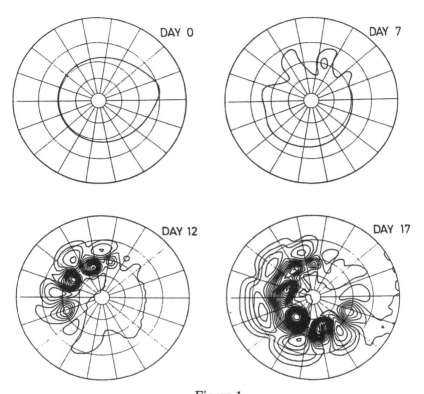

Figure 1
Downstream propagation and amplification of an initially very small per-
turbation on a zonally symmetric unstable baroclinic flow (Simmons and
Hoskins, 1979)

expand their resolution to fill the available computer. Current operational global
NWP models have a physical mesh of 1.125° and about 20 levels in the vertical.

The forecast field from the earlier data provides the background field (or first-
guess field) for the analysis of the new data. In many current operational systems
the method used to combine the two information sources is the method of optimum interpolation (Gandin, 1965; Lorenc, 1981). This procedure is illustrated
schematically for a three component system in Fig. 5 (Rodgers, 1976). The *a priori* information is that the system is at the point x_0 with uncertainty covariance
S_x. The measurement only measures two components of the system with error

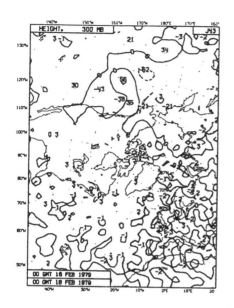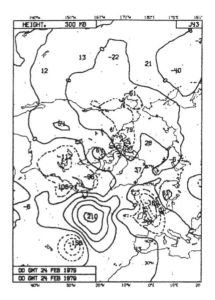

Figure 2
Perturbations on a realistic flow were initially confined to the north-east
Pacific, right. Six days later they are affecting much of north-west Europe,
left. (Hollingsworth et al., 1985)

covariance S_ϵ and so has infinite uncertainty along the third component. The two measurements are then combined according to the principles of estimation theory, and the final estimate x has its uncertainty covariance S given by

$$S^{-1} = S_x^{-1} + S_\epsilon^{-1} \qquad (1)$$

In many current operational systems the data is grouped into 6-hour groups, and a data assimilation process of the type illustrated in Fig. 6 is carried through, using three modules: analysis, initialization and forecast.

3. THE ANALYSIS MODULE

The method of optimal interpolation (Gandin, 1963; Lorenc, 1981) is widely used in operational work on analysis. The algorithm can be viewed as a two step procedure, which first filters the observational data and then interpolates it.

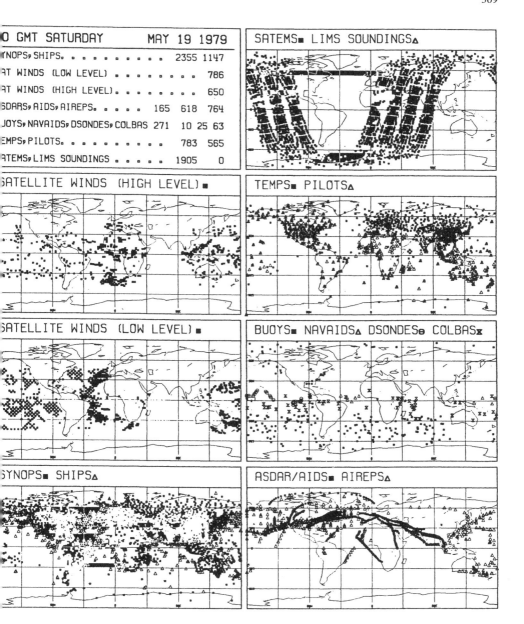

Figure 3
Typical data distribution in a 6-hour period bracketing 0000 UTC on May 19th, 1979.

310

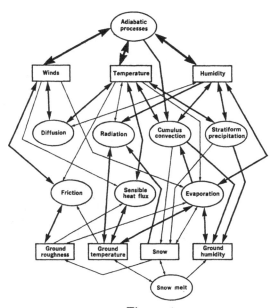

Figure 4
Schematic representation of the processes included in the ECMWF model.

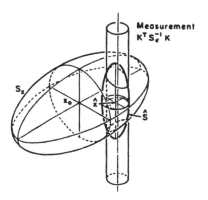

Figure 5
Illustration of the schematic relation between the *a priori* error covariance,
the measurement error covariance, and the final analysis error covariance.
(*Rodgers, 1976*)

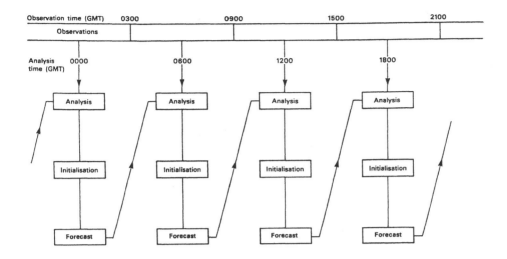

Figure 6
The 6 hourly intermittent data assimilation.

3.1 The O/I Algorithm

Let A, T, F, D denote the Analysed, True, Forecast, and Observed data at a point r where r indexes both position and variable. We introduce non-dimensional quantities:

$$\alpha(r) = \frac{A(r) - T(r)}{\sigma_f(r)}$$

$$f(r) = \frac{F(r) - T(r)}{\sigma_f(r)}$$

$$\delta(r) = \frac{D(r) - T(r)}{\sigma_f(r)}$$

and pose the following interpolation formula

$$\alpha(r) - f(r) = \sum_i W_i(r)(\delta_i - f_i) \tag{2}$$

where $\delta_i = \delta(r_i)$, $f_i = f(r_i)$, r_i indexes observation point and observed variable, and $\sigma_f(r)$ is the standard deviation of the forecast error (used as a convenient way to non-dimensionalise a multivariate problem).

We now require that the weights $W_i(r)$ be chosen so as to minimize the expected analysis error $\langle \alpha^2 \rangle$. If we calculate the expected value of (2) we find

$$\langle \alpha^2(r) \rangle = \langle f^2(r) \rangle + \sum_i \sum_j W_i W_j \{ \langle f_i f_j \rangle + \langle \delta_i \delta_j \rangle \} + 2 \sum_i W_i \langle f(r) f_i \rangle \qquad (3)$$

where we make the reasonable assumption that $\langle \delta_i, f_j \rangle = \langle \delta_i, f_i \rangle = 0$ and we assume that the observation and forecast errors have zero mean.

Taking $\partial/\partial W_i$ of Eq. (3) and setting the results to zero, to get the minimizing weights, we find

$$[W_i(r)] = [\langle f(r), f_i \rangle]^T (\mathbf{P} + \mathbf{D})^{-1}$$

where

$$\mathbf{P} = [\langle f_i, f_j \rangle]$$

$$\mathbf{D} = [\langle \delta_i, \delta_j \rangle]$$

are the prediction error and observation error covariance matrices associated with (and depending only on) the observed variables and the positions at which they are observed.

The analysis Eq. (2) then becomes

$$\alpha(r) - f(r) = [\langle f(r), f_i \rangle]^T (\mathbf{P} + \mathbf{D})^{-1} [\delta_i - f_i] \qquad (4)$$

or

$$a(r) = [F_i(r)]^T (\mathbf{P} + \mathbf{D})^{-1} [d_i]$$

where $a(r) = \alpha(r) - f(r)$ is the analysis increment, $[d_i] = [\delta_i - f_i]$ is the vector of data increments and $F_i(r) = \langle f(r), f_i \rangle$ is the correlation of forecast error (and variable) at the analysis point and the forecast error for the observation points (and variables).

To demonstrate the two step nature of the O/I algorithm we re-write Eq. (4) as

$$a(r) = [F_i(r)]^T \mathbf{P}^{-1} \{ \mathbf{P}[\mathbf{P} + \mathbf{D}]^{-1} \} [d_i] \qquad (5)$$

where we shall show that the first step

$$[a_i] = \mathbf{P}(\mathbf{P} + \mathbf{D})^{-1} [d_i] \qquad (6)$$

produces the analysed values for the observed quantities and the second step

$$a(r) = [F_i(r)]^T \mathbf{P}^{-1}[a_i] \tag{7}$$

is a straight forward interpolation of the analysed values of the observed quantities. It is clear that the error characteristics of the background field have a profound effect on how best to use the observed data. We now discuss the two steps separately, and then discuss the relations between them, following Hollingsworth (1987).

3.2 The O/I Filter

We consider first the simple case where the observation errors are uniform and uncorrelated, $\mathbf{D} = \sigma^2 \mathbf{I}$. In this case $\mathbf{P}$ and $\mathbf{D}$, which are both symmetric and positive definite, are also commutative, so they have the same eigenvector matrix $\mathbf{E}$, i.e.

$$\mathbf{P} = \mathbf{E}^T[\lambda^2]\mathbf{E}, \quad \mathbf{E}^T\mathbf{E} = \mathbf{I} \quad \text{and} \quad \mathbf{D} = \mathbf{E}^T[\sigma^2]\mathbf{E}$$

where $[\lambda^2]$ and $[\sigma^2]$ are diagonal matrices. It follows that

$$\mathbf{P} + \mathbf{D} = \mathbf{E}^T[\lambda_i^2 + \sigma^2]\mathbf{E}$$

$$(\mathbf{P} + \mathbf{D})^{-1} = \mathbf{E}^T \left[\frac{1}{\lambda_i^2 + \sigma^2}\right] \mathbf{E}$$

$$\mathbf{P}(\mathbf{P} + \mathbf{D})^{-1} = \mathbf{E}^T \left[\frac{\lambda_i^2}{\lambda_i^2 + \sigma^2}\right] \mathbf{E} = \mathbf{E}^T \left[\frac{1}{1 + \nu_i}\right] \mathbf{E} \tag{8}$$

where $\nu_i = \sigma^2/\lambda_i^2$ is the ratio of the observation error to forecast error in component i. Daley (1985) provides a discussion of the response characteristics of the O/I algorithm. For a univariate problem for which the auto-correlation of background error is bell-shaped, one finds that the background error is large on large scales and small on smaller scales. As a result, the O/I filter takes a lot of information from the data on large scales but not on small scales.

In the general case where $\mathbf{P}$ and $\mathbf{D}$ do not commute, we define $\mathbf{Q} = \mathbf{E}[\lambda^{-\frac{1}{2}}]$, so that $\mathbf{Q}^T\mathbf{P}\mathbf{Q} = \mathbf{I}$. Now $\mathbf{Q}^T\mathbf{D}\mathbf{Q}$ is symmetric and positive definite, so that

$$\mathbf{Q}^T\mathbf{D}\mathbf{Q} = \mathbf{R}^T[\nu]\mathbf{R}$$

where $\mathbf{R}$ is unitary, so that

$$[a_i] = (\mathbf{R}^T\mathbf{Q}^T)^{-1} \left[\frac{1}{1 + \nu}\right] (\mathbf{R}^T\mathbf{Q}^T)[d] \tag{9}$$

It is readily shown (Hollingsworth, 1987) that ν is a measure of the ratio of observation error to background error in a convenient 'information basis' which is complete but non-orthogonal.

Thus from both Eqs. (8) and (9) we see that the effect of the O/I filter is to selectively damp observational error in the data increment, according to the relative accuracy of the prediction error and observation error components relative to some suitable basis.

3.3 The O/I Interpolator

To discuss the interpolation properties of the O/I algorithm, we consider the following problem:

Let $\mathbf{F}(r) = \{F_j(r)\}$ be any set of n functions $j = 1, \ldots n$.

Let $\mathbf{A} = \{A(r_i)\}$ be any set of n data points $i = 1, \ldots n$.

Find an interpolation formula to interpolate the data $\mathbf{A}$ using the functions $\mathbf{F}$.
Solution:

Define $\mathbf{A}(r) = \sum_j c_j F_j(r) = \mathbf{F}^T(r)\mathbf{c}$ and let $\mathbf{A} = [A(r_i)]$

Since $\mathbf{F}$ interpolates $\mathbf{A}$, i.e.

$$\mathbf{A} = [\mathbf{F}^T(r_i)]\mathbf{c},$$

it follows that $\mathbf{c} = [\mathbf{F}^T(r_i)]^{-1}\mathbf{A}$, provided the inverse exists, so that

$$A(r) = \mathbf{F}^T(r)[\mathbf{F}^T(r_i)]^{-1}\mathbf{A}$$

We now specialise to the case when $\mathbf{F}(r)$ are the forecast error correlations, and we see that

$$a(r) = [\langle f(r), f_i \rangle]^T [\mathbf{P}]^{-1}[a_i]$$

as in Eq. (7). The well-known Lagrangian interpolation formula is one example of this formula, for which $\mathbf{P} = \mathbf{I}$.

3.4 The Relationship Between the Filter and Interpolator

Consider the case where $\mathbf{D} = \sigma^2 \mathbf{I}$.

We expand the data increment in terms of the eigenvectors $\mathbf{e}_i$ of $\mathbf{P}$ i.e.

$$\mathbf{d} = \sum d_i \mathbf{e}_i \tag{11}$$

Now the analysed values at the observation points are given by

$$[a(r_i)] = \mathbf{P}(\mathbf{P} + \mathbf{D})^{-1}\mathbf{d} = \mathbf{E}^T \left[\frac{1}{1+\nu}\right] \mathbf{E}^T \sum d_i \mathbf{e}_i$$

i.e.

$$[a(r_i)] = \sum_i \frac{1}{1+\nu_i} d_i \mathbf{e}_i$$

and the analysed value at any point is given by

$$a(r) = [F_j(r)]^T \mathbf{P}^{-1} [a_i]$$

or

$$a(r) = \sum \frac{1}{1+\nu_i} d_i \varepsilon_i(r) \tag{13}$$

where $\varepsilon_i(r) = [F_j(r)]^T \mathbf{P}^{-1} \mathbf{e}_i$ are the continuous interpolators of the eigenvectors $\mathbf{e}_i$, using the prediction error correlation. Eqs. (11), (12) and (13) provide a neat summary of the effect of the O/I algorithm in this case.

3.5 General Comments

The O/I algorithm uses the data at a single time to correct the background field in as efficient a way possible. An operational implementation of the O/I approach must resolve a number of practical issues. It is impossible to invert a matrix corresponding to a global dataset. One must make a series of local calculations which involve a number of compromises on data selection, continuity between adjacent analysis volumes, multivariate relationships and so on (Lorenc, 1981; Hollingsworth, 1987). Some of these limitations may be relaxed with a variational approach which seeks to minimize directly an expression analogous to Eq. (3). (Le Dimet and Talagrand, 1986).

An important aspect of the O/I approach is that it is a multivariate algorithm which can exploit linear multivariate relations between different variables. The final analysis is a linear combination of the 'structure functions' or forecast error correlations. Thus any linear constraint imposed on the structure functions is satisfied by the final continuous analysis within its domain of validity. Current operational systems impose global constraints of approximate non-divergence and hydrostatic balance, and a constraint of approximate geostrophy in mid-latitudes. Empirical justification for these constraints is provided by Hollingsworth and Lönnberg (1986) and Lönnberg and Hollingsworth (1986).

Finally we note that the O/I algorithm can only use data which is linearly related to the model variables. It has great difficulty in using data such as radiances which are non-linearly related to the model variables. Such data must be transformed or 'retrieved' to variables such as temperature or humidity which are linearly related to the model variables.

4. NON-LINEAR NORMAL MODE INITIALIZATION

The non-linear normal mode initialization (NNMI) algorithm plays an important role in current assimilation systems. It has little if any synoptic effect on the forecasts beyond day one or two. However it plays an important role in ensuring a smooth start to the forecast by controlling the amplitude of rapidly propagating gravity wave 'noise' which is introduced in the course of the analysis step. This control on the 'noise', Fig. 7, is critical for the quality control procedures at the next analysis time, as it ensures that the next background field is accurate and noise free.

The principle of the approach is rather simple (Machenhauer, 1977; Baer and Tribbia, 1977). The most troublesome noise is caused by gravity waves with large phase speeds (order 100–300 m/s). The propagation of such waves is not much affected by the spatial variations in the flow. Hence it is convenient to calculate the normal modes of the model for a resting basic state. These may be partitioned into two sets $[\eta_R]$ and $[\eta_G]$ consisting of Rossby waves η_R and gravity waves η_G.

For the gravity wave (or Rossby wave) components the model equations may be written in modal form as

$$\frac{\partial}{\partial t}\eta_G = i\lambda_G\eta_G + NL + F \tag{14}$$

where $i\lambda_G\eta_G$ consists of the linear terms in the equation for that mode and NL, F correspond to non-linear and forcing terms. For practical purposes the term $i\lambda_G\eta_G$

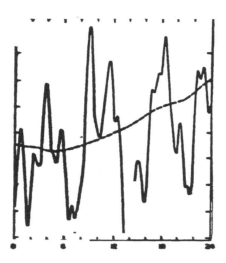

Figure 7
Time evolution over 24 hours of the surface pressure at a typical point
in forecasts from an uninitialised (solid) and from an initialised analysis
(Temperton and Williamson, 1981).

changes much more rapidly than the non-linear and forcing terms, so we group these terms together, in a term H, evaluate them at the initial time H_0, and approximate Eq. (14) by

$$\frac{\partial}{\partial t}\eta_G = i\lambda_G\eta_G + H_0$$

This has the solution

$$\eta_G = \underbrace{\exp(i\lambda_G t)\left(\eta_{G/t=0} + \frac{H_0}{i\lambda_G}\right)}_{\text{fast}} - \underbrace{\frac{H_0}{i\lambda_G}}_{\text{slow}}$$

which consists of a rapidly varying (fast) term and a slowly varying (slow) term. The first step is to modify the initial gravity wave amplitudes so as to set the fast term equal to zero, i.e.

$$\eta_{G/t=0} = -\frac{H_0}{i\lambda_G}$$

Because of the non-linearities, the initial tendency for the fast components will still be non-zero. However the procedure can be iterated and it converges rapidly for the modes of most concern — two iterations on the gravest modes in the vertical is usually sufficient. Recent developments covering diabatic effects and tides are discussed by Wergen (1988).

5. THE FORECAST MODEL

Current operational four-dimensional data assimilation systems provide high quality initial datasets for NWP, Bengtsson *et al.* (1982). If the atmosphere were a linear system then current procedures would be closely identified with the Kalman-Bucy filter approach to the interpretation of data (Ghil *et al.*, 1981). The forecast model plays a big role in any such procedure.

An important indicator of the quality of the initial datasets is the accuracy of short-range forecasts. Since the random components of the errors of the forecasts and the verifying observations are uncorrelated, the short-range forecast verifications provide a useful upper-bound on the accuracy of the analyses. Fig. 8 (Hollingsworth *et al.*, 1986) compares the 6 hour prediction error with the observation error for wind in two data rich areas. The observation error is defined to include instrumental error and representativeness error, so it includes the atmospheric variance on scales not resolved by the model. The figure shows that on the resolved scales the prediction error is comparable with the observation error. This provides a resource for reliable real time quality control in data rich areas. Even in data sparse areas the forecasts are good enough to enable detection of important errors in many different observing systems. Exploitation of these capabilities has led to distinct improvements in data quality, in analysis quality, and in forecast quality.

Another useful indicator of the way in which the state of the art has progressed in NWP is the following comparison. Fifteen years ago the hemispheric rms difference between different centres' analyses of the 500mb height field was about 30m. Today the rms hemispheric rms error of a two-day forecast is about 30m. This marked improvement has come about through better data, through better assimilation methods, through better models, and through the availability of the computer power needed to run today's sophisticated systems.

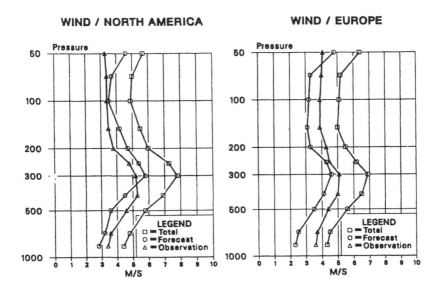

Figure 8
The perceived 6-hour forecast error for the vector wind (m/s) using quality
controlled data for North America and Europe, along with the calculated
prediction error on the resolved scales, and observation error which includes the variance of the unresolved scales. (Hollingsworth et al., 1986).

6. QUALITY CONTROL AND DATA MONITORING

Any operational assimilation system needs a suite of quality control procedures to
identify and reject erroneous data. Many checks are used including checks against
adjacent data, checks against the first guess, and checks against an analysis which
does not use the data (Lorenc, 1981). The availability of an accurate first guess
is of great assistance in quality control. Even in data sparse areas, the forecasts
provide a great deal of valuable quality control information.

6.1 Aireps

Fig. 9 shows a histogram of the magnitude of the vector wind differences between
aircraft data and forecast for a recent one month winter period over the Atlantic
and Pacific (Rubli, pers. comm.). The histogram consists of two parts (a) a 'regular'
part where the departures follow a 'regular' law for departures less than 30 m/s (the
histogram is not gaussian in this part, as it would then be parabolic rather than

linear); and (b) a flat part consisting of gross errors with magnitudes in excess of 30 m/s (Lorenc, 1987). If the magnitude of the vector wind departure is about 30 m/s then there is a 50-50 chance that the report has a gross error. The current operational tolerance in the first guess check for Aireps is set to about 28 m/s. Tests with the tolerance set to 20 m/s resulted in the rejection of obviously good data, as judged from synoptic evaluations (Lönnberg, pers. comm.).

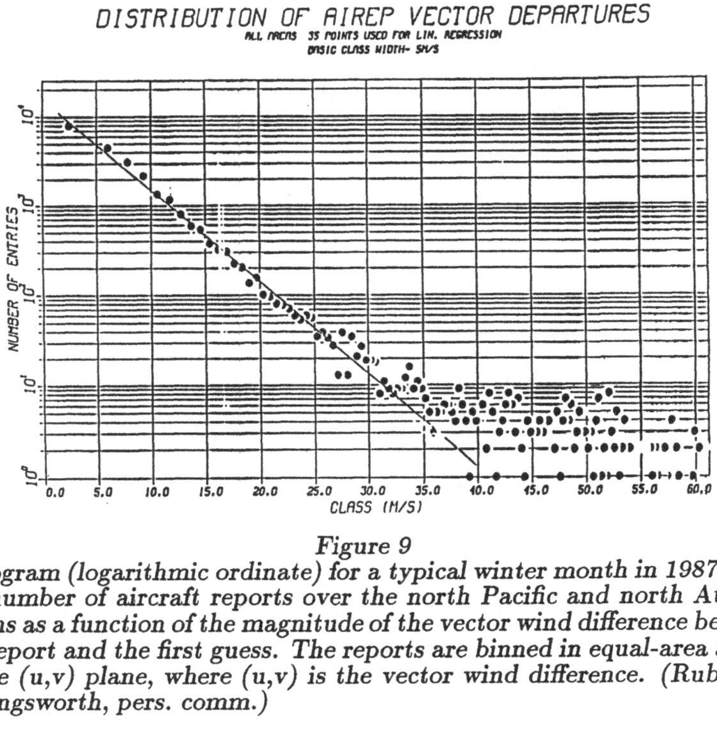

DISTRIBUTION OF AIREP VECTOR DEPARTURES

Figure 9
Histogram (logarithmic ordinate) for a typical winter month in 1987/88 of the number of aircraft reports over the north Pacific and north Atlantic oceans as a function of the magnitude of the vector wind difference between the report and the first guess. The reports are binned in equal-area annuli of the (u,v) plane, where (u,v) is the vector wind difference. (Rubli and Hollingsworth, pers. comm.)

An interesting example of an error that got through the quality checks before being detected in the operational evaluations is shown in Fig. 10 (Böttger, pers. comm.). The plot shows the reported position (marked wrong position) of a series of Aireps from an ASDAR unit on board an aircraft. There was a programming error in the ASDAR software which reversed the longitude relative to Greenwich and the latitude relative to the equator. The correct position is also shown in highlights. Although the data from the incorrectly positioned aircraft was not detected on

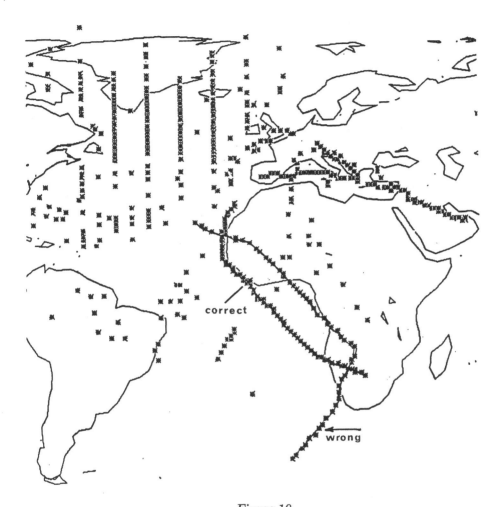

Figure 10
Distribution of aircraft reports within 3 hours of 1200 UTC on October 5 1984.
The tracks of a series of reports from an ASDAR unit are indicated as reported
(incorrect) and as they should have been reported (correct track). (Bottger and
Humphreys, pers. comm.)

322

the first flight of the ASDAR unit, the error was detected by the subjective (*ex post facto*) evaluation, and the software error was corrected rapidly by the Aircraft operator.

6.2 Radiosonde Monitoring

The fact that the assimilation system synthesizes many different data streams, as well as the dynamics and thermodynamics of the atmospheric system, means that biases between observations and the first guess frequently reveal important biases in the data, even in data sparse areas.

Fig. 11 shows the mean 250 mb wind bias between analysed and 6 hour forecast field over a remote area of the southern ocean in May 1981. The largest bias (9 m/s) occurs over a sonde on Marion Island. Hollingsworth *et al.* (1986) estimated that the station had a 12° bias in direction. The operators made local investigations and the reports have been of good quality since about 1983.

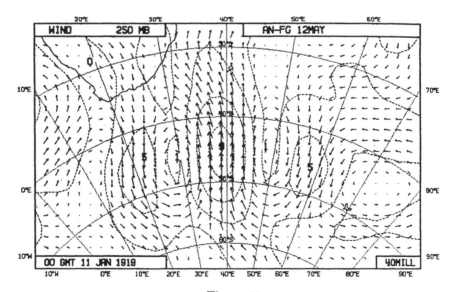

Figure 11
The mean analysis increment in the 250mb wind field near 40°E, 45°S during May 1981. The isotach interval is 2m/s.

This monitoring capability has been exploited systematically for several years. Figs. 12 and 13 shows three year time series of monthly mean biases relative to the first guess height at 500, 100 and 50 mb at two isolated stations (Bermuda Fig. 12, Diego Garcia Fig. 13) both managed by the same operator. Large diurnally varying biases are evident at both stations from 1985 until mid 1988. Investigations in May 1988 identified shortcomings in the observational practice at one of the stations. These shortcomings were remedied at both stations with evident benefit to the quality of the data (Radford, pers. comm.). All radiosonde stations are now monitored in this way.

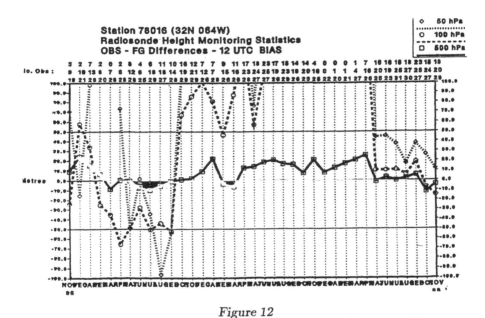

Figure 12
A three-year record (Nov 1985 – Nov 1988) of the monthly mean difference (observation minus first guess) between the 1200 UTC report and the corresponding first-guess for the geopotential at 500mb, 100mb and 50mb at Bermuda. Note the change in behaviour after May 1988. The numbers at the top give the number of reports at each level, with fewest reports at the 50mb level. (Radford, pers. comm.)

6.3 Remotely Sensed Wind Data

Remotely sensed data from satellites provides most of the available data for the North Pacific and North Atlantic, for the tropics and for the Southern Hemisphere. The use and quality control of this data presents special problems.

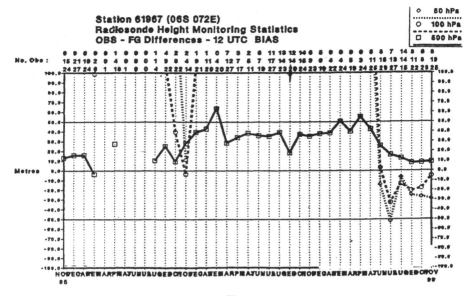

Figure 13
As Fig. 12 for Diego Garcia (Radford, pers. comm.)

a) Low-level cloud track winds

As discussed by Reed *et al.* (1988) low level cloud track winds can provide a vital means of documenting the location and intensity of tropical disturbances. More generally, the low level cloud track winds provide a vital means of documenting the intensity of the large scale tropical circulation, a matter of great concern for WCRP programs such as TOGA and GEWEX.

There are important practical difficulties in how to use single level data of this kind (Barwell *et al.*, 1986; Baede *et al.*, 1987) because one does not know how best to spread it out in the vertical. There are also indications that the data may be biased in some areas (Hollingsworth *et al.*, 1989). However the biggest problem with the low level cloud data is that we do not get enough of it. Fig. 14 shows a (rather poor) analysis of the 850 mb wind field over the tropical Atlantic associated with an easterly wave that developed into hurricane Gloria. The Meteosat picture for this time shows a well developed vortex with many possible low level targets (Reed *et al.*, 1988). As shown in Fig. 14, there are practically no low level winds near the disturbance but there are many CTW reports in the inactive area south of the ITCZ. This peculiar situation arose because of an administrative decision by the data producer not to produce a low level wind even if it were possible, provided an upper level wind could be produced in the same area.

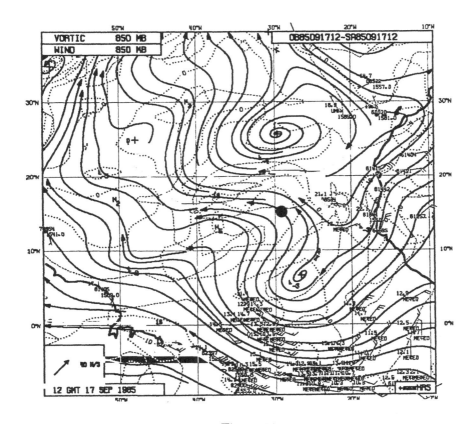

Figure 14
The operational ECMWF 850mb operational wind analysis over the tropical Atlantic at 1200 UTC on September 17 1985. Note the abundance of low-level cloud track winds south of the equator, and the paucity of reports north of the equator, even though there was in the area a very active system which eventually became hurricane Gloria (position marked by black dot). (Reed et al., 1988).

This anomaly has been pointed out to all the CTW producers, and the supply of data is slowly improving. However the situation is still far from being satisfactory. Fig. 15 shows the average number of cloud-track winds per day in each 5 degree box over the globe in August 1988. The low yields in the synoptically active tropical

326

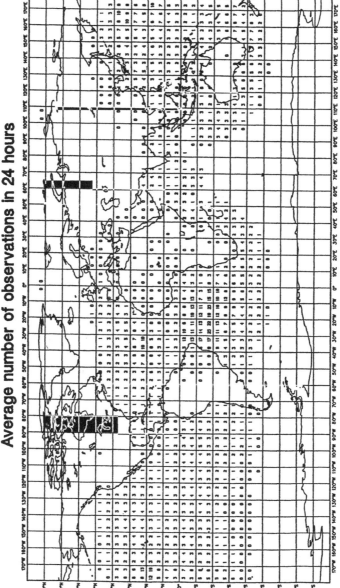

Figure 15

Average number of low-level cloud track winds available per day during August 1988 in each 5-degree box. Note the low yields along the main tropical zones of activity centred on about 15° N in the Atlantic and Pacific.

areas are quite unsatisfactory for NWP purposes and for climate purposes.

b) Upper level cloud track winds

The upper level cloud track winds have large speed dependent biases in strong wind situations, especially near the sub-tropical jet streams. This has been documented by Pierrard (1986) and by Kallberg and Delsol (1986). The latter paper, shows how bad the situation was in the FGGE data of ten years ago. Colocations with radiosondes showed that there were serious problems with all the data producers (see Fig. 16).

Recent studies of operational data from 1987 (Lönnberg, pers. comm., 1988) show that the situation has hardly changed in ten years. Fig. 17 shows the mean bias of the upper level winds relative to the first guess in December 1988. The worst biases are found for the GMS winds, both in the Northern Hemisphere (where the 6 hour forecast is accurate when judged against *in situ* measurements from sondes and aircraft) and in the Southern hemisphere (where the background is less accurate). Substantial biases are also evident in the data from the GOES and Meteosat satellites.

Some work is going on to improve the upper level data (Hayden, pers. comm.; Schmetz, pers. comm.). In view of the importance of these data for NWP and for climate purposes, it is essential that there be a thorough view of the procedures used to produce the winds. Until satellite-borne lidar wind sounders become available in the next decade or two, the cloud track winds will be the only source of wind data at upper and lower levels in many critical areas of the tropics. It is essential that the long-standing biases in these data be removed as soon as possible.

6.4 Scatterometer winds

The ERS-1 satellite, scheduled for launch in late 1990, will provide global coverage of the oceanic surface wind field. The measurements will actually be measurements of the normalised radar cross-section (NRCS) of the surface. This is a function of the radar polarisation, of the low-level wind speed, of the sea surface temperature, of the low level static stability (Woiceshyn *et al.*, 1987), as well as of the state of sea (Hasselmann, 1985). To interpret the NRCS measurement to geophysical quantities, such as wind direction and wind speed, a single model function will be used. Obviously there will be a need to determine if the model function is capable of coping with a wide range of atmospheric conditions. The basic validation of the

328

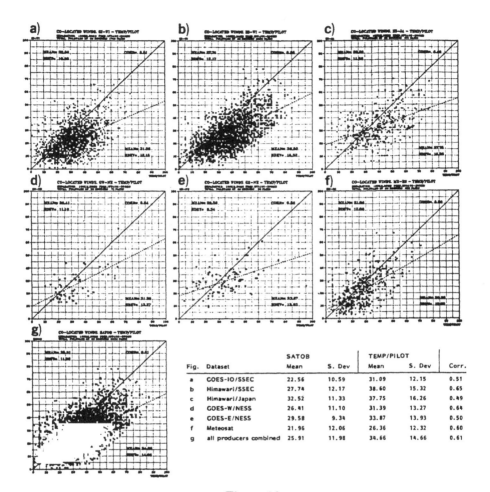

Figure 16
Scattergrams of colocated SATOB data and TEMP/PILOT data for
different producers. a) GOES-10/SSEC, b) Himawari/SSEC, c) Hi-
mawari/Japan, d) GOES-W/NESS, e) GOES-E/NESS, f) Meteosat/ESA,
g) all producers combined. The table gives the means and standard deviations as well as the correlation for each system in each figure. The dashed
line is the linear regression. The time window is ±3 hours, separation
100km and 20hPa. Only data between 100hPa and 300hPa, poleward of
20°N/S are included. Pairs differing more than 40 ms⁻¹ are excluded.
(Kallberg and Delsol, 1987)

329

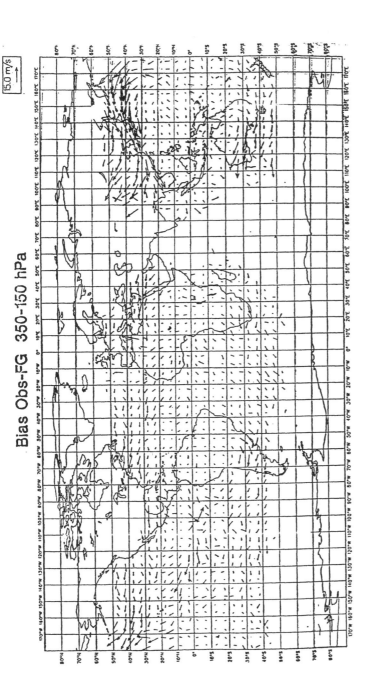

Figure 17

Mean difference between the upper tropospheric reported cloud track winds and the corresponding first guess wind (interpolated to the same position and pressure) during December 1988. The arrow at the top right gives a scale for 15m/s. From calculations with radiosondes, we know that the first guess is rather accurate, so these results demonstrate large biases in the upper-tropospheric cloud-track winds near the main jets.

geophysical products will be made in short-term calibration/validation campaigns. These are expensive exercises of short duration, and they cannot be undertaken globally.

There is a need for global monitoring of the quality of the ERS-1 data. As an example of a simple useful calculation that requires very little CPU time (but a great deal of data handling capacity), Fig. 18 shows results from an assimilation using Seasat scatterometer (SASS) data (Anderson *et al.*, 1989) when the SASS data were colocated with ship data. The upper plot (Fig. 18a) shows the reported wind direction at the ship relative to the forward pointing direction of the radar antenna. With 40,000 ship reports over a 5 day period we see the expected uniform distribution. For Seasat observations at the same times and places we see a very non-uniform distribution (Fig. 18b). This indicates significant inadequacies in the upwind-downwind-crosswind discrimination (Woiceshyn *et al.*, 1987). This problem took several years to be identified in the SASS data. It can be documented by comparison with a few days of ship data or even with a single analysis (Anderson *et al.*, 1988). This and other results in Anderson *et al.* indicate the potential of current data assimilation systems to intercompare observations with each other and with earlier observations, and the value of such procedures for quality assurance of observational data.

6.5 Temperature Soundings from Satellites

Recent studies by Pailleux, Kelly and Andersson (pers. comm.) have identified the occurrence of marked biases in the TOVS temperature retrievals generated by NESDIS. They found that the soundings at low levels tended to over estimate the low-level temperature in cold air outbreaks, and to under estimate the low-level temperature in areas of warm advection. The magnitude of the errors in the 1000–800 mb layer mean temperature could be as large as 8 or 10 degrees. Associated with the low level errors, they also found compensating errors in the vertical, leading to substantial errors in the static stability. These errors in the TOVS data contributed to a negative impact of the data on forecast skill, as not all of the errors were detected by the quality control procedures in the assimilation system.

Since cold air advection is most common on the west side of the mid latitude oceans and since warm air advection is most marked on the east side of the oceans,

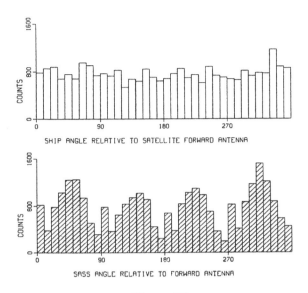

Figure 18
This figure compares the reported wind direction from ship data and from collected Seasat wind data: Top: Histogram of number of ship reports of wind direction relative to forward antenna beam direction, in 10 degree bins. Bottom: Similar result for the Seasat winds. The latter results illustrates problems in the Seasat data. (Anderson et al., 1988)

it is interesting to see one of the ways in which Pailleux, Kelly, Andersson and Radford documented the problems using some simple statistics.

Fig. 19 shows the monthly mean difference between the path-3 (cloudy, mainly MSU) retrievals and the first guess for the 1000–850 mb layer mean temperature for December 1988, averaged over 5 degree boxes. There are large positive biases (TOVS warmer than first guess) of up to 6 degrees off the east coasts of Asia and North America. There are also large negative biases (TOVS colder than the first guess) over North West Europe. We know from the radiosonde data that the first guess bias contributes less than 1 K to these results. For the path-1 (clear) retrievals the patterns are quite similar but the amplitudes of the bias is about half that shown in Fig. 19. Apart from the mid-latitude biases, there are also very large biases of up to 4 K in the sub-tropical highs.

The biases in the vertical temperature structure reported in the TOVS data are documented in Fig. 20, which shows the bias (TOVS minus first guess) in

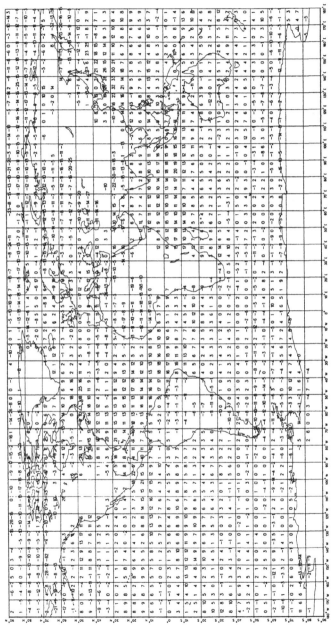

Figure 19

December 1988 mean difference (in tenths of a degree) between reported path-3 NESDIS retrievals for the 1000–850mb layer mean temperature and the first guess (observation minus first guess). Radiosonde data shows that the first guess is quite accurate near the east coasts of N America and Asia. These results demonstrate large biases in the retrieved temperatures. (Pailleux, Anderson, Kelly, Radford, pers. comm.)

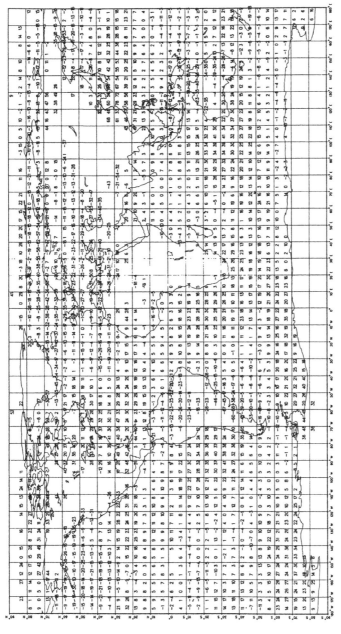

DEC 1988
NOAA10 = MSU SOUNDINGS
BIAS OBS-FG 860-1000 HPA

Figure 20

As Fig. 19. for a crude measure of the tropospheric static stability: $(T1 - Th)$ from clear (path 1) retrievals, where $T1$ is the layer mean temperature for the 1000–700mb layer and Th is the layer mean temperature for the 500–300mb layer. Note the large biases due to the retrieval procedure on both sides of the north Atlantic and north Pacific. (Pailleux, Anderson, Kelly, Radford, pers. comm.)

clear (path 1) retrievals for the difference $T_L - T_H$ between the (low level) layer mean temperature for the 1000–700 mb layer (T_L) and the (upper level) layer mean temperature (T_H) for the 500–300 mb layer. Large biases of up to 6 degrees are evident in Fig. 20 off the coasts of Asia and North America. The path 3 retrievals show biases in some cases of up to 6 degrees. We know from the radiosonde data that almost all this bias is attributable to the TOVS data rather than the first-guess. The reasons for these biases are still under investigation. However the TOVS minus first guess biases differ little between the statistical retrievals produced before September, 1988, and the physical retrievals produced since then. Thus the biases are not attributable to the retrieval method, but to some other aspect of the processing.

7. THE VALUE AND LIMITATIONS
OF GLOBAL NWP DATASETS FOR CLIMATE STUDIES

Global NWP real-time analyses of the mass wind and humidity fields from the surface to 10mb have been archived routinely from the FGGE year (1979) to the present. High-resolution fields have been made available to the community for the FGGE year and for the TOGA experiment. Lower resolution analyses are available for the intervening years (1980–84).

7.1 Advantages

The advantages of such global NWP datasets for climate studies are considerable. They present a coherent synthesis of large and heterogeneous data streams which could not possibly be handled, quality controlled, or synthesised, by a solitary scientist or by a small group. It is a formidable problem to form a coherent view of the atmospheric flow based on noisy data from say, a temperature and humidity sounder in one orbit, a scatterometer in another, and randomly distributed winds from cloud track vectors from geostationary satellites, as well as data from ships, aircraft, and radiosondes. The NWP datasets present a coherent global synthesis of the atmospheric observations, on the resolved scales, which is of known accuracy.

The global NWP datasets have found widespread application in many climate studies. Detailed structural studies of phenomena such as tropical easterly waves have shown that despite the great data sparsity in the Tropics, the structures in the NWP analyses are remarkably similar to what can be inferred from special field experiments such as GATE (Reed *et al.*, 1988). Studies of the 30–60 day tropical oscillation, based on the NWP analyses, have contributed greatly to the elucidation of the structure and mechanism of the oscillation (Murakami, 1988). The NWP analyses have been used in studies of blocking and low-frequency variability in mid-latitudes in both hemispheres, leading to important insights on the interaction of high-frequency transients with low frequency motions (Hoskins and Sardeshmukh, 1987). The global NWP real-time datasets are the subject of intensive study by oceanographers to determine their suitability for forcing their models of the equatorial ocean, of the global ocean, and for forcing models of ocean surface waves.

Both the atmospheric and oceanographic studies have demonstrated a number of problems in the earlier analyses. The NWP centres have responded constructively to criticism, and many of the earlier problems have been corrected. As the NWP analyses find wider application, further problems will undoubtedly be found with them. That is in the nature of scientific work. The NWP centres welcome this interaction, and benefit considerably from it, as it leads to distinct improvements in the quality of their analyses.

7.2 Limitations

One must of course take care in using the NWP data-sets for climate studies, as the resulting picture of the atmosphere may be affected by limitations in the NWP models or in the assimilation procedures. Many of the analysed quantities are quite robust, and results based on them will be reliable. Some of the inferred (but unobserved) fields produced by the NWP systems are sensitive to the model parameterisations and are clearly labelled with 'health warnings' for research workers. Typical examples are the diabatic heating field, the humidity field and the divergent wind field. A principal goal of WCRP is to provide the observations needed to document these fields.

NWP assimilation systems have an inherent truncation, due the grid used in the model. Information in an observation on scales of motion shorter than the truncation limit is regarded as sampling error. The NWP assimilation procedures are explicit and avowed filters of the observations (Hollingsworth, 1987). The analysis module behaves like an elaborate pattern-recognition device which knows the dynamical properties of typical forecast errors for the synoptic patterns. Information matching these characteristics is extracted from the observations, and everything else is damped as noise. Thus the NWP assimilation systems will attempt to represent the grid-averaged convective heating within a tropical convective system, but will smooth out information on individual cumulus clouds. Workers interested in such phenomena with very short space and time scales will not wish to use the NWP datasets.

Evolutionary changes in the procedures which generate the NWP datasets must also be taken into account by climate scientists. For example, the vertical velocity field in data sparse areas of the tropics is almost entirely controlled by the convective parameterisations and the assimilation procedures. The intensity of the tropical circulation in the ECMWF datasets has shown marked secular change as a result of improvements in the parameterisations and procedures (Wergen, 1988; Bengtsson and Shukla, 1988). The analysed inter-annual variability in this field is mainly due to changes in assimilation procedures.

This has led to calls for the re-analysis of the observations of the last decade with a homogeneous assimilation system. Such a re-analysis would be a large and expensive undertaking. The results would be affected by important fluctuations in the availability of polar orbiter sounding data, of geostationary cloud track wind data, and of buoy data in the Southern Hemisphere. It would be necessary to reproduce extensive real time quality control information on all the observing systems. Operational forecast centres have a strong vested interest in doing thorough quality control on all observations before use, but they do not necessarily archive all this information. Data monitoring activity is being systematised between the large centres under the aegis of WMO. Archiving of complete real-time quality control information would be of considerable value in any planned re-analysis.

8. INVERSION AND QUALITY CONTROL
OF REMOTELY SENSED DATA

Remotely sensed radiometric measurements need to be inverted in some way if they are to be synthesised with or compared with direct measurements. Thermal radiance measurements can provide estimates of atmospheric temperature and composition, using the radiative transfer equation. Winds can be inferred from scatterometer measurements of normalised radar back-scatter cross-sections. To do so one needs an empirical relation (called a model function) to relate the radar returns to several possible surface winds, and one may then need an ambiguity-removal algorithm to choose the most likely wind direction. Similar inversion problems are found with many types of remotely sensed data. Since the inversion problems are usually not well-posed, it is essential to use *a priori* information to achieve a reasonable solution. Many heuristic approaches have been tried with greater or lesser success. Recent theoretical developments have proposed a new approach to the problem which is more complete and rigorous.

8.1 Unified Variational Retrieval/Analysis Procedures

Experience with remotely sensed data in meteorology indicates that the best approach to the inversion problem is to work with the observed radiometric quantities, and with *a priori* estimates for them derived from a 'forward' calculation (Menke, 1984). The 'forward', or *a priori* estimate, of the observed radiometric quantities is derived by an explicit algorithm of known accuracy. The inputs to the forward calculation are the relevant geophysical dynamic or thermodynamic quantities to be retrieved. The *a priori* estimates for the inputs may come from an assimilating model for the dynamic and thermodynamic fields.

The forward calculation is generally quite non-linear, as in the case of the radiative transfer equation. Once the forward calculation is available, general variational formalisms can be used (Le Dimet and Talagrand, 1986; Lorenc, 1986) to derive a maximum-likelihood estimate of the geophysical quantities. These begin by using the observations, together with the 'forward' estimate of the observations, and the error statistics of both, to get a first estimate for the observed quantity. An iterative minimisation follows to find the variational maximum-likelihood solution. The variational formulation usually includes penalties on non-meteorological features of the solutions. The descent algorithm will usually require functional differentiation of the non-linear forward calculation, or linearisation about the current iterate, in order to find the descent direction. Early results are encouraging (Eyre, 1989; Eyre and Lorenc, 1989).

This approach fits naturally into the variational approach to 4-dimensional data assimilation discussed earlier. In such a unified approach, the best possible use can be made of all earlier, and indeed some later data. The unified variational approach is being developed for operational use in the 1990's at a number of meteorological and oceanographic research centres. It offers a solid theoretical foundation for optimal use of remotely sensed data.

8.2 Coupled Assimilation Systems

In some cases the use of remotely sensed data will require *a priori* information about several geophysical systems such as wind, sea-surface temperature, low-level static stability, and state of sea. For such problems, a coupled model embedded in an assimilation system has obvious advantages. As an example of how coupled models will be exploited in this context, one may cite the work of the WAM group on the development of a coupled wind-wave assimilation system at ECMWF. One of the main motivations for this is the evidence that the radar returns measured by scatterometers over the oceans are affected not just by the near-surface wind and the short gravity-capillary waves, but also by the longer $O(10s)$ waves. A correct interpretation (or 'retrieval') of the scatterometer measurements may require prior information on the winds from the atmospheric model, and prior information on the waves from the wave model (Hasselmann, 1985). An example of some of the potential benefits of this approach are discussed in Janssen *et al.* (1989). They show how a coupled system can be used to cross-validate wind and wave measurements.

A partly-coupled approach is being implemented for the assimilation of wind data and ocean circulation data for TOGA. There is substantial evidence that ocean circulation models can be used to assess the quality of estimates of surface heat and momentum fluxes derived from atmospheric assimilation. Again a variational 4-dimensional assimilation using wind, altimeter, current, and BT data is likely to provide the most complete synthesis of the data.

8.3 Quality Control of Remotely-Sensed Data

Apart from its role in interpreting the observational data, the four-dimensional assimilation process plays a vital role in quality controlling the observational data. Murphy's Law ('anything which can go wrong will go wrong') applies as much to complex heterogeneous observation systems as to mundane everyday affairs. It is essential that data problems be identified as quickly as possibly so that long series of data are not corrupted by errors which could have been identified and corrected in near-real time.

Verifications of the NWP analyses and short-range forecasts with the observations provide powerful tools for quality control of the observational data, and for documenting the most important short-comings of the model. The value of this capability of modern data assimilations has been demonstrated extensively in meteorological applications of both remotely sensed and *in situ* data. Studies of this kind are of particular benefit to data producers if they demonstrate special types of problems in special synoptic situations. Fast detection and correction of data problems is as valuable for climate studies as for NWP.

9. REAL-TIME INTEGRATION AND SYNTHESIS OF WCRP OBSERVATIONS

Even though meteorologists use the word 'analysis' to describe their initial datasets, the word 'synthesis' would be a more accurate description of the way in which the current data are combined with earlier data and with all our *a priori* information to produce as accurate a picture of the atmosphere as possible. The value of the global NWP datasets to climate studies is increasingly appreciated for the coherent synthesis they draw from the observations. This synthesis is a valuable additional tool for climate studies, and has stimulated many new insights.

The NWP experience in quality control, interpretation and synthesis of heterogeneous *in situ* and remotely sensed data is a useful guide to the problems involved in full utilisation of the satellite systems to be deployed for WCRP studies in the 1990's. Remotely sensed geophysical data from the TOGA and other WCRP experiments will be used most efficiently if they can be provided in real time to the operational weather/ocean centres producing the level III-a analyses.

In the next decade, operational NWP centres will develop and implement rather sophisticated and unified retrieval/assimilation systems. The III-a centres can provide beneficial feed-back to the data producers. Regional and global problems with the observational data can be identified quickly and put right. The fact that the 'quick-look' III-a data has used all the remotely sensed data will be a considerable stimulus to the scientific community to work on the WCRP data early in the experiment. Otherwise there will be a long lead time before the data from the special instruments are made generally available.

In processing experimental data one needs two or more passes through the data, for quality control, before final conclusions can be drawn. If the remotely sensed data is used in real time, and if the quality control information from the III-a assimilations is archived, then we will have an invaluable resource for any delayed-mode III-b analyses.

All this activity has clear relevance to the wider field of climate studies concerned with the interpretation and synthesis of remotely sensed data on the climate system — atmosphere/vegetation/ocean waves/ocean circulation/biosphere/ice. There is an enormous potential to improve our understanding of the climate system if we develop the tools to interpret remotely sensed data in an effective way. For example measures of the chlorophyll content of the mixed layer of the ocean provides information on both the marine biology and marine dynamics. Our understanding will be considerably sharpened if we build quantitative models of the processes affecting the climate system, and then use the models as a part of an assimilation system. The confrontation between the models and the data will be an effective stimulus to improve the quality of the models, of the data, and of our understanding of the system.

REFERENCES

Anderson, D., A. Hollingsworth, S. Uppala, P. Woiceshyn (1989): A study of the feasibility of using sea and wind information from the ERS-1 Satellite, Part I: Quality Control. Submitted to *J. Geophys. Res.*

Baede, A.P.M., S. Uppala and P. Kallberg (1987): Impact of Aircraft Wind Data on ECMWF Analyses and Forecasts during the FGGE period 8–19 November to appear in *Quart. J. Roy. Meteor. Soc.,* July 1987.

Baer, F. (1977): Adjustment of initial conditions required to suppress gravity oscillations in non-linear flows. *Beitr. Phys. Atmos.,* **50**, 35–366.

Barwell, B.R., and A.C. Lorenc (1985): A study of the impact of aircraft wind observations on a large scale analysis and numerical weather prediction system. *Quart. J. Roy. Meteor. Soc.,* **111**, 103–129.

Bengtsson, L., M. Kanamitsu, P. Kallberg and S. Uppala (1982): FGGE 4-dimensional data assimilation at ECMWF. *Bull. Am. Meteor. Soc.,* **63**, 29–43.

Bengtsson, L., J. Shukla (1988): Integration of Space and in-situ observations to study global climate change. *Bull. Am. Meteor. Soc.,* **69**, 1130–1143.

Daley, R. (1985): The analysis of synoptic scale divergence by a statistical interpolation procedure. *Mon. Wea. Rev.,* **113**, 1066–1079.

Eyre, J. (1989): Inversion of cloudy satellite sounding radiances by nonlinear optimal estimation; theory and simulation for TOVS. Submitted to *Quart. J. Roy. Meteor. Soc.*

Eyre, J., A.C. Lorenc (1989): Direct use of satellite sounding radiances in numerical weather prediction. *Meteor. Mag.,* **118**, 13–16.

Gandin, L.S. (1963): Objective analysis of meteorological fields. Translated from Russian by the Israeli Program for Scientific Translations (1965).

Ghil, M., S. Cohn, J. Tavantzis, K. Bube and E. Isaacson (1981): Applications of estimation theory to numerical weather prediction. In Dynamical Meteorology Data Assimilation Methods, Ed. L. Bengtsson, M. Ghil, E. Källén, pub Springer, pp. 139–224.

Hasselmann, K. (1985): Assimilation of microwave data in atmospheric and wave models. Proc. ESA Alpbach Conference on use of satellite data in wave models, pp. 47–52.

Hollingsworth, A. (1987): Objective analysis for numerical weather prediction. Special Volume *J. Met. Soc. Jap.* 'Short and Medium Range Numerical Weather Prediction' ed. T.Matsuno, pp. 11–60.

Hollingsworth, A., A.C. Lorenc, M.S. Tracton, K. Arpe, G. Cats, S. Uppala and P. Kallberg (1985): The response of Numerical Weather Prediction Systems to FGGE II-b Data Part I: Analyses. *Quart. J. Roy. Meteor. Soc.*, **111**, 1–66.

Hollingsworth, A, D.B. Shaw, P. Lönnberg, L. Illari, K. Arpe and A.J. Simmons (1986): Monitoring of observation quality by a data assimilation system. *Mon. Wea. Rev.*, **114**, 861–879.

Hollingsworth, A. and P. Lönnberg (1986): The statistical structure of short range forecast errors as determined from radiosonde data. Part I: The wind errors. *Tellus*, **38A**, 111–136.

Hollingsworth, A. J. Horn and S. Uppala (1989): Verification of FGGE assimilations of the tropical wind-field: The effect of model and data bias. *Mon. Wea. Rev.*, **117**, April Issue.

Hoskins, B.J., P.D. Sardeshmukh (1987): A diagnostic study of dynamics of the northern hemisphere winter of 1985/86. *Quart. J. Roy. Meteor. Soc.*, **113**, 759–778.

Janssen, P., P. Lionello, M. Reistad, A. Hollingsworth (1989): Hindcasts and data assimilation with the WAM model during the SEASAT period. To appear in *J. Geophys. Res.*

Kallberg, P. and F. Delsol (1986): Systematic biases in cloud-track-wind data from jet stream regions. Programme on Short and Medium Range Numerical Weather Prediction Research, Tokyo pp. 15–18. WMO PSMP Rept 19, available from WMO, Geneva.

Le Dimet, F.X. and O. Talagrand (1986): Variational algorithms for analysis and assimilation of meteorological observations: theoretical aspects. *Tellus*, **38a**, 97–110.

Lönnberg, P. and A. Hollingsworth (1986): The statistical structure of short range forecast errors as determined from radiosonde data. Part II: Covariance of height and wind errors. *Tellus*, **38A**, 137–161.

Lorenc, A.C. (1981): A global three-dimensional multivariate statistical interpolation scheme. *Mon. Wea. Rev.*, **109**, 701–721.

Lorenc, A.C. (1986): Analysis methods for numerical weather prediction. *Quart. J. Roy. Meteor. Soc.*, 1177–1194.

Machenhauer, B. (1977): On the dynamics of gravity oscillations in a shallow water model, with application to normal mode initialisation. *Contrib. Atmos. Phys.*, **50**, 253–271.

Menke, W. (1984): Geophysical Data Analysis: Discrete Inverse Theory. Academic Press pp. 260.

Murakami, T. (1988): Intraseasonal atmospheric teleconnection patterns during the northern hemisphere winter. *J. Climate*,I, 117–131.

NASA (1988a): Earth Observing System, Tech.Memo. 86129.

NASA (1988b): Earth System Science — A closer view.

Pierrard, M.C. (1985): Intercomparison between cloud-track-winds and radiosonde winds. In Proceedings of Fifth Meteosat User Conference, Rome, pub. European Space Agency.

Reed, R.J., A. Hollingsworth, W.A. Heckley, F. Delsol (1988): An evaluation of the performance of the ECMWF operational forecasting system in analysing and forecasting tropical easterly Wave disturbances. *Mon. Wea. Rev.*, **116**, 824–865.

Rodgers, C.D. (1976): Retrieval of atmospheric temperature and composition from remote measurement of thermal radiation. *Rev. Geophys. Space PHys.*, **14**, 609–624.

Simmons, A.J. and B.J. Hoskins (1979): The downstream and upstream development of unstable baroclinic waves. *J. Atmos. Sci.*, **36**, 1239–1254.

Temperton, C. and D.L. Williamson (1981): Normal mode initialisation for a multilevel grid-point model. Part I: Linear Aspects. *Mon. Wea. Rev.*, **109**, 729–743.

Wergen, W. (1988): Diabatic non-linear normal mode initialisation for a spectral model. *Beit. Phys. Atmosph.* **61**, 274–302.

Woiceshyn, P.M., M.G. Wurtele, D.H. Boggs, L.F. McGoldrick and S. Peteherych (1986): 'The necessity for a new parameterisation of an empirical model for wind/ocean scatterometry'. *J. Geophys. Res.*, **91**, 2273–2288.

INTRODUCTION TO CHEMICAL TRACERS
OF THE OCEAN CIRCULATION

J.-F. MINSTER
UM39/GRGS,
18, av. E. Belin,
31055 Toulouse Cedex,
France.

INTRODUCTION

My task in this set of lectures was to provide some background information on chemical tracers of the ocean circulation. Thus, I present in a first section a dimensional analysis of the conservation equation for oxygen in the North Atlantic ocean. Section 2 gives an introduction to some of the most useful tracers, as well as of the species frequently measured: oxygen, nutrients and carbon species. Section 3 describes in more detail the processes at play at the air-sea interface and how they affect the tracer concentrations at the surface. Finally in section 4 two recent studies on how tracers constrain diapycnal mixing in the deep ocean are discussed.

D. L. T. Anderson and J. Willebrand (eds.), Oceanic Circulation Models: Combining Data and Dynamics, 345–376.
© 1989 by Kluwer Academic Publishers.

1. DIMENSIONAL ANALYSIS OF A
TRACER CONSERVATION EQUATION

In dynamical oceanography, one generally starts by estimating the relative importance of the various terms of the equations. Similarly, it would be desirable, when studying tracers of the ocean circulation, to estimate the relative importance of the different terms of the diffusion-advection equation which describes their behavior. This is rarely done, mainly because of insufficient data. Also, tracers have been used in the past to evaluate the intensity of some processes. The best example is the estimation of the average intensity of vertical transports, deduced from ^{14}C data and a one-dimensional vertical diffusion-advection model (e.g. Munk, 1966; Craig, 1969). Nevertheless, two attempts have been made recently in this direction, using dissolved oxygen concentration in the North Atlantic ocean.

It is useful to rewrite the basic equation used in these attempts. With classical notations:

$$\frac{\partial C}{\partial t} + \mathbf{u}.\nabla C = \nabla.(K\nabla C) + J \tag{1}$$

Here, C is a tracer concentration, $\mathbf{u}$ is the velocity vector, K stands for diffusion and J for the source-sink terms.

In general, this equation is applied on a relatively large scale, i.e. larger than a few hundreds of kilometers. K is thus a parametrisation of processes below the mesoscale, or even including the mesoscale. Also, in general, K is considered constant. This may not be true, in particular near the boundaries. Indeed, typical values of K_v, the vertical diffusion coefficient, are of the order of $1 \times 10^{-4}\mathrm{m}^2\mathrm{s}^{-1}$ in the interior of the ocean basins, but two orders of magnitude larger near the boundaries (e.g. Sarmiento $et\ al.$, 1976; Armi, 1978). As a consequence, terms due to the gradients of the diffusion coefficient are neglected, which are difficult to estimate but could be significant. Finally, in most cases, mixing in the ocean is calculated along isopycnal levels and vertically. However, lateral mixing should rather occur along neutral surfaces (e.g. McDougall, 1987), which may be quite different from isopycnal levels in the deep ocean. With these limitations in mind, let us consider the above-mentioned analyses.

1.1 Oxygen in the Deep Northwest Atlantic Ocean

Using the TTO (Transient Tracers in the Ocean) data, I recently estimated the various terms of Eq. (1) near the boundary and in the interior of the deep Northwest Atlantic basin (Minster, 1985). The results are given in Table 1. Information concerning the TTO expedition can be found in a special issue of the Journal of Geophysical Research (Brewer *et al.*, 1985).

Table 1

WESTERN BOUNDARY (TTO 11)				
v	w	K_i	K_v	J
m/s		m^2/s		
1×10^{-1}	1×10^{-4}	1×10^{2}	1×10^{-4}	
		1×10^{4}	1×10^{-2}	
dC/dy	dC/dz	d^2C/dy^2	d^2C/dz^2	
μmol/kg/m		μmol/kg/m^2		
1×10^{-6}	5×10^{-3}	2×10^{-11}	1×10^{-5}	
		products		
		μmol/kg/s		
1×10^{-7}	5×10^{-7}	2×10^{-9}	1×10^{-9}	4×10^{-9}
		2×10^{-7}	1×10^{-7}	
INTERIOR (TTO 38)				
v	w	K_i	K_v	J
m/s		m^2/s		
1×10^{-3}	1×10^{-6}	1×10^{2}	1×10^{-4}	
		1×10^{3}	1×10^{-6}	
dC/dy	dC/dz	d^2C/dy^2	d^2C/dz^2	
μmol/kg/m		μmol/kg/m^2		
1×10^{-5}	2×10^{-3}	7×10^{-12}	2×10^{-6}	
		products		
		μmol/kg/s		
1×10^{-8}	2×10^{-9}	7×10^{-10}	2×10^{-10}	4×10^{-9}
		7×10^{-9}	2×10^{-12}	

TTO station 11 was used for the boundary, and station 38 for the interior. In the table, the meridional velocity of the Western Boundary Undercurrent has been taken as 0.1 ms^{-1} (Warren, 1981). The horizontal velocity in the interior was chosen such that the 2000 km of the basin would be ventilated in less than 70 years (see e.g. Broecker, 1981). The zonal velocity across the boundary is not known and was not considered. The vertical velocity was taken as 1×10^{-3} of the horizontal one. Different values of the diffusion coefficients are taken in the table. The zonal and meridional isopycnal turbulent coefficients were taken as equal, though this may not be true. A typical value of oxygen consumption of the order of 0.12 μmol/kg/y was used (Broecker and Peng, 1982). The gradients and curvatures were calculated on the vertical profiles and along the $\sigma_4 = 45.89\%_0$ isopycnal level. Six levels on the vertical and six stations along the isopycnal were used. However, these values should only be considered as orders of magnitudes.

For the boundary, one finds that the north-south advection term is large, even though the oxygen gradient is small. In order to balance this term, it seems necessary that the horizontal diffusion coefficient be as large as 1×10^4m^{2}s^{-1}, or the vertical one of the order of 1×10^{-2}m^{2}s^{-1}. Interestingly, the latter value is close to those derived for the benthic boundary layer. Unless oxygen consumption is two orders of magnitude larger than taken here, it is not likely to play a significant role.

In the interior, a number of terms are of comparable order of magnitude. However, for vertical diffusion to play a significant role, it is necessary that the diffusion coefficient be larger than 1×10^{-4}m^{2}s^{-1}. If it is as small as 1×10^{-6}m^{2}s^{-1} (Osborn and Cox, 1972), mixing is essentially isopycnal. On the contrary, in this domain, even if oxygen consumption is 10 times less than the world average (which is likely the case in these deep waters), it contributes to changing the oxygen concentration.

1.2 Oxygen near the Surface Northeast Atlantic Ocean

Jenkins (1987) also compared the relative importance of the various terms of Eq. (1) for the oxygen and tritium data from the so-called beta-triangle area, centered near 30°N, 30°W. The analysis was made for $\sigma_\theta = 26.20\%_0$, near 200 m. Here, the horizontal velocity was taken as 0.01 ms^{-1}, as derived from a T-^{3}He age (see below). The isopycnal diffusion coefficient was $K_i = 400$ m^{2}s^{-1}, as estimated by Armi and Stommel (1983) in this area. The vertical one was estimated as $K_v = 1 \times 10^{-5}$m^{2}s^{-1} using Garrett's formula (1979). Then:

$$\mathbf{u}.\nabla_H O_2 = 0.5 \quad \text{pmol/kg/s}$$

$$K_i\nabla^2 O_2/\mathbf{u}.\nabla_H O_2 < 0.1$$
$$K_v\nabla^2 O_2/\mathbf{u}.\nabla_H O_2 > 0.1$$
$$w(\partial O_2/\partial z)/\mathbf{u}.\nabla_H O_2 = 1$$

Thus, Eq. (1) reduces to:

$$\mathbf{u}.\nabla_H O_2 + w\partial O_2/\partial z - J = 0$$

In order to balance this equation, oxygen consumption has to be of the order of 10 μmol/kg/y, an estimate in fair agreement with others at this depth.

1.3 Conclusion

A number of features of tracers are illustrated by these analyses. First, it is clear that, like for dynamical quantities, the relative importance of the various terms depends on where they are considered, as well as on the space/time scale of the analysis.

For oxygen on the scale of the above studies, that is, not the basin scale, but rather a 100 km horizontal scale, mixing is not significant in surface layers, but plays a role comparable to advection in the deep ocean, even near the boundaries. In addition, in both studies, the non-conservation of the element is a term comparable to transport.

This is precisely where the interest of tracers in the ocean circulation can be found: while it is clear that the advection field can be best constrained by a dynamical description, tracers provide a measure of mixing processes. This capacity is enhanced by the non-conservativity of the elements: it is well-known that it is consumption which makes oxygen a tracer of the world water masses. Similarly, and to keep with the same element, Thomas et al. (1989) could demonstrate that because of the strong oxygen gradient created by consumption below the photic zone, this element provides much more severe constraints on vertical mixing below the mixed layer than temperature, in the North-East Pacific ocean.

Finally, because the boundary conditions and source-sink processes are different for each tracer, the relative importance of the various terms of Eq. (1) will differ for each. They will thus provide constraints in different places and on different scales. The next section provides some of the necessary background information on the tracers of ocean circulation.

2. GENERAL INFORMATION ON TRACERS OF OCEAN CIRCULATION

Providing general background information on the various tracers is an extensive task, which exceeds the volume of these notes. Most of the necessary knowledge is given in the textbook by Broecker and Peng (1982). More specific information related to tracing the ocean circulation can also be found in the documents prepared for the World Ocean Circulation Experiment (WOCE). However, I frequently found it useful to recall some of the basic features of tracers. Here, I wish to provide particularly first order information on the boundary conditions and on the source-sink terms. A list of the tracers is given in Table 2. I will discuss only a few of these.

2.1 Dissolved Oxygen and Nutrients

It is taken for granted that oxygen and nutrients should be part of the suite of tracers to be measured during WOCE. Yet, the behavior of these tracers is far from simple.

Oxygen concentration varies from about 350 μmol/kg in surface waters at 0°C to 0 μmol/kg near 500 m in the Northeast Pacific ocean. Boundary conditions are dominated by oxygen exchange with the atmosphere, near equilibrium being reached on a monthly time scale. Due to the change of solubility of oxygen with temperature, the surface concentration varies from about 200 μmol/kg in warm tropical waters to 350 μmol/kg in polar waters. However, because of a number of effects to be described in the next section, the surface concentration is generally supersaturated by a seasonally variable amount of the order of 5%. This is comparable to the effect of consumption near the surface in one year, or to 10% of the effect of mixing of warm water with cold water.

Below the depth of the photic zone, oxygen is consumed by the oxidation of organic matter. The average rate of consumption is 350 μmol/kg/1000y, i.e. the observed total variation of concentration over the turnover time scale of deep waters. However, as already seen in the previous section, this value varies with depth and geographical location. In subsurface waters, J is of the order of 10 μmol/kg/y, whereas in deep waters, it can be as low as 0.01 μmol/kg/y. Presently, one does not have a map of J for the world ocean. As seen in the previous section,

Table 2

TRACERS OF THE OCEAN CIRCULATION

Tracer	Chemical form	Biogeochem. activity	Time scale	Source function
Small Volume Tracers				
Chlorofluoro-carbons	gas	none (?)	< 30 y.	exponential increase
Tritium	water	none	< 30 y.	variable
^{3}Helium				
— radiogenic	gas	none	1 mo.–30 y.	tritium decay
— primordial	gas	none		ocean ridges
Radiocarbon (bomb)	CO_2 carbonates	active	< 30 y.	exponential decrease
Stable isotopes	water	none		polar water
Large Volume Tracers				
Radiocarbon (natural)	CO_2 carbonates	active	20–50000 y.	stationary
^{226}Radium	metal ion	active	20–5000 y.	stationary (sediments)
^{39}Argon	gas	none	20–1000 y.	stationary
^{228}Radium	metal ion	active	30–500 y.	stationary (sediments)
^{85}Krypton	gas		< 30 y.	exponential increase
^{90}Strontium	metal ions	negligible	< 30 y.	variable
^{137}Caesium		(?)		(point source)

Notes: — *Tracers are separated into large volume and small volume depending on whether they can be measured from 10 liter bottles or require larger volume samples.*
— *Inputs are from the atmosphere, unless otherwise indicated.*
— *Radiocarbon is separated into a 'small volume' component, due to anthropogenic inputs, and a large volume, natural and stationary component.*
— *Time scale of applicability is limited to 30 years for tracers of anthropogenic origin, and depends on the decay constant for natural ones. Of course, sensitivity also depends on the measurement precision relative to variations in the ocean.*

J and mixing processes contribute about equally to the variations of the oxygen concentration.

For nutrients, the picture is similar. For example, nitrate concentrations vary from 11 μmol/kg in the surface North Atlantic to 45 μmol/kg in the intermediate Northeast Pacific waters. The values at the surface are mostly determined by a competition between the residence time of the waters at the surface and a subtraction by biologic activity. For nitrate, these so-called preformed values are 11 μmol/kg in the North Atlantic and 19 μmol/kg in the surface Antarctic waters. As a consequence, as for oxygen, mixing and *in situ* regeneration processes contribute about equally to the variations of the nutrient concentrations, in most of the deep ocean.

It is common practice to assume that the J terms for oxygen and nutrients are proportional, the ratios being named the Redfield ratios (Redfield *et al.*, 1963). It is then possible to eliminate J in the corresponding Eqs. (1), and to define a conservative tracer, like temperature or salinity (Broecker, 1974). Unfortunately, there is now clear evidence that the Redfield ratios are not constant in the ocean (Minster and Boulahdid, 1987; Boulahdid and Minster, 1988; Fig. 1). For phosphate, the variations are of the order of 50%. This result is in agreement with the observation that the composition of particulate matter, as collected in sediment traps, varies with depth (e.g. Honjo *et al.*, 1982).

Finally, recent measurements of dissolved organic species (organic complexes of phosphorus, nitrogen and carbon) suggest that their concentrations are much higher than previously thought (Suzuki *et al.*, 1985; Sugimura and Suzuki, 1988). The idea that oxygen consumption and nutrient regeneration are linked with the oxidation of particulate organic matter may prove too simple. Actually, Sarmiento *et al.* (1988) demonstrated that in their model the observed repartition of phosphate in the tropics was more easily reproduced if half of the organic material production is put in the form of dissolved organic matter with a long life time, which can be transported over long distances from the point of formation.

It is thus clear that oxygen and nutrients are difficult to use as quantitative tracers of ocean circulation though they are clear qualitative indicators of water masses and mixing processes. Similar statements can be made for dissolved silica (Kawase and Sarmiento, 1985; Rintoul, 1988). In my view, their main interest lies in the estimation of J, an integrated measurement of biogeochemical activity.

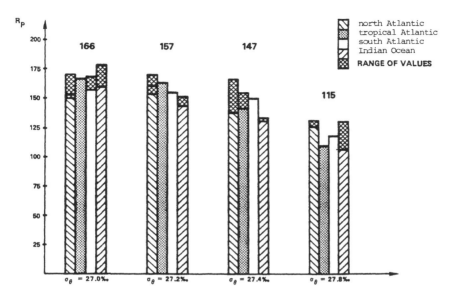

Figure 1
Variation with depth of the oxygen-phosphate Redfield ratio, calculated
from a mixing and consumption model of the isopycnal dissolved concentrations. Adapted from Minster and Boulahdid (1987).

2.2 Carbon Species

Because ^{14}C is a good tracer of ocean circulation (see Table 2), it is useful to
understand and quantify the processes which control the carbon species. The main
reason is, of course, the CO_2 problem. The total inorganic carbon concentration
varies from about 1900 μmol/kg in surface waters to 2400 μmol/kg in deep North
Pacific waters. In the simplest description, this carbon is present as dissolved CO_2,
and as carbonate and bicarbonate ions. They are in chemical equilibrium:

$$CO_2 + CO_3^{--} + H_2O \leftrightarrow 2HCO_3^-$$

As a rule of thumb, CO_2 makes 1%, CO_3^{--} 10% and HCO_3^- 90% of the total. In
principle, the measurement of two combinations of these species allows a calculation of the concentrations of the three. Most frequently, one measures carbonate
alkalinity (Alk = 2 CO_3^{--} + HCO_3^-) and total inorganic carbon (C_{tot} = CO_2 +
HCO_3^- + CO_3^{--}), by titration (complications related to the presence of other weak

acids in the ocean do not matter here). However, dissolved CO_2 can also be measured directly by other techniques, and the results do not agree, at the level of 5% of dissolved CO_2. This may seem small for C_{tot}, but is quite significant compared to the dissolved CO_2 gradient at the air-sea interface, which controls gas exchange there. This could be due to the effect of organic complexes in the titration technique (Bradshaw and Brewer, 1987). Thus, the measurement of the carbon species is not yet fully controlled.

As for oxygen, the boundary conditions are dominated by gas exchange with the atmosphere and biologic activity in the trophic layer. However, dissolved CO_2 is constantly re-equilibrated with the carbonate species, so that the ocean presents a large reservoir to equilibrate with the atmosphere. This excess capacity of the ocean is measured by a factor named the Revelle factor, which varies from 9 in warm waters to 14 in cold waters. As a consequence, the time scale for equilibration with the atmosphere is of the order a year, i.e. larger than the seasonal time scale of surface temperature evolution (a prime factor for the solubility of gases). Thus, CO_2 is rarely in equilibrium between the ocean and the atmosphere. In many places of the ocean, the net exchange is the difference of seasonal fluxes of opposite signs. Evaporation-precipitation is a minor factor in determining the CO_2 boundary conditions. This is the main control for alkalinity changes and thus affects the relative abundances of the carbonate ions. However, this propagates onto variations of the dissolved CO_2 concentrations which are one order of magnitude smaller than the other effects.

In the classical view, oxidation of particulate matter consumes oxygen and releases C_{tot} in a Redfield ratio close to $dO/dC = -1.4$ (see Takahashi *et al.*, 1985). This explains most of the observed variations of C_{tot} between the surface waters and the deep Pacific waters (compare the variations for oxygen and for C_{tot}). An additional effect, of the order of 20% of the previous one, is due to redissolution of the carbonate tests of plankton. This creates still more difficulties for deriving a map of J for carbon, because oxidation of organic matter occurs mainly in intermediate waters while redissolution of carbonate tests occurs mostly at the sediment-water interface. Finally, because of the observation of 'oceanographically consistent' large concentrations of dissolved organic carbon, this classical view may be too simple.

A further complication is invasion of anthropogenic CO_2. The atmospheric increase of the CO_2 partial pressure both increases invasion into the ocean (mostly in high latitude areas) and decreases degassing (mostly in the tropics) (see Broecker *et al.*, 1986). At present, the effect is a little less than 0.1% of C_{tot} in the areas of deep water formation. This is measurable as C_{tot} can be determined to a precision of the order of 0.1%. Further, on the path of deep water transport, the effect is not determined precisely, because of the uncertainty in J. As for degassing in the tropics, one is just getting an idea of the natural interannual variability (e.g. Andrie *et al.*, 1986), which is dominated by the variability of the current system (Garçon *et al.*, 1988).

This relative complexity of the carbon behavior in the ocean partly explains why one does not yet have a clear understanding of the uptake of anthropogenic CO_2 by the ocean. A second reason is the lack of an accurate description of transport by ocean currents. Great hopes are put on the WOCE and J-GOF (Joint Global Ocean Flux study) programs to improve the situation. This will probably make extensive use of models of the ocean circulation (Maier-Reimer and Hasselmann, 1987).

2.3 ^{14}Carbon

^{14}C has probably been the first radioactive tracer in use for studying ocean fluxes (Broecker *et al.*, 1960). It is now conveniently considered under two separate terms: a natural steady-state distribution resulting from exchange with the atmosphere of ^{14}C produced by cosmic rays, and an anthropogenic component, mainly produced in the early sixties by nuclear tests.

2.3.1 *Natural ^{14}Carbon*

^{14}C data are expressed in Δ^{14}C units, which correct for natural variations of ^{14}C abundance relative to the other carbon isotopes not related to radioactive decay. Δ^{14}C varies by $-10\%_{00}$ every 83 yrs. The natural values would range from $-110\%_{00}$ in the North Atlantic ocean to $-240\%_{00}$ in the deep North Pacific ocean (Fig. 2), an apparent aging of 1000 years. For the measurements to provide useful constraints on the residence time in water masses, a relative precision of about $1\%_{00}$ is needed. Until now, this can only be achieved by a beta-counting technique, from 200 liter water samples. Natural ^{14}C is thus a 'large volume' sample.

The boundary conditions are determined by CO_2 gas exchange at the surface. For ^{14}C, it is not only necessary to equilibrate the chemical reaction between dissolved CO_2 and the carbonate ions, but also to equilibrate the isotopic composition of each of the species. As a consequence, the equilibration time scale for this tracer

356

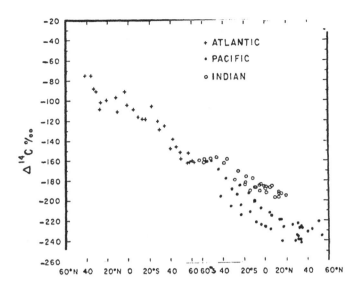

Figure 2
Observed $\Delta^{14}C$ values in the deep ocean. Adapted from Stuiver et al., (1983).

is of the order of 10 years. This explains why, on Fig. 2, the Antarctic surface waters have a $\Delta^{14}C$ value of $-160‰$: they are not equilibrated with the atmosphere. In addition, because most of the existing measurements were made after the anthropogenic increase had started, the natural boundary conditions are poorly known (see Broecker, 1981, for a discussion). However, because of the time scale of equilibration, the surface values are not sensitive to seasonal effects or to temperature. The main variations are along the vertical. ^{14}C should thus be a sensitive indicator of mixing across the main thermocline.

The J term for ^{14}C is similar to that for C_{tot}, with the additional effect that mass fractionation between the isotopes varies with the type of material (organic or mineral). Craig (1969) found that, in the deep Pacific ocean, input of ^{14}C from the particulates approximately balanced radioactive decay. However, if one works with the isotope ratio $^{14}C/C_{tot}$ (this is the case for $\Delta^{14}C$), the effect of J is much less sensitive: particulate inputs increase both ^{14}C and C_{tot}, but in a ratio similar to that of the surface water. J is frequently omitted in calculations using $\Delta^{14}C$. Errors are less than 10% (Fiadeiro, 1982).

On Fig. 2, the trend between the North Atlantic and Antarctic waters is mostly a mixing line. Deviations from this line can be observed, which are of the order of $10\%_0$, corresponding to a time scale for renewing of the Atlantic waters of less than 100 years (Broecker, 1981). The aging of the Pacific deep waters is then rather of the order of 500 years. It is for this ocean that natural ^{14}C measurements are most useful. It is also the least sampled. In particular, the existing data (mostly from Geosecs) do not provide an adequate description of the boundary currents (see Toggweiler et al., 1988a).

2.3.2 Anthropogenic $^{14}Carbon$

Man has been modifying the atmospheric $^{14}C/C$ ratio in two opposite ways: between 1850 and 1950, injection of CO_2 by fossil fuel burning decreased $\Delta^{14}C$ by $30\%_0$ in the atmosphere. This is the so-called Suess effect (Suess, 1955). Nuclear explosions increased that ratio up to $750\%_0$ in 1963. Since then, this ^{14}C has been invading the ocean. Because of the slow exchange rate, anthropogenic ^{14}C in the atmosphere is homogeneously distributed and decreases slowly ($250\%_0$ at present) (Fig. 3).

The concentration in the surface ocean is still increasing. For example, in the Greenland sea, the value changed from $-55\%_0$ in the fifties, to $+10\%_0$ at the time of Geosecs (1972) and $+20\%_0$ at the time of TTO (1981) (Smethie et al., 1986). The situation of this anthropogenic ^{14}C in the ocean is thus quite different from that of the natural one. The variations are large and mostly affect the surface ocean and the areas of deep water formation. A large number of measurements rather than an excellent precision is thus necessary. This can be achieved by measuring ^{14}C by accelerator mass spectrometry, which requires a sample of less than 100 ml for a 1% precision (e.g. Bard et al., 1988).

Inventories of the anthropogenic tracers provide a good picture of their fate in the ocean (Fig. 4; Broecker et al., 1985). Because these inventories are made with a much too limited sampling, they are probably very imprecise. Yet, they clearly show that the ocean is transporting tracers from high latitudes to temperate areas, while tropical upwelling prevents invasion from the surface in low latitudes. These inventories are sensitive tests of the behavior of world ocean models (Toggweiler et al., 1988b).

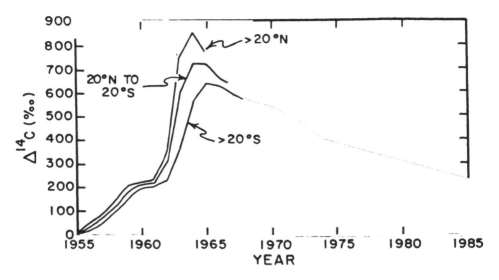

Figure 3a
Input functions of some of the tracers of the ocean circulations: ^{14}C in the troposphere from Broecker and Peng (1982)

2.4 Tritium and ^{3}Helium

Tritium is almost uniquely anthropogenic. It is generally expressed in Tritium units (TU), such that the isotopic ratio $T/H = 1 \times 10^{-18}$, which amounts to 7.088 disintegrations/kg water/minute. Its decays into ^{3}He, with a half-life of 12.43 years. It is measured by degassing first ^{3}He present in the water sample, then waiting for the decay of T for about a year, and finally measuring the resulting ^{3}He by mass spectrometry (Jenkins, 1981). The present detection limit is 0.001 TU.

Tritium is exchanged very fast between the ocean and the atmosphere, both by gas exchange and by precipitation (Weiss and Roether, 1980). Its input function to the ocean is thus very narrow in time and varies strongly with latitude (see Roether, this volume). This input function can generally be considered as well known, though its accuracy is limited by that of precipitation over the ocean. On the other hand, since this input is so limited in time, the seasonal variations of the ocean mixed layer probably play a role in the injection of tritium at the thermocline level (Sarmiento, 1983; Jenkins, 1988). In the ocean, T, being part of water molecules (HTO), is only affected by radioactive decay. Note that its decay, being of comparable time

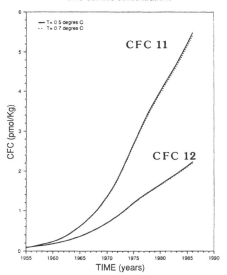

CFC surface concentrations

Figure 3b
Input functions of some of the tracers of the ocean circulations: F-11 and
F-12 calculated by Mantisi in cold surface water (unpublished)

scale to the anthropogenic input time scale, must be corrected for, if tritium is to
be considered as a dye, like bomb ^{14}C (e.g. Broecker, 1981). In the surface ocean,
the tritium concentration is steadily decreasing, with present values of the order of
5 TU in the northern hemisphere. Below the surface, it is still increasing, mostly
by lateral transport (e.g. Jenkins, 1982). Like bomb ^{14}C, the inventory of tritium
shows transport from high to temperate latitude, particularly in the Indian ocean,
and depletion in the tropics (Broecker, 1986).

Its daughter product, ^{3}He, is also passive but exchanges easily with the atmosphere, when the water gets to the surface. A most efficient way to use the T-^{3}He
information is to calculate a tritium-helium age:

$$ t = \frac{1}{a} \ln \left(1 + \frac{^{3}\mathrm{He}}{\mathrm{T}} \right) $$

where a is the decay constant of tritium. Thanks to the precision of the measurements, the resolution of this clock is of the order of a month. Of course, this 'age' is
affected by mixing, but the latter effect is not significant on short time scales. This

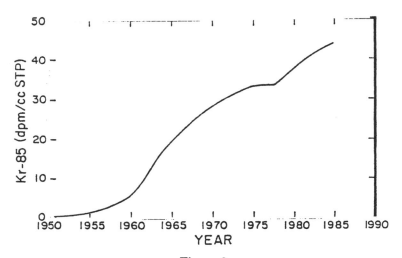

Figure 3c
Input functions of some of the tracers of the ocean circulations: ^{85}Kr in
the troposphere, from Smethie et al. (1986).

makes it a very good tracer of ventilation of the main thermocline (e.g. Jenkins,
1988).

Besides its production by tritium decay, ^{3}He is also injected into the oceans
along mid-oceanic ridges. It comes from degassing of helium contained in mantellic
rocks during the formation of the ocean floor. This Helium presents an isotopic
ratio ^{3}He/^{4}He which is eight times larger than the atmospheric ratio (Jenkins *et
al.*, 1978). This creates plumes of excess ^{3}He which spread from the ridge source in
the deep ocean, and which are measurable on distance scales of the order of several
thousand kilometers (Lupton and Craig, 1981). In principle, these plumes provide
an integrated picture of the diffusion-advection field in the deep ocean, much like
water masses. Its interest should be greater in the deep Pacific ocean, because the
^{3}He signal is larger, and because water masses are less well defined.

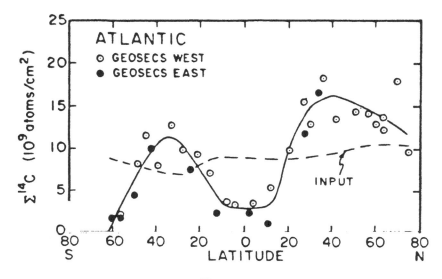

Figure 4
Inventory of bomb ^{14}C in the Atlantic ocean at the time of Geosecs. Adapted from Broecker et al. (1985).

2.5 Chlorofluorocarbons

Chlorofluorocarbons (CFC's or freons) are anthropogenic, chemically stable, passive gases in the ocean. Their atmospheric concentration is well documented, and has increased nearly exponentially since 1940 (Fig. 3b). Their life time in the atmosphere is of the order of 75 years. They equilibrate with the surface ocean on a time scale of the order of a month. The solubility of the two most frequently used freons is very different: at $-1.5°C$, the equilibrium concentration in the surface ocean is 5 pmol/kg for freon-11 (CCl3F), and 2 pmol/kg for freon-12 (CCl2F2). The concentration in warm waters is 3 to 4 times less. They can be measured onboard a ship, once serious contamination problems are overcome, from 40 cc samples, at a 1 to 4% precision and with a 0.02 pmol/kg sensitivity. Finally, because of their different industrial history, the relative abundance of these two freons changed with time, which provides a clock of their injection time in the ocean (as for all these clocks, it is sensitive to mixing, which can be estimated from the observed concentrations). Freons are thus very promising transient tracers of the ocean circulation. Because they are not radioactive, they follow an accumulation evolution, which makes them particularly useful for describing invasion of the deep ocean (e.g. Weiss *et al.*, 1985).

2.6 Conclusion

I (subjectively) consider that the above tracers are the most promising for the ocean circulation. Information on other tracers of Table 2 can be found in Broecker and Peng (1982) for stable isotopes, Smethie and Mathieu (1986) and Smethie *et al.* (1986) for [85]Kr, Schlitzer *et al.* (1985) and Smethie *et al.* (1986) for [39]Ar and Moore *et al.* (1985) for [226]Ra and [228]Ra.

In order to illustrate the relative behavior of transient tracers, I calculated their evolution in a box of volume V, renewed by a flux F of surface water. The residence time of the water in the box is thus: $t_w = V/F$. The concentration of the tracers, tritium, freon-11 and [85]Kr, were given in the input water in the form of: $C = C_0 \exp(t - t_0)/a$, where the initial concentrations C_0, time t_0 and the time constants a were selected in order to fit approximately the surface concentrations in cold northern hemisphere water (say, the surface Greenland Sea). Starting from 1963, the selected values are:

$$C = 12 \exp(-0.065t) \qquad \text{in TU for tritium;}$$
$$C = 0.5 \exp(0.11t) \qquad \text{in pmol/kg for freon} - 11;$$
$$C = 0.065 \exp(0.06t) \quad \text{in dpm/100kg for } [85]\text{krypton}$$

The calculated evolution of the tracer concentrations are given in Fig. 5a, for two different values of t_w.

In practice, one is generally looking for an estimate of t_w. A first approach consists in using the estimated concentration in the box at one time, and adjusting t_w using the known input function. A second approach consists in repeated surveys. The time evolution in the reservoir is *a priori* less sensitive to the input function. Fig. 5b shows the relative sensitivity of t_w on the knowledge of the time constant a, for the two techniques. For the second technique, I assumed that dC/dt was known, and t_w was taken as 40 years. Note that this sensitivity corresponded to a sensitivity on the input function or to that on the historical evolution in the surface water. For the two tracers F-11 and [85]Kr, the input functions of which increase with time, this parameter increases almost linearly with time for the two techniques. For these tracers, this test is in favour of [85]Kr compared to freon.

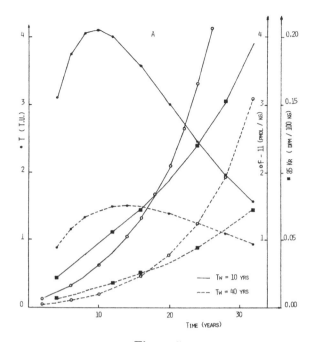

Figure 5a
Calculated evolution of tritium, F-11 and ^{85}Kr in a box in contact with cold northern hemisphere surface water. The calculation is made for residence times, t_w, of 10 and 40 years.

By comparison, for T, the input function of which is decreasing with time, t_w is less sensitive to a. Also, between 20 and 30 years after the beginning of the experiment, the second technique is much preferable. There, the tritium radioactive decay is the dominant source of information.

Because the different tracers cannot be measured with the same precision or as easily, the estimation of C or of dC/dt cannot be as precise for all tracers. Fig. 5c

364

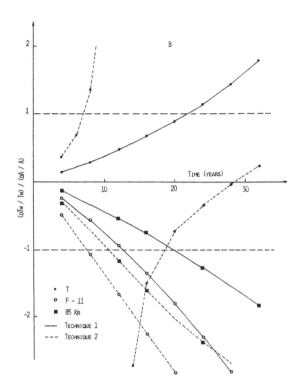

Figure 5b
Evolution with time of $(dt_w/t_w)/(da/a)$ for t_w = 40 years. The calculation is made when t_w is calculated from the estimated concentration (Technique 1) and when it is derived from the evolution of the concentration (Technique 2).

shows the relative precision on t_w for a given relative precision on C or on dC/dt. The asymptotic value is about the same for the two tracers freons and ^{85}Kr and for the two techniques. For tritium, the error propagation degrades with time in the two techniques, but is significantly better in the second approach.

As freons can be measured more precisely and much more frequently than ^{85}Kr, it will be much easier to estimate and monitor their values. The above simple calculation suggests that this more than compensates for the radioactive decay of ^{85}Kr, not to speak of the evolution of the F-11 to F-12 ratio. However, repeated

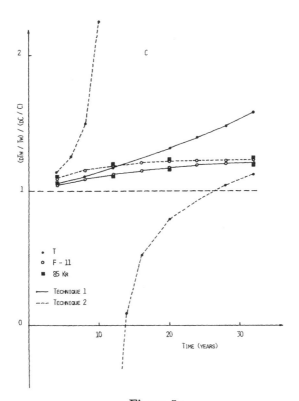

Figure 5c
Relative error on t_w for a given relative error on the estimated concentra-
tion (Technique 1) and on the time evolution of the concentration (Tech-
nique 2).

surveys of T-^{3}He are apparently still more useful.

3. SURFACE BOUNDARY CONDITIONS

In the previous section, I briefly described the input function of some transient tracers into the ocean. However the actual transfer from the atmosphere is a complex
process. Sarmiento (1983), for example, demonstrated that the observed pattern
of tritium in the North Atlantic ocean was reproduced better by his model, if the
winter time convection in the mixed layer was parameterized. Here, I briefly mention two types of studies which tend to describe better the seasonal evolution of gas

exchange through the air-sea interface and the ocean mixed layer.

3.1 Seasonal Evolution of the CO_2 Air-Sea Gas Exchange Coefficient

Gas exchange at the air-sea interface depends on the product of a gas transfer coefficient by the partial pressure gradient of the gas between the ocean and the atmosphere (see, e.g., Liss and Merlivat, 1986). Both vary seasonally and with geography. In this paragraph, I deal with the gas transfer coefficient.

The gas transfer coefficient, k, depends on water temperature, T, and wind speed, V. A review of these dependences and their estimates is made by Liss and Merlivat (1986). They propose an algorithm for CO_2,:

$$k = 0.17V \left(\frac{S_c(T)}{S_c(20)} \right)^{-\frac{2}{3}} \qquad\qquad V < 3.6 \text{ ms}^{-1}$$

$$k = (2.85V - 9.65) \left(\frac{S_c(T)}{S_c(20)} \right)^{-\frac{1}{2}} \qquad 3.6 < V < 13 \text{ ms}^{-1}$$

$$k = (5.9V - 49.3) \left(\frac{S_c(T)}{S_c(20)} \right)^{-\frac{1}{2}} \qquad V > 13 \text{ ms}^{-1}$$

The wind speed is given at 10 m above the sea surface. S_c is the Schmidt number, and 20 is a reference temperature, for which $S_c = 595$ for CO_2. Though this parametrisation is intended to describe all physical processes, Jenkins (1988) and Spitzer and Jenkins (1988) suggest adding another formula for describing air injection at wind speeds, which creates gas supersaturation in winter times. They propose a wind speed dependence of the order of:

$$V^{2.2\pm0.2}$$

This parametrisation allows one to map the gas transfer coefficient, given values of the wind speed and of the temperature. With the Liss and Merlivat formulae, this has been done using climatological winds by Erickson and Duce (1987), using meteorological winds by Thomas *et al.* (1988) and using Seasat scatterometer data by Etcheto and Merlivat (1988). In particular, Thomas *et al.* provide numerical values of the gas transfer coefficient for each season and each $7.5° \times 7.5°$ latitude-longitude square.

One of the striking features of these maps is the zonal variability of the coefficient: it is comparable to the meridional variability, so that calculating net gas fluxes from zonal averages of the CO_2 partial pressure gradient might give strongly biased results. Secondly, the seasonal variability of the gas transfer coefficient tends to be correlated with that of the gas partial pressure gradient so that calculating net fluxes from yearly averages of the two values is inappropriate. Thirdly, the interannual variability is of the order of 30 to 100% depending on the latitude. It is thus necessary to monitor this coefficient on a regular basis.

3.2 Seasonal Evolution of the Surface Partial Pressure

An increasing number of data demonstrates a strong seasonal variation of the surface CO_2 or oxygen partial pressure, including inversion of the ocean-atmosphere gradient (e.g., Andrie et al., 1986; Goyet, 1987; Takahashi et al., 1986). It will probably not be feasible to measure these values seasonally over enough of the ocean for calculating reliable net fluxes. It is thus required to understand and model the processes controlling gas concentrations in the ocean mixed layer, at least to interpolate in space and time between the existing data. Progress in this direction has been made recently (Musgrave et al., 1988; Spitzer and Jenkins, 1988; Thomas et al., 1989).

The method consists in describing the evolution of the gas concentration using a one-dimensional ocean mixed-layer model, including production and consumption by biogeochemical activity, and comparing with observations. This could be done for Ocean Weather Station P and near Bermuda. Fig. 6 illustrates some results for oxygen. The seasonal evolution of the concentration is dominated by the temperature evolution. However because of the kinetics of gas exchange, oversaturation is created during the warming periods, and undersaturation during the cooling periods. If air injection is added (which is not the case on the figure), oversaturation, of the order of 1%, is created in winter, as suggested by noble gas data (which are not affected by biogeochemical activity) (Spitzer and Jenkins, 1988). Due to production in the mixed layer, oxygen supersaturation is increased by a factor of 3 to 4 in spring time. During winter, deepening of the mixed layer entrains undersaturated water created by consumption below the photic zone. Thus the gas exchange flux is inverted with seasons. As this deepening occurs during episodic storm events, this creates a short time scale variability of the mixed layer concentration which is much larger than measurement precision (see Emerson, 1987).

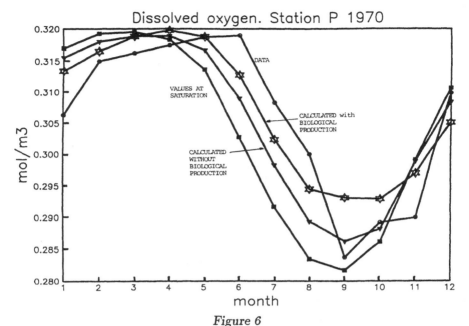

Figure 6
Evolution with time of the oxygen concentration at the surface at Ocean
Weather Station P. The figure shows the evolution of the concentration at
saturation, which depends on the temperature, the effect of the kinetics
of gas exchange, which requires about a month to equilibrate the mixed
layer with the atmosphere, the supersaturation created in spring by biological activity, and the undersaturation created in winter by entrainment
of undersaturated water from below the trophic layer into the mixed layer.
Adapted from Thomas et al. (1989).

3.3 Discussion

These results illustrate the complexity of the exchange process between the ocean
and the atmosphere. How much of this complexity should be incorporated in the
boundary limit description of tracers of the ocean circulation is not clear. In principle, the variations of the gas transfer coefficient can be monitored using satellite
data. It is conceivable that parametrisation of mixed layer processes as a function
of measurable physical parameters will be feasible in the future. As for the biogeochemical part, it is possible that the models be constrained by the ocean colour
determination of the plankton surface biomass. The recent studies open an interesting perspective for the description of the ocean-atmosphere chemical transfers.

4. MIXING IN THE DEEP OCEAN

In the conclusion of section 1, I stated that one of the main interests of tracers of the ocean circulation was to provide constraints on mixing processes. This was demonstrated for the surface ocean, using T-^{3}He data, by Jenkins (see Jenkins, 1988). In the deep ocean, it is not clear whether diapycnal mixing occurs in the interior of the basins, or whether it results from enhanced vertical mixing along the boundaries (Armi, 1979). This could have serious implications on the dynamics of the deep ocean (Gargett, 1984). Here, I wish to describe briefly two studies dealing with vertical mixing of tracers in the deep ocean.

4.1 The Two-Degree Discontinuity as Explained by Boundary Mixing

The two-degree discontinuity (TDD) can be described as a break in the temperature-salinity relationships of the deep Atlantic water. Broecker and Peng (1982) demonstrated that such a break could be formed in a one-dimensional vertical column at the level of introduction of a less dense water (the North Atlantic Deep Water, NADW) into a vertically advected bottom water (the Antarctic Bottom Water, AABW). More recently, using the TTO hydrographic and nutrient data, I could describe more precisely some properties of the TDD (Minster, 1985): the TDD seems to follow an isopycnal level which encounters the western boundary at the level of the Western Boundary Undercurrent (WBUC). It is better defined near the western boundary than near the mid-oceanic ridge. It is a level of larger proportion of northern water. Along the western boundary, the composition of the undercurrent evolves and entrains more than half of the AABW entering the western basin. Below the TDD, the water properties can be described by mixing of northern waters from the level of the discontinuity and of AABW. Above this level, the water composition can be explained by a mixture of three components: Northern waters, AABW, and shallower water the properties of which are close to those of waters from the African coast. The proportion of AABW is larger near the ridge axis than along the western boundary.

These properties are well explained if vertical mixing is dominated by boundaries (Fig. 7). Below the TDD, mixing would occur along the bottom between AABW and waters from the undercurrent. At this level, AABW is entrained southwards by the current. Above the TDD, mixing would again occur along the boundary, between WBUC water, dominated by NADW, and waters from shallower

370

levels. These two mixing lines would be propagated into the interior by isopycnal or neutral mixing, which would create the temperature-salinity break. Boundary mixing would also occur along the ridge axis. This would smooth the TDD and introduce AABW above it. Of course, diapycnal mixing could contribute to this.

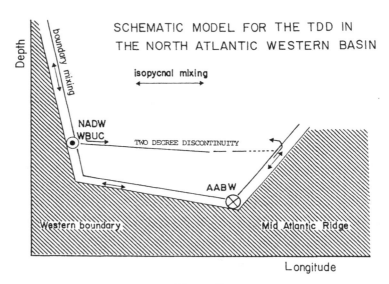

Figure 7
Schematic model of the formation of the two degree discontinuity by intensified boundary mixing at the level of the Western Boundary Undercurrent. Adapted from Minster (1985).

4.2 Purposeful Tracer Experiment

A purposeful release of tracers is probably one of the best way to estimate a macroscopic value of diapycnal mixing. A pilot experiment of such a release has recently been performed in the Santa Monica basin off the coast of California (Ledwell *et al.*, 1986; Watson and Ledwell, 1988). Sulphur hexafluoride (SF6) and perfluorodecalin (PFD) were released in 1985, near 800 m, below the sill depth of this basin. This release was confined both vertically, within 0.01°C of the 5.15°C isotherm, and horizontally, inside a circle of about 10 km. The initial streaky dispersion was sampled with slowly filling syringes. After 5 months, this could be replaced by spot, Niskin bottle sampling.

Careful sampling near the boundaries demonstrated that little of the tracers went close to them. Yet, their vertical profiles showed a spreading with time. If interpreted in terms of Fickian diffusion, this would correspond to a vertical diffusivity of the order of 0.3 cm^2s^{-1}, a value close to the typical values required to explain natural tracer distributions in the open ocean. It is also in agreement with the value obtained from Gargett's (1984) formula describing the dependence of diapycnal diffusivities as a function of buoyancy frequency (0.51 cm^2s^{-1} for the Santa Monica basin).

4.3 Discussion

The two results are not necessarily contradictory. As stated by Watson and Ledwell (1988), one should be cautious about extrapolating the results of the Santa Monica basin experiment to the open ocean as the basin is quite small, with dimensions of the order of the Rossby radius of deformation. On the other hand, the interpretation proposed for the TDD only requires that mixing along the western boundary be dominant over internal mixing. Boundary mixing might well be intensified by the WBUC. A large scale open-ocean purposeful tracer experiment is being planned for the WOCE program. Some of its requirements are discussed by Watson and Ledwell (1988). Such an experiment should be an important step towards a better parametrisation of diapycnal mixing.

5. CONCLUSION

In section 1, I indicated why tracers can help provide information on the circulation of the ocean, mostly of the intensity and geography of large scale (that is, larger than mesoscale) mixing processes. Then, I tried to provide some background information on those tracers which I consider most useful or most measured. In particular, in the case of a very simple model, I provided indications on the relative usefulness of tritium, freons and ^{85}krypton. In the third section, I described two studies which used tracers to analyse mixing in the deep ocean. Whereas the two degree discontinuity can be explained if mixing is intensified in the western boundary undercurrent relative to the interior, the purposeful tracer experiment of Ledwell *et al.* (1986) can be explained by interior diapycnal mixing, with values in agreement with Gargett's parametrisation (1984). Finally, the last section indicates how the various mixed-layer processes affect the surface concentrations of the tracers, and thus their boundary conditions.

I certainly do not pretend that these notes provide an overview of chemical tracers of the ocean circulation. In particular, I did not address the problem of using these data to estimate the advection and diffusion fields. I assumed that this was more adequately discussed in the notes of other speakers.

ACKNOWLEDGMENTS

I wish to thank the two organizers of this workshop, D. Anderson and J. Willebrand, for the very pleasant conditions they created. Discussions with the speakers and participants were most stimulating.

REFERENCES

Andrie, C., C. Oudot, C. Genthon and L. Merlivat (1986): CO_2 fluxes in the tropical atlantic during FOCAL cruises. *J. Geophys. Res.*, **91**, 11741–11755.

Armi, L. (1978): Some evidence for boundary mixing in the deep ocean. *J. Geophys. Res.*, **83**, 1971–1979.

Armi, L. (1979): Effects of variations in eddy diffusivity on property distributions in the ocean. *J. Mar. Res.*, **37**, 515–530.

Armi, L. and H. Stommel (1983): Four views of a portion of the North Atlantic gyre. *J. Phys. Oceanogr.*, **13**, 828–857.

Bard, E., M. Arnold, G. Östlund, P. Maurice, P. Monfray and J.C. Duplessy (1988): Penetration of bomb radiocarbon in the tropical Indian ocean measured by means of accelerator mass spectrometry. *Earth Planet. Sci. Lett.*, **87**, 379–389.

Boulahdid, M. and J.F. Minster (1988): Oxygen consumption and nutrient regeneration ratios along isopycnal horizons in the Pacific ocean. Marine Chem. (in press).

Bradshaw, A.L. and P.G. Brewer (1988): High precision measurements of alkalinity and total carbon dioxide in seawater by potentiometric titration: presence of an unknown protolyte(s)? Marine Chem. (in press).

Brewer, P.G., J.L. Sarmiento and W.M. Smethie (1985): The transient tracers in the ocean (TTO) program: The North atlantic study, 1981; The tropical Atlantic study, 1983. *J. Geophys. Res.*, **90**, 6903–6905.

Broecker, W.S. (1974): 'NO', a conservative water-mass tracer. *Earth Planet. Sci. Lett.*, **23**, 100–107.

Broecker, W.S. (1981): Geochemical tracers and ocean circulation. In: Evolution of physical Oceanography, B.A. Warren and C. Wunsch, eds, MIT press, Cambridge, 434–460.

Broecker, W.S. (1986): The distribution of bomb tritium in the ocean J. Geophys. Res., 91, 14331–14344.

Broecker, W.S., R. Gerard, M. Ewing and B.C. Heezen (1960): Natural radiocarbon in the Atlantic Ocean. J. Geophys. Res., 65, 2903–2931.

Broecker, W.S., J.L. Ledwell, T. Takahashi, R. Weiss, L. Merlivat, L. Memery, T.H. Peng, B. Jähne and K.O. Munnich (1986): Isotopic versus micrometeorologic ocean CO_2 fluxes: a serious conflict. J. Geophys. Res. 91, 10517–10527.

Broecker, W.S. and T.H. Peng (1982): Tracers in the sea. Eldigio Press, New York, pp. 690.

Broecker, W.S., T.H. Peng, H.G. Östlund and M. Stuiver (1985): The distribution of bomb-radiocarbon in the ocean. J. Geophys. Res., 90, 6953–6970.

Craig, H. (1969): Abyssal carbon and radiocarbon in the Pacific. J. Geophys. Res., 74, 5491–5506.

Emerson, S. (1987): Seasonal oxygen cycle and biological new production in surface waters of the pacific ocean. J. Geophys. Res., 92, 6535–6544.

Erickson, D.J. and R.A. Duce (1987): On the global transfer velocity field of gases with a Schmidt number of 600. Searex Newslett., 10, 7–10.

Etcheto, J. and Merlivat, L. (1988): Satellite determination of the carbon dioxide exchange coefficient at the ocean atmosphere interface. J. Geophys. Res., (in press).

Fiadeiro, M.E. (1982): Three-dimensional modeling of tracers in the deep Pacific ocean II. Radiocarbon and the circulation. J. Marine Res., 40, 537–550.

Garçon, V., L. Martinon, C. Andrie, P. Andrich and J.F. Minster (1988): Kinematics of CO_2 fluxes in the tropical Atlantic ocean during the 1983 northern summer. J. Geophys. Res. (in press).

Gargett, A.E. (1984): Vertical eddy diffusivity in the ocean interior. J. Marine Res., 42, 359–393.

Garrett, C. (1979): Mixing in the ocean interior. Dyn. Atmos. Oceans, 3, 239–265.

Goyet, C. (1987): Variations saisonnières de pCO_2 dans les eaux de surface du Sud Ouest de l'Océan Indien. Thèse de Doctorat de l'Université Pierre et Marie Curie, Paris, France.

Honjo, S., S.J. Manganini and J.J. Cole (1982): Sedimentation of biogenic matter in the deep ocean. *Deep Sea Res.,* **29**, 609–625.

Jenkins, W.J. (1981): Mass spectrometric measurement of tritium and 3Helium. Proceedings of consultation group on low level tritium measurement. IAEA, Vienna, 179–189.

Jenkins, W.J. (1982): On the climate of a subtropical gyre: decade timescale variations in watermass renewal of the Sargasso sea. *J. Mar. Res.,* **40**, 265–290.

Jenkins, W.J. (1987): ^{3}H and ^{3}He in the beta triangle: observations of gyre ventilation and oxygen utilization rates. *J. Phys. Oceanogr.,* **17**, 763–783.

Jenkins, W.J. (1988): The use of Anthropogenic tritium and helium-3 to study subtropical gyre ventilation and circulation. *Phil. Trans. R.Soc. Lond.,* **A 325**, 43–61.

Jenkins, W.J., J.M. Edmond and J.B. Corliss (1978): Excess ^{3}He and ^{4}He in Galapagos submarine hydrothermal water. *Nature,* **272**, 156–158.

Kawase, M. and J.L. Sarmiento (1985): Nutrients in the Atlantic thermocline. *J. Geophys. Res.,* **90**, 8961–8979.

Ledwell J.R., A.J. Watson and W.S. Broecker (1986): A deliberate tracer experiment in Santa Monica basin. *Nature,* **323**, 322–324.

Liss, P.S. and L. Merlivat (1986): Air-sea gas exchange rates: introduction and synthesis. In: The Role of Air-Sea Exchange in Geochemical Cycling, P. Buat-Menard, ed. NATO A.S.I. series, **185**, 113–128.

Lupton, J.E. and H. Craig (1981): A major helium-3 source at 15°N on the east Pacific rise. *Science,* **214**, 13–18.

Maier-Reimer, E. and K. Hasselmann (1987): Transport and storage of CO_2 in the ocean — An inorganic ocean circulation carbon cycle model. *Climate Dynamics,* **2**, 63–90.

McDougall, T.J. (1987): Neutral surfaces in the ocean: implications for modelling. *Geophys. Res. Lett.,* **14**, 797–800.

Minster, J.F. (1985): The two degree discontinuity as explained by boundary mixing. *J. Geophys. Res.,* **90**, 8953–8960.

Minster, J.F. and M. Boulahdid (1987): Redfield ratios along isopycnal surfaces. A complementary study. *Deep Sea Res.,* **34**, 1981–2003.

Moore W.S., R.M. Key and J.L. Sarmiento (1985): Techniques for precise mapping of ^{226}Ra and ^{228}Ra in the ocean. *J. Geophys. Res.,* **90**, 6983–6994.

Munk, W.H. (1966): Abyssal recipes. *Deep Sea Res.* **13**, 707–730.

Musgrave, D.L., J. Chou and W.J. Jenkins (1988): Application of a model of upper-ocean physics for studying seasonal cycles of oxygen. *J. Geophys. Res.* (in press).

Osborn, T. and C. Cox (1972): Oceanic fine structure. *Geophys. Fluid Dyn.*, **3**, 321–345.

Redfield, A.C., B.H. Ketchum and F.A. Richards (1963): The influence of organisms on the composition of sea water. In: The Sea, vol 2, M.N. Hill ed., Interscience, New-York, 26–77.

Rintoul, S.R. (1988): Mass, Heat and nutrient fluxes in the Atlantic ocean determined by inverse methods. PhD thesis, MIT/WHOI.

Sarmiento, J.L. (1983): A simulation of bomb tritium entry into the Atlantic ocean. *J. Geophys. Res.*, **13**, 1924–1939.

Sarmiento, J.L., H.W. Feely, W.S. Moore, A.E. Bainbridge and W.S. Broecker (1976): The relationship between eddy diffusion and buoyancy gradient in the deep sea. *Earth Planet. Sci. Lett.*, **32**, 357–370.

Sarmiento, J.L., J.R. Toggweiler and R. Najjar (1988): Ocean carbon-cycle dynamics and atmospheric pCO_2. *Phil. Trans. R. Soc. Lond.*, **A 325**, 3–21.

Schlitzer R., W. Roether, U. Weidmann, P. Kalt and H. Loosli (1985): A meridional ^{14}C and ^{39}Ar section in Northeast Atlantic deep water. *J. Geophys. Res.*, **90**, 6945–6952.

Smethie, W.M. and G. Mathieu (1986): Measurement of krypton-85 in the ocean. *Marine Chem.*, **18**, 17–33.

Smethie, W.M., H.G. Östlund and H.H. Loosli (1986): Ventilation of the deep Greenland and Norwegian seas: evidence from krypton-85, tritium, carbon-14 and argon-39. *Deep-sea Res.*, **33**, 675–703.

Spitzer, W.S. and W.J. Jenkins (1988): Rates of vertical mixing, gas exchange, and new production: estimates from seasonal gas cycles in the upper ocean near Bermuda. *J. Marine Res.* (submitted).

Stuiver M., P.D. Quay and H.G. Östlund (1983): Abyssal water carbon-14 distribution and the age of the world oceans. *Science*, **219**, 849–851.

Suess, H.E. (1955): Radiocarbon concentration in modern wood. *Science*, **122**, 415–417.

Sugimura, Y. and Y. Suzuki (1988): A high temperature catalytic oxidation method of non volatile dissolved organic carbon in sea water by direct injection of a liquid sample (preprint).

Suzuki, Y., Y. Sugimura and T. Itoh (1985): A catalytic oxidation method for the determination of total nitrogen dissolved in sea water. *Marine Chem.,* **16**, 83–92.

Takahashi, T., W.S. Broecker and S. Langer (1985): Redfield ratios based on chemical data from isopycnal surfaces. *J. Geophys. Res.,* **90**, 6907–6924.

Takahashi, T., J. Goddard, S. Sutherland, D.W. Chipman and C.C. Breeze (1986): Seasonal and geographic variability of carbon dioxide sink/source in the oceanic areas: observations in the North and equatorial Pacific ocean, 1984–1986 and global summary. Final report to the CO_2 research division. Office of Energy Research, USA, pp. 52.

Thomas, F., C. Perigaud, L. Merlivat and J.F. Minster (1988): World-scale mapping of the CO_2 ocean-atmosphere gas-transfer coefficient. *Phil. Trans. R. Soc. Lond.,* **A 325**, 71–83.

Thomas, F., V. Garçon and J.F. Minster (1989): Modeling the seasonal cycle of dissolved oxygen in the upper ocean at station P. *Deep Sea Res.* (in press);

Toggweiler, J.R., K. Dixon and K. Bryan (1988a): Simulations of radiocarbon in a coarse resolution, world ocean model I: Steady-state, pre-bomb distribution. *J. Geophys. Res.* (submitted);

Toggweiler, J.R., K. Dixon and K. Bryan (1988b): Simulations of radiocarbon in a coarse resolution world ocean model II: distribution of bomb-produced [14]C. *J. Geophys. Res.* (submitted).

Warren, B.A. (1981): Deep circulation of the world ocean. In: Evolution of Physical Oceanography, B.A. Warren and C. Wunsch, eds, MIT press, Cambridge, 6–41.

Watson, A.J. and J.R. Ledwell (1988): Purposeful released tracers. *Phil. Trans. R. Soc. Lond.,* **A 325**, 189–200.

Weiss, R.F., J.L. Bullister, R.H. Gammon and M.J. Warner (1985): atmospheric chlorofluoromethanes in the deep equatorial Atlantic. *Nature,* **314**, 608–610.

Weiss, W. and W. Roether (1980): The rates of tritium input to the world oceans. *Earth Planet. Sci. Lett.,* **49**, 435–446.

ON OCEANIC BOUNDARY CONDITIONS FOR TRITIUM, ON TRITIUGENIC ³He, AND ON THE TRITIUM-³He AGE CONCEPT

Wolfgang Roether

Universität Bremen, Fachbereich 1

Postfach 330440, 2800 Bremen 33,

F.R. Germany

ABSTRACT

Time dependence of tracer boundary conditions is a critical item in oceanic modelling of transient tracer data. For tritium, delivery to the ocean has occurred by flux imposed from the atmosphere, and both a flux and a concentration surface-ocean boundary condition can be formulated. Previous accounts of these are outlined, and ways to obtain boundary conditions needed in a given modelling context are suggested. ³He from tritium decay allows tritium-³He 'dating', and this approach largely circumvents the tracer boundary condition problem. However, the tritiugenic component of oceanic ³He is small, so that its separation from the natural ³He background can pose problems. The lowest achievable error in determining tritiugenic ³He is about ± 0.04 TR (one Sigma); this limit is related to availability of high-precision helium and neon data. The transport equation for tritium-³He age is explored and compared with that for regular transient tracers. A tendency of age distributions to develop towards stationarity is found. A case is made that tritium-³He age distributions are suitable for evaluation by inverse modelling. The case is discussed for tritium-³He age distributions in the lower main thermocline of the Northeast Atlantic, and aspects requiring further attention are noted.

377

D. L. T. Anderson and J. Willebrand (eds.), Oceanic Circulation Models: Combining Data and Dynamics, 377–407.
© 1989 by Kluwer Academic Publishers.

1. INTRODUCTION

Transient tracers (tritium, tritiugenic ^{3}He, chlorofluoromethanes or freons, bomb $-^{14}$C, ^{85}Kr) have a role in assessing oceanic transport on time scales between about a year and a few decades. Their potential is related to their time-dependent boundary conditions at the ocean surface. They may be looked upon as dyes that have — inadvertently and globally — been added to ocean surface water, and whose subsequent penetration into deeper ocean layers can be tracked. Their inputs have been given occurrences in time, for which the term 'input histories' is aptly used. Tracer observations give independent oceanographic information and may induce new views on ocean circulation and mixing.

The oceanographic potential of transient tracer data is utilized through incorporating the tracers into ocean circulation models. Inverse modelling (Wunsch and Minster, 1982) might look attractive, but, owing to the transient nature of the distributions, this requires quantifying the time derivative of tracer concentration, and observations to obtain this are generally not at hand. Therefore, it is more common to run models forward in time, the information being extracted by comparing simulated and observed tracer distributions (Sarmiento, 1983; Thiele *et al.*, 1986). This approach evidently implies coping with time dependent tracer boundary conditions, which can be difficult to establish. A general approach to this problem is discussed by Wunsch (1987). He shows that missing parts of tracer boundary conditions can be calculated back from interior ocean observations. It is clear, however, that the more one knows about the boundary conditions *a priori*, the more useful tracer data will be in constraining ocean circulation and mixing.

Ocean surface boundary conditions are relatively straightforward to obtain for the gaseous tracers freons and ^{85}Kr. Their environmental releases accumulate in the atmosphere. Atmospheric concentrations, due to relatively fast mixing, are fairly homogeneous worldwide, and they essentially impose a solubility equilibrium concentration in ocean surface water. Therefore, a single record of their tropospheric concentration history is sufficient to estimate their ocean surface water concentrations world-wide over the length of the record. In contrast to this, any tritium released into the troposphere enters the ocean (or, to a smaller fraction, the continental hydrosphere) within weeks. Consequently, ocean surface tritium concentrations arise in a dynamic balance between addition from the atmosphere and dilution by interior ocean mixing; this balance, naturally, is subject to appreciable regional variation. Special efforts are therefore required to establish ocean-surface boundary conditions for tritium.

Tracer boundary conditions are generally needed also for interior ocean boundaries. The problem one faces there is similar for all transient tracers: these boundary conditions can be based only on observations, and suitable observations, over extended periods of time, are often lacking. An extreme case has been treated by Wunsch (1988). The problem is not addressed further here, but reference is made to the discussion by Thiele *et al.* (1986). These authors placed their interior ocean boundaries such that tracer concentrations on part of the boundaries were either very small or near to surface water concentrations, so that a 'perturbation approach' was possible. The remaining part had flow outward from the model domain, and because the model was run forward in time from initial concentration zero, no tracer boundary condition had to be specified there.

Tritium decay in ocean waters has produced detectable quantities of ^{3}He, which allows a tritium-^{3}He 'age' to be calculated. Age distributions can be inverted into circulation information more directly (Jenkins, 1987) than is possible for regular transient tracers. Methodical questions remain, however, an example being non-linear effects of mixing on the age distributions (Jenkins and Clarke, 1976). Moreover, separation of tritiugenic ^{3}He from the natural ^{3}He background can pose problems, because the tritiugenic component typically amounts to only a few percent of the background.

In the following, the first topic is ocean surface boundary conditions for tritium. It will be seen that for establishing boundary conditions in a given modelling context various approaches are possible, which, however, will usually have to be elaborated for the specific application. Thereafter, various aspects relevant to using tritium-^{3}He age as an oceanic 'tracer' are considered. This part starts with the problem of separating tritiugenic ^{3}He from the natural background, continues with features of the transport equation for tritium-^{3}He age and its difference to that for regular transient tracers, and ends with the presentation of an empirical set of tritium-^{3}He age data. The conclusion will be that it is advantageous to use tritium and ^{3}He data in the combined form of tritium-^{3}He age, and that age distributions are attractive for evaluation by inverse modelling.

2. TRITIUM OCEAN SURFACE BOUNDARY CONDITION

Ocean surface boundary conditions for tritium can be formulated either as a flux or as a concentration condition. In principle, it would be straightforward to construct a *concentration condition* by objective interpolation of available surface water tritium observations. This has been done for the Pacific (Fine and Östlund, 1977), but, in general, observations are quite limited, which restricts such a procedure to certain oceanic regions as well as in time. Particularly sparse are observations prior to about 1965, during which time surface water tritium concentrations rose to their maximum. If surface water data are lacking, other information such as concerning the tritium addition/dilution balance for the ocean surface layer mentioned above has to be introduced. For the North Atlantic, for example, surface water observations have been extrapolated back in time from 1965 by introducing information on bomb fallout delivery from the troposphere into a simple oceanic mixing model (Dreisigacker and Roether, 1978); the surface concentration history so obtained has recently been extended forward in time beyond 1972 by interpolation of the since-plentiful surface water observations (Doney and Jenkins, 1988). In remoter ocean areas, the situation may be even more difficult to handle.

A worldwide tritium *flux condition*, on the other hand, has been constructed by Weiss and Roether (1980). As shown below, the procedure essentially requires atmospheric tritium data only, so that the problem of lack of oceanic observations is minimized. A flux condition may be regarded as more natural than a concentration condition, considering that, as mentioned, tritium is added to the ocean essentially as a flux imposed from the atmosphere. A further aspect is that combining a tritium flux condition with surface water tritium concentration data should provide constraints on upper ocean mixing. For these reasons, availability of a tritium flux condition is desirable for oceanic tritium modelling. The choice between a concentration and a flux condition is a special feature of tritium. For the freons, which in surface waters essentially have imposed concentrations (see above), a freon flux boundary condition cannot be formulated.

The basis of the Weiss and Roether (1980) flux condition is as follows. Oceanic tritium input was computed as the sum of three fluxes, i.e. net troposphere to ocean mixed-layer tritium deposition for open-ocean conditions, excess tritium deposition due to horizontal tropospheric tritium inflow from continental areas, and tritium addition by river run-off. The latter two fluxes are appreciable in the North Atlantic and North Indian oceans (deposition of about 25% and 38% of the total, respectively), but much smaller elsewhere. The open ocean net tritium deposition flux d was computed as

$$d = \left(P + f \frac{E}{s} \frac{h}{(1-h)} \right) c_p - \frac{E}{s} \frac{1}{(1-h)} c_s \qquad (1)$$

where,

$P =$ preciptation;

$E =$ evaporation (m/year);

$f =$ average ratio of actual tritium concentration in near-surface water vapour relative to that in isotopic equilibrium with precipitation;

$s = 1.13 =$ ratio of tritium concentration in water to that in water vapour in thermodynamic equilibrium;

$h =$ actual air humidity expressed as fraction of humidity in thermodynamic equilibrium with water surface;

$c_p =$ tritium concentration in precipitation (TR);

$c_s =$ tritium concentration in ocean surface water (TR).

d is obtained in TR m/year (the tritium concentration unit TR, synonymous to TU, represents a ^{3}H/H ratio of 10^{-18}; Taylor and Roether, 1982); it is a function of position and time. The first term describes tritium deposition by precipitation, as well as by air humidity being incorporated into surface water by interfacial exchange (so-called vapour exchange); the latter portion usually dominates. The second term represents the re-evaporation portion of vapour exchange; this term does require information on oceanic tritium (c_s), but is small at least up to 1967 because $c_s \ll c_p$. P, E, h were taken from the literature (climatological data); uncertainties are amplified in the case of h owing to the form in which it enters Eq. (1). A critical part is f, for which a value of unity was used, based on observations which showed tritium in air humdity at ship's height to be in near-equilibrium with that in collected precipitation samples. However, these observations were limited regionally (see Koster et al., 1988, Fig. 2) as well as in time. Thus, a somewhat smaller, or variable, value can not be excluded.

Distributions of c_p and c_s were constructed on the basis of compilations of available tritium data. c_p was factorized into temporal and areal distribution functions, based on tritium records in monthly precipitation samples from stations all over the globe, and the areal distribution for the open ocean was taken as simply meridional. Figs. 1 and 2 reproduce the temporal and meridional distributions used. The

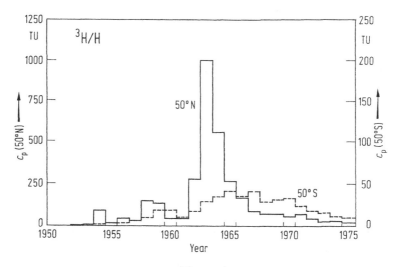

Figure 1
Mean annual tritium concentrations 1952 to 1975 in marine precipitation
for 50°N (solid line) and 50°S (broken line; right-hand ordinate). From
Roether and Rhein (1988).

point to make is that the temporal distributions, dissimilar in both hemispheres,
are quite well defined (although in tropical latitudes time dependence must be intermediate to those shown in Fig. 1 for 50°N and 50°S; the southern hemispheric
distributions generally have larger uncertainties than the northern ones), while the
meridional distribution function has considerable uncertainty. This arises because
areal correlation between stations is much lower than the temporal one, mostly
due to tritium enrichment over continents (concentrations up to a factor of about
4 higher than for corresponding open ocean conditions), which is caused by continental net tritium deposition being low. Continental influence sometimes makes it
even difficult to pick among the available stations those that might represent open
ocean conditions.

The tritium enrichment over continents leads to the second flux considered,
i.e. excess tritium deposition near continents. The flux was computed as continent-
to-ocean water vapour outflow (climatological) times appropriate tritium concentration, which amount was then distributed over an estimated exchange length for

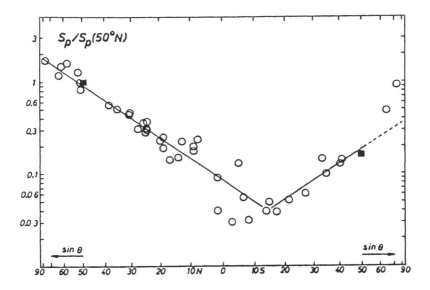

Figure 2
Latidudinal variation of open-ocean tritium delivery relative to 50° N.

relaxation of the excess by the oceanic deposition. The resulting flux has considerable uncertainty, mediated by the fact that it is of substantial magnitude in a
few regions only. Partly, continental influence was accounted for implicitly, i.e. by
residual continental influence on the distribution of Fig. 2. The third, and least
problematic, flux, due to river runoff, was computed from runoff (climatological)
and observed or estimated tritium concentration, estimates being based on simple
delay models between precipitation and continental runoff. Time-integrated total
tritium depositions for the three major oceans as a function of latitude are shown
in Fig. 3.

A check on the deposition fluxes was made by integrating deposition over ocean
basins as well as in time and comparing the resulting total with ocean-basin tritium
inventories. These inventories can be computed fairly reliably. Agreement within
about 20% was found, the depositions being a little larger than the inventories. The
comparison has since been repeated, with similar results (Broecker *et al.*, 1986).
The agreement is certainly satisfactory in view of the uncertainties in the flux
calculations. It can also be taken as proof for the relevance of vapour exchange,
and that f could hardly be much below unity, as had been assumed previously

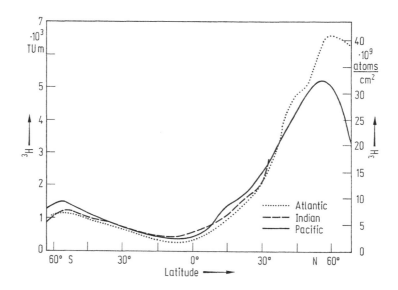

Figure 3
Latitudinal variation of total amount of tritium delivered to the major
oceans up to 1973. From Roether and Rhein (1988).

(Bolin, 1958; see also Koster *et al.*, 1988). The agreement constrains the oceanic
tritium deposition deduced to be correct on average to about 20% (the rates might
even be renormalized by means of the inventories). On smaller scales, deposition
uncertainties undoubtedly are larger, and they are considerable in enclosed seas.
These uncertainties are mostly a result of large error margins in the areal tritium
distribution function. Therefore, if improvement of the flux condition was sought, a
promising step would be a reassessment of this distribution function, possibly with
the help of an atmospheric circulation model.

If one needs a tritium ocean surface boundary condition in a specific context, one has the following options: (i) to construct one's own source function,
presumably using one of the procedures outlined above; (ii) to accept, renormal-
ize or otherwise modify one of the existing concentration source functions; (iii)
to accept, update or otherwise modify the existing flux source function. Renormalization means employing regional observations to adjust amplitudes. Ways of
modification include mixing concentration and flux information; such an approach
should lead to improved boundary conditions because it introduces additional ob-

servational constraints. As examples of how to obtain specific source functions in this way, the deductions of tritium concentration histories in the surface water of the Norwegian and Arctic Seas (Heinze *et al.*, 1988), as well as in the northern Red Sea (Kuntz, 1985), may be referred to. Boundary condition information for the southern hemisphere is sparser than for the northern hemisphere, and special efforts will be required there.

3. SEPARATION OF TRITIUGENIC ^{3}He

That a tritiugenic component of ^{3}He is at all measurable in the ocean, and consequently tritium-^{3}He dating feasible, is due not only to the small natural abundance of this nuclide and to low helium solubility, but also to the fact that man's tritium release into the environment just happened to be adequate. The component has remained, however, but a small fraction of oceanic ^{3}He (generally less than 10%), so that, had the release been one order of magnitude less, oceanic tritium-^{3}He dating would be a rather marginal tool. Quantifying the tritiugenic component, therefore, is often limited by one's ability to separate it from the natural portion. The separation is complicated by the fact that natural oceanic He is not simply determined by equilibration with atmospheric He at the ocean surface, but additionally has an appreciable terrigenic component, which generally increases with depth and 'water age'. The significance of this component has the same cause that renders tritiugenic ^{3}He measurable, i.e. low atmospheric He concentration and low ^{3}He/^{4}He ratio owing to the loss of He to space.

Tritiugenic ^{3}He stands out clearly in northern hemisphere central waters. In Fig. 4, for example, tritiugenic ^{3}He is bounded by a ^{3}He minimum layer below it (δ^3He $\leq$ +1.5%), that slopes down to the north to about 2000 m depth. The ^{3}He increase below the minimum must be due to terrigenic ^{3}He because tritium is virtually absent, but the terrigenic ^{3}He will decrease upward, so that sufficiently far up, terrigenic ^{3}He can be expected to be small. Near to the ^{3}He maximum in Fig. 4 (> 9%), the tritiugenic component can therefore be determined well. The situation is less favourable near the minimum, as well as in many other parts of the ocean particularly in the southern hemisphere.

The question of separating tritiugenic ^{3}He quantitatively can be attacked as follows. One considers three oceanic He components, i.e. atmospheric, terrigenic (concentrations He_a, He_t, in cc(STP) per kg of seawater), and tritiugenic (^{3}He*,

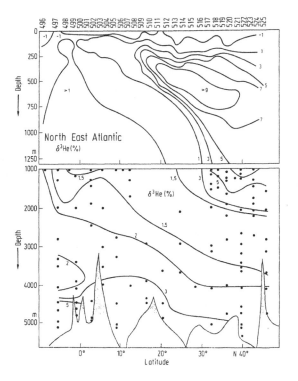

Figure 4
$\delta^3 He$ section (in %), 8°S – 46°N, in the Northeast Atlantic. From Roether
and Rhein (1988).

$^4\mathrm{He}^* = 0$). Denoting the isotopic ratio $^3\mathrm{He}/^4\mathrm{He}$ of any component as R, one has
for total He and total $^3\mathrm{He}$

$$\mathrm{He} = \mathrm{He}_a + \mathrm{He}_t$$
$$^3\mathrm{He} = R_a\mathrm{He}_a + R_t\mathrm{He}_t + {}^3\mathrm{He}^* \qquad (2)$$

A good starting point is to consider that oceanic He, to lowest order, is determined
by solubility equilibrium values with atmospheric He (atmospheric $^3\mathrm{He}/^4\mathrm{He}$ ratio
$= R_0 = 1.384 \times 10^{-6}$; the equilibrium value for oceanic He (Weiss 1971; 3.97×10^{-5}
cc(STP)/kg for 0°C and 35 PSU) will be denoted He_{eq}). This means that, in

such approximation, oceanic He is given by He_{eq} and ^3He by R_0He_{eq}. In the treatment to follow, we shall use the customary notations $\delta^3He = R/R_0 - 1$, and $\Delta He = He/He_{eq} - 1$ (similarly for Ne), and we shall scale He by He_{eq}, and ^3He by R_0He_{eq}. The latter scaling factor can be converted to tritium concentration units (TR), in order to allow direct comparison with tritium later on. The meaning of oceanic $^3He^*$ of 1 TR is the concentration formed when tritium at 1 TR has decayed fully. The conversion (Jenkins, 1987) is

$$R_0He_{eq}(TR) = \frac{10^{18}m_{H_2O}}{2V_{mol}(1 - S/1000)}R_0He_{eq} \quad (cc \; ^3He(STP)/kg \; seawater) \quad (3)$$

where m = molar mass, V_{mol} = ideal gas molar volume in cc (STP), and S = salinity in PSU. The numerical value of R_0He_{eq} for 0°C and 35 PSU is 22.9 TR.

The atmospheric component of He in sea water (He_a, R_aHe_a) is different from the scaling values because of isotopic separation in the air-water equilibration, and because it is affected by dissolution of, and exchange with, air bubbles, that are present through wave action. This action leads to He_a being a few percent above solubility equilibrium (Jenkins and Clarke, 1976), and it also introduces a slight further shift in R_a relative to R_0 (Fuchs et al., 1987). The latter authors present a simple theory of the bubble effects, which leads to

$$R_a = R_0\left(1 + \delta^3He_{eq} + q\Delta He_a\right) \quad (4)$$

where δ^3He_{eq} (typical value -0.017) represents the air-seawater thermodynamic equilibrium separation according to Benson and Krause (1980), and the following term represents the bubble correction ($q = -0.043 \pm .016$); its value typically amounts to -0.002 only. The theory has been disputed by Jenkins (1988), who presented a different bubble correction. Time will decide which of the treatments is more adequate; at this stage it is sufficient to state that the bubble effect can be represented by a small term in Eq. (4), such as, or similar to, that shown.

He_a can be assessed by introducing measurements of oceanic Ne concentration. The point is that oceanic Ne has only an atmospheric component, so that Ne deviations from solubility equilibrium (Ne_{eq}; Weiss, 1971) will characterize the bubble-derived Ne portion, from which the corresponding He portion is then to be derived. A relation between the two will exist, because it is the solubility of the species and its molecular diffusivity in water (Jähne et al., 1987) which determine the portion, and these parameters are not very different for He and Ne. In fact, the Fuchs *et al.* (1987) theory gives for the ratio He_{bubble}/Ne_{bubble} a value of $p = 0.92 \pm 0.02$.

A complication arises, however, because, as a result of finite atmosphere-ocean gas exchange rates, gases in the mixed-layer are subject to a time lag in their equilibration with the atmosphere. The concentration shifts induced by this lag are larger for Ne than for He, owing to about three times larger temperature dependence of solubility, and to somewhat slower gas exchange. Solubility decreases with increasing temperature, and therefore there will be a mixed-layer gas excess relative to a non-lag situation near to the seasonal temperature maximum and a deficiency near the minimum. The latter case applies here because the concern is the characteristics of subsurface waters, which waters are usually formed near to the temperature minimum. The magnitude of the deficiency was estimated using the convective model of a seasonal thermocline of Fuchs *et al.* (1987). This model simulates the seasonal development of the depth of the convective surface layer (under annually periodic conditions), and computes its gas concentration under gas exchange with the atmosphere (Roether, 1986) and entrainment from (during mixed layer deepening), or detrainment into (during ascent), a seasonal thermocline. A model feature added was to prescribe an exponential temperature decrease with depth (scale length = seasonal amplitude of depth of convection) in the seasonal thermocline while ignoring salinity variations. The model was run in time until annually periodic gas concentrations were obtained. Model outputs are the He and Ne deficiencies relative to solubility equilibrium, at the time of deepest convection. Values were found to be essentially proportional to the seasonal amplitude in mixed layer temperature, and inversely to the gas exchange coefficient, and those for Ne to be approximately threefold those for He. The considerations that led Fuchs *et al.* (1987) to choosing a rather small gas exchange coefficient, apply only partly in the present context, and therefore a larger coefficient was chosen (5 m/day for He). To simulate conditions such as those addressed in section 5 below, the seasonal temperature range was taken to be 7° to 12°C (Robinson *et al.*, 1979). The resulting difference between the Ne and the He deficiency amounted to 0.4% for a seasonal-maximum depth of convection of 200m, rising to 0.7% for 800m depth.

Combining the effects, one may express the atmospheric He component in terms of excess Ne (ΔNe), and of the Ne minus He deficiency difference owing to the surface-layer seasonal temperature variation, $\Delta(Ne - He)_T$, neglecting a small term, as

$$He_a = He_{eq}(1 + p\Delta Ne + \Delta(Ne - He)_T)$$
$$\equiv He_{eq}(1 + p\Delta(Ne)_{cor}) \tag{5}$$

The correction $\Delta(Ne - He)_T$ is expected to be on the order of $+0.005$. The actual value has to be obtained from a model such as the one just sketched. It is noted that observations for heavier gases, such as Ar (GEOSECS, 1987), will allow one to tune the model, because these gases are subject to a larger effect of temperature variation (Craig and Weiss, 1971), and at the same time to a smaller effect of bubbles. p also is a model value, to which the same considerations apply that were made in connection with Eq. (4). Some modification of p, as well as of R_a, may be required for certain water masses that contain noticeable amounts of He and Ne, introduced by subsurface release of air bubbles from melting continental ice (Schlosser, 1986).

Eqs. (4) and (5) quantify the atmospheric component of oceanic He in terms of observable quantities, laboratory values, and values to be obtained from models. The model-dependent values require further study, but are not very critical as they do not exceed measurement errors much (see below). It should therefore not be too difficult to reduce the error in R_a to $\pm.001$, and that of He_a essentially to that of ΔNe. It must be mentioned, however, that, so far, published records of He isotope data do not include Ne data on a routine basis, and even if so, precision is often limited. While this is changing now, the equations at present, consequently, represent a concept rather than a recipe to treat data at hand. The present practice mostly is to ignore terrigenic He and replace $p\Delta(Ne)_{cor}$ by ΔHe (Eq. 5 and below), or even to ignore variations in the atmospheric component He_a. These approximations are tolerable if $^3He^*$ is sufficiently large.

The terrigenic He component has the problem that R_t varies appreciably in ocean waters, depending on whether the He is crust or mantle derived (typical R_t/R_0 about 0.1 resp. 8; Craig and Lupton, 1981). Because of its high R_t value, mantle He, if present in substantial amounts, will generally mask tritiugenic ^3He. Furthermore, no additional constraint that might serve to separate terrigenic and tritiugenic ^3He is in sight. As the next step, we shall therefore combine terrigenic and tritiugenic He, using Eqs. (2), (4) and (5):

$$He_t = He_{eq}(\Delta He - p\Delta(Ne)_{cor})$$

$$^3\mathrm{He}^* + R_t\mathrm{He}_t = R_0\mathrm{He}_{eq}\Big(\delta^3\mathrm{He}(1 + \Delta\mathrm{He})$$
$$- \delta^3\mathrm{He}_a(1 + p\Delta(\mathrm{Ne})_{cor}) + \Delta\mathrm{He} - p\Delta(\mathrm{Ne})_{cor}\Big) \qquad (6)$$

Present measurement errors are commonly $\pm0.15\%$ for $\delta^3\mathrm{He}$, and $\pm0.5\%$ for He and Ne. That these have larger errors is mostly because they are sensitive to minute inclusions of air in the sampling process at sea (Fuchs $et\ al.$, 1987). With dedicated effort one might be able to reduce the errors in $\Delta\mathrm{He}$ and in $\Delta\mathrm{Ne}$ (prior to correction) to $\pm0.25\%$; it is evident that, even then, these will be dominating errors when evaluating Eqs. (6).

In practice, tritium is mostly present in rapidly ventilated parts of the ocean were He_t is rather small, so that $^3\mathrm{He}_t$ can be treated as a correction. A simple case holds when the waters in question can be approximately taken to be a mixture of two components, of which one contains no He_t and the other no $^3\mathrm{He}^*$. For the former component, Eqs. (6) then give R_t, knowing which, allows one to calculate $^3\mathrm{He}^*$ for all water mixtures. The equation to be used is

$$^3\mathrm{He}^* = R_0\mathrm{He}_{eq}\Big(\delta^3\mathrm{He}(1 + \Delta\mathrm{He})$$
$$- \delta^3\mathrm{He}_a(1 + p\Delta(\mathrm{Ne})_{cor}) - \delta^3\mathrm{He}_t(\Delta\mathrm{He} - p\Delta(\mathrm{Ne})_{cor})\Big) \qquad (7)$$

The approach is now illustrated for the waters above the $^3\mathrm{He}$ minimum layer in Fig. 4. One purpose will be to demonstrate the extreme He and Ne data precision needed, which arises from $\delta^3\mathrm{He}_t$ being large, so that the error in $^3\mathrm{He}^*$ is dominated by the last term in Eq. (7). For the waters in question, it appears justified to assume them to be a two-component mixture, of water just below the minimum layer with negligible $^3\mathrm{He}^*$ (see above), with upper-ocean water free of terrigenic He. Table 1 summarizes the tentative calculation. Only He data of moderate quality (Fuchs, 1987), and no Ne data, exist parallel to the data of Fig. 4, but reasonable $\Delta\mathrm{He}$ and $\Delta\mathrm{Ne}$ numbers were picked taking into account Atlantic GEOSECS He and Ne data (GEOSECS, 1987). They were chosen such that terrigenic He at the $^3\mathrm{He}$ maximum is a factor of three lower than just below the minimum layer (in reality it is probably lower by more than a factor of three).

One notes that the calculated error of tritiugenic $^3\mathrm{He}$ at the maximum is appreciable, and that it exceeds those in $^3\mathrm{He}^* + ^3\mathrm{He}_t$, which holds because the sum is obtained first and the subsequent subtraction of $^3\mathrm{He}_t$ to obtain $^3\mathrm{He}^*$ adds uncertainty; the error is essentially proportional to $\delta^3\mathrm{He}_t$ times the uncertainty in

Table 1
Estimated terrigenic and tritiugenic ^{3}He for section of Fig. 4. δ^3He$_a$ = −0.02; tritiugenic ^{3}He below minimum layer assumed zero; errors given use lowest measurement and model errors; for these and for the choice of ΔHe and $p\Delta$(Ne)$_{cor}$, see text.

	δ^3He	ΔHe	$p\Delta$(Ne)$_{cor}$	R_t/R_0	He$_t$ + ^{3}He* (TR)	^{3}He* (TR)
Below ^{3}He-min	0.02	0.065	0.05	3.82±.8	1.43±.10	—
At ^{3}He-max	0.09	0.055	0.05	as above	2.77±.10	2.34±.27

ΔHe − $p\Delta$(Ne)$_{cor}$. The errors in Table 1 are pessimistically large, however, as they are simply those for isolated samples. In reality, He$_t$ will be quite smooth so that uncertainties can be reduced by averaging. Moreover, water mass analysis will probably allow one to estimate the terrigenic He remaining at the maximum, better than does the first Eq. (6). For example, in the case that water mass analysis would exclude a contribution of the deep component at the maximum, this would (by Eq. 6) immediately impose ΔHe = $p\Delta$(Ne)$_{cor}$ in Eq. (7), which would reduce the error in ^{3}He* to ± 0.04 TR. The error is not much larger if some deep-component water is still present but the amount can be determined by water mass analysis with small uncertainty. Thus, a value of ± 0.04 TR (one Sigma error) represents about the minimum error with which, under the most favourable circumstances, tritiugenic ^{3}He can be separated. That the value is larger by one order of magnitude than the state-of-the-art detection limit for tritium, is not surprising considering that one has to separate tritugenic ^{3}He from a large background.

A further effect of finite gas exchange to note, is that ^{3}He* can be non-zero in deep mixed layers, so that water masses formed have a non-zero initial tritium-^{3}He age. This is of little consequence, however, because the age excess is on the order of 0.5 years only (Fuchs et al., 1987), which, in relative age, is generally small, and moreover, the interest is mostly in age differences (see below), so that a non-zero initial age hardly matters anyway.

4. THE TRITIUM-^{3}He AGE CONCEPT

Tritium-^{3}He age, t_a, is defined as

$$t_a = t_{1/e} \ln(1 + {}^3\text{He}^*/{}^3\text{H}) \tag{8}$$

where $t_{1/e}$ = mean life of tritium (17.93 years; Unterweger et al., 1980) and ^3H = tritium concentration (TR). Jenkins and Clarke (1976) were the first to point to non-linear effects of mixing on t_a, which are obvious in the case of two-component mixing of younger, tritium-rich with old, tritium-free water. In such an extreme case, all water mixtures, even if predominantly old, will formally show the same age, i.e. that of the younger component. As Jenkins (1987) has shown, a transport equation holds for t_a, being obtained when transport equations for tritium and tritiugenic ^{3}He are entered into dt_a/dt following (8), which equation may be written as

$$D \, t_a = 1 - (t_a)_t + \text{ correction terms} \tag{9}$$

where,

$$D = \text{advection} - \text{diffusion operator} = \mathbf{u}.\nabla - K_h \nabla^2 - K_z \frac{\partial^2}{\partial z^2}$$

K_h, K_z = horizontal, vertical diffusivity; $\mathbf{u}$ = velocity vector, ∇ = horizontal gradient operator; subscripts t and z denote partial differentiation (formulation in isopycnic-diapycnic coordinates is also possible). Only time-independent advection and mixing is considered. A unity 'source term' appears, as t_a increases explicitly one year per each calendar year. The correction terms, representing non-linear mixing, are

$$K_h \nabla \ln({}^3\text{H}^2 + {}^3\text{H}{}^3\text{He}^*) \nabla t_a + K_z \left(\ln({}^3\text{H}^2 + {}^3\text{H}{}^3\text{He}^*) \right)_z (t_a)_z$$

They naturally are proportional to the diffusivities, and they contain products of gradients. The terms were large while tracer gradients were large early during the tritium transient, but they will since have gradually become smaller. In fact, Jenkins (1987) reported the terms to be rather small in 1979 observations from the Beta triangle area, for $t_a < 20$ years. In this age range the time derivative generally is also small (see below), so that one will often have $D \, t_a = 1$, to first order.

By comparison, regular transient tracers have a negative source (representing decay), and their transport equation equivalent to Eq. (9) is

$$t_{1/e} \frac{D\,c}{c} = -1 - t_{1/e} \frac{c_t}{c} \qquad (10)$$

Apparently the transport terms $D\,c$ and the time derivative are now scaled up by the tracer mean life $t_{1/e}$. The quantity $t_{1/e}$ represents a large factor (for example, freon lifetimes in the ocean are thought to be on the order of 100 years or more), so that the unity source term is no longer prominent, in contrast to the situation in Eq. (9). (It is noted in passing that, inasmuch as differentiating the logarithm in Eq. 8 to obtain $D\,t_a$ leads to a differential of the argument divided by the argument itself, there is correspondence between the form $(D\,c)/c$ and $D\,t_a$.) Specifically, $t_{1/e}$ is generally at least comparable to the time scale of evolution of the tracer cloud (c/c_t), and therefore non-stationarity, which, as mentioned, is difficult to determine from observations, will virtually always represent a large term in Eq. (10). One does not even gain much if a transient tracer cloud has evolved to a stage that its time dependence has become relatively small (which can be expected to be true for tritium over part of the upper ocean), because then its spatial derivatives will generally also be small, which in turn can induce significant data noise in the transport terms. In contrast to this situation, spatial derivatives of t_a cannot become too small, because $D\,t_a$ must remain of order unity.

The foregoing comparison should have illustrated that Eq. (9) indeed has desirable properties. A hint of caution is due, however, as D and tracer boundary conditions can in principle act together to avoid spatial derivatives becoming small eventually. An example is an upwelling situation in which downward mixing of tracer counteracts its advection (Roether et al., 1970). Tracer transport terms then remain large and are of opposite sign, so that Eq. (10) will work well because, to first order, $D\,c = 0$ (this evidently implies stationarity to the same order). At the same time, Eq. (9) will then be problematic to use, because gradients, and hence the correction terms, stay large. Strictly, therefore, one has to make a distinction between situations where spatial transient-tracer derivatives tend to decrease with time, and those where this is essentially not the case. It is judged that the former will generally be the norm.

In such a 'normal' situation, non-linear mixing will produce, as has been pointed out above, a distorted t_a field during the early evolution of the tritium and ^{3}He clouds, while later on, when the tracer clouds are becoming smoother, distortion will decrease. The balance of terms in Eq. (9) will thus change, and therefore, a time dependence of the t_a field is to be expected. On the other hand, observational evidence of a t_a field with low non-stationarity will be presented in section 5. It will now be shown that this observation is not unexpected, inasmuch as Eq. (9) implies a tendency towards stationarity, provided only that the correction terms have become small. This can be seen from an expansion of t_a in terms of a stationary distribution t_{a0}, in the form

$$t_a = t_{a0}(1 - re^{-t/t_0}) + \text{higher terms} \tag{11}$$

which represents, to first order, relaxation of t_a from a fraction $1 - r$ of t_{a0} to t_{a0}. Entering Eq. (11) into (9) with neglected correction terms, one indeed obtains $t_0 = t_{a0}$, without any further restriction. The result is reasonable because it means that, to first order, it takes a period t_{a0} to flush out an initial disturbance and to approach a stationary age distribution. As for a quantitative estimate, one might assume that presumably $0 < r < 1$, because, as mentioned above, t_a is expected to have started with low values early in the tritium transient period. One might furthermore assume the relaxation to have started ($t = 0$) at about the time of the tritium maximum in ocean surface water, i.e. 1965 in the northern hemisphere. With these values, Eq. (11) then predicts that for TTO data, for example, i.e. in 1982, the remaining non-stationarity for $t_{a0} = 10$ years should be $(t_a)_t < 0.18$. The limit will necessarily decrease with later time of observation. The relaxation expansion demonstrates that the transport and source terms in Eq. (9) in fact act to restore the age distribution towards stationarity.

Tritium-^{3}He age studies reported so far (Jenkins, 1987; 1988) treated dominantly advective transport, in which case observed t_a distributions can be particularly easy to interpret: if all mixing terms are negligible, and t_a is stationary, one is simply left with the advective balance $\mathbf{u}\nabla t_a = 1$, and t_a will represent water travel time along trajectories of flow. In this situation, t_a has been termed 'advective age' (Jenkins, 1988). A more general case is that the other terms are 'small', but not negligible. It is then often difficult to interpret t_a as an 'age'. It is noted, however, that this difficulty does not impair at all the usefulness of t_a distributions to constrain ocean circulation and mixing in models. This argument also applies to the correction terms, provided that observational data are at hand that enable one to quantify them. In inverse modelling, for example, it is straightforward to incorporate the terms, and the same argument holds for a remaining non-stationarity. The magnitude of the 'small' terms is still important, however, because, naturally, the smaller a value, the less relative precision will be required in its determination. A practical case will be presented below. It will be seen that extended regions exist where the terms in question are < 0.2, and also that this condition is accompanied by $t_a \leq 15$ years. In regions where these limits are substantially surpassed, the use of t_a distributions has yet to be systematically explored. Investigations of t_a are also lacking for the southern hemisphere, because of the problem involved in separating tritiugenic ^{3}He (see above).

5. TRITIUM-^{3}He AGE DISTRIBUTIONS ON ISOPYCNAL SURFACES IN THE LOWER NORTHEAST ATLANTIC MAIN THERMOCLINE

As an example for treatment of actual data, this section comments on a study (Fuchs, 1987) that used an extensive tritium and ^{3}He data set from various cruises, 1977 – 1984, in the Northeast Atlantic. Distributions on isopycnal surfaces $\sigma_0 = 26.85, 27.0, 27.15, 27.3, 27.45, 27.6,$ and 27.75 were constructed by interpolation of station data to these surfaces followed by hand-contouring. Tritiugenic ^{3}He was computed according to Eq. (7) but with the terrigenic component ignored (see above), to yield tritium-^{3}He ages. Fig. 5 presents age distributions (in years) on the isopycnals 27.0, 27.3, and 27.6, together with, for the purpose of illustration, the ^{3}He distribution (δ^3He in %) on the middle horizon. The horizons correspond to depth horizons of about 200 – 400, 700, and 1000 m, but they ascend northward and outcrop at roughly 45°, 53°, and 62°N, respectively. Ages increase, smoothly, primarily southward, and a comparison of the two 27.3-horizon fields demonstrates that the age field is much different from that of ^{3}He. Ages in excess of 15 years are only present on the two lower horizons, south of about 28°and 32°, respectively. On the deepest horizon, which is within the Mediterranean water layer, isochrones bend back southeastwards off the Iberian peninsula. This is interpreted as Mediterranean water influence, because the ages observed in the region in question are about those expected for Mediterranean water layer source water (Schlosser, 1985), and concentrated flow along the peninsula in the layer has been shown to exist (Reid, 1978).

Fig. 6 is an example of observational evidence on stationarity of the age fields. The age profiles shown are for neighbouring stations taken 3.5 years apart, and

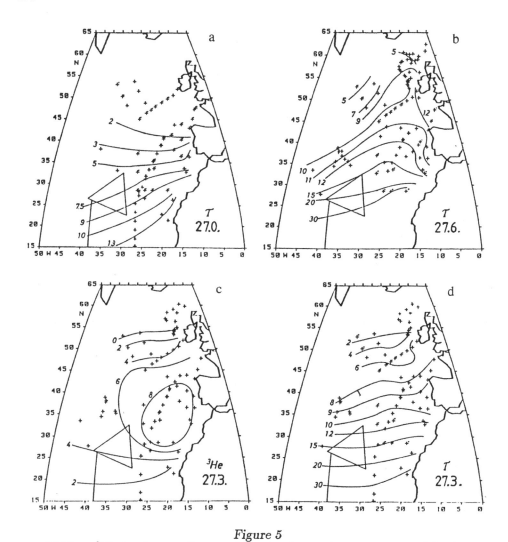

Figure 5

Tritium-³He isochrones (in years) on density surfaces $\sigma_0 = 27.0$ (a), 27.3
(b), and 27.6 (c), and ³He isolines (in %) on 27.3 (d) in the Northeast
Atlantic. From Fuchs (1987).

they do not reveal any significant time trend. The finding is particularly important as the study uses observations that span a period of some years. The evidence was interpreted as giving an upper limit for $(t_a)_t$ of 0.15. Estimated isopycnal mixing term magnitudes, both for non-linear and true mixing, are presented in Fig. 7 for two horizons intermediate to those of Fig. 5; the corresponding values for diapycnal mixing were estimated to be about one order of magnitude smaller than these. It appears that the non-linear mixing term is somewhat larger than that for 'true' mixing, and that the former only becomes appreciable (> 0.2) in the southern regions, about where the ages exceed 15 years. Vertical advection was also shown to be small.

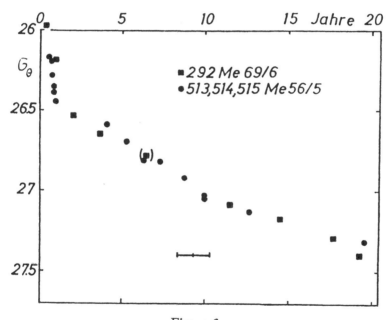

Figure 6

Profiles of tritium-^{3}He age (in years) versus density for stations near 28° N, 25° W, in April 1981 ('Meteor' cruise 56/5) and in November 1984 ('Meteor' cruise 69/6). From Fuchs (1987).

It thus appears that over much of the study area vertical advection, the time derivative, and all the mixing terms are small. Therefore, the simplified equation $\mathbf{u}_I \nabla t_a = 1$ should provide meaningful isopycnal velocities $\mathbf{u}_I$. Fig. 8 gives velocity

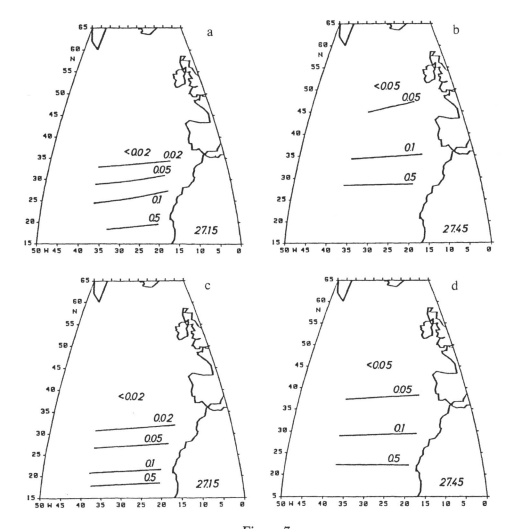

Figure 7

Magnitudes of non-linear mixing terms (upper panels) and 'true' mixing terms (lower panels), on the isopycnals indicated, see text. From Fuchs (1987).

vectors so obtained from the data in Fig. 5 for the upper two horizons. It might be mentioned that if one extrapolates the northernmost velocity vectors back to zero age (Jenkins, 1987), one arrives at about the areas where ventilation of the respective isopycnal horizons is expected. This finding lends support to the velocities derived.

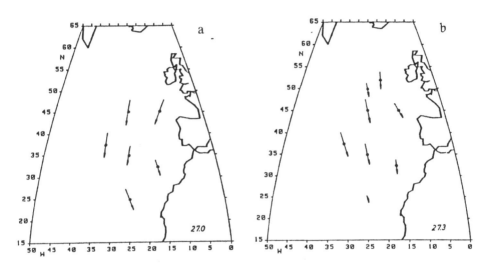

Figure 8
Isopycnal velocities derived from tritium-³He age gradients on the isopycnals indicated, see text. From Fuchs (1987).

The data between the σ_0 horizons 27.15 and 27.6 were further evaluated by inverse modelling. The model domain was from 10° to 30°W and 25° to 50°N, with 5° × 5° resolution, and the σ_0 range was subdivided into three layers. The deepest of these layers, 27.45 – 27.6, incorporates some Mediterranean layer water, which layer is centered between about 27.5 and 27.8 (Zenk, 1975). The age distribution was taken as stationary, and conservation of mass as well as of potential vorticity was included in the model. Diapycnal transport was neglected, but both true and non-linear isopycnal mixing were included, with the diffusivity prescribed (see below). Solutions were obtained using singular value decomposition. Fig. 9 presents flow fields obtained for the two lower layers. The circulation deduced is anticyclonic through both layers, north of 30°N. The region further south is that of large ages, where stationarity, and hence the model result, becomes questionable. Typical flows are about 0.25×10^6 m^3/s per box, which corresponds to approximately 0.4 cm/s velocity, which is the expected order of magnitude (Saunders, 1982). The transports were found to be quite stable when the isopycnic diffusivity, and a damping coefficient introduced in the usual way to suppress the effect of small singular values (Menke, 1983), were varied.

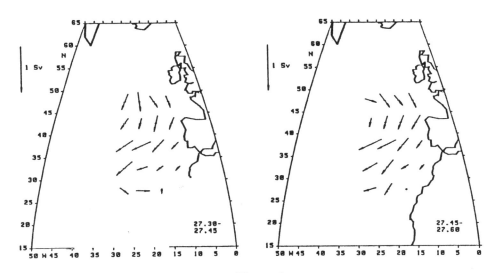

Figure 9

Transports from inverse model of tritium-^{3}He ages, potential vorticity, and mass, for density layers indicated, with $K_I = 400$ m^2/sec (same for potential vorticity) and damping coefficient 0.01, see text. From Fuchs (1987).

The foregoing results can be taken as evidence that tritium-^{3}He ages in principle can be determined adequately and evaluated successfully. However, not much experience exists so far on evaluating age distributions, so that, naturally, question marks remain. Some are addressed below, in connection with a certain discrepancy of the above results, noted previously (Roether and Fuchs, 1988), with an accepted oceanographic finding: while Figs. 8 and 9 indicate the circulation to be anticyclonic up to or even slightly beyond 50°N (Fig. 8), Saunders (1982) reports for the depth range in question a sluggish flow divide at roughly 42°N with cyclonic flow to the north of it, changing to anticylonic circulation throughout the domain only below 1250 m depth. He basically used geostrophic shear data, which he converted into absolute flow. At 48°N, his geostrophic current profile has northward maximum flow in the depth range in question. Reid's (1978) map of 1000 dbar geopotential anomaly relative to 2000 dbar similarly indicates a general northward flow starting even somewhat further south, although with some structure superimposed. Reid's signal is very clear off the Iberian peninsula. Reid (1981; Fig. 3.11 B) reports Russian geopotential anomaly maps 500 dbars relative to 1500 dbars, that show a similar flow pattern. In the light of this, a search for weak points in the above evaluation will now be made.

The assumed stationarity of the age distributions will certainly stand, to the extent that the circulation is stationary, because in the relevant region ages are low, and as shown above, stationarity should be approached fast. The scale analysis for the mixing terms necessarily depends on the diffusivities chosen; the isopycnal value used in Figs. 7 and 9, $K_I = 400$ m^2/s, should be valid in the southern part of the region investigated (Thiele et al., 1986; Jenkins, 1987), while further north, evidently, a substantially larger value would still be tolerable (Fig. 7); the latter argument holds for the vertical diffusivity throughout, because the vertical mixing terms (for $K_z = 10^{-5}$m^2/s) were found, as mentioned, to be considerably smaller than the isopycnal ones. This makes it improbable that the discrepancy could be caused by too small diffusivities being chosen. The diffusivity question can be addressed also in a different way: the observational fact is that ages increase evenly southward (Fig. 5). If indeed isopycnal flow was against the age gradient, then the observed age pattern would have to be dominated by mixing, as implicitly suggested, for example, by Sarmiento et al. (1982). This possiblity was explored, and it was found that it would require isopycnal diffusivities larger than 5×10^4m^2/s. This finding supports the model results, unless isopycnal mixing coefficients are very much larger than currently accepted. As vertical advection was found to have little effect for as much as 2×10^{-6}m/s vertical velocity, it is likewise improbable that ignoring vertical advection has caused a problem.

The apparent smoothness of the tracer distributions (Fig. 5) might be somewhat exaggerated owing to the involved hand contouring. While this might have affected second derivatives, and hence estimated magnitudes of the 'true' mixing terms, it should have had less effect on the non-linear mixing terms, which imply first derivatives only (Eq. 9). Non-uniform data coverage across the model domain exists and might have caused problems: a certain data gap is apparent with a center near 43°N, 30°W, and moreover, the data west and southwest from that gap originate from the earliest observations used, i.e. from 1977–78. This might have introduced a bias in the age gradients possibly extending far enough north and northeast to lead to a biased flow field in the critical region. Another point is that the mentioned divide reported by Saunders (1982) is not far from the northern model boundary. To what extent this vicinity might have influenced the result is not clear, but if there is an effect, it might similarly have acted to mask the expected Mediterranean Water layer flow off the Iberian peninsula.

No suspected single cause of the discrepancy can thus be named, but concerted action of several smaller problems cannot be excluded. Perhaps the weakest parts are data coverage and the choice of the model boundaries. It can be argued that improved data coverage, and availability of data allowing an extension of the model northwestward, would help to resolve the question. Model checks to find out in detail the robustness of the deduced circulation through the model domain might also be called for.

6. CONCLUSIONS

There is no unique answer to the problem of obtaining an adequate ocean-surface tritium boundary condition in a given modelling context. One has to choose between a concentration and a flux boundary condition, and one will often have to do one's own adaption of the boundary conditions that exist in the literature, which are outlined above; tritium boundary conditions in the southern hemisphere will require special efforts. The best constrained boundary conditions might be obtained by a mixed approach, i.e. by employing information on both concentration and flux. At intermediate depths, tritium will carry information mostly on comparatively large scales. For those, adequate assessment of tritium ocean surface boundary conditions is judged as feasible. In the upper ocean, on the other hand, tritium presumably will mostly be used in the form of tritium-^{3}He dating, which does not require much boundary information.

Combining tritium and ^{3}He observations in the form of tritium-^{3}He age is advantageous because one obtains a 'tracer' that carries oceanographic information rather independent of tracer boundary conditions, and is well suited for evaluation by inverse modelling. A main point is that non-stationarity usually is but a small term in the transport equation for age. It is judged that modelling of tritium-^{3}He age distributions, an example of which is presented above, so far has not received the attention it deserves. The example is interpreted as showing that observed age distributions in the northern hemisphere upper ocean can be evaluated successfully.

Some flags for future attention were placed in the discussion of the above example, however: there is foremost a call for adequately dense and uniform tracer data coverage over coherent, and preferably basin-scale, regions. Combining the above data with those from the TTO program (Jenkins, 1988), and with French data (Andrie *et al.*, 1988), for example, would go a long way towards producing a data base from which indisputable results could be expected. WOCE data, to become available over the next several years, should give worldwide coverage of the desired nature, provided that sampling scales for both tritium and ^{3}He will be chosen adequately.

In practice, as in the example above, observations available will only be synoptic in parts. Care should be taken to obain data that will allow one to determine the non-linear mixing terms in Eq. (9), as well as time-series information to estimate non-stationarity. In general, problems from non-synoptic data will be much less for tritium-^{3}He age than for the regular transient tracers. A challenging task is to expand the use of tritium-^{3}He ages over as much of the upper and intermediate ocean as possible. In the northern hemisphere this refers to the regions beyond about the 20 year isochrones, while in the southern hemisphere there is challenge throughout. The main problem is the required correction for terrigenic ^{3}He. To allow such correction, a verification of Eqs. (4) and (5), and high-precision routine measurement of He and Ne concentrations will be required. It is noted that the correction procedure proposed may be problematic in enclosed seas where tritiugenic ^{3}He is non-vanishing throughout (Mediterranean, Norwegian Sea), so that determination of the isotopic ratio of terrigenic He becomes difficult. A further problem is that, beyond the 20 year isochrones, non-stationarity of, and non-linear mixing effects on, the age distributions become substantial. This does not exclude successful modelling, but does require particularly detailed observations to enable one to determine adequately the respective terms in the transport equation for tritium-^{3}He age.

ACKNOWLEDGEMENTS

The present article has been made possible by efforts of my former research group at Heidelberg — at sea, in the laboratory, and in thought. Much thank is due for this work to many. I acknowledge the support of Karl-Otto Münnich, head of the Heidelberg institute. I am grateful for discussions with Gerhard Fuchs, Peter Schlosser, and Reiner Schlitzer. Figs. 1, 3 and 4 are reproduced by kind permission of Springer Verlag. Much of the work enabling this study was supported by the Deutsche Forschungsgemeinschaft.

REFERENCES

Andrie, C., P. Jean-Baptiste and L. Merlivat (1988). Tritium and helium-3 in the Northeastern Atlantic Ocean during the 1983 TOPOGULF cruise, *J. Geophys. Res.*, in press.

Benson, B. B., and D. Krause, Jr. (1980). Isotopic fractionation of helium during solution: a probe for the liquid state, *J.Solution Chem.*, 9, 895–909.

Bolin, B. (1958). On the use of tritium as a tracer of water in nature, in: *Proc. of the Second United Nations International Conference on the Peaceful Uses of Atomic Energy, Geneva, 1958*, Paper A/Conf., 15/P/176, United Nations, New York, 1958, p. 336–343.

Broecker, W. S., T. H. Peng, and G. Östlund (1986). The distribution of bomb tritium in the ocean, *J.Geophys. Res.*, 91, 14331–14344.

Craig, H., and R. F. Weiss (1971). Dissolved gas saturation anomalies and excess helium in the ocean, *Earth Planet. Sci. Lett.*, 10, 289–296.

Craig, H., and J. E. Lupton (1981). Helium-3 and mantle volatiles in the ocean and the oceanic crust, in: *The Sea, Vol. 7: Oceanic Lithosphere*, edit. C. Emiliani, Wiley and Sons, p. 391–428.

Doney, S. C., and W. J. Jenkins (1988). The effect of boundary conditions on tracer estimates of thermocline ventilation rates, *J. Mar. Res.*, submitted.

Dreisigacker, E., and W. Roether (1978). Tritium and Sr-90 in North Atlantic surface water, *Earth Planet. Sci. Lett.*, 38, 301–312.

Fine, R. A., and H. G. Östlund (1977). Source function for tritium transport models in the Pacific, *Geophys. Res. Lett.*, 4, 461–464

Fuchs, G. (1987). Ventilation der Warmwassersphäre des Nordostatlantiks abgeleitet aus ³Helium- und Tritium-Verteilungen, Doct. Dissertation, Univ. of Heidelberg, 207 pp. and tables.

Fuchs, G., W. Roether and P. Schlosser (1987). Excess ³He in the ocean surface layer, *J. Geophys. Res.*, **92**, 6559–6568.

GEOSECS (1987). *GEOSECS Atlas of the Atlantic,Pacific, and Indian Ocean Expeditions 7: Shorebased Data and Graphics*, Washington, U.S. Government Printing Office, 200 pp.

Heinze, C., P. Schlosser, K. P. Koltermann and J. Meincke (1988). A tracer study of the deep water renewal in the European Polar Seas, *Deep Sea Res.*, submitted.

Jähne, B., G. Heinz and W. Dietrich (1987). Measurement of the diffusion coefficients of sparingly soluble gases, *J.Geophys. Res.*, **92**, 10767–10776.

Jenkins, W. J. (1987). ³H and ³He in the Beta triangle; observations of gyre ventilation and oxygen utilization rates, *J. Phys.Oceanogr.*, **17**, 763–783.

Jenkins, W. J. (1988). The use of anthropogenic tritium and helium-3 to study subtropical gyre ventilation and circulation, *Phil. Trans. R. Soc. Lond.*, A, **325**, 43–61.

Jenkins, W. J., and W. B. Clarke (1976). The distribution of ³He in the western Atlantic Ocean, *Deep Sea Res.*, **23**, 481–494.

Koster, R., W. S. Broecker, J. Jouzel, R. Suozzo, G. Russell, D. Rind, and J. W. C. White (1988). The global geochemistry of bomb produced tritium; general circulation model compared to the real world, preprint.

Kuntz, R. (1985). Bestimmung der Tiefenwasserzirkulation des Roten Meeres anhand einer Boxmodellauswertung von Tritium- ³He- und Salinitätsdaten, Doct. Dissertation, Univ. of Heidelberg., 85 pp.

Menke, W. (1983). *Geophysical Data Analysis: Discrete Inverse Theory*, Academic Press, 260 pp.

Reid, J. L. (1978). On the middepth circulation and salinity field in the North Atlantic Ocean, *J. Geophys. Res.*, **83**, 5063–5067.

Reid, J. L. (1981). On the middepth circulation of the world ocean, in: *Evolution of Physical Oceanography*, ed. B. A. Warren and C. Wunsch, Cambridge, Mass., p. 70–111.

Robinson, M. K., R. A. Bauer and E. H. Schroeder (1979). *Atlas of North Atlantic-Indian Ocean Monthly Mean Temperature and Mean Salinities of the Surface Layer,* U. S. Naval Oceanographic Office Ref. Pub. 18, Washington D. C.

Roether, W. (1986). Field measurements of gas exchange, in: *Dynamic Processes in the Chemistry of the Upper Ocean,* ed. J. D. Burton, P. G. Brewer, R. Chesselet, Plenum Press, p. 117–128.

Roether, W., K.-O. Münnich and H. G. Östlund (1970). Tritium profile at the North Pacific (1969) Geosecs intercalibration station, *J. Geophys. Res.,* **75**, 7672–7675.

Roether, W., and G. Fuchs (1988). Water mass transport and ventilation in the Northeast Atlantic derived from tracer data, *Phil. Trans. R. Soc. Lond., A,* **325**, 63–69.

Roether, W., and M. Rhein (1988). Chemical tracers in the ocean, in *Landolt-Börnstein,* New Series, Group V, Vol. 3b, ed. J. Sündermann, Springer, Berlin-Heidelberg-NewYork-Tokyo, chapt. 4.3, in press.

Sarmiento, J. L. (1983). A simulation of bomb tritium entry into the Atlantic Ocean, *J. Phys. Oceanogr.,* **13**, 1924–1939.

Sarmiento, J. L., C. G. H. Rooth and W. Roether (1982). The North Atlantic tritium distribution in 1972, *J. Geophys. Res.,* **87**, 8047–8056.

Saunders, P. M. (1982). Circulation in the eastern North Atlantic, *J. Mar. Res.,* **40**, Supplement, 641–657.

Schlosser, P. (1985). Ozeanographische Anwendungen von Spurenstoffmessungen im Mittelmeerausstrom und im Europäischen Nordmeer, Doct. Diss., University of Heidelberg, 206 pp.

Schlosser, P. (1986). Helium: a new tracer in Antarctic oceanography, *Nature,* **321**, 233–235.

Taylor, C. B., and W. Roether (1982). A uniform scale for reporting low-level tritium measurements in water, *Int. J. Appl. Radiat. Isotopes,* **33**, 377–382.

Thiele, G., W. Roether, P. Schlosser, R. Kuntz, G. Siedler and L. Stramma (1986). Baroclinic flow and transient-tracer fields in the Canary-Cape Verde Basin, *J. Phys. Oceanogr.,* **16**, 814–826.

Unterweger, M. P., B. M. Coursey, F. J. Schima, and W. B. Mann (1980). Preparation and calibration of the 1978 National Bureau of Standards tritiated-water standards, *Int. J. Appl. Radiat. Isotopes,* **31**, 611–614.

Weiss, R. F. (1971). Solubility of helium and neon in water and seawater, *J. Chem. Eng. Data*, **16**, 235–241.

Weiss, W., and W. Roether (1980). The rates of tritium input to the world oceans, *Earth Planet. Sci. Lett.*, **49**, 435–446.

Wunsch, C. (1987). Using transient tracers: the regularization problem, *Tellus*, **39** B, 477–492.

Wunsch, C. (1988). Eclectic modelling of the North Atlantic. II. Transient tracers and the ventilation of the Eastern Basin thermocline, *Phil. Trans. R. Soc. Lond.*, A, **325**, 201–236.

Wunsch, C., and J.-F. Minster (1982). Methods for box models and ocean circulation tracers: Mathematical programming and nonlinear inverse theory, *J. Geophys. Res.*, **87**, 5747–5762.

Zenk, W. (1975). On the origin of the intermediate double-maxima in T/S profiles from the North Atlantic, *'Meteor' Forsch.-Ergebnisse*, A, **16**, 35–45.

OCEAN CARBON MODELS AND INVERSE METHODS

Berrien Moore III
Institute for the study of Earth, Oceans and Space
University of New Hampshire
Durham, New Hampshire 03824
USA

Bert Bolin and Anders Björkström
Department of Meteorology
University of Stockholm
Arrhenius Laboratory
S-106 91 Stockholm
Sweden

Kim Holmén
Rosentiel School of Marine and Atmospheric Science
University of Miami
4600 Rickenbacker Causeway
Miami, Florida 33149
USA

Chris Ringo
Institute for the study of Earth, Oceans and Space
University of New Hampshire
Durham, New Hampshire 03824
USA

D. L. T. Anderson and J. Willebrand (eds.), Oceanic Circulation Models: Combining Data and Dynamics, 409–449.

1. INTRODUCTION

Any theoretical treatment of a problem concerning our environment, such as our present concern, namely that of deducing the parameterization of an ocean-carbon model by inverse methods, must be based on some kind of model. In reality of course, the natural phenomena are so complex and our data so few that all such problems are indeterminate (i.e. have more than one solution that is consistent with the data). However, only by adopting models, and thereby, by definition, imposing a reduction on the complexity of reality do overdetermined systems arise, and also only by adoption of models can we hope to make efficient use of the available data and information. It is obvious that whether or not the results will be of interest depends on how well our model captures the essence of the phenomena in nature which it seeks to describe.

Assuming that we are able to design a physically-chemically 'reasonable' model, we face the question of how detailed we may assume it to be. Basically, the number of degrees of freedom of a model should not exceed the degrees of freedom that the data contain, which is used to determine the set of unknowns that describe the circulation, the transfer processes, and the biochemical processes. Wunsch (1985) has argued however, that because of nature's complexity and the sparsity of data, a model should be parameterized as an underdetermined problem, and rather than search for a single solution, one should explore the range of values for solutions and their implication for other geophysical phenomena. It should be recognized, from one viewpoint, that in such an analysis one does not consider explicitly imperfections or inconsistencies in the model. Rather, one assumes implicitly that all available data fit the model exactly and that the range of uncertainty is dependent exclusively on insufficient data. Even if direct measurements from which model data are derived are accurate, model variables often represent mean values in space and/or time and cannot necessarily be obtained accurately from available measurements.

It therefore seems reasonable to design a diagnostic model with sufficient simplicity that it formally becomes overdetermined. Since, in fact, any number of data is insufficient to describe reality in all its complexity and since the problem in reality is not overdetermined, it seems preferable to call such a model incompatible. That is, the model and the data are incompatible because of an inadequate model or inaccurate data or both.

There is actually a possible middle-ground (suggested by Wunsch) between these two approaches. Namely, even if the initial formalization is incompatible (overdetermined), one could cast it in an indeterminate (underdetermined) mode by associating with each equation an additional unknown with an *a priori* error bound reflecting, in a sense, the error associated with the forcing term. We believe that this is an important suggestion; however, it does raise the dimensionality of the system considerably, and because of this we have not had sufficient opportunity to pursue the approach.

Consequently, it is the incompatible framework that we have chosen to focus on here. In Section 2 we give a detailed account of one method of formulating a tracer-based box model of the ocean. For further details and a specific application of the methods to an Atlantic Ocean model, consult Bolin *et al.*(1987). In Section 3, we will briefly survey a handful of inverse methods that one may use to parameterize the model. Since Wunsch (this volume) contains a detailed review of several linear techniques, this section will serve to complement that work, as well as to give an introduction to a few nonlinear schemes, an area not treated by Wunsch. All the techniques we review, both linear and nonlinear, are outgrowths or generalizations of ordinary least squares. Hence, many other useful types of inverse methods are neglected, most noticeably perhaps being linear programming. For a clear account of this method's use in an oceanographic setting, see e.g. Schlitzer (1987, 1988). We further note that the linear programming method is not that dissimilar to Wunsch's suggestion of associating a constrained error term with each equation.

2. DIAGNOSTIC EQUATIONS

2.1 The General Setting.

We consider an ocean basin and seek to determine its pattern of circulation, the rates of turbulent transfer and biochemical processes. Our approach is to develop a system of diagnostic equations, each of which either exploits the geostrophic structure of the oceans, or employs data on the bulk properties of ocean water (specifically the spatial, quasi-steady distributions of temperature, salinity and a set of biologically active tracers). As a consequence, we need to prescribe boundary conditions for the ocean domain that is being considered, relative to the exchange of heat, water, and chemical properties with the atmosphere and adjacent aquatic bodies.

As a starting point, we adopt the classical assumption that outside a rather narrow zone around the equator, the horizontal motions in the ocean are quasi-geostrophic; furthermore, conservation of water within the oceanic domain under consideration remains a fundamental requirement. In an incompatible framework, one may in principle accept solutions that do not satisfy the requirement of water continuity exactly. This possibility was discussed by Wunsch and Minster (1982); however, incompatibilities in the water continuity equations implicitly introduce artificial sources and sinks in the other tracers which is particularly troublesome when considering the time-evolution of a tracer through the system (e.g. carbon dioxide). We accordingly develop a method for satisfying the water continuity requirement exactly. In light of the uncertainties in the boundary conditions for water however, it must be recognized that there are serious difficulties with this approach and that it may certainly be questioned. The linear programming approach would be particularly appropriate for exploring the consequences of this requirement. Furthermore, simply by explicitly including an unknown error term for each forcing, we could improve upon our current methodology while avoiding the issue of fictitious sources and sinks. An alternative would be to treat external exchanges of water as constrained unknowns.

By adding the tracer continuity equations (including equations for the conservation of water), one both supplements the geostrophic equations in the determination of horizontal velocities, and determines vertical velocities, turbulent transfers, and the biological rates such as new primary production or decomposition. (New primary production is the net primary production that leaves the photic zone; hence, it is a measure of the rate at which carbon is removed from the photic zone and falls to deeper layers in the form of detrital material. See Eppley and Peterson, 1979).

Finally, the analysis is formulated through a set of finite-element equations or, in other words, in terms of a set of equations for specific subregions of the domain (i.e., 'boxes').

In order to illustrate the handling of a large number of boxes and for a discussion of the concepts of indeterminacy and incompatibility, it is useful to be able to refer to a structurally simple model that is similar to the one which we have used in our analysis of the Atlantic Ocean (Bolin *et al.*, 1987). Consider an ocean domain which consists of R (=12) horizontal regions and S (=8) layers in the vertical, i.e., the total number of boxes is $I = RS$ (=96); see Fig. 1. It is readily seen that this configuration has $T_h = WS$ (=136) vertical interfacial surfaces through which there is horizontal transfer and $T_v = R(S-1)$ (=84) horizontal interfacial surfaces through which there is vertical transfer; W (=17) is the number of surfaces between the R regions. There is accordingly $T = T_h + T_v$ (=220) interfacial surfaces. The diagnostic equations are obtained by assuming quasi-geostrophic flow (Section 2.2), water continuity (Section 2.3) and tracer continuity (Section 2.4).

2.2 The Condition of Geostrophy.

We consider first the condition imposed by geostrophy in the abstract, ignoring for the moment that such conditions do not hold near the equator. In Fig. 2, the parallelogram ABCD denotes one of the W interfacial surfaces in the model. The lines z_{s-1} and z_s define one of the S model layers. The velocity in the x-direction u, and the density ρ, vary from place to place in ABCD. The product of these two, $m(y,z) = u(y,z)\rho(y,z)$, is a mass flux. We seek an expression for the integral of m taken over the shaded area Ω_s in layer s between the separating levels z_{s-1} and z_s. For any two horizontal directions x and y, where y is 90 degrees to the left of x, the balance between Coriolis acceleration and pressure gradient is

$$-fu - \frac{1}{\rho}\frac{\partial p}{\partial y} = 0 \qquad (2.1)$$

where f is the Coriolis parameter. Since hydrostatic conditions prevail, we have $\partial p/\partial z = -g\rho$ so that differentiation of (2.1) with respect to z gives

$$\frac{\partial}{\partial z}(fm) = g\frac{\partial \rho}{\partial y} \qquad (2.2)$$

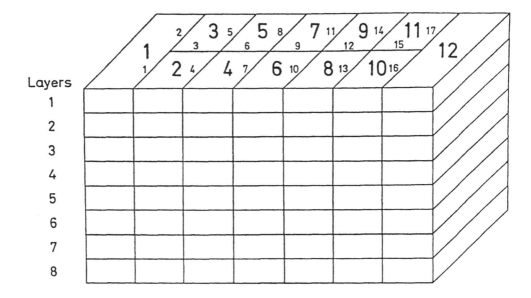

Figure 1
Geometrical arrangement of boxes in the domain under consideration. The
bold-face numerals indicate the regions ($R=12$) and the thin numerals
in the uppermost layer (1 to 17) refer to the interfaces between regions.
($W=17$). There are S ($=8$) layers in the vertical.

We assume a reference level z_0 (which can be a function of y) where the mass flux
is $m_0(y) = m(y, z_0(y))$. Since f is independent of z we can integrate (2.2) in the
vertical and get

$$m(y, z) = m_0(y) + \frac{g}{f} \int_{z_0(y)}^{z} \left(\frac{\partial \rho}{\partial y} \right) dz \tag{2.3}$$

The mass flux through the area Ω_s will be

$$m_s = \int \int_{\Omega_s} m \, dy \, dz = \int \int_{\Omega_s} m_0 \, dy \, dz + \int \int_{\Omega_s} \frac{g}{f} \int_{z_0(y)}^{z} \left(\frac{\partial \rho}{\partial y} \right)_{\zeta=z} d\zeta \, dy \, dz \tag{2.4}$$

The last term on the right side can be evaluated numerically, given data on ρ in
the wall and having made a choice of reference level (in our case, the level $z_0 = 0$
has been used). The first term on the right side can be integrated, using the fact
that m_0 is independent of z:

$$\int \int_{\Omega_s} m_0 \, dy \, dz = m_A \Omega_s \tag{2.5}$$

where m_A is the average value of the water flux (ρv) along the reference level z_0. The S mass fluxes m_s, $s = 1,\ldots,S$ are thus tied together by S equations of the form

$$m_s = m_A \Omega_s + M_s , \qquad s = 1,\ldots,S \tag{2.6}$$

where Ω_s and M_s can be computed from hydrographic data. The average value m_0 becomes an additional unknown. Each of the W walls thus supplies S equations and one unknown.

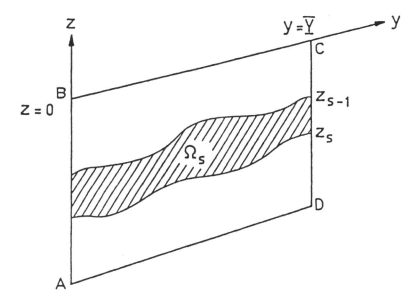

Figure 2
Vertical section along a wall between regions. For notation see text.

Alternatively, the mass flux m_0 can be algebraically eliminated between successive Eqs. (2.6), in which case we obtain $S-1$ equations at each wall for the flux differences between successive layers:

$$\frac{m_s}{\Omega_s} - \frac{m_{s+1}}{\Omega_{s+1}} = \frac{M_s}{\Omega_s} - \frac{M_{s+1}}{\Omega_{s+1}} , \quad s = 1,\ldots,S \tag{2.7}$$

We shall use the latter formulation.

2.3 Water Continuity; Definition of 'Loops' as Advective Variables.

The continuity equation for water, i.e., the equation

$$\nabla(\rho\mathbf{v}) = F_b \qquad (2.8)$$

where F_b denotes the local supply of water from external sources, can be integrated over an arbitrary box, giving

$$\sum_{t=1}^{T} \text{sign}\,(i,t)m_t = b_i^m\,, \qquad i = 1,\dots,I \qquad (2.9)$$

Here, T is the number of box interfaces, and hence the number of horizontal and vertical fluxes to or from the box under consideration. The technical function $\text{sign}(i,t)$ is either plus one or minus one, depending on sign conventions that can be adopted arbitrarily. (If we further define $\text{sign}(i,t) = 0$ when the flux m_t does not involve box i, the summation over t is formally correct). The term b_i^m stands for the net source of water to box i from external regions. It may consist of exchange with adjacent ocean basins, or net precipitation, or both.

We combine the T mass fluxes m_t into a T-dimensional vector $\mathbf{m}$, and also combine the I net sources b_i^m to an I-dimensional vector $\mathbf{b}^m$. We then summarize the I continuity equations (2.9) in matrix form (where again I is the total number of boxes in the domain):

$$\mathbf{A}_0\mathbf{m} = \mathbf{b}^m \qquad (2.10)$$

where the $(I \times T)$-dimensional matrix $\mathbf{A}_0$ will be defined by

$$(A_0)_{it} = \text{sign}(i,t).$$

Since one of the water continuity equations is redundant, the rank of $\mathbf{A}_0$ is $I - 1$. With the aid of (2.10) we may derive by using the pseudo-inverse, $\mathbf{A}_0^+$, of $\mathbf{A}_0$ (also called the Moore-Penrose inverse or the generalized inverse; see Wunsch, 1978; Fiadeiro and Veronis, 1982; or Bolin *et al.*, 1983 (the Appendix)), a particular solution $\mathbf{m}_0$ with minimum norm. This solution $\mathbf{m}_0$, describes an advective field that satisfies exactly the boundary conditions $\mathbf{b}^m$ for the water exchange with the exterior.

Eq. (2.10) constitutes a set of I equations that are to be part of the total set that defines our system. Treating these equations as simply part of the system would imply, however, that the water requirement would be only approximately satisfied since the overall system is incompatible. It, would, of course, be possible to weight these equations heavily in the search for a best fit solution and thereby reduce the errors in the water continuity equations. For several reasons (e.g. numerical), it is desirable, however, to ascertain perfect water continuity but also to avoid introducing weights.

Our approach to this problem is as follows. We first observe that any solution m to Eq. (2.10) can be written uniquely as

$$\mathbf{m} = \mathbf{m}_h + \mathbf{m}_0 \qquad (2.11)$$

where $\mathbf{m}_h$ is a solution to the homogeneous water continuity equations, i.e., to the system

$$\mathbf{A}_0\mathbf{m}_h = \mathbf{0} \qquad (2.12)$$

There is an alternative way to describe a solution to (2.12), namely by replacing the T advections by $T - (I - 1)$ new variables defined as closed circulations of water 'loops'. The technique is most simply illustrated by an example. The nine boxes in the upper part of Fig. 3 are connected by 12 advective fluxes and they have no external exchanges. The 12 numbers inserted show a flow pattern where water continuity is fulfilled in all boxes. In other words, these 12 numbers would define a vector $\mathbf{m}_h$. It is clear that the same information can also be expressed by only 4 numbers using the method in the lower part of the figure. It is easily realized that every flux pattern, without external gains and losses, can be reduced from 12 to 4 numbers in the same way.

In terminology from linear algebra, the observation we just made can be stated as follows. Since Eq. (2.12) is homogeneous, its solutions form a subspace of the space of all vectors m. The dimension of this space is $L = T - (I - 1)$, which is $12 - (9 - 1) = 4$ in the case of Fig. 3. The 4 loops chosen constitute a basis in that subspace, i.e., any solution $\mathbf{m}_h$ to (2.12) can be written as linear combinations of loops in a unique way.

When the configuration of boxes is less regular than in Fig. 3, the way to establish a set of L loops may be less obvious. However, it can always be done except in degenerate cases (e.g. when the system is composed of decoupled regions) of no interest in the present context. It is important to ascertain that the loop basis

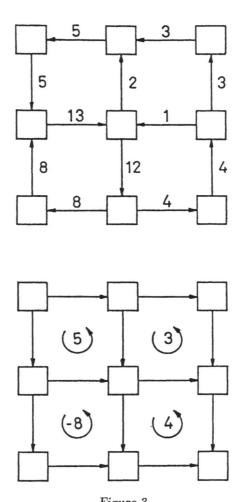

Figure 3
A water-conserving motion pattern for nine boxes and the representation
of the same pattern in terms of 'loops'.

is wide enough for all water advection patterns to be represented, since otherwise interesting solutions might not be expressible, and therefore be lost.

The problem of avoiding this is equivalent with the problem of making sure that no loop is a linear combination of the others: if one is able to find L linearly independent loops, these must, by definition, span an L-dimensional space. In a two-dimensional case, such as Fig. 3, it is easy to show that the circulations around the L enclosed, finite areas make up a 'natural' basis, with the wanted linear independence. We should point out that the resulting solution is of course independent of the choice of basis.

We note from Fig. 3 that there will be linear relations between the T advections m_t, $t = 1, \ldots, T$, that are components of the vector $\mathbf{m}_h$, and the loops, Y_j, $j = 1, \ldots, L$. For example,

$$m_1 = -y_1, \qquad m_3 = y_1$$
$$m_2 = -y_2, \qquad m_4 = y_1 + y_2.$$

In matrix form,

$$\mathbf{m}_h = \mathbf{L}_0 \mathbf{y}, \tag{2.13}$$

where $\mathbf{L}_0$ is a $T \times L$ matrix.

Inserting (2.13) in (2.11) we get the following expression for the general solution to the inhomogeneous water continuity equations:

$$\mathbf{m} = \mathbf{L}_0 \mathbf{y} + \mathbf{m}_0. \tag{2.14}$$

2.4 Continuity Equations for Tracers.

Tracer continuity was dealt with in detail in our previous papers (Bolin et al., 1983 and Bolin et al., 1987). We consider the tracer distributions q_i^n, where $n = 1, \ldots, N$ identifies the N tracers being considered.

For salinity and the biochemical tracers q_i^n is expressed as the amount of tracer material per unit mass of sea water. Because of adiabatic heating of sea water when pressure is increased, the continuity equation for internal energy is somewhat more complicated and will be dealt with later. Slightly modifying the derivation given by Bolin et al., (1983) with regard to the units of q, we have

$$\sum_j m_{ij} \int_{\Omega_{ij}} q_i^n d\Omega - \sum_j K_{ij} \int_{\Omega_{ij}} \frac{\partial q^n}{\partial \nu} d\Omega - \gamma_o^n F_{oi} - \gamma_c^n F_{ci} = -\lambda^n Q_i^n + F_{bi}^n \tag{2.15}$$

where the summations are extended over the values j that refer to adjacent boxes. m_{ij} is the water flux in the direction perpendicular to the surface Ω_{ij} between boxes i and j; K_{ij} is the average eddy diffusivity across this surface; F_{oi} is the withdrawal of carbon from box i due to new primary production or release from net decay, and F_{ci} is similarly the withdrawal or release of carbon due to new carbonate formation or dissolution; γ_o^n and γ_c^n are the Redfield ratios for tracer $n = 1, \ldots, N$ as related to these two processes, respectively (i.e., the ratios of carbon to various nutrients in organic material and of carbon to alkalinity in inorganic material); Q_i^n is the total amount of the nth tracer in box i; λ^n is the radioactive decay rate for the nth tracer, and finally F_{bi}^n is the source or sink for the nth tracer at the external boundary for box i.

We note that the eddy diffusivity term in our formulation supposedly accounts for transfer due to motion on scales less than those directly captured by the advective flux described by the first term in Eq. (2.15). K_{ij} accordingly may be considered as transfer by subgrid scale motions and transient (e.g. seasonal) flow patterns. In the case of heat flux, it should be recognized that temperatures will change as a result of vertical displacement due to adiabatic compression or expansion. Since compressibility of sea water depends on both temperature and salinity, pressure changes affect temperature differently for different salinities, but an approximate treatment can be accomplished by the use of potential temperature obtained by considering the compressibility of standard sea water. In the upper parts of the oceans, the spatial variations of temperature are comparatively large and the pressure correction plays a rather insignificant role. In the deeper layers, where small temperature changes due to the compressibility may be of importance also salinity variations are small and the use of potential temperature accounts approximately for this effect.

As before we evaluate the integrals in (2.15) by employing a finite difference formulation:

$$\int_{\Omega_{ij}} q^n d\Omega = \frac{1}{2}(q_i^n + q_j^n)\Omega_{ij}$$

$$\int_{\Omega_{ij}} \left(\frac{\partial q}{\partial \nu}\right) d\Omega = \frac{1}{\Delta_{ij}}(q_i^n - q_j^n)\Omega_{ij} \tag{2.16}$$

where Δ_{ij} is the distance between center points of adjacent boxes. It should be noted that this formulation of centered differences in our formulation of the advective exchanges might be inappropriate (see Fiadeiro and Veronis, 1977). As seen from (2.15) and (2.16), the transfer is partly recipient-controlled. This implies that the time integral of q_i^n in (2.15) might be unstable when we use the given tracer field as the initial condition to derive the changes of this field as dependent on a solution of ocean circulation and biochemical processes derived with the present method. The instability can be avoided by imposing constraints between the advective and turbulent unknowns (see e.g. next section).

When adopting the loop formulation for the advective velocities, the tracer continuity equations are modified. We notice from Eqs. (2.15) and (2.16) that the advective flux of tracer n into box i is given by

$$f^n = \sum_i f_{ij}^n = \sum_i m_{ij}\left(\frac{q_i^n + q_j^n}{2}\right) \tag{2.17}$$

where the summation is over all adjacent boxes j. We replace m_{ij} by a set of loops y_{jk} which enter the box i from j and leave it for some box k. The flux of tracer material transported into box i (i.e. f_{jk}) due to loop y_{jk}, is given by

$$f_{jk} = \tfrac{1}{2}y_{jk}\left((q_j^n + q_i^n) - (q_i^n + q_k^n)\right) = \tfrac{1}{2}y_{jk}(q_j^n - q_k^n) \tag{2.18}$$

Since the flow of m_{ij} is nondivergent, all terms of the kind given by (2.17) can be transferred into expressions as given by (2.18). In other words,

$$\sum_j f_{ij}^n = \sum_{jk} \tfrac{1}{2}y_{jk}(q_j^n - q_k^n) \tag{2.19}$$

It is important to note here that the replacement of the advective fluxes m_{ij} by loops y_{jk} means that the average concentrations $(q_i + q_j)/2$ that enter as coefficients in Eq. (2.15) are changed to one half of the concentration difference, $(q_j - q_k)/2$, between boxes adjacent to the one for which tracer continuity is being considered (i.e. box i). This makes the coefficients for loops and eddy diffusivities to be of the same magnitude. This simplifies the procedure required to ensure that the advective and turbulent terms in (2.15) are given the same weight in deriving an optimum solution.

Further we assume (Bolin et al., 1983) that there is no horizontal transfer (from one region to another) of organic and inorganic detrital matter, i.e. whatever is formed in the surface layers of the sea is decomposed below. There is no net accumulation at the bottom of the sea. This provides $2R (= 24)$ additional continuity equations

$$\sum_{s=1}^{S} F_{os} = 0 \quad \text{for all } r = 1, \ldots, R \tag{2.20}$$

and

$$\sum_{s=1}^{S} F_{cs} = 0 \quad \text{for all } r = 1, \ldots, R \tag{2.21}$$

Here, as for water continuity, it is desirable that these equations are satisfied exactly, which can be achieved by a variable substitution.

2.5 The Set of Diagnostic Equations and the Inequality Constraints

From our conceptual point of view geostrophy has been the basis for the development thus far, but from a computational viewpoint water continuity is fundamental. We organize our matrix and the discussion in the next two sections to reflect this latter view. Specifically, we combine the Eqs. (2.14) for water continuity, (2.7) for geostrophy, (2.15) for tracer continuity (including heat), and (2.20) and (2.21) for detrital continuity and obtain a system of linear equations that will be denoted

$$\mathbf{Ax} = \mathbf{b} \tag{2.22}$$

where $\mathbf{A}$ is a matrix that can be determined from hydrographic data, data on steady-state tracer distributions and the Redfield ratios. Fig. 4 shows the organization of the matrix. The column vector of unknowns is given by

$$\begin{aligned}
\mathbf{x} = ((m_t, t = 1, \ldots, T), (K_t \Omega_t / \Delta_t, t = 1, \ldots, T), \\
(F_{oi}, i = 1, \ldots, I), (F_{ci}, i = 1, \ldots, I))^T \tag{2.23}
\end{aligned}$$

We note that this system is based on advections (m_t) and not on loops. The 'forcing' of our system, $\mathbf{b}$, is the vector of inhomogeneous terms in Eqs. (2.14), (2.7) and (2.15), and they are:

— in (2.14) externally imposed water flow defined by $\mathbf{b}^m$;
— in (2.7) specified change of geostrophic flow between layers as determined by $M_s/\Omega_s - M_{s+1}/\Omega_{s+1}$;
— in (2.15) externally imposed flow of tracers and, in the case of ^{14}C, radioactive decay.

To this set we add a set of inequality constraints

$$\mathbf{Gx} \geq \mathbf{h}. \tag{2.24}$$

For the moment, this inequality refers to two kinds of constraints.

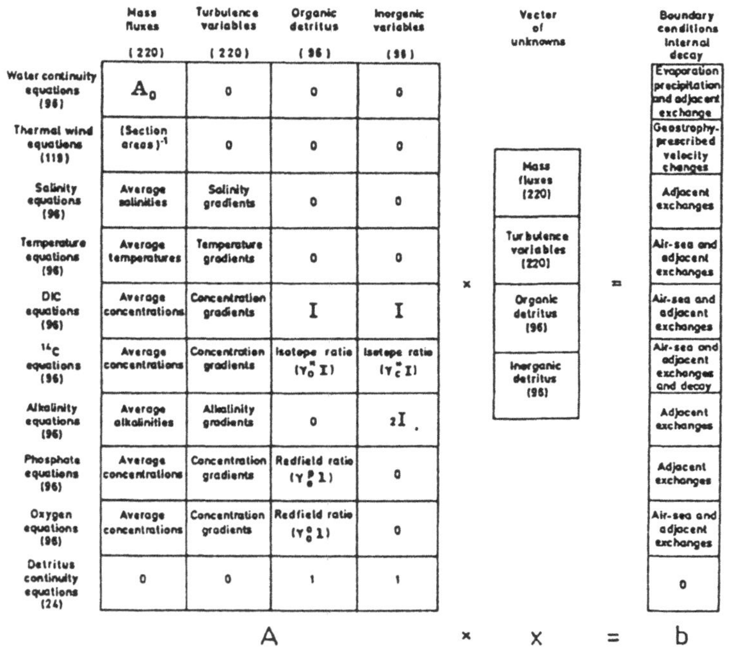

	Mass fluxes (220)	Turbulence variables (220)	Organic detritus (96)	Inorganic variables (96)		Vector of unknowns		Boundary conditions Internal decay
Water continuity equations (96)	A_0	0	0	0				Evaporation precipitation and adjacent exchange
Thermal wind equations (118)	(Section areas)$^{-1}$	0	0	0				Geostrophy-prescribed velocity changes
Salinity equations (96)	Average salinities	Salinity gradients	0	0		Mass fluxes (220)		Adjacent exchanges
Temperature equations (96)	Average temperatures	Temperature gradients	0	0		Turbulence variables (220)		Air-sea and adjacent exchanges
DIC equations (96)	Average concentrations	Concentration gradients	I	I	$\times$	Organic detritus (96)	$=$	Air-sea and adjacent exchanges
^{14}C equations (96)	Average concentrations	Concentration gradients	Isotope ratio ($\gamma_0^n\, I$)	Isotope ratio ($\gamma_c^n\, I$)		Inorganic detritus (96)		Air-sea and adjacent exchanges and decay
Alkalinity equations (96)	Average alkalinities	Alkalinity gradients	0	$2I$				Adjacent exchanges
Phosphate equations (96)	Average concentrations	Concentration gradients	Redfield ratio ($\gamma_e^p\, 1$)	0				Adjacent exchanges
Oxygen equations (96)	Average concentrations	Concentration gradients	Redfield ratio ($\gamma_e^o\, 1$)	0				Air-sea and adjacent exchanges
Detritus continuity equations (24)	0	0	1	1				0

$$A \qquad \times \qquad x \qquad = \qquad b$$

Figure 4

Arrangement of the matrix **A**, the vector of unknowns **x** and the right hand side vector **b** of Eq. (3.22).

a) Turbulent constraints. In the definition of the coefficients of eddy diffusivity (implicit in (2.15); see further Bolin *et al.* (1983) we generally wish to impose the condition that $K_{ij} \geq 0$. We have noted, however, that the present formulation of advective transfer using partly recipient-control in the finite element formulation may lead to unstable sets of equations for time integration. It is formally possible to avoid this by making use of the stabilizing property of the eddy diffusivities K_{ij}. The condition, $K_{ij} \geq |m_{ij}|/2$, will ensure stable time integration; however, it is stronger than necessary (i.e., sufficient but not necessary) and there is no physical justification for imposing such a condition.

b) Decomposition of new primary production and the dissolution of carbonate formation. We need to ascertain by using inequality constraints that:

— new primary production and carbonate formation only take place in boxes reached by sunlight;

— decomposition and dissolution in a box in a region can never exceed the production that has occurred in the surface layer above minus decomposition and dissolution in layers above that of concern.

2.6 Basic Characteristics of the System

Our purpose for this section is to clarify the characteristics of the common base that our approach shares with traditional inverse methods for the analysis of oceanic data that have been developed by others (particularly Wunsch, 1978; Fiadeiro and Veronis, 1982; Wunsch and Minster, 1982; Wunsch, 1984; Fiadeiro and Veronis, 1984; and Wunsch, 1985), who generally have used inverse methods in an indeterminate (underdetermined) setting. We wish to elaborate further on the reasons why we introduce an incompatible (or overdetermined) inverse framework and establish the characteristics of that framework. In the end, we hope that this way of building an inverse structure, piece by piece, will remove some of the confusion about the application of this methodology in constructing models of oceanic processes.

Consider first a purely advective abiotic ocean box-model as defined in Section 2.1 that uses I boxes, arranged in a network of R regions within which there are vertical flows between S layers and horizontal flows between R regions at each layer (*cf.* Fig. 1).

Suppose we seek to determine the water flow, i.e. the $T = T_h + T_v$ different terms, from the imposition of the $I - 1$ independent water continuity constraints alone. Alternatively one simply replaces the systems of T advective fluxes with the

$I - 1$ continuity constraints by the system of L loops, noting that $L = T_h + T_v - (I - 1) = T - (I - 1)$.

We now add the geostrophic constraints where we assume for the moment that they can be applied at each of the S layers in each of the W walls. There are thus WS additional constraints on the T advections, but we need to introduce W unknown reference velocities; alternatively, there are $W(S - 1) = T_h - W$ constraints on just the T advective fluxes (*cf.* Section 2.2). Accordingly the system of L unknown loops satisfying the $W(S - 1)$ geostrophic constraints has an indeterminacy of $L - W(S - 1) = T_h + T_v - (I - 1) - (T_h - W) = T_v + W - (I - 1) = R(S - 1) + W - (R \cdot N \cdot S - 1) = W + 1 - R$.

In the particular case we chose in Section 2.1, we find that the resulting indeterminacy for computing the advective field is six-dimensional $(17 + 1 - 12 = 6)$, which corresponds precisely to the requirement of determining additionally six loops (*cf.* Fig. 5) to define uniquely the two-dimensional flow in one horizontal layer. This, in turn, is equivalent to the necessity of defining a set of reference velocities in order to determine completely the flow by using the conditions of geostrophy and water continuity. There are W reference velocities that need to be determined, and water continuity for each region gives $R - 1$ independent requirements. We note that by prescribing a set of reference level velocities or by applying a minimization procedure (e.g. pseudo-inverse; see Wunsch, 1978; Fiadeiro and Veronis, 1982 or Wunsch, this volume), we can derive a solution that satisfies water continuity and the geostrophic condition exactly.

It is interesting in this context also to recall the method to analyze hydrographic data proposed by Stommel and Schott (1977): the β-spiral. This method is based on the assumption that the geostrophic relation is satisfied exactly and that accordingly the horizontal divergence and vertical change of vertical velocity can be derived from the north-south component of the current field and the variation of the Coriolis parameter (β). Although the horizontal velocity can be obtained reasonably well in this way, we also know that the large-scale oceanic circulation patterns are maintained by surface winds and thermohaline processes at the ocean surface which impose dynamic forcing on the oceanic system, and by basin configuration and bottom topography which constrain possible motions. They all create nongeostrophic components in the flow and are reflected in the field of vertical velocities and presumably in the large scale distribution of the tracers in the sea. It is therefore desirable not to impose the geostrophic condition too severely. Also,

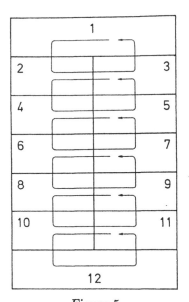

Figure 5
Same as in Fig. 1, viewed 'from above'. If mass fluxes must be exactly
geostrophic, the vertically integrated flux has only one degree of freedom at
each of the W (=17) walls. Mass continuity for the R (=12) regions gives
R−1 independent constraints for the W free parameters. The total number
of free parameters that govern the mass fluxes thus is W − R+1 = 6. The 6
free parameters can be represented by the rates of the vertically integrated
circulations around each of the 6 loops.

inaccuracy of hydrographic data is another reason for not requiring that geostrophy
be exactly satisfied. Finally, the indeterminacy will also be larger than $W + 1 - R$
(i.e. six-dimensional) for the particular system that we have adopted above for
illustration, because the geostrophic condition in an actual application must not be
applied close to the equator.

We return to our example abstract ocean basin model as described in Section
2.1. If we also make use of hydrographic information by imposing the continuity
relations for salinity and enthalpy (temperature), we obtain $2I$ (= 192) additional
equations and the system would shift from being indeterminate to being incompatible. In this setting however, considering only purely advective solutions is
inadequate, since turbulent transfer of salt and enthalpy is essential (*cf.* Wunsch
and Minster, 1982). Including the $T_h + T_v = T$ (= 220) turbulent transfers as
unknown eddy diffusivities shifts the system back to being indeterminate, with an

indeterminacy in the illustrative case of at least 34 dimensions, when applying the geostrophic condition in each vertical surface between boxes.

At this stage, our objective in building an ocean model that contains the biochemical interplay in the ocean becomes important. We are interested in questions of the kind: What are the rates of transfer of CO_2 between the atmosphere and various oceanic regions? Exploring this question implies a need to determine the rate and distribution of various biochemical processes such as primary production and decomposition. As already mentioned, we need to employ tracers which carry information about these processes. In the 84-box model developed in Bolin et $al.$ (1987), we used five additional tracers: total dissolved inorganic carbon (DIC), ^{14}C, alkalinity, oxygen and phosphorous. We also determined two additional processes for each box: 1) the rate of new primary production of organic tissue and its decay in deeper layers and 2) the rate of new carbonate formation and its decomposition.

In the present idealized case, these additional unknowns give a total of $J_x = 2T + 2I$ ($2 \times 220 + 2 \times 96 = 632$) unknowns. The additional constraints raise the dimension of the forcing space (the right hand side) to $J_b = T_h - W + (I - 1) + N \times I + 2R$, ($136 - 17 + 95 + 7 \times 96 + 2 \times 12 = 910$), where N ($= 7$), is the total number of tracers. The $N \times I$ equations are the continuity equations for the N tracers; the $2R$ equations arise from detrital continuity. Thus the system we have defined is of dimension 910×632, and it would appear to be formally incompatible. In the application of Bolin et $al.$ (1987), an analysis of the rank of A confirmed that it was in fact an incompatible system.

It is generally this formulation that we have applied in our development of a model for the Atlantic Ocean. The details and results of this application can be found in Bolin et $al.$ (1987). Rather than repeat those results here, we now briefly outline some possible inverse techniques that are available to parameterize the model.

3. PARAMETERIZATION OF THE MODEL:
METHODOLOGICAL ISSUES FOR INVERSE METHODS

There are many inverse techniques, both linear and nonlinear, that one may now use to parameterize the model we have just developed. We do not attempt to review them all here, for that subject would easily fill many volumes in itself (Wunsch, this volume, contains an excellent review of several linear techniques; see also Wunsch and Minster, 1982). Rather, we briefly outline some of the more easily implementable linear schemes that we have considered for our studies of both the 84-box model and the formalism introduced by Fiadeiro and Veronis (1984), and then conclude with a short discussion of some hopeful nonlinear techniques.

3.1 Linear Inversion

Whether one chooses a linear or a nonlinear inversion technique will depend on what assumptions one wishes to make about the variances of the various data one is using, versus the ease of implementation and the extent of computing power one has available.

If one chooses a linear scheme, then one is implicitly assuming that the entries of the matrix A are known exactly, or at least that their variances are negligible compared with those associated with the right hand side b. How well suited a linear inversion technique is to the particular problem under consideration depends in part on the degree to which this assumption is valid. Linear schemes generally have the advantage of being easier to implement than nonlinear ones, mainly because software is often more readily available, the techniques are more well-developed, and they are computationally simpler.

3.1.1 Ordinary Least Squares with Convex Constraints

This is the approach that we initially used to parameterize the Atlantic Ocean model in Bolin et al. (1987). Simply stated, this method solves the problem:

'over all vectors x satisfying $Gx \geq h$, choose the one which minimizes $\|Ax - b\|$; if there is more than one such x, choose the one such that $\|x\|$ is minimum.'

As mentioned, in this case, the system had full column rank and was incompatible, so that the solution x was (formally) unique.

This method's major advantages are that it is easy to include *a priori* information on the sizes of solution components in the solution procedure (in the form of convex constraints), and that reliable, efficient software is available for its implementation (we have had success with the algorithm developed by Haskell and Hanson, 1981). In hindsight we now see however, that its disadvantages probably outweigh its advantages. For one, the great sensitivity of the least squares method in general is well-established (see e.g. Sprent, 1969; Hoerl and Kinnard, 1970a; Marquardt, 1970). Although the convex constraints do in general serve to reduce this inherent instability, the high variances associated with small singular values can still be troublesome (*cf.* Bolin *et al.*, 1987).

In addition, estimates of the statistical uncertainty of the solution, as well as estimates of how well the solution components are resolved, are not easily obtained using this method. Much more information may be extracted from the singular value decomposition of the matrix $\mathbf{A}$, however, upon which the least squares solution is based. By manipulating the singular value decomposition, one is also able to reduce the sensitivity of the least squares solution.

3.1.2 Singular Value Decomposition: A Classical Approach

Because of noise in the data, the information in the matrix $\mathbf{A}$ is only partly significant. A standard way to identify and perhaps remove the less certain information (and hence decrease the variance of the solution) is to decompose $\mathbf{A}$ in the form

$$\mathbf{A} = \mathbf{U}\mathbf{S}\mathbf{V}^T \tag{3.1}$$

where $\mathbf{U}$ and $\mathbf{V}$ are orthogonal matrices of sizes $(m \times m)$ and $(n \times n)$, respectively, and $\mathbf{S}$ is of the form

$$\mathbf{S} = \begin{pmatrix} \mathbf{D} \\ \mathbf{0} \end{pmatrix} \tag{3.2}$$

where $\mathbf{D}$ is an $n \times n$ diagonal matrix of singular values s_i. This is called the singular value decomposition (SVD) (see Lanczos, 1961; Wunsch, 1978; Wunsch and Minster, 1982; or Fiadeiro and Veronis, 1982). The least squares solution to the unconstrained problem can then be written

$$\mathbf{x} = \mathbf{V}\mathbf{S}^{+}\mathbf{U}^T\mathbf{b}$$

or, in component form,

$$x_i = \sum_j \frac{U^{(j)}\mathbf{b}}{s_j} V_{ij} \tag{3.3}$$

where $U^{(j)}$ denotes the jth column of $\mathbf{U}$. Obviously, if s_j is a 'small' singular value and if there is a substantial component of $\mathbf{b}$ in the 'direction' $U^{(j)}$, then the jth term would form a large component in the solution; whereas, if the singular value were zero, then the jth term would contribute nothing to the solution. The closeness between these two extremes is the essence of the difficulties associated with small singular values. When there are no convex constraints, it is in principle straight-forward to resolve the difficulties. Simply, establish a logical definition of 'small', and eliminate in the sum (3.3) all the singular values that are less than this smallest acceptable value.

In determining a potential 'cut-off' point for the singular values, one should consider noise in the data. One approach is to establish a new matrix $\mathbf{A}_1$ by slightly altering the data set, wherein, each element is perturbed within preset confidence intervals about each box average. Having determined this new matrix $\mathbf{A}_1$, which is in a sense not different from the matrix $\mathbf{A}$ insofar as the data are concerned, we compute the norm of $\mathbf{A} - \mathbf{A}_1$ (i.e. its largest singular value) and this is used to establish the lower bound for the singular values. Of course, for the purpose of sensitivity studies one should test higher and lower values for this bound as well as a variety of perturbations. In pursuing the approach, however, we met a difficulty in simultaneously satisfying the convex constraints $(\mathbf{G}, \mathbf{h})$. The complication occurs because the calculation requires the full application of the SVD prior to the onset of the calculation of the solution. Specifically, one begins with the creation of a nonfull rank matrix $\mathbf{A}_{\mathrm{tr}}$ which is formed through the SVD of $\mathbf{A}$ by a truncation of the diagonal matrix $\mathbf{S}$. Specifically, set $\mathbf{A}_{\mathrm{tr}} = \mathbf{U}\mathbf{S}_{\mathrm{tr}}\mathbf{V}^T$ where $\mathbf{S}_{\mathrm{tr}}$ is the diagonal matrix of singular values except that below the cut-off point, all singular values have been set to zero, and hence the associated V-singular vectors form a basis for the null space (kernel) of $\mathbf{A}_{\mathrm{tr}}$.

One can now treat this truncated system by using the techniques that we used in Bolin et al. (1983) for meeting constraints in an indeterminate setting (see also Wunsch and Minster, 1982). We use the vectors from the null space to modify the pseudo-inverse solution so that the resulting solution meets the constraints. The appealing character of this approach is that we use only the directions associated with the small singular vectors to form the final solution, but we do not weight highly these directions as one implicitly does when using also the inverse of the small singular values. The disadvantage is that the method does not necessarily lead to a solution. In other words, there may be no linear combination x of vectors from the null space of $\mathbf{A}_{tr}$ such that $\mathbf{x} + \mathbf{A}_{tr}^{+}\mathbf{b}$ meets the constraints imposed by $(\mathbf{G}, \mathbf{h})$. This circumstance can be avoided by rephrasing the least squares concept in a less restrictive manner. In order to realise this rephrasing we state again the constrained least squares problem in this context. Given $(\mathbf{A}_{tr}, \mathbf{b})$ and $(\mathbf{G}, \mathbf{h})$ find the vector x such that

i) $\mathbf{Gx} \geq \mathbf{h}$;

ii) $\mathbf{A}_{tr}\mathbf{x} - \mathbf{b} = \mathbf{A}_{tr}\mathbf{A}_{tr}^{+}\mathbf{b} - \mathbf{b}$; and

iii) $\|\mathbf{x} - \mathbf{A}_{tr}^{+}\mathbf{b}\|$ is minimal (this condition is equivalent to demanding that $\|x\|$ is minimal).

In this context one relaxes conditions (ii) and (iii).

We do this by introducing the augmented full rank system

$$\begin{pmatrix} \mathbf{A}_{tr} \\ \alpha\mathbf{I} \end{pmatrix} \mathbf{x} = \begin{pmatrix} \mathbf{b} \\ \mathbf{0} \end{pmatrix} \tag{3.4}$$

where α is a positive number. Since this system is full rank we can use our standard procedures with $(\mathbf{G}, \mathbf{h})$ as before. In the case where the previously discussed non-full rank (indeterminate) system has a constrained solution, the augmented technique will converge to the same solution if we consider the limit as alpha goes to zero.

The disadvantages shared by each of these approaches to small singular values is that they require a reconstruction of $\mathbf{A}$, or a creation of $\mathbf{A}_{tr}$, before beginning any analysis. Noise through round-off errors enters the process during this reconstruction and before beginning the analysis. This is particularly dangerous, since the reconstruction demands the full SVD using all three matrices. With large systems $(\mathbf{A}, \mathbf{b})$, this full application of the SVD is to be treated with caution. Thus far, using advanced algorithms and 64 bit precision, we have been dissatisfied with the results. Further advances in the SVD algorithms perhaps hold more promising results.

There is, in addition, a further problem that plagues many of these approaches, which has not received enough attention in the literature. It is the alteration of the 'zero structure' of the matrix $\mathbf{A}$ (i.e., the pattern of zeroes and non-zeroes), that reflects the topology of the model (i.e., what boxes are connected to what boxes). In this context, $\mathbf{A}_{tr}$ will not in general have the same zero structure as $\mathbf{A}$ and therefore there is some concern with the interpretation of these new, nonzero elements.

3.1.3 Ridge Regression

Another technique that has been developed to deal with the sensitivity of the least squares solution is called ridge regression, which received widespread attention in the statistical literature after the idea was introduced by Hoerl and Kinnard (1970a). The basic idea is to add a small positive quantity k to the diagonal elements of S in (3.2); k is called the ridge parameter. One can see from the component form (3.3) that replacing the singular values s_j by $s_j + k$ may greatly reduce the effect that the smaller singular values have on inflating the variance of the least squares solution. In statistical language, this means that by introducing a small bias to the solution (via k), we are generally able to greatly reduce the variance of the solution, and hence reduce the mean-squared error.

The fundamental problem with ridge regression is that one does not, in general, know what k to choose. Obviously, one would like to reduce the solution variance as much as possible by making k as large as possible, but in so doing, one faces the danger of destroying smaller-scale structure in the solution (that may be statistically significant) by introducing an unacceptable bias into the solution. Many mechanical procedures have been introduced for selecting k, and their effectiveness in general regression problems have been extensively tested (see e.g. Lawless and Wang, 1976; Hocking et al., 1976; Dempster et al., 1977; Wichern and Churchill, 1978; Hoerl et al., 1986). Hoerl and Kinnard (1970b) proposed a graphical method for choosing k called the ridge trace, but the method is not practical for systems of the size we are dealing with here. For lack of a better method, one could choose k so that the solution variance is near an a priori estimate, if one is available.

A chief difficulty with ridge regression for the present application is that, like truncation of singular values and other biased techniques, the zero structure of A is destroyed. Techniques do exist however, to incorporate convex constraints into the procedure (see Hemmerle and Brantle, 1978). Marquardt (1970) provides a good discussion which unifies the generalized inverse and the biased linear estimator approaches.

3.2 Nonlinear Inversion

As Wunsch (this volume, e.g.) has pointed out, one rarely (if ever) has complete knowledge of the data, or complete ignorance of the parameters to be estimated, when performing an inversion. The degree to which an inverse procedure is appropriate to the particular problem at hand then, is reflected by the degree to which the state of knowledge of both the parameters and the data is incorporated into the procedure. While convex constraints do provide a means by which we may utilise *a priori* information on the parameters in the inversion procedure, linear procedures are unequipped to adjust the input data according to any prior information we may have concerning them. Procedures are being developed to determine whether or not the variances of the entries of $\mathbf{A}$ contribute significantly to the variances of the solution parameters (see Ringo *et al.*, 1989). If the tracer data from which $\mathbf{A}$ is derived have substantial variability (and therefore contribute significantly to the solution variance), then a nonlinear inverse procedure may be more appropriate than a linear one.

In addition, if one is working with an incompatible system such as the one described in Section 2, one may find that the incompatibilities (i.e., residuals) resulting from a linear inversion are too large to accept. Rather than reject the model structure, one may wish to try a nonlinear inversion technique in which the parameters and the data are adjusted simultaneously. We outline here just a few of the many techniques available.

3.2.1 Total Least Squares

Incompatibility of equations is caused by two factors: The data are inaccurate and the equations are approximate. Both factors are significant, but let us assume here that all apparent contradictions between equations are due to inadequate input data. In this context it becomes meaningful to search for an exact solution, and in the search process simultaneously adjust the elements a_{ij} and b_i in computing a solution vector $\mathbf{x}$.

Let us first consider the least squares problem for the $m \times n$ system $(\mathbf{A}, \mathbf{b})$ where we remove all convex inequality constraints. The unconstrained problem for the system $(\mathbf{A}, \mathbf{b})$ is simply the problem of finding the vector $\mathbf{x}$ (unique since we still assume that $\mathbf{A}$ is of full column rank) such that $\|\mathbf{Ax} - \mathbf{b}\|$ is minimum. An equivalent statement of the problem is to find the vector $\mathbf{r}$ such that $\|\mathbf{r}\|$ is minimal and $\mathbf{b} + \mathbf{r}$ is in the range of $\mathbf{A}$. The solution is then the vector $\mathbf{x}$ for which $\mathbf{Ax} = \mathbf{b} + \mathbf{r}$. This statement is the one in which it is easiest to formulate the total least squares concept. In this rephrasing, we see that a least squares problem for an incompatible system is simply the determination of the minimal perturbation of the boundary conditions (i.e. the vector $\mathbf{b}$) such that the perturbed system is no longer incompatible. This, in effect, is to assume that $\mathbf{A}$ is well known that all 'incompatibilities' originate from inadequate boundary conditions.

The total least squares method allows consideration of uncertainty in $\mathbf{A}$ as well as $\mathbf{b}$. We perturb not only $\mathbf{b}$ but also the n column vectors in $\mathbf{A}$. In order to define a minimal perturbation we need an $(n + 1)$-dimensional weighting matrix:

$$\mathbf{W} = \text{diag}\,(w_1, \ldots, w_{n+1})$$

The total least squares problem for $(\mathbf{A}, \mathbf{b})$ with distribution weight W (but still without convex constraints) is to

$$\text{minimize } \|[\mathbf{E} : \mathbf{r}]\mathbf{W}\|_F \tag{3.5}$$

over all matrices $\mathbf{E}$ and vectors $\mathbf{r}$ of the appropriate dimensions and subject to $\mathbf{b} + \mathbf{r}$ being in the range of $\mathbf{A} + \mathbf{E}$. Here $\| \cdot \|_F$ denotes the Frobenius norm (i.e. the square root of the sum of the squares of the matrix elements) and $[E : r]$ represents the $m \times (n + 1)$ augmented matrix. Having determined a minimizing pair $(\mathbf{E}, \mathbf{r})$, the x that satisfies $(\mathbf{A} + \mathbf{E})\mathbf{x} = \mathbf{b} + \mathbf{r}$ is said to solve the least squares problem $(\mathbf{A}, \mathbf{b})$ with distribution weight $\mathbf{W}$.

It is important that $\mathbf{W}$ is on the right of $[\mathbf{E} : \mathbf{r}]$ in (3.5) since matrix multiplication from the right affects the columns on $[\mathbf{E} : \mathbf{r}]$ and hence $\mathbf{E}$ and $\mathbf{r}$ can be weighted separately. By adjusting the weights in $\mathbf{W}$, one can shift the balance between adjustments of $\mathbf{A}$ by $\mathbf{E}$ and of $\mathbf{b}$ by $\mathbf{r}$. For instance, if one wants to minimize the perturbations of $\mathbf{b}$ and allow greater perturbations of the data matrix $\mathbf{A}$, one would increase the penalty (measured by $\mathbf{W}$) when perturbing $\mathbf{b}$ (by $\mathbf{r}$) by increasing the weight w_{n+1} while lowering the weights $w_1, \ldots, w_n$.

The method of total least squares is essentially equivalent to a statistical procedure known as linear functional regression (see Sprent, 1969; or Kendall and Stuart, 1973). In attempting to deduce a functional relationship between 'regressor' variables however, it is assumed that the problem under consideration is set up as a standard regression problem. That is, the columns of the matrix $\mathbf{A}$ correspond to the regressor variables and the rows are repeated observations of each variable. In this case, each column of **A** may be scaled by the variance of the elements of that column, giving all elements of A unit variance; then the functional regression approach is appropriate. But in the present context, the matrix **A** has a definite pattern of zeroes and non-zeroes; the zeroes have zero variance and the non-zeroes have non-zero variance. Therefore, one cannot hope to scale **A** to give all entries unit variance. The result of applying total least squares to the present problem then, would be to adjust the zeroes of **A** along with the non-zero entries. In terms of the perturbation matrix **E** described above, this means that the minimal **E** will not in general have the same zero structure as **A**, and hence the 'adjusted' matrix **A** + **E** will not. Unless one can interpret these new non-zero entries in some physical manner then, it would be hard to justify the use of the method. As should be apparent by now, this problem of preserving the zero structure of **A** is a pervasive problem.

Gerhold (1969) gives a method by which individual variances can be specified for each element of **A**, which generalizes functional regression. Unfortunately though, the method reduces to solving a large system of nonlinear equations, and the method Gerhold proposes for solving them (Newton-Raphson) is not guaranteed to converge for arbitrary initial guesses for the solution. Also, one cannot impose convex constraints. Nevertheless, it may be worthwhile to test the method in the present setting, although we have not done so.

3.2.2 Total Inversion

Perhaps the most 'complete' solution of the nonlinear inverse problem to appear in the geophysical literature to date was provided by Tarantola and Valette (1982) (see also, Wunsch and Minster, 1982). In this approach, it is assumed that one has covariance information on both the data and on the unknowns. An iterative procedure is then used to adjust the data field while simultaneously adjusting the parameters. Although we have no experience with the method, others have had great success with it, particularly in the area of seismology. See Wunsch and Minster (1982) and particularly Mercier (1986) for oceanographic applications.

The routine's attraction lies in its ability to easily incorporate *a priori* statistical information on both the data and the parameters into the inversion procedure, something lacking in other methods. Also, estimates of the statistical uncertainty and resolution of the individual solution components are readily obtainable as by-products of the solution procedure. One drawback is that all *a priori* information on the unknowns must be Gaussian, and hence convex constraints are not implementable. Also, the method is very time-consuming, as is the case with most nonlinear schemes. To our knowledge, it has not been tested on a system of the size with which we are working here.

We conclude our discussion of inverse methods with an outline of an experimental technique that is quite similar to, but computationally less demanding than, the total inversion approach. It is an iterative minimization scheme in which the zero structure of **A** is preserved, and constraints on both the unknowns and the data may be set. Although the method will be described here in terms of the incompatible framework we developed in Section 2, it is general enough to be used in a variety of settings. For example, it could be used with the indeterminate structure suggested by Wunsch; wherein, unknown but constrained error terms are included in each equation. It could also be used in conjunction with a linear programming scheme, and it also probably has connections with control theory.

3.2.3 Separable Least Squares: An Experimental Approach

The elements of the matrix **A** are uncertain because they are functions of concentration data q_i^n, the correct values of which are unknown. Using the symbol $\mathbf{A}(q)$ to denote that **A** is a function of q_i^n, our problem is to

$$\text{minimize } \|\mathbf{A}(q)\mathbf{x} - \mathbf{b}\| \text{ over all } \mathbf{x} \text{ and } \mathbf{q} \tag{3.6}$$

This is known as the separable least squares problem. Algorithms for its solution have been developed by Golub and Pereyra (1973), and Kaufmann (1975); see also Björk (1981). Unfortunately, the technique becomes of little use in our application, unless constraints can be set for the numbers q_i^n; this is for the following reason. There are as many concentration data q_i^n as there are tracer equations (both numbers are equal to the number of boxes times the number of tracers, or 96 7 = 672). The matrix **A** has 672 rows for tracer balance equations and another 80 rows for the geostrophy equations. The combined set of unknowns, **q** and **x**, has 672 components q_i and 429 independent components **x**. Since 672 + 429 is greater than 672 + 80, the system becomes grossly underdetermined. It has many exact solutions; however, these solutions may include negative concentrations as well as conflicts with the constraints $(\mathbf{G}, \mathbf{h})$ on $\mathbf{x}$. Thus for the approach to be interesting, constraints must also be imposed both on the concentrations $\mathbf{q}$ and on the solution $\mathbf{x}$. However, the need for bounds on $\mathbf{q}$ further hampers the employment of algorithms devised to resolve the separable least squares problem, in addition to the two problems with the total least squares technique (inability to treat constraints on $\mathbf{x}$ and inability to conserve the element structure of $\mathbf{A}$ as $\mathbf{A}(q)$).

As an approach to this methodological issue we have been considering the possibility of iteratively minimizing over $\mathbf{x}$ and over $\mathbf{q}$ in Eq. (3.6). The approach is still heuristic but might be valuable.

For simplicity in notation we sketch this algorithm for a general constrained system $[(\mathbf{A}, \mathbf{b}), (\mathbf{G}, \mathbf{h})]$ based on I boxes and N chemical tracers:

Step 1. Establish highest and lowest values acceptable for each tracer in each box and express these constraints by developing N matrices $\mathbf{G}^n$, each of dimension $2I \times I$, and $2I$-dimensional vectors $\mathbf{h}^n$, $n = 1, \ldots, N$. In the following scheme Step 1 is not repeated, i.e. it is not part of the subsequent iteration process.

Step 2. Solve the constrained least squares problem for the system $[(\mathbf{A}, \mathbf{b}), (\mathbf{G}, \mathbf{h})]$.

Step 3. Establish the $I \times I$ transfer matrices $\mathbf{K}^n$ for each of the N tracers. By a transfer matrix $\mathbf{K}^n$ we mean the square matrix $\mathbf{K}^n$ that arises if the I continuity equations for tracer n are algebraically reformulated to express $\mathbf{q}$ instead of $\mathbf{x}$ as a vector:

$$\mathbf{A}^n \mathbf{x} - \mathbf{b}^n = \mathbf{K}^n \mathbf{q}^n + \mathbf{k}^n,$$

where $\mathbf{A}^n$ is the submatrix of $\mathbf{A}$ containing the continuity equations for tracer n, $\mathbf{b}^n$ is the corresponding right hand side, and $\mathbf{k}^n$ is a vector of boundary conditions, which are independent of $\mathbf{q}$.

Step 4. Solve the N constrained least squares problems for the (square) systems $(\mathbf{K}^n, \mathbf{k}^n)$, $(\mathbf{G}^n, \mathbf{h}^n)$, $n = 1, \ldots, N$, where $\mathbf{k}^n$ indicates the boundary conditions for the nth tracer and $(\mathbf{G}^n, \mathbf{h}^n)$ represents the interval constraints on the data (as given in Step 1).

Step 5. With the data set found in Step 4, define a new matrix $\mathbf{A}(q)$ and to the extent that $\mathbf{b}$ depends upon the tracer concentrations, define a new boundary matrix $\mathbf{b}$.

We now repeat the sequence beginning at Step 2 again. In practice, we find that iteration needs only to be repeated two or three times, depending in part on the size of the 'error' intervals in the data set (Step 1). Note that this approach does not produce a unique solution. If one started Step 2 with a matrix other than A but still one formed from data consistent with error intervals, it is not certain that the resulting two iterative paths would lead to the same solutions. We do not consider this a major difficulty since we begin with the mean values for each box which seems the most reasonable, but the concern does increase if the data intervals are large. Further exploration using this iterative approach to reducing the incompatibilities will be pursued in a subsequent publication.

4. SOME PRELIMINARY EXPERIMENTS

As has been outlined in the previous section there are a number of methodological issues that arise, when we wish to apply the set of basic equations (2.22) to the real ocean. Although many of these have not been resolved, it may be of interest to consider some preliminary experiments that have been attempted (detailed accounts are given in Bolin et al., 1987, 1989).

In our first attempts to apply the present methodology we have used the ordinary least square method with convex constraints (see section 3.1.1; the algorithm used has the name 'Subroutine LSEI' and has been described by Haskell and Hanson, 1981). It should be emphasized, however, that because of insufficient data, the matrix A in reality is not well determined, nor is the vector b. The coarse box configuration and the approximate algorithms that are used to evaluate the different terms in the tracer equation (2.15) lead to a considerable uncertainty in the solution x.

Oceanic data are most plentiful for the Atlantic Ocean, which we accordingly choose for initial experimentation. We adopt a configuration of boxes as shown in Figs. 6a and 6b. Note the similarity between this structure and the simple model used as an illustration in section 1 (cf. Fig. 1). For further details about the geometry of the model and the extraction of the necessary data, cf. Bolin et al. (1987).

The model configuration defines $I=84$ boxes with $T=184$ interconnecting surfaces. The number of loops, therefore becomes $T-I+1=101$. The 536-dimensional vector x is composed of the following elements:

184 mass fluxes, 'advections', m_t;

184 eddy diffusivities, 'turbulences', i.e. expressions $K_{ij}\Omega_{ij}/\Delta_{ij}$;

 84 organic production terms, as defined in Eq. (2.15);

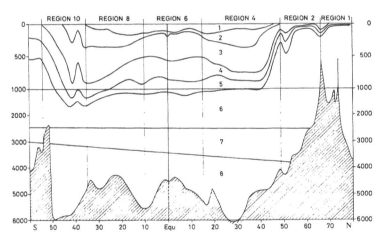

Figure 6
a) *Definition of regions 1–12 (denoted by bold numerals in the lower right hand corners of the boxes) as used in the analysis of the Atlantic Ocean. The areas where the layers 2–6 reach the ocean surface north and south of layer 1 are also shown, b) Vertical cross-section along the set of GEOSECS observations in the western basin of the Atlantic Ocean. The density surfaces shown are chosen as boundaries between layers 1–6. Two planes separate the layers 6–8. In this configuration altogether 84 boxes are defined. The definitions of the separating surfaces are given in Table 1.*

Table 1
Vertical Structure of layer configuration (ϕ denotes latitude).

Layer	Separating surface
1	
	$\sigma_\theta = 25.80$
2	
	$\sigma_\theta = 26.60$
3	
	$\sigma_\theta = 27.20$
4	
	$\sigma_\theta = 27.40$
5	
	$\sigma_\theta = 27.60$
6	
	$z = 2500$ m
7	
	$z = 3000 + 1200(\phi + 60)/140$
8	$-60° \le \phi \le 80°$

84 calcium carbonate production terms, as defined in Eq. (2.15).

There are 80 equations that define the thermal wind equation (it cannot be applied close to the equator), and there are 84 equations for each of the seven tracers. An annual average of the enthalpy field is not appropriate in the present analysis and enthalpy has therefore not been used as a tracer in the experiment to be presented below. The water continuity equation can be exactly satisfied by replacing the 184 advections by 101 loops. In addition there are 337 constraints as defined by Eq. (2.24). Obviously the system is grossly overdetermined. This problem of overdetermination requires careful investigation. Work is in progress in which the application of nonlinear inversion techniques is being attempted permitting ranges of uncertainty of the basic data to be included in the analysis.

Since there are about 500 tracer equations and merely 80 geostrophic conditions, the latter must be upweighted in order to achieve a balance between the impact on the solution of the tracer fields on the one hand and the geostrophic condition on the other. A weight of 4 has been used in the following experiment. The mean ageostrophic flow component then becomes about 16% as compared with

51% in the case when the same weight for all equations is used; see further Bolin *et al.* (1987).

Due to the complexity of the model and accordingly the size of the vector **x**, we shall describe the major features of the solution in a few simplified graphs. The complete solution is given in Bolin *et al.* (1987). In the following figures we display the average horizontal field of motion in

(i) *Surface Waters*, i.e. layers 1–3 at midlatitudes and in equatorial regions, including also layers 4–5 in polar regions where these approach the ocean surface; water above about 600 m depth is being included (Fig. 7a).

(ii) *Intermediate water*, i.e. layers 4–5 at midlatitudes and in equatorial regions (i.e. regions 4–11); water between about 600 and 1300 m depth is included (Fig. 7b).

(iii) *Deep water*, i.e. layers 6–8; water below about 1300 m at low and middle latitudes is included; the layers approach the ocean surface in regions 1–3 and 12 (Fig. 7c).

A large number of solutions has been derived by varying the data fields, boundary conditions, weighting of the equations, etc. In a broad sense they are all similar, but a careful comparison is required to permit firm conclusions. We notice the following features of a basic solution.

The North Atlantic *surface currents* are controlled by the flow through the Caribbean Sea and the Gulf of Mexico with a prescribed intensity of 810×10^{12} t/yr = 26 Sverdrups through the Florida Straits. A major part of this flow continues northward in the western basin (16 Sverdrups) and across the Atlantic south of Greenland and Iceland (21 Sverdrups). If a smaller geostrophic weight is applied a weaker anticyclonic circulation is obtained. Still the circulation deduced is weaker than in reality.

The northward flow of surface water in the southern Atlantic takes place in the eastern basin. The water crosses the Atlantic south of the equator and continues northward outside the Brazilian coast to feed into the the Caribbean Sea. An intense eastward current prevails through regions 10 and 11, caused by the inflow through the Drake Passage and leaving the Atlantic into the Indian Ocean. Because of the uncertain boundary conditions this current is not accurately determined.

The flow of intermediate water in the northern Atlantic is largely a reflection of the flow in the surface layers, but is weaker. The northward flow of intermediate water from the convergence zone in the southern Atlantic is only found in the

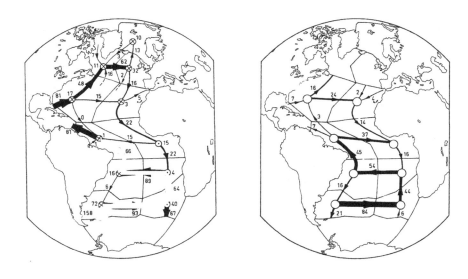

Figure 7

a) Advective flow (in 10^{13} t/yr) in the surface layers (layers 1–3 in regions 4–11; layers 4–5 in regions 1–3 and 12) as deduced for a reference case with geostrophic weighting being 4. Flow to and from intermediate or deep waters is shown by × and · respectively. The transfers through the Caribbean Sea and the Mexican Gulf, flow to and from the Mediterranean Sea in layers 1 and 2–5 respectively and from the Amazon river have been prescribed as boundary conditions and are also shown.

b) Advective flow (in 10^{13} t/yr) in the intermediate layer (layers 4–5 in regions 4–11) as deduced for a reference case. The exchanges with the surface water and the deep water are shown in Figs. 7a and 7c respectively.

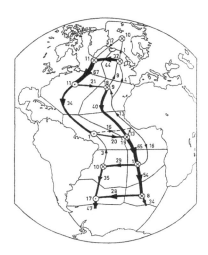

Figure 7 (ctd)
c) Advective flow (in 10^{13} t/yr) in the deep waters (layers 6–8) as deduced for a reference case. Flow to and from intermediate or surface water is shown by · and × respectively. Flow through the Romanche Fracture Zone (dashed line 16 × 10^{13} t/yr) and the flow through the bottom layer 8 in the eastern basin (regions 3, 5, 7 and 9) is shown separately.

eastern basin and does not penetrate beyond the equatorial region. The flow is southward in the western basin.

Deep water is formed in the three northernmost regions at a total rate of 470×10^{12} t/yr = 15 Sverdrups. Strong westward flow from region 3 to region 2 is obtained, which is in qualitative agreement with the penetration of tritium and CFC's into the western basin of the deep Atlantic (*cf.* Jenkins and Rhines, 1980; Weiss *et al.*, 1985). The southward flow is most intense in the western basin, although some water that is sinking into the deep layers in region 5 (presumably caused by the inflow of saline and comparatively heavy water from the Mediterranean Sea) also moves southward. In the equatorial region the flow through the Romanche Fracture Zone in layer 8, attains the minimum permissable value, 160×10^{12} t/yr = 5 Sverdrups, that is given as a constraint. Water is also supplied into layer 8 from above in this region of the eastern basin. This water spreads southward and northward and rises into layer 7 elsewhere in the basin in a similar way as has been derived by Schlitzer (1987). In addition the solution is characterized by eastward flow in layers 6 and 7 of the equatorial region and the southward flow in the deep sea of the northern Atlantic thereby shifts from the western to the eastern basin.

The advective flow pattern as deduced is characterised by a major meridional circulation cell as shown in Fig. 8. Its intensity is about 600×10^{12} t/yr = 20 Sverdrups, which rate, however, is dependent upon the weighting of the geostrophic equations. Superimposed on this cell there is upwelling in the equatorial region of the intermediate and surface layers. A weak cell in the opposite direction is found in the subtropical region of the deep sea, the existence of which is not well established by observations.

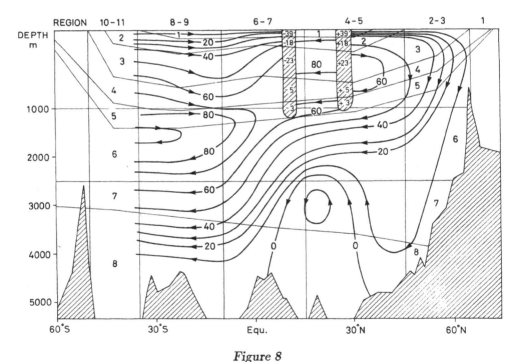

Figure 8

Meridional circulation shown by stream lines for each 10×10^{13} t/yr flow of water. Water in the layers 1–5 is leaving the Atlantic Ocean from region 6, flowing into the Caribbean Sea and returning into region 4 from the Gulf of Mexico.

Turbulence is primarily of importance for exchange between boxes at the sea surface and adjacent boxes below. The average magnitude of the exchange of water due to turbulence is on the average about half of that due to advective flow. We notice further that intense turbulence sometimes appears spotwise in regions where the tracer gradients are small and accordingly the magnitude of the turbulence exchange is unreliable.

We deduce further that total new primary production of the Atlantic Ocean is 2.5×10^{15} g C/yr of which 2.2×10^{15} g C/yr is in organic form and the rest in inorganic form. Because of the uncertainty of the solution in Antarctic regions this total production may be an overestimate. It is hardly meaningful to consider the vertical distribution of decomposition of organic detritus and dissolution of inorganic detritus for each region separately. The average distribution is, however, shown in Fig. 9. We find that about 80% of the organic detritus that leaves the surface layers (which is about 100m thick) is decomposed above 1000 m depth, while the corresponding figure for inorganic detritus is about 65%. The ratio of organic to inorganic detritus flux, being about 5.4 at 100 m, has been reduced to 2.8 at 1000 m and is less than 2.0 below about 3000 m.

The experiment described above should be considered as an example of the use of inverse methods with the aim of combining hydrographic and chemical data for analysis of oceanic processes. We impose requirements which constrain the interpretation of the data and permit the extraction of a consistent answer to a hypothesis that has been posed. The lack of an exact solution and accordingly errors in the way the equations that we have formulated have been satisfied, is a characteristic of overdetermined problems and warrants close analysis. We wish to be able to distinguish between inadequate hypotheses, inappropriate spatial resolution in the model, insufficient accuracy in the numerical formulation, inaccurate and insufficient data as a cause for the incompatibility. The method of least squares as used in the present example is not well suited for such an analysis, but other methods as outlined in section 3 show promise.

In conclusion we note, however, that the solution we present shows resemblance with reality and the approximately correct location of a 'layer of no motion' as is probable from the meridional cross-section (Fig. 8), is actually determined with the aid of the tracer data.

446

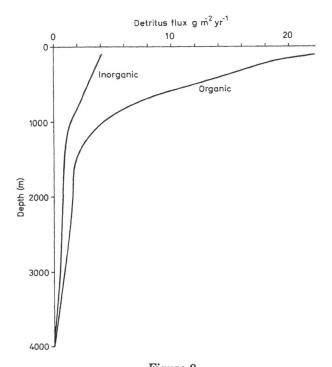

Figure 9

Average vertical flux of organic and inorganic detrital matter (in units of g
$C\,m^{-2}\,yr^{-1}$*) as a function of depth and deduced for the regions 1–11 in the*
reference case. Each box in the regions considered has been located with
regard to its depth and vertical extension assuming that its horizontal area
is that of the surface box of the region. The mean depth of the Atlantic
Ocean deduced in this way becomes about 4000 m. The decomposition and
dissolution rates as obtained from the model solution have been assigned
to the appropriate layers and the downward flux determined.

ACKNOWLEDGMENTS

This work has been supported by the Swedish Natural Science Research Council
(NFR) under contracts E-EG 0223-113, E-EG 0223-115, E-EG 0223-117 and E-
EG 0223-118. The investigation at the University of New Hampshire has been
supported by a grant from the National Aeronautics and Space Administration
(NASA): NAGW-848. Finally, we would like to acknowledge helpful discussions
with Å. Björk, A. Copeland, R. Hansson, N. Metzl, J. F. Minster, and C. Wunsch,
which have greatly aided in furthering our understanding of inverse methods.

REFERENCES

Björk, Å. (1981): Least squares methods in physics and engineering. Lectures given in the Academic Training Programme of CERN 1980-1981. CERN 81–16, 61 p. CERN, Geneva.

Bolin, B., A. Björkström, K. Holmén, and B. Moore (1983): The simultaneous use of tracers for ocean circulation studies. *Tellus* **35B**, 206–236.

Bolin, B., A. Björkström, K. Holmén, and B. Moore (1987): On Inverse Methods for Combining Chemical and Physical Oceanographic Data: A Steady-state Analysis of the Atlantic Ocean. Tech. Rep., Meteorologiska Institutionen (MISU), Stockholm Universitet, Arrheniuslaboratoret, 134pp.

Dempster, A.P., M. Schatzoff, and N. Wermuth (1977): A simulation study of alternatives to ordinary least squares. *J. Amer. Stat. Assoc.*, **72**, 77–106.

Eppley, R.W. and B.J. Peterson (1979): Particulate organic matter flux and planktonic new production in the deep ocean. *Nature*, **282**, 677–680.

Fiadeiro, M.E. and G. Veronis (1984): Obtaining velocities from tracer distributions. *J. Phys. Oceanogr.*, **14**, 1734–1746.

Fiadeiro, M.E. and G. Veronis (1982): On the determination of absolute velocities in the ocean. *J. Mar. Res.*, **40** (Suppl.), 159–182.

Fiadeiro, M.E. and G. Veronis (1977): On weighted-mean schemes for the finite-difference approximation of the advection diffusion equation. *Tellus* **29**, 512–522.

Gerhold, G.A. (1969): Least-squares adjustment of weighted data to a general linear equation. *Amer. J. Phys.*, **37**, 156–161.

Golub, G.H. and V. Pereyra (1973): The differentiation of pseudo-inverses and nonlinear least squares problems whose variables separate. *SIAM J. Numer. Anal.* **10**, 413–432.

Golub, G.H. and C.F. Van Loan (1980): An analysis of the total least squares problem. *SIAM J. Numer. Anal.* **17**, 883–893.

Haskell, K.H. and R.J. Hanson (1981): An algorithm for linear least squares problems with equality and nonnegativity constraints. *Math. Programming*, **21**, 98–118.

Hemmerle, W.J. and T.F. Brantle (1978): Explicit and constrained generalized ridge estimation. *Technometrics*, **20**, 109–120.

Hocking, R.R., F.M. Speed, M.J. Lynn (1976): A class of biased estimators in linear regression. *Technometrics*, **18**, 425–437.

Hoerl, A.E. and R.W. Kinnard (1970a): Ridge regression: biased estimation for nonorthogonal problems. *Technometrics*, **12**, 55–67.

Hoerl, A.E. and R.W. Kinnard (1970b): Ridge regression: applications to nonorthogonal problems. *Technometrics*, **12**, 69–82.

Hoerl, R.W., J.H. Schuenemeyer, and A.E. Hoerl (1986): A simulation of biased estimation and subset selection regression techniques. *Technometrics*, **28**, 369–380.

Jenkins, W.J. and P.B. Rhines (1980): Tritium in the deep North Atlantic Ocean. *Nature*, **286**, 877–880.

Kaufman, L. (1975): Variable projection methods for solving separable nonlinear least squares problems. *BIT*, **15**, 49–57.

Kendall, M.G., and A. Stuart (1973): The Advanced Theory of Statistics, Vol. II, 3rd. edition, 723 pp.

Lanczos, C. (1961): Linear Differential Operators, Van Nostrand, Princeton, 564 pp.

Lawless, J.F. and P. Wang (1976): A simulation study of ridge and other regression estimators. *Comm. in Stat.* (Part A), **5**, 307–323.

Marquardt, D.W. (1970): Generalized inverses, ridge regression, biased linear estimation, and nonlinear estimation. *Technometrics*, **12**, 591–612.

Mercier, H. (1986): Determining the general circulation of the ocean: a nonlinear inverse problem. *J. Geophys. Res.*, **91(C4)**, 5103–5109.

Ringo, C., A. Copeland, and B. Moore (1989): Random perturbations of pseudoinverses and least squares solutions. (in review)

Schlitzer, R. (1988): Modeling the nutrient and carbon cycles of the North Atlantic 1. Circulation, mixing coefficients, and heat fluxes. *J. Geophys. Res.*, **93**, 10, 699–10, 723.

Schlitzer, R. (1987): Renewal rates of east Atlantic deep water estimated by inversion of ^{14}C data. *J. Geophys. Res.*, **29**, 2953–2969.

Sprent, P. (1969): Models in Regression and Related Topics. Methuen, London, 173 pp.

Stommel, H. and F. Schott (1977): The β-spiral and determination of the absolute velocity field from hydrographic station data. *Deep Sea Res.*, **24**, 325–329.

Tarantola, A. and B. Valette (1982): Generalized nonlinear inverse problems solved using the least squares criterion. *Rev. Geophys. Space Phys.*, **20**, 219–232.

Weiss, R.F., W.S. Bullister, R.H. Gammon, and M.J. Warner, (1985): Atmospheric chlorofluoromethanes in deep equatorial Atlantic, *Nature*, **314**, 608–610.

Wichern, D.W. and G.A. Churchill (1978): A comparison of ridge estimators. *Technometrics*, **20**, 301–311.

Wunsch, C. (1978): The general circulation of the North Atlantic west of 50°W determined from inverse methods. *Rev. Geophys. Space Phys.*, **16**, 583–620.

Wunsch, C. (1984): An estimate of the upwelling rate in the equatorial Atlantic based on the distribution of bomb radiocarbon and quasigeostrophic dynamics. *J. Geophys. Res.* **89**, 7971–7978.

Wunsch, C. (1985): Can a tracer Field be Inverted for Velocity?, *J. Physical Oceanogr.*, **15**, 1521–1531.

Wunsch, C. and J.-F. Minster (1982): Methods for box models and ocean circulation tracers: Mathematical programming and non-linear inverse theory. *J. Geophys. Res.*, **87**, 5647–5662.

MODEL OF THE NUTRIENT AND CARBON CYCLES
IN THE NORTH ATLANTIC
AN APPLICATION OF LINEAR PROGRAMMING METHODS

Reiner Schlitzer
Universität Bremen FB-1
Postfach 330440, Bremen
F.R.Germany

ABSTRACT

As example for the use of Linear Programming methods in the context of oceanographic tracer models a model of the nutrient and carbon cycles in the North Atlantic is presented. The strategy is to make use of large hydrographic- and tracer data-sets to investigate the various processes that are involved in shaping the tracer distributions and to obtain numerical values for the respective rate constants. Linear Programming is seen to be a powerful tool because calculation of extreme solutions allows one to examine the whole range of possible solutions. The method provides diagnostics to identify well and poorly determined model parameters and the important (or redundant) parts of the input data. Often it is possible to infer what additional information is needed to better determine certain model parameters.

451

D. L. T. Anderson and J. Willebrand (eds.), Oceanic Circulation Models: Combining Data and Dynamics, 451–463.
© *1989 by Kluwer Academic Publishers.*

1. INTRODUCTION

The distributions of dissolved nutrients (phosphorous, nitrogen, silicate, etc.) and carbon in the ocean are affected by a large number of simultaneously occuring processes. Like all substances in sea water they are transported by ocean circulation and mixing. In addition, nutrients and carbon take part in the marine biological cycle. In the euphotic zone organisms extract dissolved nutrients and carbon from the sea-water and use them for the formation of particulate organic and inorganic matter (organic tissue and shells). To a great extent the organic matter is recycled (redissolved) in the upper layers, but a fraction of the particulates falls out of the euphotic zone. These particles are dissolved in the intermediate and deep waters or they reach the sediment. The distribution of carbon is not only influenced by circulation, mixing and the biological cycle but also by the exchange of CO_2 between atmosphere and the ocean.

All the mentioned processes leave signatures in the distributions of nutrients and carbon. The well known tongue-shaped features in sections of silicate, phosphate, nitrate and ΣCO_2 (see e.g. GEOSECS west Atlantic sections, Bainbridge, 1981) indicate major (and persistent) oceanic currents. Tracer gradients are relatively strong in the source regions of these currents and are continuously weakened downstream. This feature is attributed to turbulent mixing processes. Biological processes result in low nutrient concentrations in surface waters and relatively high values in the deep water. CO_2 gas-exchange keeps surface water pCO_2 values close (but not equal) to the atmospheric concentrations.

The fact that the various processes (circulation, mixing, biology, gas-exchange) leave structures in the tracer distributions encourages quantitative evaluation of tracer data. The hope is to gain a better understanding of the ongoing processes and to obtain numerical estimates of the respective rate constants. In this paper a model of the nutrient and carbon cycles in the North Atlantic is described that makes use of large collections of hydrographic, nutrient and carbon data and yields estimates for circulation and mixing rates, net productivity rates, particle fluxes, and ocean atmosphere water, heat, oxygen and CO_2 fluxes. Linear Programming methods are used to solve the model.

2. THE MODEL

The North Atlantic is divided into a series of layers which are bounded by the sea surface by isopycnal surfaces and the ocean floor (see Fig. 1). These layers are further divided into boxes by zonal sections at 52.5°N, 12.5°N, and 9.5°S (see Fig. 2). The model provides high vertical resolution near the surface where the biological processes and the ocean-atmosphere gas-exchange occur (layer thickness approximately 100 m). In the deep ocean only the major water masses AAIW (Antarctic Intermediate Water), NADW (North Atlantic Deep Water) and AABW (Antarctic Bottom Water) are resolved.

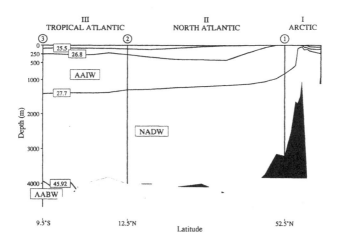

Figure 1

Meridional section through the north Atlantic from 9.5°S to the Bering Strait. Indicated are the zonally averaged depths of the isopycnal surfaces that are used as vertical boundary surfaces in the model and the sections that divide the north Atlantic into the three model regions. The horizontal axis is not linear in latitude but is scaled such that the area of the different layers in the figure is approximately proportional to the volume of the corresponding boxes of the model.

There are three seperate regions: (a) the Arctic Ocean including the Norwegian and Greenland Seas, (b) the subtropical Atlantic, and (c) the tropical Atlantic. These regions represent distinctly different oceanographic regimes. The Arctic Ocean and especially the Greenland and Norwegian Seas are the areas where

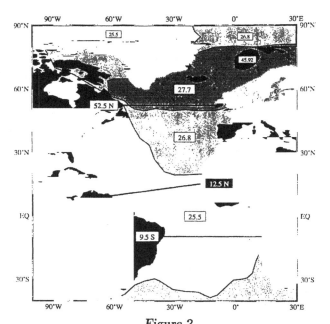

Figure 2

Map of the North Atlantic showing the sections of the model and the annual mean outcrop regions of the model layers.

formation of NADW takes place. It is believed that the heat loss to the atmosphere and the uptake of atmospheric CO_2 are especially large in these regions. Also, because of strong vertical exchange nutrient concentrations in surface waters and the biological productivity are high. The subtropical Atlantic includes the entire Gulf Stream and eastern basin recirculation. Nutrient concentrations in the surface waters and biological productivity are relatively small. In the tropical Atlantic upwelling leads to high pCO_2 and nutrient concentrations at the surface. There is strong evidence that in the tropical Atlantic — in contrast to the subtropical and subpolar Atlantic — CO_2 is lost to the atmosphere.

Concentration data for a suite of tracers are used as input to the model and the water transports between the boxes of the model, the iso- and diapycnal mixing coefficients, the biological rate constants and the air-sea exchange rates are estimated. The basic equation in the model is continuity of tracer within the boxes of the model:

$$\sum_i c_i J_i + \sum_i \rho F \frac{\partial c_i}{\partial n} K_i + Q = \rho V \frac{\partial c}{\partial t} \tag{1}$$

Eq. (1) provides the link between the unknown rate constants (J_i = water fluxes across the boundary surfaces of the boxes; K_i = turbulent mixing coefficients; Q = source/sink terms) and the measured tracer concentrations c_i and concentration gradients $\partial c / \partial n$. Gas-exchange, formation and dissolution of particulate matter are included in the source terms Q of the nutrient balance equations. Eq. (1) is linear in the unknowns (model parameters) J, K and Q and serves as constraint equation for the unknowns.

The principle of geostrophy provides an additional constraint on the horizontal water fluxes J_i across the sections of the model. Geostrophic flow calculations yield the vertical shear of the flow and leave only the reference velocities unknown. Estimates of the wind-driven Ekman flows are added to the geostrophic flows in the surface layers of the model.

Formulation of balance Eqs. (1) for the boxes of the model for mass, heat, salt, oxygen, phosphorous, nitrogen, silicate, carbon and alkalinity results in a set of linear constraints for the unknown model parameters in the vector $\mathbf{x}$:

$$\mathbf{A}\mathbf{x} = \mathbf{b} \pm \mathbf{e} \qquad (2)$$

Because of data-errors and approximations in the model the balance equations are only required to be satisfied within certain tolerances e_i. The model is set up as a Linear Programming problem, i.e. the model Eqs. (2) are transformed into a set of linear inequalities

$$\mathbf{b} - \mathbf{e} \le \mathbf{A}\mathbf{x} \le \mathbf{b} + \mathbf{e} \qquad (3)$$

and the Simplex Algorithm is used to calculate solutions $\mathbf{x}_e$ that satisfy the set of inequalities (3) (model constraints), a set of a priori inequalities for the unknowns

$$\mathbf{x}_- \le \mathbf{x} \le \mathbf{x}_+ \qquad (4)$$

and minimize (or maximize) a conveniently chosen objective function f

$$f = \sum w_i \, x_i \qquad (5)$$

If model constraints and a priori constraints cannot be satisfied simultaneously (the system is then called infeasible) the Simplex Algorithm returns the vector $\mathbf{x}_b$ that leads to the smallest misfits (i.e. satisfies (4) and (5) 'best').

A variety of interesting oceanographic parameters can be written in the form (5) [1] and can be chosen as objective function f. Running the model with these objective functions then yields the smallest and largest numerical values for the particular oceanographic parameters that are consistent with the model constraints (i.e. satisfy the tracer constraints (3) and the *a priori* constraints (4)). In this manner minimal and maximal values for water, heat and nutrient fluxes across the sections and isopycnals of the model, iso- and diapycnal mixing coefficients, net productivity and air-sea gas-exchange rates can be calculated. The method reveals which oceanographic parameters are well determined by the model constraints (narrow range between minimal and maximal value) and which are only poorly determined.

Most implementations of the Simplex Algorithm also solve the dual problem (Luenberger, 1984). By inspection of the dual it is possible to identify the important model constraints; e.g. those constraints that actually limit the values of the objective function (oceanographic parameter). Thus Linear Programming not only allows calculation of solutions to the model constraints (3) and (4) but in addition allows identification of well (and poorly) determined parameters and identification of the important constraints and data in the model.

3. RESULTS FOR THE 'INORGANIC' CASE

For a first set of numerical experiments only the inorganic components phosphate, nitrate and ΣCO_2 of the nutrients phosphorous, nitrogen and carbon are considered. This approach is analogous to 'traditional' nutrient cycle models in which organic components of the nutrients are ignored (this model setup is referred to as case 'I'). The distributions of phosphate, nitrate and carbon exhibit similar structures. As examples, Figs. 3a, b and c show average nitrate profiles at 9.5°S, 12.5°N, and 52.5°N. In the tropical and subtropical Atlantic (9.5°S and 12.5°N) the surface waters are depleted in nutrients (almost zero concentrations) whereas the deep waters have high concentrations. Vertical gradients in the transition zone in about 100 m to 400 m depth are large. Deep and intermediate waters of southern origin (AAIW and AABW) can be distinguished from waters of northern origin (NADW) through their higher nutrient values. At 52.5°N, nutrient concentrations in the surface water are relatively high, the concentration difference between surface water

---

[1] Wagner (1984) shows that the objective function can be generalized to include absolute values.

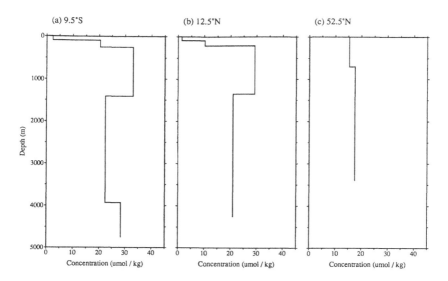

Figure 3
Average layer nitrate concentrations ($\mu mol/kg$) at 9.5° S, 12.5° N, and 52.5° N.

and deep water is small. Again the deep water nutrient concentrations are higher than surface water concentrations.

Model calculations show that the set of tracer-constraints (3) and (4) with only the inorganic components considered in the nutrient and carbon balances is feasible; e.g. that a solution exists that satisfies constraints (3) and (4). Thus the climatological flow field of Levitus[2] is consistent with large-scale, steady-state nutrient and carbon budgets in the North Atlantic. The solution to (3) and (4) is not unique. Fig. 4 shows meridional water flows and iso- and diapycnal mixing coefficients for one particular solution. Many features of the circulation pattern in Fig. 4 are observed in all solutions that have been calculated. In the top three layers (upper 1000 m) the water flow across 9.5°S and 12.5°N is northward. A mixture of subtropical surface water and AAIW flows northward into the Greenland and Norwegian Seas where NADW is formed and subsequently is flowing southward

---

[2] Levitus' (1982) climatological temperature and salinity data have been used for the calculation of geostrophic flows and for deriving temperature and salinity values on the box boundaries.

458

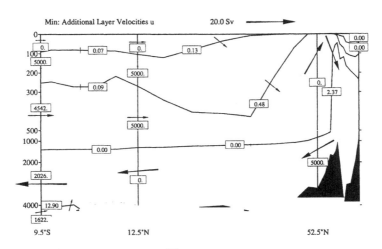

Figure 4
Meridional and diapycnal water transports (proportional to lengths of arrows) and isopycnal and diapycnal mixing coefficients (numbers in boxes;
isopycnal: m^2/s; diapycnal: cm^2/s) for solution that minimizes the deviation of the additional layer velocities from Ekman velocities in the surface
layers of the sections at 9.5°S, 12.5°N, and 52.5°N, and from zero in the
deeper layers. Note the scale change of the vertical axis at 500 m. (From
Schlitzer, 1988).

across 52.5°N, 12.5°N, and 9.5°S. Diapycnal mixing coefficients are small (0 to 0.5
cm^2/s) except in the Greenland and Norwegian Seas and at the interface between
NADW and AABW in the tropical Atlantic.

Fig. 5 shows the fluxes of particulate silicate (biogenic opal), organic carbon,
and $CaCO_3$. Particle fluxes are largest in areas with high biological productivity
(subpolar regions, tropical Atlantic) and very small in the subtropical gyre region.
The flux of particulate organic carbon in the tropical Atlantic decreases rapidly
with depth. The numerical values for the particle fluxes agree well with independent
estimates or measurements (Eppley and Peterson, 1979; Martin *et al.*, 1986). By
calculating solutions which minimize and maximize the net productivity in the
model area it is found that in the Atlantic north of 10°S the net productivity
amounts to between 2.3×10^{14} and 10.1×10^{14} gC/yr. The productivity rates
depend on the supply of nutrients into the euphotic zone and are sensitive to and
closely correlated with upwelling rates and diapycnal mixing coefficients.

Particle Fluxes

S: 3.E+05 mol SiO2/s ➤
O: 1.E+06 mol Corg/s ➤
I: 1.E+06 mol Carb/s ➤

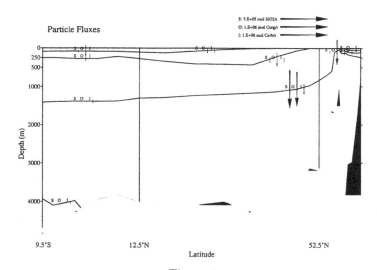

Figure 5
Downward fluxes of particulate silicate (S), organic carbon (O), and
CaCO₃ (I) across the vertical layer boundaries for the solution in Fig. 4.
(From Schlitzer, 1988).

The exchange rates of CO_2 between the atmosphere and the ocean are found to be only loosely constrained in the model. Model solutions are possible for which the net uptake of CO_2 by the ocean is almost zero or at the maximum amounts to about 80% of the CO_2 production due to burning of fossil fuel. This indetermination of CO_2 gas-exchange rates arises because of the uncertainty of the ΣCO_2 data due to sparse data coverage and lack of information about the temporal increase of ΣCO_2-concentrations in the ocean.

4. RESULTS FOR THE 'INORGANIC + ORGANIC' CASE

In addition to the inorganic components phosphate, nitrate and ΣCO_2, significant amounts of dissolved organic phosphorous, nitrogen and carbon compounds are found in sea water. However, measurement techniques and concentration data are controversial and the distributions of these organic compounds in the ocean are not well known quantitatively. Although it is suspected that the organic components are involved in the nutrient cycles they are usually ignored in models because of the large uncertainty of the concentration data. Use of these data in models presently cannot lead to final answers. The main purpose of such efforts is to examine different scenarios for the organic components and to assess the question whether the results for new production and particle fluxes change when organic nutrients are included in the model.

Two main features seem to evolve from the existing data: (a) surface water concentrations are higher than concentrations in deep water; and (b) there seems to be no geographical variation of the concentrations (Menzel, 1974). Reported concentrations differ widely. Traditional estimates (see Menzel, 1974) are ca. 6.5 μmol N_{org}/kg in surface water and ca. 3.5 μmol N_{org}/kg in deep water; Japanese workers (Sugimura and Suzuki, 1987) recently reported values 3 to 7 times as high. In both data-sets surface water concentrations are higher than deep water values.

Three sets of model-experiments with dissolved organic nutrients have been run: (a) using traditionally accepted concentrations ('T1'), (b) implementing Sugimura's values in a simple way (using his surface value for the first layer of the model and his deep water value for all other layers) ('T2'), and (c) constructing concentration values from the correlation between organic nutrients and apparent oxygen utilization (AOU) that has been found by Sugimura and Suzuki (1987) ('T3').

The vertical profiles of total nitrogen that result in the three cases are shown in Fig. 6. For case 'T1' the profiles of total nitrogen (solid line) are similar to the nitrate profiles in Fig. 3 except that the overall concentrations are higher and that the vertical concentration-gradient in about 200 m is smaller than in the nitrate profiles. At 52.5°N concentrations in the surface and deep layers are nearly equal. Using Sugimura's values leads to profiles of total nitrogen that are considerably different. For case 'T2' (dashed line) surface concentrations at 9.5°S and 12.5°N are slightly larger than in the deep water; at 52.5°N surface values are about twice the deep water values. At 9.5°S and 12.5°N there is a distinct minimum in about 100 to 200 m depth. When the concentrations of organic nitrogen are derived from the correlation with AOU (case 'T3'; dotted line) concentrations of total nitrogen are nearly constant below about 500 m depth; surface concentrations are about 10% smaller than deep water values and the shallow minimum in about 150 m (see case 'T2') is less pronounced. At 52.5°N the profile for case 'T3' resembles the profile for case 'I' but the overall concentrations are a factor of 2.5 higher in case 'T3'.

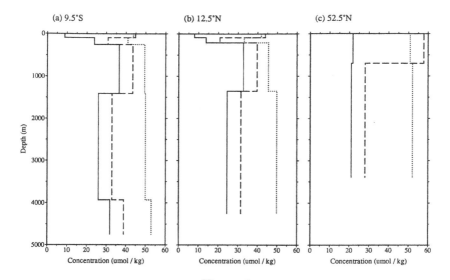

Figure 6
*Average layer concentrations of total nitrogen ($\mu mol/kg$) at 9.5°S, 12.5°N,
and 52.5°N. Solid line: using values from Menzel, (1974); dashed line: using values from Sugimura et al., (1988); dotted line: constructing concentrations of organic components from correlation with AOU (Sugimura et
al., 1988).*

Model calculations show that with traditional values of organic nutrients (case
'T1') the set of model constraints is still feasible. The circulation pattern in the
solutions is similar to the one that was found in the model without dissolved organic nutrients (case 'I' in section 3) and values for the iso- and diapycnal mixing
coefficients lie in the same range as the mixing coefficients for case 'I'. However,
the maximal values for the fluxes of particulate organic matter are about 20 to
50% smaller than for case 'I' with the largest reduction observed in the subpolar North Atlantic. Overall new production in the North Atlantic lies between
4.1×10^{14} and 8.0×10^{14} gC/yr. Obviously, the reduction of the maximal particle
fluxes is due to the reduced transport of nutrients into the euphotic zone (mixing
coefficients about the same as in case 'I' but vertical nutrient gradients smaller).
The geographical pattern of particle fluxes remains unchanged. Particle fluxes are
largest in regions with high biological productivity (subpolar North Atlantic, see
e.g. Koblentz-Mishke, 1970) and are smallest in the subtropical gyre region.

The models with Sugimura's values for the dissolved organic nutrients (cases

'T2' and 'T3') are infeasible; e.g. there exists no solution that satisfies all tracer balances. The violations of the model constraints are largest in case 'T2' (simple way of implementing Sugimura's values). Higher nutrient concentrations in surface water compared with deep water are incompatible with the meridional circulation cell that evolves from the hydrographic data. At 52.5°N northward flow in the surface layer and southward flow in the deep layer as indicated by geostrophic calculations would lead to a net import of nutrients into the subpolar and Arctic Ocean which (if steady-state nutrient distributions are assumed) could only be balanced by an unrealistically large deposition of organic material in the sediments (ca. 220 $gC_{org}m^{-2}a^{-1}$).

In case 'T3' (organic components correlated with AOU) the model is also infeasible but the misfits of the model constraints (3) and (4) are much smaller than in case 'T2'. Some of the mixing coefficients are negative and heat, salt and phosphorous balances in the tropical Atlantic are violated. Fluxes of particulate organic matter are small (net productivity in model region ca. 2×10^{14} gC/yr) but fluxes of $CaCO_3$ and opal are comparable with results in cases 'I' and 'T1'.

5. SUMMARY

Linear Programming is seen to be a powerful tool for obtaining estimates of oceanographic circulation rates, mixing coefficients, and biological and geochemical rate-constants. Not only does the method yield solutions to the set of model constraints, it also naturally leads to an examination of the whole range of possible solutions. Well and poorly determined parameters and important model constraints and data can easily be identified.

Four different scenarios for the nutrients in the North Atlantic have been studied. The models with inorganic nutrients only (case 'I') and inorganic nutrients plus traditionally accepted concentrations of organic components (case 'T1') are feasible. Circulation rates in the model solutions are consistent with independent investigations. Diapycnal mixing coefficients on the shallow isopycnals (ca. 100 to 200 m depth) are small (0 to 0.3 cm^2/s) but span a large range in the deep water (0 to 13 cm^2/s). Particle fluxes depend on the supply of nutrients to the euphotic zone and are consistent with sediment trap measurements. The geographical pattern of the particle fluxes agrees with the geographical pattern of primary productivity. Numerical values for the particle fluxes are significantly reduced if organic components are considered in the nutrient budgets.

Use of recently reported data for the organic components (Sugimura and Suzuki, 1987) leads to infeasible model constraints. However, it has to be kept in mind that these data were obtained in the western Pacific and probably are not representative for the North Atlantic. More data must be gathered before the controversy about the organic components of phosphorous, nitrogen and carbon can be resolved and before the effect on calculated particle fluxes can be quantified.

REFERENCES

Bainbridge, A. E. (1980). *GEOSECS Atlantic Expedition*, vol. II, *Sections and Profiles 1972-1973*, U.S. Govt. Print. Office, Washington D.C.

Koblentz-Mishke, O. J., V. V. Volkovinsky, and J. G. Kabanova (1970). Plankton primary production of the world ocean. In: *Scientific Exploration of the South Pacific*, W. S. Wooster (ed.), 183–193, Washington, D.C., National Academy of Science.

Luenberger, D. G. (1984). *Linear and Nonlinear Programming*, Addison-Wesley, Reading, Mass.

Menzel, D. W. (1974). Primary productivity, dissolved and particulate organic matter, and the sites of oxidation of organic matter, in *The Sea, Vol. 5*, ed. M.N. Hill, pp. 659–678, Interscience, New York.

Schlitzer, R. (1988). Modeling the nutrient and carbon cycles of the north Atlantic. Part I: Circulation, mixing coefficients, and heat fluxes, *J. Geophys. Res.*, in press.

Sugimura, Y., and Y. Suzuki (1987). A high temperature catalytic oxidation method of non-volatile dissolved organic carbon in seawater by direct injection of liquid sample, *Mar. Chem.*, in press.

Wagner, H. M. (1969). *Principles of Operations Research: With Applications to Managerial Decisions*, 937 pp., Prentice-Hall, Englewood Cliffs, N.J.

THE DESIGN OF NUMERICAL MODELS OF THE OCEAN CIRCULATION

K. Bryan

Geophysical Fluid Dynamics Laboratory/NOAA
Princeton University, P.O. Box 308
Princeton, N.J. 08542
U.S.A.

ABSTRACT

The design of numerical models is introduced through an analysis of the stability of finite difference approximations of the shallow-water equations. An outline of nonlinear instability, and methods to control it, is followed by a discussion of vertical coordinate systems, and examples of the application of models of various design to the investigation of the large scale potential vorticity pattern in the main thermocline.

1. FUNDAMENTALS OF MODEL DESIGN

A widely held viewpoint has been that the relatively sparse data sets available in oceanography did not justify the development of detailed numerical models similar to those used for research and applications in meteorology. Recently, however, there has been a growing appreciation of the power of computer simulation in nearly all fields of science, and this trend has influenced oceanography as well. Special properties of the ocean put severe demands on models which are based on the fundamental equations of fluid motion. One property is the very wide span of energy-containing spatial scales that are important for transport and mixing. These

465

D. L. T. Anderson and J. Willebrand (eds.), Oceanic Circulation Models: Combining Data and Dynamics, 465–500.
© *1989 by Kluwer Academic Publishers.*

scales range from the diameter of mesoscale eddies to the dimensions of ocean basins. A second property that put special demands on models is the very great range of time-scales that are important in ocean circulation, which range from the two to three week time-scale of mesoscale eddies, to the thousand year time-scale for the ventilation of the deep Pacific.

The numerical methods that have been used in models of ocean circulation are not unique to oceanography. Nevertheless it may be useful to gather some of this material, scattered through the literature, in a more convenient form. The reader is referred to 'Numerical Models of Ocean Circulation' (National Acad. Science, U.S., 1975) and a more recent work, 'Advanced Physical Oceanographic Numerical Modeling' (O'Brien, 1986). The goal of these lectures is not to make a comprehensive survey, but only introduce some of the concepts needed to understand the design of the different ocean circulation models.

1.1 Linearized Equations

In the preliminary design stage, linear theory is a convenient guide, keeping in mind that the ultimate goal is a very general model that can be used to simulate the nonlinear behaviour of the ocean circulation as well. Let us consider the appropriate equations for small perturbations in a uniformly stratified ocean at rest, ignoring friction and diffusion. Let $\mathbf{u}$ be the horizontal velocity, w the vertical velocity, p the pressure and N the Brunt-Vaissalla frequency, where

$$N^2 = -g\partial_z\bar{\rho}/\rho_0 \qquad (1.1)$$

$\bar{\rho}$ is the horizontally averaged density, and b is the buoyancy defined as

$$b = -(\rho - \bar{\rho})g \qquad (1.2)$$

f is the Coriolis parameter. With this notation the governing linearized equations are:

$$\partial_t\mathbf{u} + f\mathbf{k} \times \mathbf{u} + \nabla p/\rho_0 = 0 \qquad (1.3)$$

$$b - \partial_z p/\rho_0 = 0 \qquad (1.4)$$

$$\nabla.\mathbf{u} + \partial_z w = 0 \qquad (1.5)$$

$$\partial_t b + N^2 w = 0 \qquad (1.6)$$

If there were no salt in the ocean, buoyancy would be simply a measure of temperature and N^2 a measure of the vertical temperature gradient.

1.2 Decomposition into Vertical Modes

A great simplification can be achieved by making a separation of variables, and expanding in terms of the vertical modes (Moore and Philander, 1977). Let $\mathbf{u}_n$ and h_n be functions of x, y and t and Z_n be a function of z only.

$$(\mathbf{u}, p/\rho_0) = \sum_{n=1}^{\infty} (\mathbf{u}_n, gh_n) Z_n \tag{1.7}$$

From Eqs. (1.3)–(1.6) the condition for separability is,

$$d_z \left(\frac{(d_z Z)}{N^2} \right) + \frac{Z}{gH_n} = 0$$

where H_n^{-1} is the eigenvalue for each vertical mode. The governing equations for u_n and h_n are,

$$\partial_t \mathbf{u}_n + f\mathbf{k} \times \mathbf{u}_n + g\nabla h_n = 0 \tag{1.8}$$

$$\partial_t h_n + H_n \nabla . \mathbf{u}_n = 0 \tag{1.9}$$

Eqs. (1.8) and (1.9) represent the governing equations for each mode. They correspond to the tidal equations for a basin of uniform depth, H_n. Many of the technical problems associated with the design of an ocean circulation model can be analyzed in terms of these very simple equations.

1.3 Dispersion Diagrams

Any two dimension velocity vector can be represented by a stream function ψ and velocity potential χ.

$$v = \partial_x \psi + \partial_y \chi \tag{1.10}$$

$$u = -\partial_y \psi + \partial_x \chi \tag{1.11}$$

Substitution of (1.10) and (1.11) in (1.8) and (1.9) gives,

$$\partial_t \nabla^2 \psi + f\nabla^2 \chi + \beta(\partial_x \psi + \partial_y \chi) = 0 \tag{1.12}$$

$$\partial_t \nabla^2 \chi - f\nabla^2 \psi - \beta(\partial_y \psi - \partial_x \chi) = -g\nabla^2 h \tag{1.13}$$

$$\partial_t h + H\nabla^2 \chi = 0 \tag{1.14}$$

where

$$\beta = \partial_y f$$

First let us consider the case of uniform rotation, $\beta = 0$. Let

$$\psi, \chi \sim \exp[i(k_x + \ell_y - \omega t)] \tag{1.15}$$

Substitution into (1.12)–(1.14) gives

$$\omega^2 = f^2 + gH_n(k^2 + \ell^2) \tag{1.16}$$

For very low wave number we have pure inertial motion. At very high wave numbers we have shallow-water gravity waves.

At low frequencies the flow is in near geostrophic balance so that the stream function approximately coincides with contours of constant depth. In that case (1.12)–(1.14) can be approximated as

$$\partial_t \nabla^2 \psi + f \nabla^2 \chi + \beta \partial_\chi \psi = 0 \tag{1.17}$$

$$-f \nabla^2 \psi + g \nabla^2 h = 0 \tag{1.18}$$

$$\partial_t h + H \nabla^2 \chi = 0 \tag{1.19}$$

If both f and β are taken to be constants, the corresponding dispersion equation is

$$\omega = -\beta k \left(k^2 + \ell^2 + \frac{f^2}{gH_n} \right)^{-1} \tag{1.20}$$

A dispersion diagram is given in Fig. 1.1. Since the frequencies are well separated, Eq. (1.16) can be used to calculate the high frequencies and Eq. (1.20) to calculate the low frequencies with little loss of accuracy. (gH_n/f) is the radius of deformation corresponding to the nth vertical mode. The radius of deformation defines a horizontal scale. For scales larger than the radius of deformation divergence effects decrease in importance in determining wave speed.

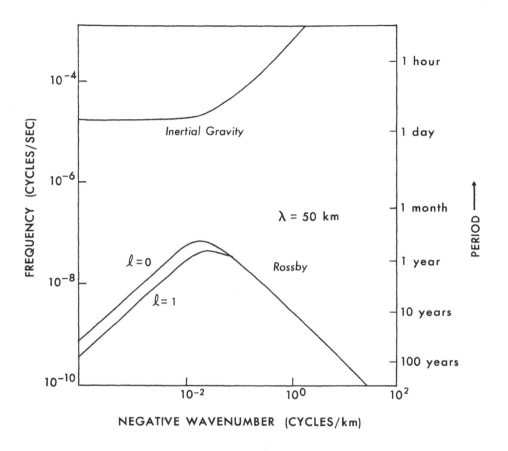

Figure 1.1
Dispersion diagram for the beta-plane at midlatitudes for a single vertical
mode with a radius of deformation of 50 km. The abscissa is the negative
east-west wave number.

1.4 Staggered Grids

Let us consider a simple, one-dimensional gravity wave,

$$\partial_t u + g\partial_x h = 0 \tag{1.21}$$

$$\partial_t h + H\partial_x u = 0 \tag{1.22}$$

Writing these equations in terms of second order finite difference equations

$$u_i^{n+1} - u_i^{n-1} + \frac{g\Delta t}{\Delta x}(h_{i+1} - h_{i-1})^n = 0 \tag{1.23}$$

$$h_i^{n+1} - h_i^{n-1} + \frac{H\Delta t}{\Delta x}(u_{i+1} - u_{i-1})^n = 0 \tag{1.24}$$

Imagine a checkerboard in a space defined by the indices n and i. In the 'red' squares (not underlined) the sum of $i + n$ would be an even number and the 'black' squares (underlined) $i + n$ would be an odd number.

4	<u>uh</u>	uh	<u>uh</u>	uh
3	uh	<u>uh</u>	uh	<u>uh</u>
2	<u>uh</u>	uh	<u>uh</u>	uh
1	uh	<u>uh</u>	uh	<u>uh</u>
n/i	1	2	3	4

Note that Eqs. (1.23) and (1.24) only establish relationships of the following kind:

i) Variables are only related to other variables in the same color square, either red or black.

ii) u's and h's in the same color square are only linked at alternate time steps. These rules define 4 independent lattices

```
        |  -   h   -   h              |  -   u   -   u
        |  u   -   u   -              |  h   -   h   -
     t  |  -   h   -   h           t  |  -   u   -   u
        |  u   -   u   -              |  h   -   h   -
        |_____                |_____
              x                             x

        |  h   -   h   -              |  u   -   u   -
        |  -   u   -   u              |  -   h   -   h
     t  |  h   -   h   -           t  |  u   -   u   -
        |  -   u   -   u              |  -   h   -   h
        |_____                |_____
              x                             x
```

This illustrates that little is achieved by defining all variables in the primitive equations at the same lattice point. Staggered grids on the $x - y$ plane are a more efficient use of computer resources. Two possible forms are the Arakawa 'B' and 'C' grids illustrated below.

'B' Grid   'C' Grid

	-	h	-	h			u	h	u	h	u
	u,v	-	u,v	-			-	v	-	v	-
t	-	h	-	h		t	u	h	u	h	u
	u,v	-	u,v	-			-	v	-	v	-
		x							x		

To represent the finite difference form of the linearized shallow-water equations, let us define the following operators,

$$\delta_x(\ ) = \left[(\ )_{i+\frac{1}{2}} - (\ )_{i-\frac{1}{2}} \right] / \Delta \tag{1.25}$$

$$\overline{(\ )}^x = \left[(\ )_{i+\frac{1}{2}} + (\ )_{i-\frac{1}{2}} \right] / 2 \tag{1.26}$$

On the 'B' grid, Eqs. (1.8) and (1.9) can be written,

$$\partial_t u - fv + g\delta_x \bar{h}^y = 0 \tag{1.27}$$

$$\partial_t v + fu + g\delta_y \bar{h}^x = 0 \tag{1.28}$$

$$\partial_t h + H(\delta_x \bar{u}^y + \delta_y \bar{v}^x) = 0 \tag{1.29}$$

On the 'C' grid

$$\partial_t u - f\bar{v}^{xy} + g\delta_x h = 0 \tag{1.30}$$

$$\partial_t v + f\bar{u}^{xy} + g\delta_y h = 0 \tag{1.31}$$

$$\partial_t h + H(\delta_x u + \delta_y v) = 0 \tag{1.32}$$

2. STABILITY

2.1 Inertia-Gravity Waves in 'B' and 'C' Grids

Mesinger and Arakawa (1976) define two different goals in numerical modelling of atmospheric flows. One is to correctly represent the geostrophic adjustment process, and the other is to correctly represent time-dependent quasi-geostrophic flow. These goals are also completely appropriate for the ocean circulation problem.

To analyze the geostrophic adjustment process, let us consider the simple case of uniform rotation. For convenience let

$$x_m = m\Delta, \quad m = 1, 2, ...M$$
$$y_n = n\Delta, \quad n = 1, 2, ...N$$

be the positions of grid points. Assuming a solution of the form

$$u, v = (U, V) \exp[i(km\Delta + \ell n\Delta - \omega t)] \tag{2.1}$$

Then the $\delta_x(\ \ )$ and $\overline{(\ \ )}^x$ operators defined by Eqs. (1.25) and (1.26) become,

$$\delta_x u = -(2iu \sin X)/\Delta \tag{2.2}$$

and

$$\bar{u}^x = u \cos X \tag{2.3}$$

where

$$X = k\Delta/2 \tag{2.4}$$

Likewise

$$\delta_y u = -(2iu \sin Y)/\Delta \tag{2.5}$$

and

$$\bar{u}^y = u \cos Y \tag{2.6}$$

where

$$Y = \ell\Delta/2 \tag{2.7}$$

Substituting (2.1)–(2.7) into the equation for Arakawa's 'B' grid (1.27)–(1.29), we obtain

$$\omega_B^2 = f^2 + \left(\frac{2f\lambda}{\Delta}\right)^2 (\sin^2 X \cos^2 Y + \cos^2 X \sin^2 Y) \tag{2.8}$$

where $\lambda^2 = gH/f^2$, the radius of deformation squared.

The equivalent expression for the 'C' grid is

$$\omega_C^2 = (\cos X \cos Y)^2 f^2 + \left(\frac{2f\lambda}{\Delta}\right)^2 (\sin^2 X + \sin^2 Y) \tag{2.9}$$

For comparison the continuous case in the same notation is

$$\omega^2 = f^2 + \left(\frac{2f\lambda}{\Delta}\right)^2 (X^2 + Y^2) \tag{2.10}$$

Plots of $|\omega|/f$ for the continuous case and grids 'B' and 'C' are shown in Fig. 2.1 for $\lambda/\Delta = 2$ (radius of deformation larger than Δ) and in Fig. 2.2 for $\lambda/\Delta = 0.2$ (radius of deformation smaller than Δ). The gradients of $|\omega|/f$ in the X and Y plane are a measure of the group velocity of gravity waves which are essential for geostrophic adjustment. A serious distortion of the geostrophic adjustment process exists where the group velocity goes to zero. For $\lambda/\Delta > 1$ the 'C' grid is clearly superior to the 'B' grid which has zero group velocity along the diagonal at $X = \pi/4$, $Y = \pi/4$.

On the other hand, for a grid size larger than the radius of deformation ($\lambda/\Delta < 1$) Fig. 2.2 shows that the 'B' grid gives much more accurate results along the x- and y-axis than the 'C' grid.

2.2 Filtering

The linearized shallow-water equations in nearly geostrophic balance can be represented by

$$\partial_t \nabla^2 \psi + f\nabla^2 \chi + \beta\partial_x \psi = 0 \tag{2.11}$$

$$-f\nabla^2 \psi + g\nabla^2 h = 0 \tag{2.12}$$

$$\partial_t h + H\nabla^2 \chi = 0 \tag{2.13}$$

corresponding to Eqs. (1.1)–(1.19). This formulation eliminates high frequency inertial-gravity waves through filtering. By elimination,

$$\partial_t \left(\nabla^2 \psi - (f^2/gh)\psi\right) + \beta\partial_x \psi = 0 \tag{2.14}$$

The quasi-geostrophic filtering approach is very useful for theoretical insight, but it has disadvantages as well. In the case of both the atmosphere and the oceans, the global circulation in the equatorial regions are best represented by the more complete primitive equations.

474

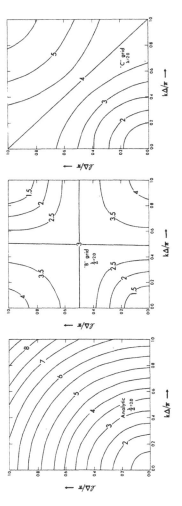

Figure 2.1

The frequency normalized by the earth's rotation, $|w|/f$, (a) the continuous case, (b) the 'B' grid, and (c) the 'C' grid. For $\lambda/\Delta = 2$.

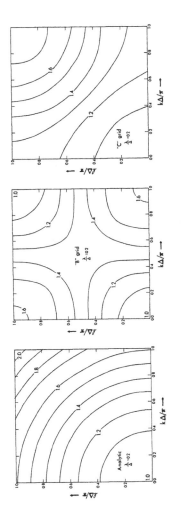

Figure 2.2

Same as (2.1), but for $\lambda/\Delta = 0.2$.

2.3 Implicit Treatment

Filtered equations eliminate unwanted high frequencies and allow a much longer time step. For practical purposes the same longer time step can be achieved for the more general primitive equations by implicit methods. Although implicit methods do not provide the theoretical insight of filtering, they have the advantage of very great flexibility. In a single formulation one can have the advantages of a very accurate model by using very short time steps, or a much more efficient, but less accurate model by using very long time steps. The disadvantage of implicit methods is that they may involve second-order partial differential equations which are time-consuming to solve in the complicated geometry of the World Ocean.

2.3.1 Inertial Motion — An Example

To illustrate implicit methods consider pure inertial motion,

$$\partial_t u - fu = 0 \tag{2.15}$$

$$\partial_t v + fv = 0 \tag{2.16}$$

Introducing the complex variable, w, where

$$w = u + iv \tag{2.17}$$

Then Eqs. (2.15) and (2.16) may be written,

$$\partial_t w - ifw = 0 \tag{2.18}$$

Using centered differences with respect to time,

$$w^{n+1} - w^{n-1} - 2(\Delta t)ifw^n = 0 \tag{2.19}$$

This is sometimes referred to as 'leap frog' differencing in the time domain. We can analyze this scheme with the Von Neumann method (Richtmyer and Morton, 1967, p. 70).

$$w^n = \theta^n \tag{2.20}$$

and

$$\theta^2 - 2(\Delta t)if\theta - 1 = 0 \tag{2.21}$$

$$\theta = (\Delta t)fi \pm (1 - (\Delta t)^2 f^2)^{\frac{1}{2}} \tag{2.22}$$

$$\theta^2 = 1 - 2(\Delta t)^2 f^2 \tag{2.23}$$

476

The Von Neumann necessary condition for stability requires that

$$|\theta| < 1 \tag{2.24}$$

therefore

$$\Delta t f < 1 \tag{2.25}$$

This places an absolute restriction on the size of the time-step to be less than some fraction of a pendulum day. For low resolution models of the ocean circulation this restriction may be very limiting.

An implicit treatment of inertial motion (2.18) is

$$w^{n+1} - w^n - \frac{i(\Delta t)f}{2}(w^{n+1} + w^n) = 0 \tag{2.26}$$

The Von Neumann approach gives

$$\theta - 1 - \frac{i(\Delta t)f}{2}(\theta + 1) = 0 \tag{2.27}$$

$$\theta = \frac{2 + i(\Delta t)f}{2 - i(\Delta t)f} \tag{2.28}$$

$$\theta = \frac{4 + 4i(\Delta t)f - (\Delta t f)^2}{4 + (\Delta t f)^2} \tag{2.29}$$

$$|\theta^2| = 1 \tag{2.30}$$

Note that $|\theta| = 1$ for all values of Δt.

2.3.2 Gravity Waves — A Second Example

As a second example let us compare explicit and implicit treatment of gravity waves. Consider the simple model,

$$\partial_t u + g\partial_x h = 0 \tag{2.31}$$

$$\partial_t h + H\partial_x u = 0 \tag{2.32}$$

Let

$$u, h = (u^n, h^n)e^{ikx} \tag{2.33}$$

An explicit, forward difference representation is

$$u^{n+1/2} - u^{n-1/2} + ig\Delta t k h^n = 0 \tag{2.34}$$

$$h^{n+1} - h^n + iH\Delta t k u^{n+1/2} = 0 \tag{2.35}$$

Combining

$$h^{n+1} - 2h^n + h^{n-1} + gH(\Delta t)^2 k^2 h^n = 0 \tag{2.36}$$

Let

$$\lambda = \frac{gH(\Delta t)^2 k^2}{2} - 1$$

Then

$$\theta^2 + 2\lambda\theta + 1 = 0$$

$$\theta = -\lambda \pm (\lambda^2 - 1)^{\frac{1}{2}}$$

$$|\theta| < 1, \quad \text{if} \quad \lambda < 1$$

Therefore

$$\frac{\sqrt{gH}\Delta t k}{2} < 1 \tag{2.37}$$

The restriction on the maximum allowable Δt can be removed by going to an implicit formulation. For example,

$$u^{n+1} - u^n + \frac{ikg\Delta t}{2}(h^{n+1} + h^n) = 0 \tag{2.38}$$

$$h^{n+1} - h^n + \frac{ikH\Delta t}{2}(u^{n+1} + u^n) = 0 \tag{2.39}$$

Combining these equations, we can obtain the finite difference form of the wave equation,

$$h^{n+1} - 2h^n + h^{n-1} + \frac{gH(\Delta t)^2 k^2}{4}(h^{n+1} + 2h^n + h^{n-1}) = 0 \tag{2.40}$$

Let

$$\lambda = (gH(\Delta t)^2 k^2)/4$$

Using the Von Neumann method,

$$\theta^2 + 2\frac{(\lambda - 1)}{(\lambda + 1)}\theta + 1 = 0$$

$$\theta = \frac{\lambda - 1}{\lambda + 1} \pm \left[\left(\frac{\lambda - 1}{\lambda + 1}\right)^2 - 1\right]^{\frac{1}{2}}$$

Since $[(\lambda - 1)/(\lambda + 1)]$ is always less than one, θ is complex and

$$|\theta^2| = 1 \tag{2.41}$$

478

Note that the implicit scheme is unconditionally stable.

The next question concerns the accuracy as a function of Δt. Assume a solution of the form,

$$h^n = \hat{h} \exp(i\omega n \Delta t)$$

Substituting in (2.40)

$$\cos(\omega \Delta t) - 1 + \lambda[\cos(\omega \Delta t) + 1] = 0$$

$$\omega \Delta t = \cos^{-1} \left(\frac{1 - \lambda}{1 + \lambda} \right) \qquad (2.42)$$

Let c be the phase speed, ωk^{-1}. In Fig. 2.3 $c/\sqrt{gH}$ is plotted as a function of λ. Note that the accuracy of the wave speed falls off rapidly as Δt is increased beyond the explicit limit given by Eq. (2.37). This will artificially slow up the geostrophic adjustment process.

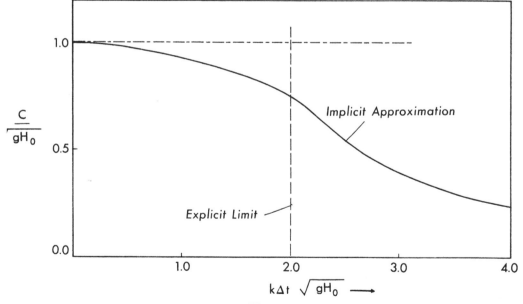

Figure 2.3
The implicit approximation of a gravity wave as a function of time step.

3. STABILITY OF ADVECTIVE SCHEMES
AND VERTICAL COORDINATES

3.1 Time Differencing Continued: A Prototype Equation

Consider the following equation,

$$\partial_t u + c\partial_z u = A\partial_z\partial_z u \tag{3.1}$$

This example contains terms representing both advection and diffusion.

Let us use a simple forward differencing scheme. It is completely explicit. We specify that $\Delta x = \Delta$.

$$u_j^{n+1} = u_j^n - \frac{c\Delta t}{2\Delta}(u_{j+1}^n - u_{j-1}^n) + (A\Delta t/\Delta^2)(u_{j+1}^n - 2u_j^n + u_{j-1}^n) \tag{3.2}$$

Substituting

$$u_j^n = \theta^n \exp(ikj\Delta)$$

we obtain

$$\theta = 1 - i\lambda - \alpha$$

where

$$\lambda = \frac{c\Delta t}{\Delta}\sin(k\Delta)$$

and

$$\alpha = (2A\Delta t/\Delta)[1 - \cos(k\Delta)]$$

Thus

$$|\theta^2| = (1 - \alpha)^2 + \lambda^2 \tag{3.3}$$

Without mixing, that is $\alpha \to 0$, the forward scheme is unconditionally unstable. Another possibility is to use centered differencing with respect to time.

$$u^{n+1} = u_j^{n-1} - \frac{c\Delta t}{\Delta}(u_{j+1}^n - u_{j-1}^n) + \frac{2A\Delta t}{\Delta^2}(u_{j+1}^n - 2u_j^n + u_{j-1}^n) \tag{3.4}$$

In this case

$$\theta^2 + (2\lambda i + 2\alpha)\theta - 1 = 0 \tag{3.5}$$

$$\theta = -(\alpha + i\lambda) \pm (\alpha^2 + 2i\alpha\lambda - \lambda^2 + 1)^{\frac{1}{2}} \tag{3.6}$$

Let $\alpha \to 0$ (no diffusion)

$$|\theta^2| = 1 - 2\lambda^2, \quad -1 \tag{3.7}$$

or

$$\frac{c\Delta t}{\Delta} < 1 \tag{3.8}$$

We have recovered the C.F.L. condition. For $\lambda \to 0$ (no advection)

$$|\theta^2| = 2\alpha^2 + 1 \pm 2\alpha(\alpha^2 + 1)^{\frac{1}{2}} \tag{3.9}$$

Note that the pure diffusion case is unconditionally unstable with centered differencing.

Thus forward differencing with respect to time is unstable for the advective term, and 'leap frog' differencing is unstable for the diffusive term. This suggests a mixed approach — centered differencing for the advective term and forward differencing for the diffusive term.

$$u^{n+1} = u_j^{n-1} - \frac{c\Delta t}{\Delta}(u_{j+1}^n - u_{j-1}^n) + \frac{2A\Delta t}{\Delta^2}(u_{j+1}^{n-1} - 2u_j^{n-1} + u_{j-1}^{n-1}) \tag{3.10}$$

The corresponding equation for the amplification factor is,

$$\theta^2 + 2\lambda i\theta + 2\alpha - 1 = 0$$
$$\theta = -i\lambda \pm (-\lambda^2 + 1 - 2\alpha)^{\frac{1}{2}}$$

and

$$|\theta^2| = 1 - 2\lambda^2 - 2\alpha.$$

In the pure advection case

$$\frac{c\Delta t}{\Delta} < 1$$

and in the pure diffusion case

$$\frac{A\Delta t}{\Delta} < \frac{1}{2}.$$

Although the form given by Eq. (3.10) requires two time levels to be stored during the calculation, the stability of the advective term makes it desirable. No artificial damping is required for linear stability. This is the preferred form for models, and is widely used in models with explicit treatment of advection and diffusion.

3.2 Nonlinear Instability

So far we have shown how the Von Neumann method can be used to analyze the stability of linear finite difference equations with constant coefficients. This analysis provides a necessary condition for stability, but it is not sufficient for the case of non-constant coefficients or nonlinear equations. 'Nonlinear instability' was first noted by Norman Phillips (1956, 1959). To illustrate this type of instability consider the simple conservation equation

$$\frac{d\theta}{dt} = 0 \tag{3.11}$$

Over a finite, closed domain, all moments of θ should be conserved. If only a discrete number of fourier components are allowed, all moments in the truncated wave number space will also be conserved. The difficulty with finite difference formulas is that they do not provide an exact truncation in wave number space. This can lead to an artificial build-up of energy through aliasing.

Consider the following example, where

$$u = \sin kx \tag{3.12}$$

$$\alpha = \sin kx$$

$$u\partial_x\alpha = k\sin kx \cos kx = \tfrac{1}{2}k\sin 2kx \tag{3.13}$$

If we approximate,

$$u\partial_x\alpha = u\delta_x\bar{\alpha}^x \tag{3.14}$$

Then

$$u\delta_x\bar{\alpha}^x = \sin kx_j \cos kx_j[\sin(k\Delta) - \sin(-k\Delta)]/(2\Delta)$$

$$= \sin(2kx_j)\sin(k\Delta)/(2\Delta) \tag{3.15}$$

To illustrate aliasing, we can write

$$\sin(2kx_j) = \sin\left(2\frac{\pi x_j}{\Delta} - \left(\frac{2\pi}{\Delta} - 2k\right)x_j\right) \tag{3.16}$$

Expanding the right hand side

$$= \sin\frac{2\pi}{\Delta}x_j \cos\left(\frac{2\pi}{\Delta} - 2k\right)x_j - \cos\frac{2\pi}{\Delta}x_j \sin\left(\frac{2\pi}{\Delta} - 2k\right)x_j$$

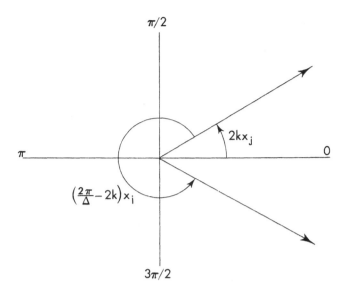

Figure 3.1
A diagram illustrating aliasing.

Since $x_j = j\Delta$

$$\sin(2kx_j) = -\sin[(2\pi - 2k\Delta)j] \tag{3.17}$$

Thus a finite grid cannot distinguish between $2k$ and the much larger wave number $2\pi/\Delta - 2k$. The wave number $2k$ might be a resolved wave number and $2\pi/\Delta - 2k$ beyond the resolvable range. Fig. 3.1 shows this in terms of angles.

The result is that an increase of amplitude that should be assigned to $(2\pi/\Delta - 2k)$ will erroneously be assigned to the smaller wave number $2k$ with the erroneous coefficient $\sin(2\pi - 2k\Delta)$ instead of $\sin(2k\Delta)$ as in Eq. (3.15).

3.3 A Variance Conserving Form

Experience has shown that nonlinear instability can be avoided by either,

1) Filtering wave numbers $k > \pi/\Delta$; or

2) Adopting a variance conserving form (Arakawa Method).

Precise filtering is rather easy to do in an idealized geometry and may be the easiest thing to do for some process studies. For the general case, however, the variance conserving forms are preferable. To illustrate this approach consider

$$\frac{d\theta}{dt} = 0, \quad \nabla.\mathbf{u} = 0 \tag{3.18}$$

Let a closed volume be divided into smaller discrete cells. For each cell,

$$\int \int \int \partial_t \theta \, dx \, dy \, dz = - \int \int \mathbf{u}\theta.\mathbf{n} \, dS \tag{3.19}$$

where $\mathbf{n}$ is a unit vector normal to the surface of the volume. The continuity equation is simply

$$\int \int \mathbf{u}.\mathbf{n} \, dS = 0 \tag{3.20}$$

A discrete approximation of Eqs. (3.19) and (3.20) is

$$V_j \partial_t \theta_j = - \sum_i^I \tfrac{1}{2} U_i (\theta_j + \theta_i) A_i \tag{3.21}$$

and

$$\sum_i^I U_i A_i = 0 \tag{3.22}$$

where V_j is the volume of the cell and A_i is the one of the $i = 1, 2, ...I$ surfaces which bound the cell. $U_i A_i$ is the total flow moving outward through the ith surface. Summing up over the J cells that form the entire volume we can see that

$$\sum_j^J V_j \theta_j = \text{constant}$$

since the entire volume is closed and individuals cells will simply exchange equal concentrations of θ with each other.

We investigate the variance by multiplying (3.21) through by θ_j and summing over j.

$$\sum_j^J \tfrac{1}{2} V_j \partial_t \theta_j^2 = - \sum_j^J \sum_i^I \tfrac{1}{2} U_i (\theta_j^2 + \theta_i \theta_j) A_i$$

This can be rewritten,

$$\sum_j^J \tfrac{1}{2} V_j \partial_t \theta_j^2 = - \sum_j^J \tfrac{1}{2} \left[\theta_j^2 \sum_i^I U_i A_i + \sum_i^I U_i A_i \theta_i \theta_j \right]$$

Using the continuity equation in discrete form (3.22) the first term on the right hand side vanishes. The second term vanishes because there is an equal and opposite contribution on either side of the cell interfaces. A flux of variance of this term is exactly balanced by a flux into an adjacent cell. Therefore

$$\sum_j^J \tfrac{1}{2} V_j \theta^2 = \text{constant}$$

We have shown that when conservation equations for a tracer in an incompressible medium are put in discrete form, it is possible to design the advection in such a way that certain constraints are maintained. In particular, if a closed volume is divided into a finite number of cells, the sum of the volume of the cells multiplied by the average concentration in each cell will remain constant, aside from errors caused by time-differencing. In addition the advection between cells can be represented in such a way that the sum of the square of the concentration in each cell multiplied by the volume will also be conserved. Note that the result is quite independent of the particular geometry or arrangement of the cells.

3.4 First and Second Order Advection Schemes

Up to this point we have discussed the instabilities arising in the numerical integration of initial value problems. In discussing Phillips's (1956) nonlinear instability we considered the case in which the solution is not accurately resolved by the numerical grid. This problem is almost universal in modelling the ocean circulation. Even though a scheme may be stable in the sense that the total variance is not increasing with respect to time, the finite difference solution will not represent a good approximation to the continuous solution, if unresolved two grid-point waves dominate. In this section we will consider a simple one-dimensional advection diffusion problem and find the minimum level of diffusion required to suppress the two grid-point waves.

Let us consider a simple one dimensional, steady state model of advection and diffusion,

$$U\partial_x \alpha - A\partial_x \partial_x \alpha = 0. \tag{3.23}$$

A second order-accurate finite difference scheme would be,

$$\frac{U}{2\Delta}(\alpha_{j+1} - \alpha_{j-1}) - \frac{A}{\Delta^2}(\alpha_{j+1} - 2\alpha_j + \alpha_{j-1}) = 0 \tag{3.24}$$

Let

$$R = U\Delta/2A \tag{3.25}$$

and

$$\alpha_j = \theta^j \tag{3.26}$$

Here θ^j plays the same role as θ^n in the initial value problems considered earlier.

Substituting in (3.23)

$$(R - 1)\theta^2 + 2\theta - (R + 1) = 0 \tag{3.27}$$

and

$$\theta = \frac{-1 \pm R}{R - 1} \tag{3.28}$$

$$= 1, \quad \frac{1 + R}{1 - R} \tag{3.29}$$

To avoid a two grid-point oscillation, both roots must be positive.

$$\theta > 0$$

Thus

$$R < 1 \quad \text{or} \quad \frac{U\Delta}{A} < 2 \tag{3.30}$$

$U\Delta/A$ may be thought of as a grid point Péclèt number. To suppress the 2 grid-point wave requires that the grid-point Péclèt number be less than two. We have only considered the one dimensional case. In practice more dimensions allow this criterion to be relaxed.

Let us now consider the case in which the advection term is represented by only first order differencing. Eq. (3.23) can be written as,

$$\left(\frac{U + |U|}{2\Delta}\right)(\alpha_j - \alpha_{j-1}) + \left(\frac{U - |U|}{2\Delta}\right)(\alpha_{j+1} - \alpha_j) - \left(\frac{A}{\Delta^2}\right)(\alpha_{j+1} - 2\alpha_j + \alpha_{j-1}) = 0 \tag{3.31}$$

Rearranging terms,

$$R'(\alpha_{j+1} - \alpha_{j-1}) - (\alpha_{j+1} - 2\alpha_j + \alpha_{j-1}) = 0 \tag{3.32}$$

where

$$R' = \frac{U\Delta}{2A + \Delta|U|} \tag{3.33}$$

Replacing R by R' in Eqs. (3.27)–(3.29) we find that a two grid-point wave is avoided if,

$$R' < 1 \tag{3.34}$$

or

$$\frac{U\Delta}{A} < 2 + \frac{\Delta|U|}{A} \tag{3.35}$$

This result tells us that the grid point Péclèt number can be infinitely large without the two-grid point wave being excited. On the other hand, Eq. (3.32) shows us that the first order scheme itself imposes a diffusion which limits effective grid point Péclèt number to the same critical value as obtained for centered differencing in Eq. (3.30).

3.5 Choice of Vertical Coordinate

There are three categories of vertical coordinate that have become popular for ocean circulation modelling. We will introduce these three systems and very briefly described their advantages and disadvantages.

3.5.1 z-Coordinate System

Perhaps the simplest coordinate system of all is based on cartesian coordinates in all three dimensions. Usually the grid size in the vertical direction is nonuniform to allow the highest resolution near the surface, and somewhat less resolution below the thermocline. A disadvantage of this representation is that the depth of the ocean is discrete, so that even slopes are represented by a series of steps. One way to alleviate this problem is by giving the lowest cell a variable depth, and this correction may be quite important in modelling high frequency barotropic waves, which are very sensitive to the configuration of the deep ocean. Another way is to use much higher resolution in the vertical, but this, of course, increases the expense of the calculation.

One of the major disadvantages of the z-coordinate system is in the treatment of horizontal mixing of momentum, heat and salinity. In the ocean, mixing by mesoscale eddies tends to be along surfaces of constant density. In regions near the western boundary isopycnal surfaces are strongly sloped, and horizontal mixing implies strong cross-isopycnal mixing. Veronis (1975) was the first to point out that horizontal mixing in regions of tilted density surfaces can cause a strong spurious upwelling in numerical models. This problem can be alleviated by going to a high enough resolution so that mesoscale eddies can be included explicitly. In that case the horizontal mixing of the model can be greatly reduced, and cross-isopycnal mixing will be correspondingly diminished. Another approach is to rotate the mixing tensor so that mixing is primarily aligned with isopycnal surfaces. This formulation requires that there be some gradients of salinity within density surfaces, a condition which is nearly always fulfilled in the real ocean. An advantage of aligning the mixing tensor along isopycnal surfaces is that the vertical component of mixing automatically becomes larger as density surfaces become very steep, and there is no requirement for a special parameterization of convection in the model (Cox, 1987).

3.5.2 Depth-Normalized Coordinate

In most meteorological models used in numerical weather prediction, pressure normalized by surface pressure is used as the vertical coordinate. The analogue in ocean circulations models is a vertical coordinate based on the depth normalized by the depth of the bottom, z/H. The advantage of this coordinate is that it allows a good resolution of the bottom boundary layer, and a more accurate representation of overflows that are so important for deep water formation. This coordinate has been used for the most part in high resolution models of near-shore circulation. One disadvantage of this coordinate system are the errors that can arise in calculating the pressure force term. Another disadvantage of the z/H coordinate for open ocean circulation models is that the coordinates would be strongly tilted in the transition region from the continental shelves to the deep sea. Mixing along the coordinate surfaces could cause a spurious cross isopycnal mixing similar to that encountered in z-coordinate models. A tilted mixing tensor may turn out to be a useful solution to this problem in the z/H-coordinate system as it is in the case of the z-coordinate models.

3.5.3 Isopycnal Coordinates

Up to this point we have discussed fixed coordinates, stationary in space and time. It has been suggested that a coordinate based on isopycnal coordinates is a more natural one for the large scale ocean circulation. In the main thermocline trajectories are believed to be largely tangential to density surfaces. Except in regions of very strong mixing, trajectories crossing isopycnal surfaces would only exist to balance the slow downward mixing of heat and salt in the main thermocline. For non-eddy-resolving models the mixing due to mesoscale eddies is easily parameterized as a diffusion of potential vorticity and water mass properties along the density surfaces.

In other types of models the results must usually be interpolated to density surfaces in order to interpret the results in terms of advection and mixing of potential vorticity. An isopycnal model is already in a very convenient form for analysis, and thus provides greater insight than other models. The disadvantage is that a semi-Lagrangian coordinate is inherently more complex to work with than fixed coordinates. This is particularly true over complex topography. However, rapid progress is being made in the development of very general isopycnal ocean circulation models.

4. THE APPLICATION OF OCEAN GENERAL CIRCULATION MODELS

Up to this point the focus has been on the technical aspects of model design. We now turn to the application of models to the study of the ocean circulation, and in particular the potential vorticity distribution in the ocean. It will be evident that no single model will provide the same insight as a hierarchy of models of increasing resolution and physical complexity. As powerful computers become increasingly available, it is a challenge to modelers to provide the oceanographic community with a variety of well tested models that will be convenient to apply to the analysis of data sets, the design of field experiments or problems of climate.

4.1 Observations

Keffer (1985) has produced a very valuable analysis of the configuration of the thermocline and the distribution of potential vorticity in the World Ocean. σ_θ is defined as $(\rho - 1) \times 10^3$, where ρ is the density referred to surface pressure. Fig. 4.1 (a) and (b) show the depth of the $\sigma_\theta = 26.15$ surface in the Pacific and the potential vorticity between the $\sigma_\theta = 26.05\text{--}26.25$ surfaces. Relative to the Atlantic, the thermocline in the Pacific is quite shallow. The depth of the $\sigma_\theta = 26.15$ surface is dominated by two great depressions, corresponding to the subtropical gyres in the Northern and Southern Hemisphere. Both gyres are strongly intensified to the west where they are bounded by western boundary currents. The gross features of

Fig. 4.1 (a) appear to fit the predictions of linear, wind-driven theory (Pedlosky, 1987).

The interpretation of the potential vorticity pattern in Fig. 4.1 (b) is more challenging. In both hemispheres the eastern part of the ocean is dominated by a high potential vorticity tongue, which extends to the west, skirting around the equatorward flank of the subtropical gyres that are evident in Fig. 4.1 (a). The central part of the gyres are dominated by water with low relative vorticity which extends to the poleward outcrop. Keffer (1985) attributes these central low potential vorticity waters to the equatorial drift of water masses which have undergone wintertime convection near the poleward outcrop. The higher vorticity tongues to the east have quite a different water mass signature. The high vorticity tongue lies just below the salinity minimum in the North Pacific. In the South Pacific the high vorticity tongue in Fig. 4.1 (b) lies well above the salinity minimum.

The configuration of the base of the thermocline is shown in another map from Keffer (1985) in Fig. 4.2 (a). The depth of the $\sigma_\theta = 27.4$ surface is nearly constant over large areas with depressions associated with the subtropical gyres greatly shrunken and displaced poleward and westward. The areas in which the isopycnal surface is nearly flat correspond to regions of uniform north-south vorticity gradient as shown in Fig. 4.2 (b). Where the density surface still retains some relief, indicating geostrophic currents, there are only very weak gradients of potential vorticity. Note that there is almost no contrast in potential vorticity between the subtropical and subpolar gyres.

4.2 A Quasi-Geostrophic Model

As an example of an eddy-resolving quasi-geostrophic model we will briefly discuss the circulation patterns found by Holland et al. (1984). A quasi-geostrophic model linearizes the vertical stratification so that it cannot explicitly include the vertical pathways of the ventilation process. However, it is ideally suited for a description of certain effects of mesoscale eddies on the ocean circulation. Fig. 4.3 (a) shows the stream function at the third level below the surface, and the potential vorticity at the same level is shown in Fig. 4.3 (b). The flow is driven by a very simple, zonally uniform wind stress pattern. The wind stress has a simple sinusoidal distribution with maximum easterlies at the northern and southern boundaries and maximum westerlies midway in between. If the fluid were at rest there would be a simple uniform gradient of potential vorticity in the y-direction. Advection and mixing by

DEPTH OF ISOPYCNAL SURFACE
$\sigma_\theta = 26.15$

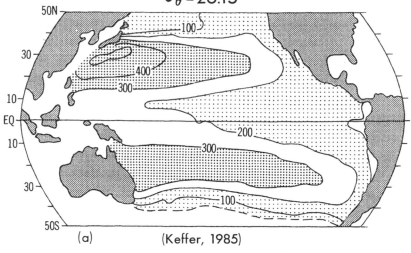

(a) (Keffer, 1985)

$-f\,\partial_z\rho$, $\sigma_\theta = 26.05 - 26.25$

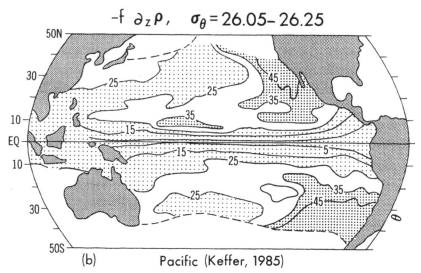

(b) Pacific (Keffer, 1985)

Figure 4.1
(a) Pressure on the $\sigma_\theta = 26.15$ surface in the Pacific. Units are decibars.
(b) Potential vorticity in the $\sigma_\theta = 26.05 - 26.25$ interval. Units are
$10^{-13}\,cm^{-1}\,sec^{-1}$. (From Keffer, 1985).

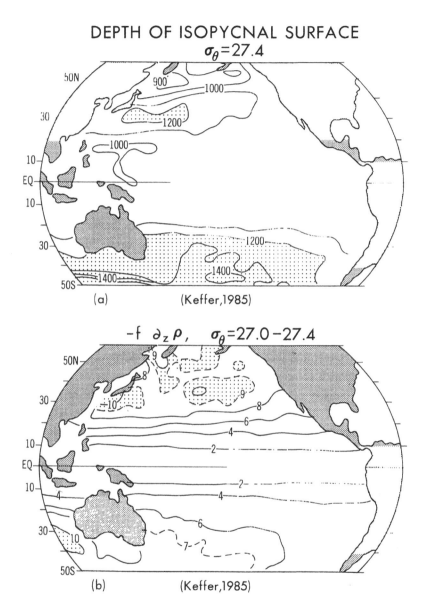

DEPTH OF ISOPYCNAL SURFACE
$\sigma_\theta = 27.4$

(a) (Keffer,1985)

$-f\ \partial_z\rho,\quad \sigma_\theta = 27.0-27.4$

(b) (Keffer,1985)

Figure 4.2
Same as Fig. 4.1 (a) Pressure, $\sigma_\theta = 27.4$ surface. (b) Potential vorticity, $\sigma_\theta = 27.0 - 27.3$ interval.

unstable baroclinic disturbances distort that pattern. The result of this stirring is to create a nearly uniform blob of potential vorticity in midlatitudes. The model appears to do an excellent job in simulating the potential vorticity of the Pacific shown in Keffer's potential vorticity map for the $\sigma_\theta = 27.0$ surface shown in Fig. 4.2 (b) which shows weak gradients of potential vorticity extending over a very large area which includes the subarctic as well as the subtropical gyre.

4.3 An Isopycnal Model

An example of an isopycnal model which was directly motivated by the ventilation theory of Luyten *et al.* (1983) is shown in Fig. 4.4 (a). The schematic representation in the y-z plane shows four layers in motion over a deep resting layer. The lower layers in motion have a fixed density, but a variable thickness. The upper mixed layer has a fixed depth, but the density is allowed to vary. The model is general enough to allow both a wind-driven and buoyancy-driven circulation, through fluxes imposed on the mixed layer. Water is exchanged between the mixed layer and the layer just below by mixing and by Ekman pumping. Water is exchanged between the moving layers below by a small cross-isopycnal mixing and a more vigorous, but local convective mixing near outcrops.

Convection is parameterized in the following way. Advection may cause a lower layer of variable depth to extend poleward, beyond the point where its density is greater than the density of the mixed layer above. This would create a statically unstable environment. If this takes place, a convective adjustment is made at each time step. The density of the mixed layer is reduced to the fixed density of the layer just below. Then the thickness of the layer below is reduced, and a corresponding thickness increase is carried out for the next layer below. The thickness change is chosen so that the buoyancy change in the mixed layer will exactly compensate the buoyancy change below.

Fig. 4.4 (b) shows the potential vorticity distribution in the third layer (second layer of variable thickness). Note that potential vorticity is a maximum near the outcrop and near the eastern boundary. There is a region of minimum potential vorticity in the central region. Upstream of the low vorticity region is a stippled area which indicates convection, with heavy stippling indicating the region of most intense convection. This is a thickness source region, where the third layer has gained thickness at the expense of the layer above. Thus the model illustrates the mechanism proposed by Keffer (1985) to explain the low potential vorticity region

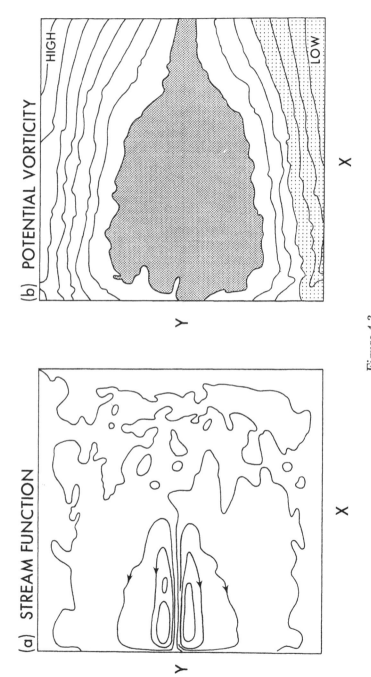

Figure 4.3
Results from the third level below the surface of an eight-level, eddy-resolving quasi-geostrophic model. (a) Stream function, (b) Potential vorticity. (From Holland et al., 1984).

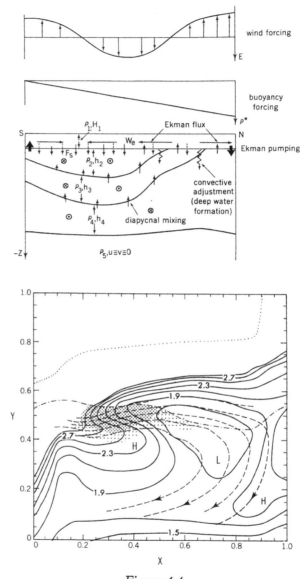

Figure 4.4

A low resolution, isopycnal-coordinate model. (a) Structure of the model in the y-z plane. (b) potential vorticity (solid lines) and flow lines (dashed) for the third layer below the surface. Cross-matching indicates a strong thickness source region due to convection. (From Huang and Bryan, 1987). .

in the central subtropical gyre shown in Fig. 4.1. It also simulates the high potential vorticity region near the eastern boundary.

4.4 A z-coordinate, Eddy-Resolving Model

As a final example, we will describe some results of Cox (1985) for an eddy-resolving, z-coordinate model. An isopycnal model may be ideal for providing insight into the ventilation process, but its semi-Lagrangian character makes it inherently complex. While we can expect rapid future developments, only limited studies with isopycnal models are now available. Much more has been done with models which have a structure that is more like those used in atmospheric weather prediction. Cox (1985) has used a z-coordinate, primitive equation model to study the ventilation process. In his study the circulation is driven by both wind and buoyancy forcing. For purposes of analysis, Cox has interpolated the results to density surfaces. The time-averaged flow on a surface of constant density which outcrops in mid-latitudes is shown in Fig. 4.5 (a). We see a subtropical gyre bounded by an outcrop to the north and a stagnant zone to the south. On the west the subtropical gyre is bounded by an intense western boundary current which flows poleward toward the outcrop. Some of the western boundary current flows off the density surface across the outcrop. This outflow is compensated by an inflow into the surface near the eastern boundary.

The time-averaged potential vorticity pattern corresponding to the same surface is shown in Fig. 4.5 (b). We see relatively high vorticity near the eastern boundary, low vorticity in the region of ventilation by surface waters flowing across the outcrop, and high vorticity to the west. The pattern is somewhat like Keffer's (1985) pattern for the Pacific, shown in Fig. 4.1 (b), if one makes allowances for the fact that the Pacific includes almost twice the span of longitude than the model does. The main difference seems to be in the character of the high vorticity tongue in the eastern part of the basin. In Keffer's (1985) pattern it appears to form in the subarctic gyre, while in Cox's case it appears to originate right at the wall.

Observations of mixing and convection in the ocean are much too sparse to allow us to calculate a potential vorticity budget, but we can get some insight from Cox's model. Fig. 4.6 (a) is the total overturning transport in megatons/s in the temperature-latitude plane. In this particular model temperature and buoyancy are equivalent. We see that poleward flow takes place at the maximum temperature. Surface cooling forces the flow towards lower temperatures as it moves poleward.

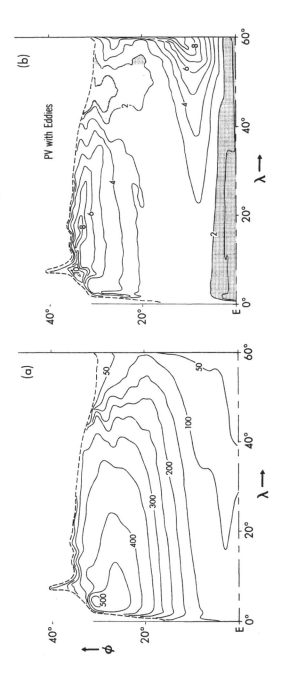

Figure 4.5

An eddy-resolving, primitive equation model based on a z-coordinate. (a) The Bernoulli function on a surface of constant density averaged with respect to time. (b) The potential vorticity pattern averaged with respect to time on the same surface. (From Cox, 1985). The units of potential vorticity are $10^{-12} cm^{-1} s^{-1}$

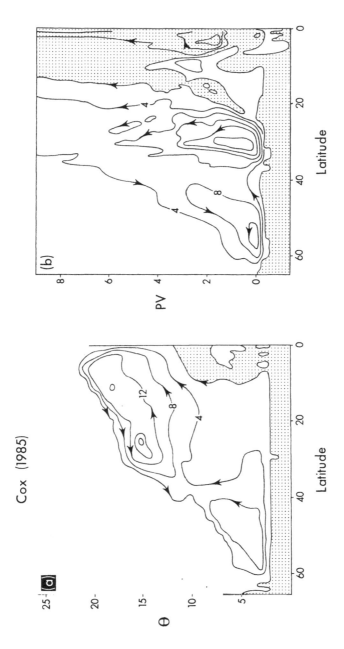

Figure 4.6

An analysis of Cox's (1985) eddy-resolving general circulation model. (a) Total transport in the temperature-latitude plane. (b) Total transport in the potential vorticity-latitude plane. Units are megatons/sec.

Some fluid makes it all the way to the poleward wall where it sinks and moves equatorward. The ventilation process of Luyten *et al.* (1983) is clearly evident in this diagram at midlatitudes. Fluid is forced southward, conserving temperature. Flow upward through the isopycnal surfaces requires nonadiabatic mixing. Note that above 10°C upward penetration is largely confined to the equator, where the thermocline is squeezed upward very close to the surface. At lower temperatures vertical mixing allows weak vertical motion through the lower thermocline at all latitudes. This diagram shows how the thermohaline circulation and the effects of wind are probably very closely coupled together in producing the overall overturning circulation of the ocean.

Potential vorticity is lost when water undergoes convection and becomes de-stratified. Heating from above and upwelling from below tends to restratify the upper part of the water column, increasing potential vorticity. Fig. 4.6 (b) for Cox's (1985) model shows the overturning transport in the potential vorticity-latitudinal plane. The transport is in units of megatons/s as in Fig. 4.6 (a). Flow towards increasing potential vorticity indicates the rate at which potential vorticity is being created. It is evident that 25°N in the subtropical gyre is a preferred site. The high vorticity waters flow poleward and are destroyed by convection and mixing in two major zones. One zone is at 35°N where the western boundary current flows out into the interior. The other site is near the poleward wall. Note that the waters flowing southward have a very low potential vorticity which closely corresponds to what Keffer (1985) concluded from the observations shown in Fig. 4.1 (b).

4.5 Discussion

The structure of the main thermocline of the World Ocean is controlled by the flux of momentum and buoyancy at the ocean surface. The classical approach to studying the ocean circulation was to trace the spreading of water mass types from source regions down into the main thermocline. The quantitative theories of ocean circulation tended to take the existence of a thermocline as a given fact, and concentrate on the effect of wind. The present trend is to consider the effects of wind and buoyancy forcing together in a more unified theory of ocean circulation. More general theories inevitably involve greater mathematical complexity, and it can be anticipated that numerical models will play an increasingly important role. Models will be needed for gaining an understanding of how the ocean circulation works, as well as for the more practical applications to data analysis and the design of field experiments.

The global models used in atmospheric numerical forecasting are often based on an elegant expansion of the velocity, temperature and moisture field based on spherical harmonics. Unfortunately the complex geometry of the World Ocean does not allow us to use the same semi-analytical approach to model the ocean circulation and the bulk of these lectures is concerned with the design of stable finite difference methods. Since the design of computers is in a constant state of flux, there is no well defined context in which a given numerical scheme can be considered optimum. The present trend toward parallel processing, instead of serial processing, would seem to favour explicit schemes with only local interactions between variables. However, technical developments may soon overcome present barriers to global operations in parallel machines.

In providing examples of the application of different types of models, we focused on the problem of explaining the distribution of potential vorticity in the subtropical gyres of the North and South Pacific. A low resolution isopycnal model, and eddy-resolving models based on the quasi-geostrophic equations and the full 'primitive equations' all provided unique insights. We have tried to show that no one model is sufficient. Understanding the ocean circulation requires a hierarchy of models of different resolution, and based on different approaches.

REFERENCES

Cox, M.D., (1985): An eddy-resolving numerical model of the ventilated thermocline. *J. Phys. Oceanogr., 15,* 1312–1334.

Cox, M.D., (1987): Isopycnal diffusion in a z-coordinate ocean model. *Ocean Modelling, 74,* 1–5.

Holland, W.R., T. Keffer, and P.B. Rhines, (1984): Dynamics of the oceanic general circulation: The potential vorticity field. *Nature, 308,* 698–705.

Huang, R.X., and K. Bryan, (1987): A multilayer model of the thermohaline and wind-driven ocean circulation.*J. Phys. Oceanogr.* 17, 1909–1924.

Keffer, T., (1985): The ventilation of the world's oceans: Maps of the potential vorticity field. *J. Phys. Oceanogr., 15,* 509–523.

Luyten, J.R., J. Pedlosky, and H. Stommel, (1983): The ventilated thermocline. *J. Phys. Oceanogr., 13,* 292–309.

Mesinger, F., and A. Arakawa, (1976): *Numerical Methods Used in Atmospheric Models. Global Atmospheric Res. Program Publ. No. 17,* World Met. Organ., Geneva, 64 pp.

Moore, D., and S.G.H. Philander, (1977): Modelling of the tropical oceanic circulation. *The Sea, 6,* Wiley, 319–361.

National Academy of Science (U.S.), (1975): *Numerical Models of Ocean Circulation.* Washington, D.C., 364 pp.

O'Brien, J.J., (1986): *Advanced Physical Oceanographic Modelling,* Edited by J.J. O'Brien, D. Reidel, 608 pp.

Pedlosky, J., (1987): *Geophysical Fluid Dynamics,* 2nd Ed., Springer-Verlag, Berlin, New York, 710 p.

Phillips, N.A., (1956): The general circulation of the atmosphere: a numerical experiment. *Quart. J. Roy. Meteor. Soc.* **82,** 123–164.

Phillips, N.A., (1959): An example of non-linear computational instability. *The Atmosphere and the Sea in Motion,* Rossby Memorial Volume, New York, Rockefeller Inst. press, 501–504.

Redi, M.H., (1982): Oceanic isopycnal mixing by coordinate rotation. *J. Phys. Oceanogr.,* **12,** 1154–1158.

Richtmeyer, R.D., and K.W. Morton, (1967): *Difference Methods for Initial Value Problems.* New York, *Interscience,* 406 pp.

Veronis, G., (1975): The role of models in tracers studies. In *Numerical models of ocean circulation.* Nat. Acad. of Sciences, Washington, D.C., 133–146.

INSTABILITIES AND MULTIPLE STEADY STATES
OF THE THERMOHALINE CIRCULATION

Jochem Marotzke
Institut für Meereskunde an der Universität Kiel
Düsternbrooker Weg 20
D-2300 Kiel 1
FRG

1. INTRODUCTION

It is well established by observations that the thermohaline circulation in the Atlantic Ocean is strongly asymmetric with respect to the equator. Salinities at high latitudes are much higher in the North Atlantic than in the South Atlantic. The deep water is mainly formed in the North Atlantic and moves southward across the equator, while warmer water in the thermocline crosses the equator northward, causing a heat flux of approximately 0.8 PW (Isemer *et al.* 1989). The question arises if this behaviour can have causes other than the obvious asymmetry in land-sea distribution. Posed differently, the question is, if ocean models with symmetric geometry and surface forcing can have several steady states, especially asymmetric ones, with deep water formation in one hemisphere only. The answer is clearly 'yes', for models of widely varying complexity. Rooth (1982), Walin (1985) and Welander (1986) investigated box models, Marotzke *et al.* (1988) a 2-dimensional model with T-S advection and diffusion, a simplified momentum balance and purely thermohaline forcing. Bryan (1986) found that the 3-dimensional GFDL model, with symmetric thermohaline and wind forcing, has an asymmetric steady state.

D. L. T. Anderson and J. Willebrand (eds.), Oceanic Circulation Models: Combining Data and Dynamics, 501–511.

The crucial feature in all these models is the difference in surface boundary conditions on temperature and salinity: The sea surface temperature (SST) is strongly coupled to the surface heat flux, SST anomalies are thus rapidly removed by enhanced heat gain or loss. The surface salinity, however, has no influence on evaporation and precipitation rates, and consequently surface salinity anomalies can persist on much longer time scales. In all models mentioned above the SST is prescribed or relaxed to prescribed values with a short time constant, and the freshwater flux is specified (mixed boundary conditions).

Whereas the existence of asymmetric steady states is well confirmed, the physical mechanism leading to the transition from the symmetric state is less clear. As shown in section 2, the arguments given up to now are incomplete. Furthermore (Bryan, 1986), in a one-hemisphere configuration, the circulation obtained by restoring boundary conditions on both temperature and salinity breaks down after switching over to mixed boundary conditions, the reason for this being unclear. Thus it is necessary, in order to gain a more complete understanding, to reconsider the stability of the thermohaline circulation in one hemisphere only. Note that this is equivalent to a symmetry condition about the equator. In section 2, box models will be shortly discussed, some results from a 2-D model are presented in section 3. Finally, a one-hemisphere version of the GFDL model with purely thermohaline forcing was employed, the experiment is described in section 4.

2. BOX MODELS

The simplest box model (Stommel, 1961) consists of a polar and an equatorial box, which are both completely mixed and held at constant temperatures with difference ΔT. They are connected by pipes at the top and the bottom, the equivalent surface salinity flux into the equatorial and polar boxes is H_S and $-H_S$, respectively. The flow is proportional to the density difference, $\alpha \Delta T - \beta \Delta S$, where ΔS is the salinity difference and α, β are the thermal and haline expansion coefficients, respectively. The surface flow is toward the polar box for $\delta \equiv \beta \Delta S / \alpha \Delta T < 1$, and reversed for $\delta > 1$ (temperature and salinity dominated, respectively).

Fig. 1 shows the equilibrium solutions in phase space. Obviously, H_S is zero for vanishing ΔS, but also for $\delta = 1$, since then the flow vanishes. Thus H_S must have an intermediate maximum which in this simple case is at $\delta = 0.5$. We see from Fig. 1 that for prescribed ΔS there exists only one solution, whereas for specified H_S there may be three. They are characterized by $\delta < 0.5$, $0.5 < \delta < 1$ and $\delta > 1$,

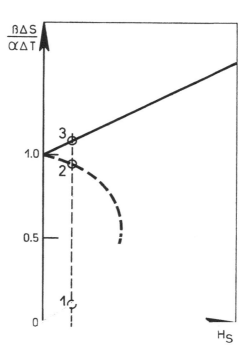

Figure 1
Equilibrium states of a 2-box model in a $\Delta S - H_S$ phase space diagram.
Points 1,2,3 denote the equilibrium states for a specified H_S. The dotted
part of the curve marks the unstable regime.

respectively (marked 1,2 and 3 in Fig. 1). 1 and 2 are thermally dominated, 3
is salinity dominated. Linear stability analysis shows that 1 and 3 are stable to
infinitesimal perturbations, while 2 is unstable. This is plausible since in the range
$0.5 < \delta < 1$ a slight enhancement (reduction) of the forcing H_S would lead to a
decrease (increase) of the equilibrium response ΔS, which is usually indicative of
an unstable system.

The advective feedback mechanisms involved are discussed by Walin (1985)
and Marotzke *et al.* (1988). Note that the stability analysis outlined here is not
valid to explain cross-equatorial instability, as it refers to one hemisphere only.
As already mentioned in the introduction, the aim of this paper is to investigate
the thermohaline circulation in one hemisphere before proceeding to more complex
geometries. Fig. 1 is taken as guidance for designing the experiments. For further

information about box models of the thermohaline circulation, see Welander (1986).

3. 2-D MODEL

The model is a meridional-plane model, with temperature and salinity advection and diffusion, and a simplified momentum balance. The strength of the circulation is controlled by the meridional density gradient. A detailed description is given in Marotzke *et al.* (1988). The system is spun up with surface temperatures and salinities prescribed, both following a cosine law in latitude, with amplitudes 25°C and 2‰ respectively. From the resulting steady state the salinity flux is diagnosed, and surface salinity flux and temperature are used as boundary conditions thereafter. Additionally, a salinity anomaly is added to the steady state solution. The results shown are for the case $\delta = 0.32$, a value reflecting the relative influence of salinity on the pole-to-equator density contrast. Fig. 2a displays the initial circulation with downwelling at the poles and upwelling everywhere else (termed 'normal' or 'positive' circulation from now on). A salinity anomaly of -0.5 ‰ in the top 750m of the poleward one third of the basin leads to a breakdown of the circulation, the resulting steady state is shown in Fig. 2b. Its strength is about half of the original one, and downwelling occurs at the equator ('reverse' or 'inverse' circulation). The state corresponds to the upper branch (salinity dominated) of Fig. 1.

There is one unphysical feature in the model used here, as occurring static instability is not removed by a convective adjustment procedure. In a second experiment, convective adjustment is present, and the situation changes dramatically. A much smaller anomaly is necessary to cause the transition (-0.02 ‰ in the top 250m of the same area as before). Before reaching equilibrium, strong convective processes occur, causing a short period of extremely strong positive circulation. Afterwards the system returns to the reverse mode. As Fig. 3 shows, this process happens periodically, with a period of approximately 16 times the travel time of a water particle around the basin. That this behaviour is not merely a peculiarity of the 2-D model and which physical processes are involved, is discussed in the next section.

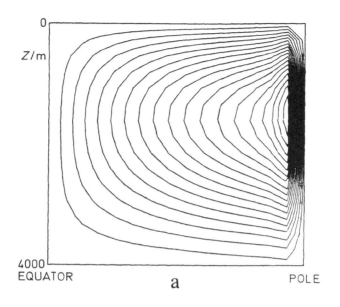

a

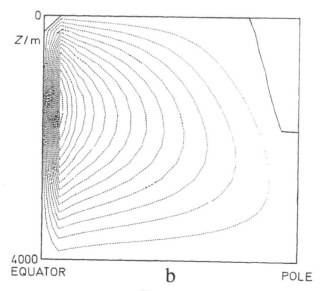

b

Figure 2

2-D model without convective adjustment. Meridional stream function for
(a) initial and (b) final state. (a): Maximum value corresponds to 11 Sv
in a basin of 6000 km width, contour interval to 0.6 Sv. (b): Minimum
−4.75 Sv, contour interval 0.3 Sv.

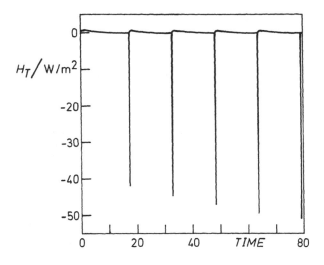

Figure 3
2-D model, including convective adjustment. Time series of basin averaged heat uptake H_T for anomaly experiment with a -0.02 %o anomaly in the top layer (250m) of the right one third of the basin. Time unit is dimensionless overturning time for a water particle.

4. GFDL MODEL

The model is a coarse resolution (3.75°zonally, 4°meridionally, 15 levels vertically), standard version of the Bryan-Cox model. Its extension is 60°in longitude and 64°in latitude, with closed walls, no bottom topography and no wind stress applied. For a detailed description see Cox (1984) and Cox and Bryan (1984). The convective adjustment mechanism employed is different from the standard one, guaranteeing complete vertical mixing (i.e. stability) after the procedure. The time step is 2 hours for all variables during periods of rapid changes of the circulation patterns (synchronous integration), during the rest of the anomaly experiment, the timestep for T, S is changed to 1 day (asynchronous integration). The spin-up is run with restoring boundary conditions on temperature and salinity, with a time constant of 30 days. The reference values follow a cosine function in latitude, the contrast between northern and southern values being 27°C and 2.5 ‰ respectively, which corresponds roughly to zonally averaged Levitus data. The resulting meridional mass transport pattern very much resembles the one from the 2-D model (Fig. 2a), with a maximum of 7.6 Sv. As in the 2-D case, the surface salinity flux is diagnosed from the equilibrium solution and used as boundary condition thereafter. A salinity anomaly of -0.01‰ is added to the top layer of the northern half of the basin. Fig. 4 shows the meridional transport 23 years afterwards. The circulation has almost completely collapsed and is even reversed at greater depths. This confirms the initial estimate that the spin-up state corresponds to the upper, thermally dominated but unstable, branch of Fig. 1. During the further development the circulation is reversed in shallower depths, too, while ceasing almost completely below. This leads to downwelling at the equator and diffusive transport of heat and salt downward, accompanied by weak poleward transport at mid-depths. Fig. 5 shows the meridional mass transport after almost 7000 years of integration. The circulation is very shallow and sluggish, and its direction determined by the extremely large surface salinity contrast (more than 6‰). Apparently, the system is close to an equilibrium, the mean surface heat flux being 0.2 W/m^2, approximately. At high latitudes, however, the stratification is only marginally stable, with cold, fresh water on top of warm, saline water. Shortly after the situation depicted in Fig. 5, static instability occurs at the western boundary at about 40°N, and within a few years the convective domain spreads over the whole northern part of the basin, with mixing from top to the bottom. An extremely strong meridional density gradient is built up, leading to an immensely violent (positive) circulation, which at times exceeds 200 Sv.

This process can be understood from the initial stratification and the surface boundary conditions: The very low surface salinity stabilizes the water column, until it is removed by a convective event. There is no fast process to build it up again. Opposed to this, the strong SST anomaly caused by the initial mixing process is rapidly removed by heat loss to the atmosphere. The column becomes strongly unstable again, convection will reach even further down, until the column is completely homogenized. Within a few decades it gives off all the heat it had accumulated within several thousand years. Fig. 6 shows that the spatially averaged heat loss during this time reaches 130 W/m^2.

508

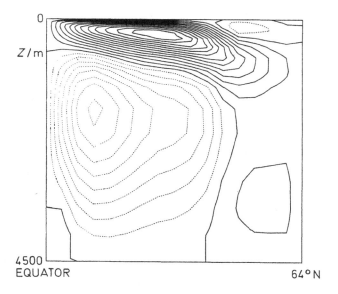

Figure 4
GFDL model, meridional mass transport, 23 years after a −0.01 ‰
anomaly has been added to the top layer (50m) of the northern half of the
basin. The range is from −1.7 Sv to 2.6 Sv. Dotted lines denote negative
values (inverse circulation). Contour interval 0.2 Sv.

The subsequent time development is characterized by very strong, low-frequency oscillations, as obvious from time series of basin averaged heat uptake and the mass transport stream function at an arbitrary point (Fig. 6). Associated with these oscillations is a strong inverse circulation emanating from the northern boundary, pulsating back and forth with maximum extension to about 40°N. Most likely it is caused by intermediate freshening of high latitude surface water that creates a negative density gradient, while convection is maintained. The oscillations gradually disappear, the circulation settles somewhat to a moderately large value (about 30 Sv), until it undergoes a transition (marked by the last peak in Fig. 6) that very much resembles the one at the beginning of the anomaly experiment. The result is again a sluggish and partly reversed circulation, and collecting the results from all models discussed here it seems safe to say that the time history is about to be repeated, as in the 2-D model. The period, which is several thousand years, should depend mainly on the vertical diffusion coefficient and is therefore of very little significance.

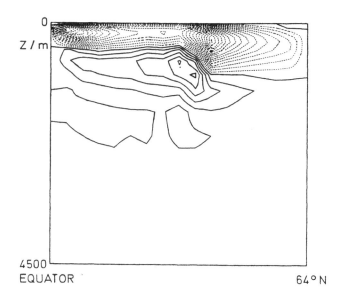

Figure 5
GFDL model, meridional mass transport after about 7000 years. The range is from −1.6 Sv to 0.5 Sv, contour interval 0.1 Sv.

5. CONCLUSIONS

The most robust common feature of the one-hemisphere models of the thermohaline circulation is the existence of several steady states, if mixed boundary conditions (freshwater flux and temperature prescribed) are applied. Box models suggest the existence of two stable circulation patterns, one temperature and one salinity dominated, with opposite meridional flow directions. Besides, there should exist a thermally dominated, unstable equilibrium solution. The 2-D model reproduces

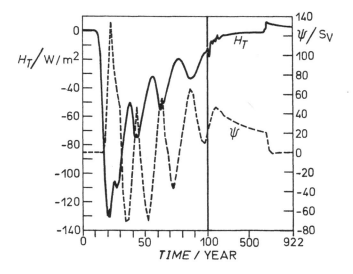

Figure 6
*Basin averaged heat uptake H_T and meridional stream function ψ at an ar-
bitrary point (54°N, 1100m depth) during the strongly convective phase.
The perpendicular line marks the transition from synchronous to asynchronous integration.*

these three steady states, if convective adjustment is absent. When convection is
incorporated in the model (i.e., sharply localized events are present), the reversed
circulation is no longer an equilibrium state, but still a very strong attractor, and
the system is close to it for most of the time. For the 3-D model without wind stress,
no stable steady state solution of the one-hemisphere thermohaline circulation has
been found.

Clearly, the system considered here is highly idealized, but exhibits how the
thermohaline part of the oceanic circulation tends to behave. A measure is gained
for how well simple models can predict overall features of the three-dimensional
thermohaline circulation. The thermohaline circulation is very sensitive to perturbations in the salinity field, if mixed boundary conditions are applied, and this

sensitivity may have a strong influence on changes of the oceanic heat transport and thus on changes of the climatic system.

ACKNOWLEDGEMENTS

I wish to thank M. Cox for making the GFDL model available to me, and Professor J. Willebrand for numerous valuable suggestions and discussions. This work has been supported by the Deutsche Forschungsgemeinschaft, Sonderforschungsbereich 133 'Warmwassersphäre des Atlantiks'.

REFERENCES

Bryan, F. (1986): High-latitude salinity effects and interhemispheric thermohaline circulations. *Nature* **323**, 301–304.

Cox, M.D. (1984): A primitive equation, 3-dimensional model of the ocean. *GFDL Ocean Group Technical Report No. 1.* GFDL, Princeton University.

Cox, M.D. and K. Bryan (1984): A numerical model of the ventilated thermocline. *J. Phys. Oceanogr.* **14**, 674–687.

Isemer, H.-J., J. Willebrand and L. Hasse (1989): Fine adjustment of large-scale air-sea energy flux parameterizations by a direct estimate of ocean heat transport. *Journal of Climate* (in press).

Marotzke, J., P. Welander and J. Willebrand (1988): Instability and multiple steady states in a meridional-plane model of the thermohaline circulation. *Tellus* **40A**, 162–172.

Rooth, C. (1982): Hydrology and ocean circulation. *Progr. Oceanogr.* **11**, 131–149.

Stommel, H. (1961): Thermohaline convection with two stable regimes of flow. *Tellus* **13**, 224–230.

Walin, G. (1985): The thermohaline circulation and the control of ice ages. *Palaeogeogr., Palaeoclimatol.,Palaeoecol.* **50**, 323–332.

Welander, P. (1986): 'Thermohaline effects in the ocean circulation and related simple models'. In: *Large-scale transport processes in oceans and atmosphere* (eds. J.Willebrand and D.L.T. Anderson). D.Reidel Publ. Co., 163–200.

SUBGRIDSCALE REPRESENTATION

by
Greg Holloway
Institute of Ocean Sciences
Sidney, B.C., Canada

1 INTRODUCTION

Among the many aspects of numerical ocean modelling, subgridscale (SGS) parameterization is nearly always the 'black sheep'. It tends to be the last thing a modeller talks about, and then only reluctantly. Why is that? Perhaps it is because most of us went to school in areas like mathematics and physics; we learned that when we are presented with a partial differential equation with boundary and initial conditions, then we are supposed to solve it by a sequence of careful, geometrically precise steps. Or we might be more sophisticated and first ask if a problem is well posed before going after 'the solution' regardless. Sometimes, even if we've somehow caused a problem to be well posed, we still might not have the mathematical power to obtain its solution explicitly, perhaps because of a nonlinearity in the equation. Yet, though we might not obtain the solution exactly, it can often be approximated by perturbation analyses. And if we can't carry out all the steps by purely analytical means, the computer can carry out numerics. Importantly, in the end we can feel that we have proceeded carefully, as mathematical, physical sorts of scientists should. Until we come to ocean modelling SGS ...

At the SGS there is an uncomfortable lack of geometrical precision about what can be said. Presumably this calls for greater courage! Only one may not escape also feeling rather clumsy and foolish. Here, at the outset, I mustn't besmirch some considerable number of colleagues who have thought deeply about SGS, certainly performing as respectable mathematical physicists. It remains a very difficult and very uncertain area. To a large extent, SGS continues to rest either upon inspired 'hand-waving' or else one just admits that it is something that makes the model 'work', where 'work' means that the code executes and numbers come out that aren't obviously ridiculous.

D. L. T. Anderson and J. Willebrand (eds.), Oceanic Circulation Models: Combining Data and Dynamics, 513–593.
© 1989 by Kluwer Academic Publishers.

Let me hasten to remark that I don't mean 'hand-waving' in a pejorative way. Honest 'hand-waving' may reflect hard thought based upon physical insight; only the steps aren't defended with strict mathematics. A greater caution should be mentioned: any idea, which may not reflect very deep insight, can be festooned with all kinds of mathematics. Unfortunately, to the extent that SGS is almost invariably a problem of averaging over nonlinear dynamics, the mathematical problem is often one of retaining only some low order terms in a divergent representation of the dynamics. What to retain in what form depends upon an assumption or, said otherwise, a guess. Then lots of mathematical symbols where they really don't matter, together with a bad guess where it does matter, still make a bad guess. An Appendix A expands on this point. Another cautionary comment at this point is that one can also bury the SGS in the numerical methodology where it is not so readily seen. There are many methods, often associated with uncentering some spatial or temporal difference operator, which may be attractive from the view of stabilizing a computational scheme. Usually this means that something (e.g., vorticity or temperature variance) is dissipated even if one hasn't explicitly included a dissipative term in the governing differential equation. That does not 'solve' the SGS problem; it only hides the SGS unless one is very clear about how consideration of the SGS in fact motivates the numerical method.

With so much bad news at the outset, one may be little inclined to read on. Indeed it would be nice to think that the parts of ocean modelling where we are most uncertain turn out to be parts that don't matter very much. Unfortunately there is little reason to think that is so and substantive reason to think it is not. Besides, I hope that it will turn out to be fun to explore a sample of SGS issues.

Concerning the topic of this School, one sometimes separates modelling into two classes: 'forward' and 'inverse', or prognostic and diagnostic. In his lectures, Prof. Wunsch indicated that the distinction between forward and inverse can be ambiguous. For the moment it may be helpful to think of each separately. In its most common usage, a forward model integrates equations forward in time in order to see how resulting flow evolution depends upon forcing or other influences. Since we are usually striving to understand reality, we hope that the model physics (as numerically implemented) can be as 'realistic' as possible. Hence one wishes to get the SGS right. The concern is especially felt when one considers ocean variability on climatic timescales for two reasons:

- Concern is global and over long enough time that model resolution is necessarily coarse compared with some regional or process model; hence more of the physics falls into SGS.

- Time duration is such that systematic misrepresentation of the SGS has opportunity to corrupt major features of the flow.

Inverse models in their archetypal use seek to estimate some properties of a system from observations of other properties. Here it may not be so clear what is the role of SGS formulation. However, the power of an inverse model to infer properties depends upon how well other properties are known. Consider a many-box model by which one may seek to infer large scale circulation from observed tracer burdens. Suppose the SGS will be represented as exchange coefficients between boxes. To the extent that these SGS coefficients are unknown and must be inferred as part of the solution, the circulation is inferred with less precision. If we can bring any *a priori* information to bear to narrow the possible bounds on SGS, then we can infer other properties within narrower bounds. As well, study of the SGS may suggest possible errors, perhaps in the assumption of an exchange coefficient formulation.

We have ample reasons from either forward or inverse modelling to be concerned about SGS. We can't ignore it or do so at our peril. And so we come to ask what can be said about the SGS. The paucity of firm answers and the wide range of uncertainty can be discouraging. Feeling a bit discouraged while trying to assemble these lectures, I sought inspiration from a bowl of soup at a favorite Chinese restaurant. The soup didn't tell so much but a fortune from the cookie afterwards somehow captured the situation:

> Having no answers to questions
> does not make them go away.

It seemed just the note on which to commence these lectures. We will look at a lot of questions without very firm answers. Perforce those remain as challenges to the students – and likely to their students in turn!

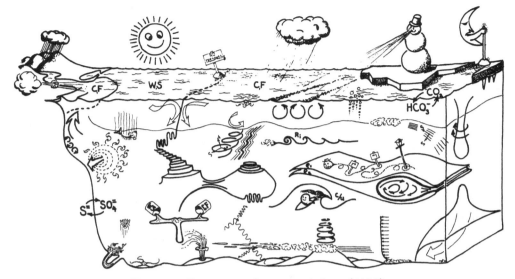

Figure 1: Processes of ocean mixing. (1981)

2 ORGANIZATION

The subject of SGS representation exceeds the material that can be presented in these lectures. A cartoon (Figure 1) suggests some of the many processes which a numerical model will not be able to resolve but which need somehow to be represented in terms of their effects upon the resolved scales.

The cartoon is incomplete. It was intended only to depict some of the processes that modify the distributions of chemical tracers, heat or salt. Further processes will affect redistributions of momentum or vorticity. The following discussions address only a few of the many processes, reflecting material with which I feel more familiar. This should not imply any priority in terms of the importance of some processes with respect to others. The discussion should be read only as a sampling among possible topics.

To the small extent that there has been effort at organizing, the subject matter is divided into three areas:

- Quasi-horizontal stirring

- Quasi-vertical mixing

- Benthic boundary processes

Perhaps the largest omission that comes to mind is the air-sea interface and mixed layer region. This is clearly of great importance. Only it is a topic with which I am not very familiar.

3 QUASI-HORIZONTAL STIRRING

The concern here will be stirring and transport by mesoscale eddies with length scales from order of 10 to 100 km. 'Quasi-horizontal' refers to some ambiguity as to whether the resulting eddy transport is best represented in the horizontal (i.e. level geopotential) or in mean isopycnal (i.e. constant potential density) surfaces. We will return to this question. 'Stirring' is an expression sometimes distinguished from 'mixing' in the sense that stirring is an advection of properties without thermodynamic modification although that stirring may cause straining, squashing, folding and drawing out of filaments giving rise to ever smaller scales leading to scales at which thermodynamic modification occurs. 'Mixing' is often thought to be the smaller scale 3-dimensional turbulence which leads directly to thermodynamic modification. Viewed this way, 'mixing' also involves a cascade across length scales due to differential advection; differences between the two terms may be more a matter of custom than of actual distinction.

Most of the material presented will be under the topic of quasi-horizontal stirring, in part to introduce ideas that are useful in other topics and also, again, reflecting my lesser familiarity with other topics. Within quasi-horizontal stirring, material is organized as follows:

1. Basis for Fickian diffusion?

2. Geophysical influences (Rossby waves)

3. Aside: Negative diffusion?

4. Role of coherent vortices

5. Statistical inhomogeneity

6. Horizontal or isopycnal?

7. SGS within partly resolved eddy field

- ∇^2, ∇^4, APV, eddy noise ... ?

3.1 Basis for Fickian diffusion?

Often the expression 'SGS parameterization' translates to "how to pick your eddy diffusion coefficients." This shows how pervasive is a kind of Fickian idea; it is the default option, often assumed without remark as a given part of the equations of motion. Specifically, this refers to the idea that SGS transport is proportional to the numerically resolved part of the property gradient,

$$\mathbf{F} = -\mathbf{K} \cdot \nabla S \qquad (1)$$

where $\mathbf{F}$ is the (vector) flux of property S, $\mathbf{K}$ is a (tensor) eddy diffusivity, ∇ is spatial gradient, and S refers to the concentration or intensity of property S on the scales that are numerically resolved.

It is easy to find fault with Fickian SGS due to its lack of systematic derivation from some averaging procedure over SGS motion. On the other side, there are some very compelling reasons why one would attempt to 'get by' with Fickian SGS so far as possible.

- The Fickian idea is understandable at an intuitive level and is relatively straightforward to implement in models.

- We are well acquainted with properties of Fickian diffusion, at least when $\mathbf{K}$ is assumed constant. Library shelves of books can be found discussing solutions to diffusion equations, including posedness considerations for boundary conditions, unusual geometries, and analytical and numerical techniques for solution. This is important. While some theoretical consideration might suggest an SGS representation other than Fickian, further difficult questions arise. What are necessary boundary conditions just from a posedness consideration? What are appropriate boundary conditions based on SGS theoretical consideration? How will this be implemented numerically (recognizing that very plausible looking difference representations often yield discrete operators very different from their intended differential forms)?

For reasons suggested, trying to think of something 'better' than Fickian is usually only the beginning of the SGS theorist's travails. Let us first consider arguments that may support a Fickian representation.

We begin with the simplest views possible, until contradictions force a more strenuous effort. One may imagine SGS turbulence in analogy to molecular agitation in a gas; little blobs of water being more-or-less randomly rearranged. Suppose that blobs tend to be exchanged over a characteristic distance l after a characteristic time τ. Suppose there's a background (uniform) gradient of mean property S such that, at the outset for each exchange, blobs are characterized by the mean S. Since the end of each exchange is the start of the next, we suppose that property S at each blob tends (perhaps incompletely) to take on the 'new' mean value of S at the blob's 'new' location before its next random exchange. One could elaborate this model with some lines of algebra but there may be little gained. If only from dimensional consistency, we are led to a Fickian transport in which the diffusivity coefficient K must be proportional to l^2/τ. Proportional maybe, but what sign? In fact the simple model is quite unambiguous in suggesting down-gradient (i.e. positive K) transport, if only from the idea that each blob is tending to carry the value of S from its starting position in each exchange. What we don't know from this simple sketch is the coefficient of proportionality between K and l^2/τ. Moreover, we've not defined l or τ sufficiently to make the idea of a coefficient meaningful. Suppose we have information on characteristic turbulent velocities, say of r.m.s. value v. Suppose also we've measured a spatial correlation scale or wavenumber peak of an eddy spectrum; letting that be l, we could try to evaluate a coefficient of proportionality between K and vl, having elected to define $\tau = l/v$. A further very important remark is that any such Fickian hypothesis depends upon assuming a separation in length scale between the large scale 'mean' (however defined!) and the small scale 'eddies'. Here we evade the issue by only speaking of a uniform gradient 'mean', hence infinitely separated in scale from any 'eddies' whatsoever. But clearly that is a cheat with respect to virtually any real application imaginable. In reality, it may firstly be highly nontrivial to decide how to distinguish 'mean' from 'eddy'; then the supposition of a spectral 'gap' between the two is often dubious if not plainly wrong. Nonetheless we plunge bravely ahead.

In the preceding paragraph, a goal was to estimate transport of some property S. There is an alternative question which turns out to give an answer as above but which clarifies the basis for defining and measuring τ. It is the Lagrangian particle dispersion consideration following Taylor (1921). By definition, a Lagrangian particle is one that is purely advected: $dx/dt = u$, where $x(t; a)$ is the location at time t of a distinct particle labelled a. u is the t-dependent speed of particle a which is identical to the Eulerian velocity $u(x(t; a), t)$ always evaluated at $x(t; a)$. Consider an ensemble of realizations of trajectories $x(t)$, each with independent realization of u. Suppose we've removed any ensemble mean u from the population and we've defined the starting point at $t = 0$ to be $x = 0$. (In this discussion, scalar notation is used. However, one may consider space to have any dimension.) For each realization

$$\frac{d}{dt}x^2 = 2x\frac{d}{dt}x = 2\int_0^t u(t')u(t)dt'$$

Averaging over the ensemble

$$\frac{d}{dt}\langle x^2 \rangle = 2\langle u^2 \rangle \int_0^t R(\tau)d\tau$$

where $\tau = t - t'$ and R is the velocity autocorrelation function following the Lagrangian particle. For sufficiently small t, $R(\tau)$ will be nearly a constant near unity. Then $\langle x^2 \rangle$ grows as $\langle u^2 \rangle t^2$. If the nature of R is such that for large t

$$\int_0^t R(\tau)d\tau \to T_L$$

where T_L is a reasonably stable limit at large t, then $\langle x^2 \rangle$ grows proportionally to t. The probability density $p(x, t)$ for finding the particle near location x at t is then given by

$$\frac{\partial p}{\partial t} = K_L \frac{\partial^2 p}{\partial x^2}$$

Now K_L is given by $K_L = \langle u^2 \rangle T_L$.

The natural question is what this apparent diffusion of probability density for a Lagrangian particle has to do with transport of S-stuff. The answer is that particles transport their labels; the a-stuff transports as S-stuff to the extent that S tends to be conserved following the fluid. Interest from the Lagrangian particle view is also stimulated by the numbers of quasi-Lagrangian floats that have been tracked over considerable time with statistical analyses of their trajectories. For a sufficiently large number of releases and for sufficient time after each release, if one tended to observe $\langle x^2 \rangle$ increasing as t, a coefficient of proportionality would be K_L which, hopefully, is the K for some S-stuff of interest. This isn't an easy observation to make with statistical confidence but it seems easier than any scheme for direct observation of eddy transports of S. As well, we recognize what the τ in the earlier blob-exchange model really is: a Lagrangian velocity integral correlation time.

Efforts to 'observe' K directly from float dispersion encounter both fundamental and practical difficulties. Depending upon the flow regime, the integral over R may not tend to converge to a stable T_L. Especially if the flow dynamics have significantly wavelike nature, persistent oscillations of R at large $t - t'$ cause slow convergence to T_L. Practically, one has ever fewer estimators at larger $t - t'$ so statistical confidence becomes a more severe problem. A second line of difficulty is that the previous discussions have ignored mean circulation while considering eddy effects. In reality it is often difficult and rather arbitrary how one goes about trying to separate 'mean' from 'eddy'. In numerical models it is easier to say what is resolved and what is not; however it is not so clear how float observations of K_L bear upon the SGS in a model of given resolution. Worse may be the supposition that K_L is simply K, i.e. a-stuff is S-stuff and one kind of S-stuff behaves like another. An illustration of the gross failure of this supposition will be given in the following subsection.

Just before ending this portion, one aside on the topic of Lagrangian dispersion may be helpful. When we saw that probability density p satisfied a diffusion equation, we might have imagined that a cloud of many particles released nearby would satisfy this diffusion equation and thus, for example, a dye spot would so disperse. Effectively, one might imagine that p is an area density of particles in a given realization. That is incorrect. p is a single-particle probability under an ensemble of independent releases. For illustration, suppose the nature of flow was one of rigid translation with amplitude and direction varying in time so that an integral over R indeed converges nicely to T_L. p would diffuse as above. However, any cloud of particles released in any realization of this 'flow' would remain 'frozen' forever while being displaced, possibly in some erratic time-dependent way. Relative to the particles which constitute the cloud, the cloud never disperses at all. For many processes of biogeochemical interest, it is the relative dispersion of the cloud rather than its erratic wandering as a whole that matters.

Approaches to relative dispersion often address two-particle joint or conditional probability density functions. In practice one may observe or wish to calculate only low order moments of the probabilities. If the distance between two particles a and b is denoted $r = |x_a - x_b|$, then observed 'neighbor-distance' laws have sometimes (Richardson and Stommel, 1948) been fit to power law forms

$$\frac{d}{dt}r^2 \propto r^\lambda$$

where values of λ have been estimated from various experimental data. Of possible interest here is the question of what can be inferred from such 'neighbor-distance' laws with respect to statistics of the eddy field which induces the relative dispersion. Here is only a very simple sketch:

Suppose the kinetic energy spectrum for the eddy field has a power law subrange $E(k) \propto k^{-\alpha}$, where k is wavenumber. For separation r in the range of k^{-1} for which the spectral power law holds, hypothesize that the rate of change of r is due only to eddies of about the scale r. Then dimensional consistency alone already gives

$$\frac{d}{dt}r^2 \propto [E(r^{-1})r]^{1/2} \propto r^{\frac{\alpha+1}{2}}$$

For illustration, $\lambda = 4/3$ implies $\alpha = 5/3$ while $\lambda = 2$ implies $\alpha = 3$, both being of considerable geophysical interest. As ever, there are cautions. How well can we justify a 'localness' hypothesis that change of r depends only upon eddies of approximate scale r? One failure clearly arises when the spectrum is so steep (essentially $\alpha > 3$) that the straining field at scale r is controlled directly by straining from very much larger scale eddies. This suggests that $\lambda = 2$, $\alpha = 3$ is a limiting case for the possible utility of 'neighbor-distance' laws with respect to inferring energy spectra. There are other defects: wave propagation which is not expressed from $E(k)$ alone will affect r^2 evolution, for example.

3.2 Geophysical influences (Rossby waves)

Let us now examine the possible basis for a Fickian SGS with more care. Suppose we include Rossby wave propagation. The dimensional consistency argument at the outset of the preceding discussion here becomes ambiguous. There is another dimensional parameter in the problem, β if we consider motion on the β-plane. Arguments following Taylor might be more useful since one could speculate on how β affects $R(\tau)$. However, this also would not anticipate how badly Fickian SGS on something so simple as a β-plane! Let's see.

3.2.1 A numerical empirical approach

A first and most straightforward approach is to perform direct numerical experiments at high enough resolution to include all the β-plane turbulence which would be SGS in some coarser model. Figure 2 shows some snapshots from experiments using a spectral transform model that is isotropically truncated at wavenumber 60.

(Very substantially higher resolution is available from modern computers, but that is not needed for present purposes.)

In Figure 2, the equations of motion are only the barotropic vorticity equation on the β-plane coupled to a passively advected tracer:

$$\frac{\partial}{\partial t}\nabla^2\psi + J(\psi, \nabla^2\psi + \beta y) = D_1\psi + \xi \tag{2}$$

$$\frac{\partial}{\partial t}\phi + J(\psi, \phi + \mathbf{G}\cdot\mathbf{x}) = D_2\phi \tag{3}$$

$\nabla^2\psi$ is relative vorticity, ψ is streamfunction, $J(,)$ denotes the Jacobian determinant with respect to $\mathbf{x} = (x, y)$, D_1 and D_2 are dissipation operators on the ψ and ϕ fields, ξ are external torques, nominally representing windstress curl, and $\mathbf{G}$ is a background gradient of total tracer concentration $\phi + \mathbf{G}\cdot\mathbf{x}$. An objective for such simulations has been to calculate the north-south and east-west components of eddy transport of ϕ-stuff, to relate these transports to components of $\mathbf{G}$ and to see if a resulting $\mathbf{K}$ bears any resemblance to the simple suggestions made previously. What may come as a bit of a pleasant surprise is that at least some of the relationships do seem to work out in a tolerably simple way. If, for example, we omit β, and consider only isotropic, homogeneous, stationary, f-plane, 2D turbulence, then one does get an isotropic K which seems to scale as vl.

When β is included then $\mathbf{K}$ becomes anisotropic for two reasons: The velocity field itself tends to become anisotropic with more east-west than north-south variance. As well, even if the velocity variances were isotropic, still north-south tracer transport would be suppressed by the more wavelike nature of north-south velocity components. Taken together, the result (based upon numerical empiricism) is that the east-west component of $\mathbf{K}$ is not much affected by β whereas the north-south component is very markedly suppressed. A more theoretical approach, to be discussed below, suggests that north-south diffusivity K_{yy} should be suppressed by a factor $(1 + (B\hat{\beta})^2)^{-1}$ where B is a bit of numerical fudge factor and $\hat{\beta}$ is β scaled relative to nonlinearity as $\hat{\beta} = \beta v/\zeta^2$ with $\zeta = \nabla^2\psi$, the relative vorticity. Numerical simulations indicate that suppression may be more marked at small $\hat{\beta}$ than theoretically suggested. Still, at least within factor of 3 uncertainty, everything 'works'. So we shouldn't ask more questions, right? But, of course, we must.

First there are minor comments. D_1 and D_2 haven't been discussed. Effectively they are the SGS *within* this model aimed at studying SGS. They are necessary to remove some ζ and ϕ variance at high wavenumber and to damp some velocity

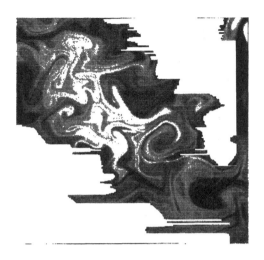

Figure 2: Snapshots are shown from computer simulation.
A passive tracer field is stirred by β-plane turbulence while a uniform background
gradient of tracer concentration is maintained with equal north-south and east-west
components. Random windstress curl sustain the statistically stationary, homogeneous eddy field. In the figure, the tracer field is coloured. Black and white contours indicate instantaneous streamfunction with the sense of cyclonic (CCW) flow
around black contours.

variance at all scales. I have not thoroughly explored the parameter space of D but an impression is that ϕ-transports are not very sensitive to this when the truncation wavenumber is 60 or so. It remains a point of caution though.

There is a more disturbing question. Two kinds of advected stuff in Eq 2 and Eq 3 are total (potential) vorticity $q = \zeta + \beta y$ and total tracer $\phi + \mathbf{G} \cdot \mathbf{x}$. Neither is purely advected; each is weakly nonconserved due to D or ξ. Eddy transport of tracer is found to be down its gradient $\mathbf{G}$ with anisotropic $\mathbf{K}$ when β is present and with numerical values of components of $\mathbf{K}$ that are tolerably consistent with simple estimates. q also has a mean gradient, i.e. β. Since q is advected by the same eddies that transport tracer, it would only 'make sense' that q should be transported southward at rate $K_{yy}\beta$ where K_{yy} is the north-south component of $\mathbf{K}$. Indeed this is just one aspect of the sometimes popular idea of potential vorticity mixing.

Regrettably, SGS by 'hand-waving' cannot be counted on to 'work' every time. *This time* we've hit a snag. Eddy transport of q is here by transport $\langle \zeta \partial \psi / \partial x \rangle$ where $\langle \rangle$ denote ensemble averaging. No q transport can be sustained by $\langle y \partial \psi / \partial x \rangle$ because this requires a mean torque $\langle \xi \rangle$ which was not supplied. Transport of ζ poses a problem because this acts as a mean body source of momentum in the fluid. That can be seen from momentum advection written in the form

$$\mathbf{u} \cdot \nabla \mathbf{u} = \zeta \times \mathbf{u} + 1/2 \nabla u^2$$

where $\zeta = \nabla \times \mathbf{u}$ is the vector vorticity. Thus southward transport of relative vorticity appears as a westward force $\langle \zeta \times \mathbf{u} \rangle$ for which neither westward acceleration nor compensating eastward force has been assumed.

We cannot get around this snag by saying "Ah but the theory used and the numerical simulations are based on quasigeostrophy. Contradictions at a momentum level are someone else's department." This dodge wouldn't be legal anyway and besides the numerical simulations employ spatial Fourier representation of ζ and of ψ so spatial average of $\zeta \partial \psi / \partial x$ does vanish. It is our 'hand-waving' SGS that has gone wrong. An eddy field that transports one kind of stuff, i.e. passive tracer, is unable to transport another kind of stuff, i.e. potential vorticity, although both are approximately advected quantities. As a practical matter we might say it wouldn't be too bad if one's SGS admitted that K for one field could be somewhat different than K for another field. Here it is more severe. K altogether vanishes for one field yet takes reasonable values for another.[1]

The difficulty apparently is that vorticity isn't a *passive* tracer but rather acts on the advecting velocity field. More precisely, vorticity is kinematically related to velocity. (Indeed that kinematic relationship is such that it is difficult, though not impossible, for velocity to transport vorticity in many circumstances.) Manipulating the vorticity equation may be a serviceable way to address SGS; only it is also hazardous. Very plausible things that one may do to vorticity can have unphysical consequences for momentum, as discussed, *e.g.* by Stewart and Thomson (1977).

There is a further caution: identifying a tracer as 'passive' can be misleading. A tracer such as CO_2 in seawater is usually deemed passive, having negligible effect on buoyancy. However, to the extent that a passive tracer may be correlated with an active tracer, perhaps on account of their distributions of sources, the passive tracer is a 'surrogate active' tracer and its transport properties may be affected by the dynamic relationships of its active 'partner'. Thus the ϕ above does not act in a simply passive way because the G_y component, corresponding to β, tends to induce some correlation between ϕ and ζ.

3.2.2 A closure assumption

Although one cannot avoid 'guessing' at some level (so far as I know), many SGS problems of the sort we just encountered on a simple β-plane may be reduced, if not eliminated, by deferring our guessing to a higher level. We would try to follow the exact dynamics while calculating eddy statistics. The problem is that nonlinear dynamics will cause second order correlations such as eddy transport

---

[1]In the sometimes desperate effort to save potential vorticity mixing, it has been suggested that the seeming contradiction that we have discussed here could really be some very peculiar artifact from talking about statistically homogenous flow on an unbounded β-plane. After all, there can't be such a thing. However, this desperate bid also fails. Consider a barotropic fluid confined to a thin shell between two concentric spheres. Suppose there is no skin friction between the fluid and the spherical surfaces. Suppose that initially the fluid is in a mean state of uniform rotation plus random barotropic eddies such that the averaged zonal flow (integrated around any circle of latitude) yields only the background uniform rotation rate. We observe that this system is, in principle, realizable to the extent that skin friction may be small enough to neglect. We further observe that the gradient of potential vorticity (local vertical component thereof) is everywhere of one sign, vanishing only at the poles. If then we hypothesize that eddies tend to mix potential vorticity, we should expect everywhere to see a tendency for meridional eddy vorticity transport of one sign. Such a vorticity transport should everywhere induce a 'westward' acceleration. That would violate conservation of the angular momentum of the fluid. So the β-plane was not a trick; potential vorticity mixing really does fail – for this case. Presumably the reader will not leap to the opposite – and entirely unwarranted – assumption that potential vorticity mixing always fails. Sometimes it's OK, sometimes not. *Caveat emptor.*

to depend upon third order correlations. The third order correlations will depend upon fourth, fourth upon fifth, and so on – indefinitely. It is this unclosed hierarchy which prevents us from just calculating the SGS and being done with it. Instead we have to guess (a 'closure assumption'), sometimes from a phenomenologically motivated basis and sometimes on a more formally structured basis (say, a 'renormalization'). In the end though, we throw away something that cannot be proven to be smaller than what we keep. Moreover, it requires a good deal more effort than simpler 'hand-waving'. That effort tends to be rewarded though, both by avoiding outright blunders such as the β-plane illustration and also by making much more detailed quantitative predictions at SGS. Those detailed predictions are often directly testable from numerical simulations so that one can discover defects in the closure assumptions, hopefully leading to their redress through clearer dynamical insight as distinct from parameter 'tuning'.

Because the algebra of any particular closure application can become quite cumbersome, let me first provide a generic overview. More thorough reading on this subject can be found in Leslie (1973) with a number of geophysically motivated applications in Holloway (1986) or Lesieur (1987). While our specific application here will be directed to the β-plane, in a later section such closure calculation will be applied to internal gravity wave breaking / mixing.

Let all of the dependent variable fields be expanded on spatial Fourier bases. This strongly prejudices us toward considering statistically homogenous (or nearly so) problems. The motivation is for tractability; travails at statistically inhomogenous turbulence will be mentioned in section 3.5. The several Eulerian field equations that we consider and will consider in these lectures may be expressed

$$\frac{d}{dt}\mathbf{y}_k + \mathbf{L}_k\mathbf{y}_k + \sum_{\Delta}^{k_*} \mathbf{N}_{klm}\mathbf{y}_l\mathbf{y}_m = \mathbf{e}_k \tag{4}$$

where $\mathbf{y}_k$ is a complex valued vector whose components are the Fourier coefficients at wavevector $\mathbf{k}$ of the dependent variable fields. Components of $\mathbf{y}$ may include expansion coefficients from vorticity, velocity, tracer concentration, buoyancy, or other variables of interest.

The $\mathbf{L}_k$ consist of matrices $\mathbf{L}$ at each $\mathbf{k}$ resulting from linear operators in the partial differential equations of motion. $\mathbf{L}$ includes both wave propagation (for example coupling velocity and buoyancy fields of internal gravity waves) and linear dissipation. That $\mathbf{L}$ is local in $\mathbf{k}$ is a consequence of assumed statistical homogeneity.

The N_{klm} are matrices of coupling coefficients by which components of y_l and y_m affect components of y_k. Summation $\sum_{\Delta}^{k_*}$ is over all l and m whose magnitudes $\mid l \mid$ and $\mid k \mid$ are each less than some specified k_* and which form a wavevector triangle $k + l + m = 0$. This latter restriction upon possible l and m is also due to statistical homogeneity. The bi-linear form Nyy reflects nonlinearity due to the advection operator. Nonlinear pressure effects (also due to advection) are included in Nyy. Finally, e_k represents any external forcing imposed upon the components of y at k.

So far this is only lists the equations of motion in a terse style. Now what to do? One can get a big computer and integrate Eq 4, retaining as large a k_* as one's computer account permits. An illustration of this approach was seen in Figure 2. Except as possible illustrations of how eddy fields might work, such detailed eddy simulations may not be our primary interest. That interest may be more directed to statistics of those eddy fields – their energy spectra, property transports, and so on. Especially we may wish to estimate $\langle yy \rangle$ where $\langle \rangle$ denotes averaging over an ensemble of realizations of Eq 4, elements of the ensemble differing by choices of initial conditions or of e_k, say. Diagonal components of yy express energy spectra and other variances while off-diagonal components include eddy transports, among other things.

How to get $\langle yy \rangle$? A straightforward, but perhaps not very elegant, approach is to numerically integrate lots of realizations of Eq 4 and average. Alternatively we may attempt to write equations for the evolution of the statistics themselves. The algebra in such attempts grows tedious, so let me here further compress to a schematic:

From Eq 4,

$$\frac{d}{dt}\langle yy \rangle + L\langle yy \rangle + N\langle yyy \rangle = \langle ey \rangle \qquad (5)$$

$$\frac{d}{dt}\langle yyy \rangle + L\langle yyy \rangle + N\langle yyyy \rangle = \langle eyy \rangle \qquad (6)$$

and so on indefinitely as each order of correlation depends upon the next higher order. (Besides omitting wavevector subscripts, L and N only refer generically to coefficient matrices that would be obtained by manipulations on L and N from Eq 4. The L and N that occur in Eq 5 will be different than in Eq 6 and so on up the hierarchy. Importantly, each of these can be carefully, if tediously, derived. There is no guessing, so far!)

Again, what to do? Some progress can be made by recognizing certain products of correlations. Suppose that we are interested purely in eddies without mean flow, hence that $\langle y \rangle = 0$. Then we could write

$$\langle yyyy \rangle = \langle yy \rangle \langle yy \rangle + \langle yyyy \rangle' \tag{7}$$

where $\langle yyyy \rangle'$ are those parts of fourth order correlations which cannot be expressed as products of second order correlations.

It is not clear that this helps since an equation for $\frac{d}{dt}\langle yyyy \rangle'$ will contain fifth order correlations. The hierarchy continues until some kind of closure assumption is invoked.

The closure with the strongest (but yet arguable) mathematical defense occurs in the limit of weak nonlinearity when L is given by wave propagation. Since each wavelet propagates almost independently of the others over many wave periods, the phases of the waves should come to be randomly unrelated. While variance statistics $\langle yy \rangle$ exist, higher statistics that depend upon phase relations among waves are presumed to vanish according to the 'random phase approximation'. Hence

$$\langle yyyy \rangle' = 0$$

Omitting also e and taking a limit $t \to \infty$, Eq 6 has

$$\langle yyy \rangle = -L^{-1} N \langle yy \rangle \langle yy \rangle$$

with

$$L^{-1} \propto \delta(\omega + \omega + \omega) \tag{8}$$

where δ denotes the δ-distribution and $\omega + \omega + \omega$ here represents the sum of the natural frequencies of three waves correlated under $\langle yyy \rangle$. Energy transfer in Eq 5 is sustained only by waves that satisfy not only $\mathbf{k} + \mathbf{l} + \mathbf{m} = 0$ but also the frequency resonance condition $\omega_{\mathbf{k}} + \omega_{\mathbf{l}} + \omega_{\mathbf{m}} = 0$. This is the 'weak wave' or 'resonant interaction' approximation.

There is continuing mathematical discussion as to whether a resonant interaction approximation is in fact valid *even in the limit* of vanishingly weak nonlinearity. For our concerns though, the SGS seem not so weak. How weak is weak? A 'rule of thumb' is to require $\omega \tau_I \gg 1$ where τ_I is some kind of interaction time scale. It may not always be clear τ_l what really is , by how much the inequality needs be satisfied, or even whether this is the correct inequality to test.

Going on to stronger interactions, one might still guess $\langle yyyy \rangle' = 0$ just to see what happens. However, it was found by Ogura (1963) that if you apply this idea to turbulence, i.e. with L given only by dissipation which is presumed weak relative to nonlinearity, then Eqs 5 and 6 lead to unacceptable expectations for negative energies. Something else needs to be done.

There have been a host of alternatives to $\langle yyyy \rangle' = 0$ proposed. A survey of those alternatives is perforce outside the scope of these lectures. The interested and courageous reader may turn to Leslie (1973) who describes carefully a number of theoretical approaches commencing from the 'direct interaction approximation' of Kraichnan (1959). Here I'll only mention one view, my preferred way of guessing. We are trying to formulate the statistical dynamics of Eq 4. We haven't got much to go on. However, one special case, which may not be very realistic but which is theoretically interesting, is the case where we set any dissipation terms in L to zero, retaining waves if we like. Forcing terms e are omitted. Then, with finite k_*, we have a closed, isolated, nonlinear system of some large but finite number of degrees of freedom. Then, though it is only a hypothesis, the *usual hypothesis* is that such a system will evolve toward a maximum entropy distribution; the system will be equally likely to be found in any of the states available. The question is what states are available. The answer is found by listing the invariants of motion. Usually there is some expression for total energy conserved by the system. Depending upon what specific dynamics we have included in Eq 4, there may be other invariants such as potential vorticity variance or a passive tracer variance. Our hypothesis is then that any state $\mathbf{Y} = \{\mathbf{y_k} \mid \forall \mathbf{k}, \mid \mathbf{k} \mid < k_*\}$ which is consistent with energy and any other invariants is equiprobable. (Salmon *et al.* (1976) argue that this principle may account for aspects of large scale ocean circulation.)

Of immediate interest is that triple correlations $\langle yyy \rangle$ vanish at maximum entropy, at least for all the systems of which I am aware. Under more realistic conditions, including forcing and dissipation, maximum entropy distributions usually will not be approached. In particular $\langle yyy \rangle$ will not vanish. Nonetheless, as we seek a closure for the hierarchy begun at Eqs 5 and 6, we impose a thermodynamic consistency condition:

If the system is considered in isolation, without forcing or dissipation, *then* the closure assumption must be consistent with evolution toward maximum entropy.

Although closure will be applied to the open system (Eq 4, i.e. with forcing and dissipation) we seek to maintain the presumed statistical properties of Eq 4 over all of its parameter space. Now we return to the question of what to do about $\langle yyyy \rangle'$. Discarding this term will not do. Then there would be nothing in Eq 6 to cause $\langle yyy \rangle$ to tend to vanish in absence of dissipative terms in L. Here we see the role played by $\langle yyyy \rangle'$: this is the term which links lower order correlations $\langle yy \rangle$ and $\langle yyy \rangle$ to all of the higher order correlations, ultimately imposing thermodynamic consistency upon $\langle yy \rangle$ and $\langle yyy \rangle$. If we propose to cut this link, we need to put something in its place. Then the question is: what?

Attempting to proceed as simply as possible, the *form* of proposed substitution is

$$N\langle yyyy \rangle' = M\langle yyy \rangle \tag{9}$$

where M is a coefficient matrix to be determined. The hypothesis is that the influence of higher order correlations can be approximated as a linear relaxation on $\langle yyy \rangle$. Physically, the picture is that real turbulence tends to 'scramble' the phase relations in triple correlations while other turbulent tendencies expressed by $N\langle yy \rangle \langle yy \rangle$ tend to restore those phase relations.

The quantitative problem becomes how actually to specify M. It is a huge matrix for which an approach from 'parameter tuning' would be quite ridiculous. One may impose certain consistency conditions such as 'realizability', e.g. that the influence of $M\langle yyy \rangle$ preserves positivity of energies. Another condition termed 'random Galilean invariance' will be mentioned in a later section 3.7.2. And one may argue phenomenologically, restricting the form of M to admit (hopefully!) no more than one or a *very* few tuneable parameters.

Fortunately, without even going into details of M, one can make certain revealing observations. Suppose the motion field is reasonably statistically stationary. (Specifically, suppose second order statistics $\langle yy \rangle$ tend to evolve slowly relative to relaxation timescales for third order $\langle yyy \rangle$. Then Eqs 6 and 9 integrate toward

$$\langle yyy \rangle = (L + M)^{-1} N\langle yy \rangle \langle yy \rangle \tag{10}$$

It is the real part of $\langle yyy \rangle$ which will provide energy transfer across the spectrum at Eq 5. Suppose that L is dominated by wave propagation; here Rossby waves, in a later section we will see internal gravity waves. M was introduced to represent the nonlinear coupling of lower order correlations to higher order correlations; thus M is a measure of the strength of nonlinearity. Denoting a particular triad by subscripts $1, 2, 3$, $Real\langle y_1 y_2 y_3 \rangle$ will be proportional to

$$Real(L + M)^{-1} = \frac{M_{123}}{(\omega_1 + \omega_2 + \omega_3)^2 + M_{123}^2} \tag{11}$$

where ω_i is the wave propagation frequency at mode i and M_{123} is a real timescale (as yet unspecified) associated with nonlinear interaction at triad $\langle y_1 y_2 y_3 \rangle$. The form in Eq 11 results from $\omega_1 + \omega_2 + \omega_3$ reflecting the imaginary part of L, supposing that the $\langle yy \rangle$ in Eq 10 are real.

What we may observe here is this: In a limit of vanishingly weak amplitude, hence vanishingly small M, Eq 11 limits upon $\pi \delta(\omega_1 + \omega_2 + \omega_3)$. We recover a weak wave resonant interaction theory. In a limit of strong motion such that background wave propagation becomes negligible, Eq 11 approaches $1/M_{123}$. This gives the form of a whole class of turbulence theories called 'eddy damped quasi-normal Markovian' (EDQNM), where the particular means of specifying M_{123} will select a particular flavour of EDQNM turbulence. Examples of EDQNM are discussed by Lesieur (1987).

Eq 11 provides a continuous bridge from weak wave interactions through fully developed turbulence. We avoid entirely the sometimes popular dichotomy in which one supposes that flow fields are wavelike until the waves 'break' and then there is turbulence. With respect to real flows, such dichotomy often leads to trying to separate velocity or other fluctuations into a 'waves part' and a 'turbulence part', sometimes suggesting 'waves' propagating through the remnant 'turbulence' from prior wave 'breaking'. In the long run, I think that such artificial separations will frustrate quantitative analyses. Eq 11 is suggested as a 'one fluid' alternative to such separations.

To close this lengthy subsection, below are given the explicit results for passive scalar stirring in β-plane turbulence following Eq 2 and 3 . Results are only quoted here, with details given in Holloway (1986) and references therein.

$$(\frac{\partial}{\partial t} + 2\nu_{\mathbf{k}} + 2\eta_{\mathbf{k}})Z_{\mathbf{k}} = \Lambda_{\mathbf{k}} + F_{\mathbf{k}} \tag{12}$$

$$(\frac{\partial}{\partial t} + 2\kappa_{\mathbf{k}} + 2\gamma_{\mathbf{k}})\Phi_{\mathbf{k}} = \Xi_{\mathbf{k}} - 2G_{\mathbf{k}}Im\Gamma_{\mathbf{k}} \tag{13}$$

$$(\frac{\partial}{\partial t} + i\omega_{\mathbf{k}} + \nu_{\mathbf{k}} + \kappa_{\mathbf{k}} + \eta_{\mathbf{k}} + \gamma_{\mathbf{k}})\Gamma_{\mathbf{k}} = -iG_{\mathbf{k}}Z_{\mathbf{k}} \tag{14}$$

where $Z_{\mathbf{k}} = \langle \zeta_{\mathbf{k}}\zeta_{-\mathbf{k}} \rangle$, $\Phi_{\mathbf{k}} = \langle \phi_{\mathbf{k}}\phi_{-\mathbf{k}} \rangle$, $\Gamma_{\mathbf{k}} = \langle \zeta_{\mathbf{k}}\phi_{-\mathbf{k}} \rangle$, $\nu_{\mathbf{k}}$ and $\kappa_{\mathbf{k}}$ are from D_1 and D_2, $F_{\mathbf{k}}$ represents any external forcing of vorticity variance, $G_{\mathbf{k}} = \mathbf{z} \cdot (\mathbf{G} \times \mathbf{k})/k^2$, $\mathbf{z}$ is the unit vertical vector, and $\omega_{\mathbf{k}} = -\beta k_x/k^2$ with

$$\eta_{\mathbf{k}} = \sum_{\Delta} \theta_{\mathbf{kpq}} b_{\mathbf{kpq}} Z_{\mathbf{p}} \tag{15}$$

$$\gamma_{\mathbf{k}} = \sum_{\Delta} \theta_{\mathbf{kpq}} c_{\mathbf{kpq}} Z_{\mathbf{p}} \tag{16}$$

$$\Lambda_{\mathbf{k}} = \sum_{\Delta} \theta_{\mathbf{kpq}} a_{\mathbf{kpq}} Z_{\mathbf{p}} Z_{\mathbf{q}} \tag{17}$$

$$\Xi_{\mathbf{k}} = 2 \sum_{\Delta} \theta_{\mathbf{kpq}} c_{\mathbf{kpq}} Z_{\mathbf{p}} \Phi_{\mathbf{q}} \tag{18}$$

with

$$a_{\mathbf{kpq}} = \mid \mathbf{k} \times \mathbf{p} \mid^2 (\frac{1}{p^2} - \frac{1}{q^2})^2 \tag{19}$$

$$b_{\mathbf{kpq}} = \mid \mathbf{k} \times \mathbf{p} \mid^2 (\frac{1}{p^2} - \frac{1}{q^2})(\frac{1}{p^2} - \frac{1}{k^2}) \tag{20}$$

$$c_{\mathbf{kpq}} = \mid \mathbf{k} \times \mathbf{p} \mid^2 /p^4 \tag{21}$$

These algebraic results may appear burdensome. Their derivation is even more tedious. However, it is important to remark none of the above results require any kind of inspired guessing. Only straightforward, if tedious, algebraic bookkeeping is required. ... except at $\theta_{\mathbf{kpq}}$.

Guessing has been focused in the quantity $\theta_{\mathbf{kpq}}$ for which my preferred prescription is

$$\theta_{\mathbf{kpq}} = \frac{\mu_{\mathbf{kpq}}}{\omega_{\mathbf{kpq}}^2 + \mu_{\mathbf{kpq}}^2} \tag{22}$$

where $\omega_{\mathbf{kpq}} = \omega_{\mathbf{k}} + \omega_{\mathbf{p}} + \omega_{\mathbf{q}}$ and $\mu_{\mathbf{kpq}} = \mu_{\mathbf{k}} + \mu_{\mathbf{p}} + \mu_{\mathbf{q}}$ with

$$\mu_{\mathbf{k}}^2 = \lambda \sum_{\mathbf{P}} \frac{k^2}{k^2 + p^2} Z_{\mathbf{p}} \tag{23}$$

The form of Eq 22 follows from Eq 11 where M_{123} is given by $\mu_{\mathbf{kpq}}$. Eq 23 is a compromise between a more difficult formulation following Kraichnan (1971) and a simpler heuristic after Pouquet *et al.* (1975). λ is a 'tuneable' coefficient. In Eq 22, I have retained only the natural or 'bare' frequencies ω. There is a theoretical question, as Legras (1980), how one may calculate the effects of nonlinearity inducing mean frequency shifts in the ω. And there are deeper questions involving higher order phase correlations as implied by 'coherent structures' – whatever *they* are. (See sect 3.4)

3.2.3 What a lot of bother!

Is the effort that goes into closure theory justified? It is certainly algebraically cumbersome and does not avoid guessing, *viz.* at Eqs 9, 22 or 23. Moreover, the computational burden to evaluate Eq 12 through Eq 23 numerically is often found to exceed the effort to evaluate Eqs 2 and 3 directly – even many times to permit ensemble averaging. So why bother?

First, one does avoid certain blunders such as vorticity transport on the β-plane. Here $\langle u\zeta \rangle$ vanishes identically while the eddy transport of ϕ evaluates as

$$\langle \mathbf{u}\phi \rangle = \sum \frac{\mathbf{k} \times \mathbf{z}}{k^2} Im\Gamma_{\mathbf{k}} \tag{24}$$

With $\Gamma_{\mathbf{k}}$ from Eq 14 and λ in Eq 23 determined – indeed overdetermined – from other considerations, Eq 24 provides an *a priori* estimate of $\langle \mathbf{u}\phi \rangle$. Magnitudes of the individual summands on the right side of Eq 24 may be examined with respect to assessing the contributions from different scales of motion to the overall transport. This latter question is important to possible experimental design aimed at direct observation of $\langle \mathbf{u}\phi \rangle$.

One may observe that near stationarity Γ, and hence $\langle \mathbf{u}\phi \rangle$, are proportional to $|\mathbf{G}|$. Therefore we could proceed in a Fickian way to express

$$\langle \mathbf{u}\phi \rangle = -\mathbf{K} \cdot \mathbf{G}$$

where it turns out that $\langle \mathbf{u}\phi \rangle$ has the sense of down-gradient with respect to $\mathbf{G}$ while $\mathbf{K}$ is anisotropic both on account of any anisotropy in $Z_{\mathbf{k}}$ and on account of the direct role of Rossby wave propagation through $\omega_{\mathbf{k}}$ in Eq 14.

Whatever its purported theoretical advantages, the apparatus set out above appears grossly too cumbersome to be of direct SGS value in a practical ocean model. However, we might look at the theoretical expressions for some guidance.

Indeed, one of the most remarkable simplifications emerges, at first unexpectedly, then as an 'obvious guess' with hindsight.

Consider the very simplest circumstance: stationary, homogenous, isotropic, 2D turbulence, i.e. with $\beta = 0$. From Eq 14 and 24, representing results in terms of isotropic diffusivity K, the result is

$$K = \sum_{\mathbf{k}} \frac{Z_k}{k^2 \xi_k} \tag{25}$$

where $\xi_k = \nu_k + \kappa_k + \eta_k + \gamma_k$ and the several terms are indicated to depend only upon scalar wavenumber k. At large k, Z_k will be decreasing with increasing k. ξ_k will be increasing with increasing k. Thus contributions to K will be dominated at smaller k.

Suppose that the eddy spectrum is broadly peaked around some wavenumber k_o. (Such a peak might be a consequence of an imposed definition of 'eddy' if we imagine some spatial filtration separating the SGS or 'eddy' component from the larger scale, resolved flow.) Suppose that explicit diffusion ν_k and κ_k are minor contributors to ξ_k near k_o. Here let me only claim (to be checked!) that contributions η_k and γ_k increase with increasing k to attain values of the order of the r.m.s. vorticity ζ_{rms} at k_o. Above k_o, these contributions continue to increase[2]. Z_k decreases above k_o. Thus, the sum in Eq 25 is really exhausted by about k_o with a value 'roughly' ζ_{rms}/k_o^2. This is only the r.m.s. eddy streamfunction ψ_{rms}. A more careful evaluation of K depends upon more detailed prescription of Z_k and actual evaluation of contributions to ξ_k. Instead we might try to collect such details 'roughly' in an order unity coefficient C so that

$$K = C\psi_{rms} \tag{26}$$

If one had a way to estimate SGS ψ_{rms} and had a value for C, then Eq 26 provides a most simple SGS prescription – applied to the right things, i.e. *not* vorticity. What about C? Say 0.4, *caveat emptor*.

Now a critic could remark that Eq 26 would have been an obvious guess, quite apart from all the closure apparatus. It is simply a mixing length guess $K = u_{rms} l'$

---

[2]In fact there's a defect here such that experts may see that my high wavenumber contributions to K fail to pass 'random Galilean invariance'. With rather more effort, this can be averted. Luckily, for present purposes it does not matter since high wavenumber contributions to K – even if estimated 'correctly' – become more negligible at successively higher wavenumbers where my 'random Galilean' failure gets worse. The present purpose is sufficiently crude that this is quite insignificant.

where a natural choice for mixing length is $l' = \psi_{rms}/u_{rms}$ and C is a fudge factor. So sometimes things do work out simply though that need not have been guaranteed. Moreover the closure apparatus provides detailed prediction for wavenumber distribution of contributions to $\langle u\phi \rangle$, which may be tested and/or included in design of field experiments.

A mixing length guess becomes less obvious when one takes only a little step beyond simple 2D turbulence by including β. In part, some guidance is given from Eq 14 from which we need $Im\Gamma_\mathbf{k}$ in Eq 24. At stationarity, $Im\Gamma_\mathbf{k}$ will be proportional to

$$\frac{\xi_\mathbf{k}}{\omega_\mathbf{k}^2 + \xi_\mathbf{k}^2}$$

Thus β induces anisotropy into the diffusivity tensor $\mathbf{K}$ through $\omega_\mathbf{k}$. By symmetry, principal axes of $\mathbf{K}$ lie along x and y unless some aspect of external forcing breaks this symmetry. x-directed transport $\langle u\phi \rangle$ is associated with small ω, hence is less affected by β. y-directed transport $\langle v\phi \rangle$ is more strongly suppressed, as seen by replacing $1/\xi$ in Eq 25 by $\xi/(\omega^2 + \xi^2)$. 'Roughly', the meridional diffusivity K_{yy} is related to previous isotropic K as

$$K_{yy} = \frac{\zeta_{rms}^2}{\omega_o^2 + \zeta_{rms}^2}K = \frac{C\psi_{rms}}{(B\hat{\beta})^2 + 1} \tag{27}$$

where $\hat{\beta} = \beta u_{rms}/\zeta_{rms}^2$ and B is another numerical fudge factor introduced here to serve two purposes. First, we are trying to estimate a representative ω_o without explicit consideration of details of the $Z_\mathbf{k}$ distribution. Second, β plays an indirect role which enhances zonal transport $\langle u\phi \rangle$ while further suppressing meridional $\langle v\phi \rangle$; this is due to β inducing anisotropy into the velocity energetics such that $u^2 > v^2$, as observed by Rhines (1975). Since we do not explicitly account for anisotropic velocity energetics in Eq 27, that is collected, in part, in B.

A formula such as Eq 27 can be calibrated to some extent against numerical experiments such shown in Figure 3. Some results from such experiments are shown in Figure 3 where measured K_{yy} are compared with Eq 27 with $B = 3$, $C = 0.4$. The fit is not so great but not so bad. There may be a stronger suppression of K_{yy} at small $\hat{\beta}$ than expected. A test against oceanic data has also been reported by Freeland (1987), comparing K from mid-depth drifter dispersion with ψ_{rms} estimated from current meter moorings.

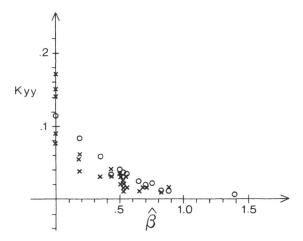

Figure 3: Measured values (x) of meridional diffusivity from numerical experiments are compared with theoretical values (o).

If one is brave, or maybe foolhardy, there is an interesting way to make use of these formulae. In ocean models, the ψ-variance at SGS is not available since it would require integrating some prognostic equation of its own. However, it may be that we can observe ψ_{rms} directly on a global basis. This is done from altimetric satellites in repeat orbits, as from Seasat or from the current Geosat mission. After correcting for a number of influences, *changes* of elevation of the sea surface can be attributed to ocean dynamics. We circumvent the problem of not knowing the geoid adequately. After removing tides and away from the equator, the remaining temporal variability of sea surface elevation h_{rms} can be associated with ψ_{rms} by geostrophy: $\psi_{rms} = g h_{rms}/f$. This is only ψ_{rms} at the sea surface, so one is left to make guesses how ψ-variance is distributed over depth. If one has some tolerance for such guessing, then the suggestion is that altimetric satellites give us a global picture of eddy diffusivity which, to the extent that it can be approximated from h_{rms} alone, is one of the very simplest things to extract from the satellite data.

There are a whole list of *caveats*. Unknown depth distribution of ψ-variance was mentioned. The relation (Eq 27) to eddy transport is based upon nearly homogeneous statistics and hence becomes suspect near strong, narrow, jetlike currents whose instabilities may be quite important to eddy transport. To estimate $\hat{\beta}$ carefully, we would need information beyond only h_{rms}. Values for B and C need be

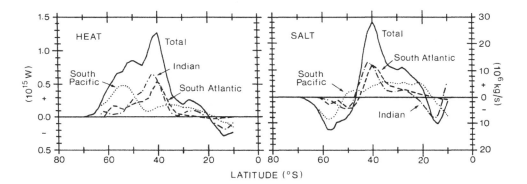

Figure 4: Zonally integrated poleward eddy heat flux in petawatts (10^{15} W) and eddy salt flux (10^6 kg/s) are shown for each ocean sector and summed over all.

assigned. To compound what may seem like folly, I tried this out for the eddy transports of heat and salt in the North Pacific above 15° latitude. Because of their effect upon buoyancy, heat and salt directly influence ψ_{rms} at the sea surface through steric anomalies; so their transport efficiency is further reduced. Put in the double negative though, results were 'not unreasonable'. This led to a more courageous calculation by Tom Keffer, bearing upon a region of much greater ocean - climate concern.

Keffer employed Eq 27 with ψ_{rms} from Seasat h_{rms} and the global ocean atlas data from Levitus (1982) to calculate meridional eddy transports of heat and salt throughout the southern hemisphere oceans away from the equator. Results (Keffer and Holloway, 1988) are shown in Figure 4.

The eddy heat transport, especially across the polar frontal latitudes 50° to 60° S, nicely resolves the heat budget dilemma identified by DeSzoeke and Levine (1981). Convergence of the eddy salt flux between 40° and about 60° S is broadly in agreement with excess precipitation over evaporation at these latitudes (Gill, 1982), recognizing the wide uncertainty in such estimates. A puzzle though is the divergent eddy salt flux poleward from 60° S. There doesn't appear to be a corresponding net fresh water loss to the atmosphere. Is the divergent eddy flux compensated by mean upwelling of more saline water? An interesting point is that calculations shown in Figure 4 suppose that K_{yy} acts upon 'horizontal' (i.e. equigeopotential) gradients. If instead we thought that K_{yy} acted along mean isopycnal surfaces, it would turn out that the slopes of isopycnals south of 60° are such that the gradient of salinity on the isopycnal is opposite in sign to the gradient of salinity on the equigeopotential.

Clearly there are puzzling curiosities and gross uncertainties in this section. However, there may also be some novel suggestions as to how satellite data might feed directly into an ocean model SGS, perhaps along lines of Eq 27.

3.3 Negative diffusion?

This topic arose in a student's question and expanded into lively discussion all around. In the previous consideration of β-plane stirring, we saw that K's can sometimes behave badly – vanishing for one field and not for another. Nonetheless, from the point of view of constraining inverse models, perhaps one may still impose an inequality that K's are everywhere non-negative. That is, can *negative* diffusion be ruled out? If not, can one guess what sort of circumstances may support negative diffusion? Three illustrations are offered for discussion. A fourth will occur when we get to small scale internal gravity wave breaking in a later section.

3.3.1 Positive diffusion in disguise

First, ordinary positive (i.e. down-gradient) diffusion can masquerade as negative diffusion. All one needs is anisotropic diffusivity in a coordinate system not aligned with the principal axes of $\mathbf{K}$. Suppose $\mathbf{K}$ is a symmetric tensor.

Aside: It is sometimes assumed that a diffusivity tensor *must* be symmetric. Not so. There is no *a priori* reason to require the antisymmetric part of $\mathbf{K}$ to vanish. Sometimes statistical symmetries in a given problem may impose symmetry upon $\mathbf{K}$. Also, flux resulting from the antisymmetric part of $\mathbf{K}$ is non-divergent, so this part of the overall flux may not be of interest in many cases. For present purpose, let us only consider symmetric $\mathbf{K}$.

Consider two dimensions of space with orthogonal coordinates (x', y'). In these primed coordinates, let components of $\mathbf{K}$ be

$$\begin{pmatrix} K_1 & 0 \\ 0 & K_2 \end{pmatrix}$$

Consider another orthogonal coordinate set (x, y) only rotated through an angle

α from the primed coordinates. In the unprimed coordinates, components of $\mathbf{K}$ are

$$\begin{pmatrix} K_1 \cos^2 \alpha + K_2 \sin^2 \alpha & (K_1 - K_2) \sin \alpha \cos \alpha \\ (K_1 - K_2) \sin \alpha \cos \alpha & K_1 \sin^2 \alpha + K_2 \cos^2 \alpha \end{pmatrix}$$

In unprimed coordinates, the x-directed component of flux is

$$F_x = -(K_1 \cos^2 \alpha + K_2 \sin^2 \alpha)G_x - (K_1 - K_2) \sin \alpha \cos \alpha G_y$$

where $\mathbf{G} = (G_x, G_y)$ is the background gradient.

Now suppose that all you have available are values of F_x and G_x. If you calculate an apparent Fickian coefficient from $-F_x/G_x$, you could get anything. It depends upon $K_1 - K_2$, α and G_y. In this way, for example, one may appear to observe negative (counter-gradient) diffusion even though, in a space of higher dimension, the diffusion is of ordinary down-gradient type.

3.3.2   Momentum transport in 2D turbulence

Here we consider 'negative viscosity'. That is, let the Fickian diffusion idea be applied to momentum transport in mean shear. This is a nice example because we can see the sign of the answer from only drawing a simple sketch. Suppose a uniform background shear $\frac{\partial U}{\partial y}$. Imagine eddies in this shear. In order not to prejudice things, let us begin our thought experiment with circular eddies at $t = 0$. That way there is no imposed eddy stress $u'v'$. To be fair, assign circulation of opposite signs to an equal number of eddies. Now imagine the eddies are simply kinematically deformed by the shear, hence tilted over into ellipses. Is that dynamically consistent? Sure. Just say that the lines we drew to indicate eddies are isolines of vorticity. Suppose also that the eddy field is narrow-banded in wavenumber. (Real eddies aren't – usually – but this is just *our* thought experiment.) By narrow-bandedness, eddy streamfunction is nearly proportional to eddy vorticity: $\psi = \nabla^{-2}\zeta \approx -l^2\zeta$. In this case, eddy advection doesn't affect eddy vorticity since $J(\psi, \zeta)$ vanishes. So only U advects vorticity and the sketch is OK.

Consider now the stress in the eddy field. By inspection we see that northward motion ($v' > 0$) tends to transport eastward ($u' > 0$) momentum. Likewise $v' < 0$ transports $u' < 0$. This works for eddies of either circulation. Hence on average $u'v' > 0$. And so an eddy viscosity $K_M = -u'v' \div \partial U/\partial y$ is negative.

Incidentally, there is no special prejudice for using closed eddies in this argument. One could imagine initial wavetrains which deform to the kind of tilted

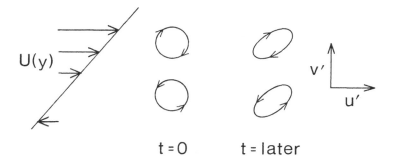

Figure 5: A sketch 'deriving' negative eddy viscosity.

ridges and troughs that also would sustain $u'v' > 0$. This and other illustrations, and implications thereof, are discussed in a thought-provoking monograph by Starr (1968).

Aside: Sometimes it will be said that one avoids peculiarities such as this eddy viscosity if one is careful to apply Fickian ideas only to (nearly) advected quantities S. Since momentum is readily transmitted through a fluid without material exchange, we ought not to attempt the Fickian representation. Here I venture a personal view: The Fickian idea is only so-so in any case. We saw that it went completely awry on the β-plane where we eddy-diffused an advected tracer but not advected vorticity. Note: *nearly* advected. Why not eddy-diffuse non-advected properties? On the one hand, we want to transport advected properties because they follow fluid blobs during random exchanges – recalling the molecular gas analogy from the outset. On the other hand, when a blob gets to its new location, we want it to take on the mean properties of that new location. So it would be nice to consider properties that are easily modified – like momentum. The one argument sounds as good (to me) as the other.

3.3.3 Upper ocean organic carbon

We go somewhat afield for another illustration. My inexpertise will show. Consider a very simple 1D (vertical) model for the distribution of primary production in the upper ocean. Suppose that plants of concentration p photosynthesize at rate λp while also suffering some kind of grazing or other mortality. λ is a depth dependent coefficient which depends upon light intensity and available nutrients. What will interest us is that the plants are advected by a varying vertical velocity w. Both p and w will fluctuate in space and time. We may wish to calculate the transport $\langle pw \rangle$ of p-stuff (organic carbon per volume, say), where $\langle \rangle$ might be time averaging over a period defining statistical stationarity.

Suppose a blob of water is fluctuating in depth about a depth z_o. Since nutrients and the grazing community (in a simplest-minded way) tend to be advected with the blob, its p tends to obey

$$\frac{dp}{dt} = \lambda_o p + p[\frac{\partial P}{\partial I}\frac{\partial I}{\partial z}\zeta] + \text{grazing, etc.} \tag{28}$$

λ_o is the equilibrium λ at z_o. As the blob undergoes vertical displacements, here denoted $\zeta(t)$, the main thing that happens is that irradiance I at the blob fluctuates. We further assume that the photosynthetic rate coefficient P responds to changes in I by moving along a '$P - vs - I$' curve. (Here I am ignoring time-dependence or 'photoadapatation' in $P - vs - I$ response. It turns out that for certain very simple representations of photoadapatation, our results to follow are not much affected. However, the present point will not concern the biological realism–or lack thereof!– of the model. If the biologically aware student will indulge me, we'll only come to a remark about Fickian assumptions.) Supposing small displacements ζ, we've kept only terms through first order in a Taylor expansion of P with respect to ζ.

Next consider w, retaining the part of total vertical velocity which is expressed by $d\zeta/dt$, e.g. the rate of vertical displacement of an isotherm. Thus we select the part of w that might be associated with internal wave activity. With p from integration of dp/dt and w from $d\zeta/dt$, form

$$\langle pw \rangle = p\frac{\partial P}{\partial I}\frac{\partial I}{\partial z}\langle(\int^t \zeta dt')d\zeta/dt\rangle = p\frac{\partial P}{\partial I}\frac{\partial I}{\partial z}\langle-\zeta^2\rangle \tag{29}$$

Suppose now we wanted to represent $\langle pw \rangle$ in a Fickian form as $-K\partial p/\partial z$. Unfortunately there's nothing in the expression above for $\langle pw \rangle$ that looks like this. (In fact I left that part out by retaining only the $d\zeta/dt$ part in w. We may imagine circumstances when turbulence is relatively weak compared with internal wave activity.) The important remark here is that if one just took the expression for $\langle pw \rangle$ and divided by mean $\partial p/\partial z$, one could get either sign for K. Although I shouldn't make too much of this simple model, there is a bio-oceanographic concern to which this may pertain. In oligotrophic waters, it is common to observe a maximum of mean p near the base of the euphotic zone. Model studies have tended to show that a *downward* transport of p-stuff *above* the p maximum is important to maintaining the mean p distribution. Hence *negative K*. Alternatively to negative K, it is sometimes supposed that negative buoyancy of the plant cells causes downward transport by sinking. Unfortunately, for the very small sizes of the plant cells, implied sinking speeds (roughly $1m/day$) appear to be too great.

3.3.4 On further consideration ...

First I must make a little confession. One of the illustrations is false. Deliberately. The others might also be wrong, but I don't know that. At the School we could take a show of hands for which one is suspect. Here the reader can peek ahead, so it is as well just to admit that the momentum transport argument (3.3.2) is wrong. However, this is an argument that comes up in lectures, at meetings, and which has appeared more than once in press. It is instructive to say what is wrong with it.

The figure as sketched is approximately right, I think. The error occurred when we looked at the tilted ellipses and 'saw' that northward flow tended to carry eastward momentum, and southward flow, westward. What we really 'saw' was that, over most of the flow domain, the orientation of flow was northeast-southwest. In two portions of each eddy, the flow also had to turn to northwest-southeast. What we could not 'see' in the figure was that in those short northwest-southeast portions, the flow intensifies. Streamfunction is not proportional to vorticity. A more careful calculation (Appendix B) shows that the shear deformation process from initially isotropic eddies leads to area-integrated stress $u'v'$ identically zero at all time. So we should have got identically zero for our eddy viscosity? No. Here is how I think the rest of the argument should go, though the student will see that this is only more 'hand-waving'. The influence of $\partial U/\partial y$ on the eddies is indeed to produce zero stress. Components of flow that contribute $u'v' > 0$ tend to have higher wavenumbers; contributions $u'v' < 0$ tend to come from lower wavenumbers. Consider next the likely effects of eddy-eddy interaction. Among the many effects possible, one usually expects a return-to-isotropy tendency (Herring, 1977). Arduous closure theory such as 3.2.2 can be used to quantify this return tendency. For the present, all we need to know (or guess) is that the return tendency is an increasing function of increasing wavenumber. Then it is the high wavenumber $u'v' > 0$ contributions which become more easily disoriented, relaxing toward $u'v' = 0$. What that leaves behind are the $u'v' < 0$ contributions at lower wavenumbers;

hence the eddy viscosity $K_M = -u'v'/\partial U/\partial y$ is positive! Sincerely (*this time*) that is what I think happens.

The foregoing discussion leading to positive K_M may be even more disturbing than the earlier suggestion of negative K_M. 2D turbulence has been actively studied for more than two decades and it has become well understood that smaller whorls make larger whorls. And so on up to mean $\partial U/\partial y$? No. What distinguishes the present discussion is that $\partial U/\partial y$ has been assumed to be spatially constant whereas the eddies started out in some finite wavenumber band. While eddy energy spills toward lower wavenumber, driven both by eddy-eddy interaction and by $\partial U/\partial y$, there remains a spectral gap between eddy scales and mean flow scales. It is this gap that permits positive K_M here despite a usual 2D turbulence tendency for 'negative viscosity'. Whether anything like a spectral gap, or even a gentle dip, exists between oceanic eddy scales and 'mean' flow isn't known. The question is more complicated because we may not know what we mean by the 'mean'. How far positive K_M would persist if we assumed only a spectral dip isn't addressed by simple 'hand-waving'. Closure theory would help but itself rests on 'hand-waving' at some level. If oceanic reality is characterized only by a gentle dip or none at all (assuming we've decided how to distinguish 'mean' from 'eddy'!) then are we back to negative K_M after all? Here I only guess and that guess is: no. We should be out of Fick and looking for something else to do about momentum transport.

To summarize this portion, it should be clear that sometimes eddy transports will be systematically counter-gradient. Simply taking the ratio of transport to gradient will yield apparent negative diffusion. However a 'rule of thumb' suggestion for such cases is that there is something else going on. One should try to figure out what that something is. Perhaps a higher dimension of space needs to be considered. Or the Fickian vehicle just isn't adequate. A personal view is this: *If one is going to suppose a Fickian representation, then negative diffusion is strictly disallowed. Clearly that's a big If.*

Before closing, let us return briefly to the plankton. Unless there was an unwitting mistake (apart from gross oversimplification of marine biology), we seemed to get negative diffusion. The 'rule of thumb' above suggests we reconsider. Indeed, the right side of Eq 29 more nearly resembles pW_* where W_* is some kind of apparent or virtual velocity

$$W_* = -\frac{\partial P}{\partial I}\frac{\partial I}{\partial z}\langle\zeta^2\rangle \tag{30}$$

This W_* is not a mean motion of the water nor even a mean motion of cells through the water. It arises as a correlation between vertical motion and photosynthetic carbon in each cell. Thus pw is not due to a gradient in mean p, as one would assume in a Fickian way, but rather is due to a gradient of P in presence of ζ^2 fluctuations. Interestingly, the sense of W_* is clearly downward so long as I is not so large as to cause photoinhibition ($\partial P/\partial I < 0$). Is it this which was mistaken for apparent cell sinking at unrealistic speeds? Just to try out some numbers, suppose $p(\partial P/\partial I)(\partial I/\partial z)$ is $1 day^{-1}/10m$ and ζ^2 is $(3m)^2$ as representative of the lower euphotic zone. Then W_* is about $1m/day$. So maybe we got lucky.

3.4 Role of coherent vortices

This will be limited to a comment only. It is a caution to bear in mind. There are circumstances, the range of which is not well known, in which a field of initially random vorticity fluctuations evolves into a set of nearly axisymmetric, largely isolated, discrete vortices. This can be seen in Figure 6 when $\beta = 0$. The figure is from some very old numerical experiments; more thorough studies on modern computers can be seen, for example, in McWilliams (1984).

A concern is that our previous discussions have not taken account of the possible tendency for vorticity collapse, seen here when $\beta = 0$. It is not clear how important this effect is. A closure theory such as 3.2.2 depends only upon variance spectra and so would be quite blind to the complicated phase relationships implied by a collection of coherent vortices. Babiano et al. (1987) have emphasized this failure, suggesting that it is the reason why simpler turbulence theories, following Kolmogorov (1941), are unable to distinguish between advected vorticity and advected passive tracer. Numerical experiments show however that vorticity variance spectra tend to have steeper slopes than do variance spectra of passive tracers advected in the same flow. A tendency for the vorticity field to collapse into coherent vortices while the tracer field does not experience this tendency may account for the different spectra of the two fields.

On the other side, the coherent vortices argument may not be right. First, it is not the case that closure theory such as 3.2.2 fails to anticipate the disparate spectra of vorticity and of tracer. Disparity results from the different expressions for b_{kpq} and c_{kpq} as these affect η_k and γ_k in Eq 12 and Eq 13. That difference results in Q_k tending to shallower spectra than Z_k as seen also in the numerical

Figure 6: Snapshots from unforced, decaying vorticity fields which were statistically identical at $t = 0$.

On the left, $\beta = 0$; on the right $\beta \neq 0$. When $\beta = 0$ the vorticity has collapsed into a set of nearly isolated, coherent vortices. When sufficient β is present, the collapse is not observed. How much β is 'sufficient' is not very well known.

experiments. Thus we don't seem to *need* coherent vortices to account for the differences. Second, consider cases when coherent vortices are not visibly present, as in the case of sufficient $\beta \neq 0$. *Still* $Q_{\mathbf{k}}$ tends to be shallower than $Z_{\mathbf{k}}$, as predicted from closure theory, but here in the absence of coherent vortices.

This comment should not be read to say that coherent vortices don't matter. Sometimes they do if, for example, we sought to predict individual realizations of the eddy fields in the manner of weather forecasting. However, coherent vortices may not be responsible for some of the effects that sometimes are attributed to them. To the extent that we seek to average over the details of SGS eddies, it remains unclear how significant is any tendency for coherent structure formation at the SGS.

3.5 Statistical inhomogeneity

From possible theoretical perspectives, the problem of SGS would be greatly simplified if only one could assume statistical homogeneity or something close to it. That is, it would help if statistics of eddy fields only changed over length scales and time scales long compared with the eddy scales themselves. Unfortunately there is little evidence that oceanic reality supports such 'quasi-homogeneity', especially in climate sensitive regions such as near strong boundary currents. What to do about it is one of the most difficult among our many unanswered questions. Here we explore just a couple of comments.

3.5.1 Inhomogenous Fickian assumption

A natural first step is to see how far we dare go with a Fickian assumption, setting aside for the moment concerns raised previously. Suppose Eq 1 applies locally insofar as $K(x)$ depends upon position x. A continuity equation for S might read

$$\frac{\partial}{\partial t}S + \nabla \cdot (uS - K \cdot \nabla S) = \frac{\partial}{\partial t}S + (u - \nabla \cdot K) \cdot \nabla S - (K \cdot \nabla) \cdot \nabla S = 0 \quad (31)$$

In the case of isotropic diffusivity, the last term contracts to more familiar $K\nabla^2 S$. Of interest here is the observation that $-\nabla \cdot K$ appears to play the role of another velocity field. It is like u and sometimes called a 'diffusion velocity'. In the case of isotropic K, it is just $-\nabla K$, i.e. a 'velocity' directed *down* the gradient of K. (Let me here hasten to remark that, while isotropic K is nearly always assumed with respect to horizontal (x,y), such assumed isotropy is hardly ever warranted.) The cautionary note to be made here is that one ought not too freely begin to treat $-\nabla \cdot K$ or $-\nabla K$ as if it *is* a kind of velocity, else one obtains paradoxes such as the following.

Suppose S is a dye spot or perhaps even a dense cluster of floats. We ask how the spot 'moves' under the influence of ∇K. Let us limit the discussion to one dimension, say x. Define the center of S as

$$r = \int^{\infty} xSdx$$

where $\int$ extends over all space but we suppose S vanishes outside some region, and we have scaled S such that $\int Sdx = 1$. Then, omitting $K\nabla^2$,

$$\begin{aligned}
\frac{\partial}{\partial t}r &= \int x\frac{\partial}{\partial t}Sdx = -\int x\frac{\partial}{\partial x}(uS - K\frac{\partial}{\partial x}S)dx \\
&= -x(uS - K\frac{\partial}{\partial x}S)\mid_{\infty} + \int (uS - K\frac{\partial}{\partial x}S)dx \\
&= \int uSdx - KS\mid_{\infty} + \int S\frac{\partial}{\partial x}Kdx \\
&= \int (u + \frac{\partial}{\partial x}K)Sdx
\end{aligned} \quad (32)$$

Terms evaluated $|_\infty$ vanish due to the finite extent of S. Apparently the center of the spot 'moves' *up* the gradient of K with some kind of 'diffusion velocity' $+\nabla K$ just equal but opposite to the previous $-\nabla K$. Did I make a sign error? That certainly should be considered! But I think not. The point really is to be cautioned from too quickly treating $-\nabla K$ (or $+\nabla K$) as if this were just another **u**. The role of $K\nabla^2$ with $K = K(\mathbf{x})$ is important.

3.5.2　The inhomogenous random walk

Here is an approach born of frustration. In fact the random walk as a 'model' for turbulent dispersion has been around at least since Taylor (1921). Taylor's careful work showed not only insight but also foresight: he avoided the statistically inhomogenous problem. Let us stumble ahead regardless.

In the random walk, a particle selects a new random step from some probability distribution of steps, doing so at a sequence of times that may either have a definite timestep increment or may themselves be random. In the very simplest walk, the timestep Δt is definite and the particle only chooses between 'forward' and 'backward' steps of length Δx. Here let us fix Δt but suppose Δx chosen from a continuous distribution. Taylor moreover took account that successive steps are not uncorrelated but rather that motion in any direction tends to persist for a while. Let us not worry about persistence by only considering $\Delta t > T_L$, c.f. sect. 3.1. Then each step can be presumed to be independent of the previous step. Imagine the particle is being continuously advected, but we will only observe its location every $\Delta t > T_L$.

The wonderful thing is that after many steps in the homogenous environment, the probability density function for the particle's location approaches a Gaussian almost regardless of what probability distribution governs Δx. (In fact there are a bunch of *caveats* here but let us not be too concerned; let us assume a central limiting tendency.) The variance $\langle x^2 \rangle$ for a particle started at $x = 0$ grows as $\langle (\Delta x)^2 \rangle t / \Delta t$, so that an effective diffusivity is $K = \langle (\Delta x)^2 \rangle / 2\Delta t$.

Intuitively it would seem straightforward to extend this model to inhomogenous turbulence. Only let the average step size $s = \langle (\Delta x)^2 \rangle^{1/2}$ be a function of x. Then a region of more intense turbulence is just characterized by larger s. (Alternatively one might take smaller Δt for more intense turbulence, or some combination of variable s and variable Δt. Let us, however, simply fix Δt, respecting that it remain larger than T_L.)

Already one might see how to evade a nagging SGS issue in ocean modelling. It is the question of boundaries. Suppose we have a usual sort of model employing constant K. Where there are boundaries, one may require no normal flow and, if the boundary is nonconducting or impermeable, no diffusive exchange. To impose the latter, we could require boundary conditions $\partial S/\partial n = 0$ where n is the normal coordinate. A critic might challenge this. How do we justify constant K in the presence of a boundary? If K is really due to eddy motion, and if the normal component of eddy flow always vanishes at the boundary, then eddy flux should vanish at the boundary without requiring any condition on $\partial S/\partial n$. We seem to be requiring a spurious boundary condition just to make up for an inappropriate SGS! Maybe we should have let K vanish near the boundary.

Random walk models can also handle boundaries in either of two ways. The constant K or constant s model just imposes that, when a particle would otherwise step through a boundary, it is instead reflected back onto the side from which it approached. Does that represent turbulent dispersion? Alternatively we could say that, as the normal component of eddy flow vanishes approaching the barrier, so s vanishes in a random walk. Which is more appropriate? (If the reader doesn't already have an answer, please pause to think about this before reading on.)

A difficulty arises because, although the usual random walk is well understood and discussed in numerous elementary textbooks, the inhomogenous, i.e. $s = s(x)$, random walk is not. (At least I have not found it. I have been variously told that this is a terrible, unsolved[3] problem or that the solution is only the same old Gaussian as before. Further insights can be found in biological literature where random walks are sometimes used to model animal behavior, as reviewed, e.g., in Okubo (1986)). We may not know what consequences follow from $s = s(x)$.

Let me suggest – tentatively – a rather simple solution which turns out to reveal a defect of the model with respect to actual turbulence. Define a coordinate

$$y = \int_{x_o}^{x} s^{-1}(\hat{x})d\hat{x}$$

where x_o is the launch point for a particle. The change of coordinates is such that in y the particle stepsize is independent of y. Therefore in y a central limiting tendency should cause the probability density function p_y to approach Gaussian.

---

[3]Unsolved in the sense of closed, analytical form. Numerical realizations are readily available.

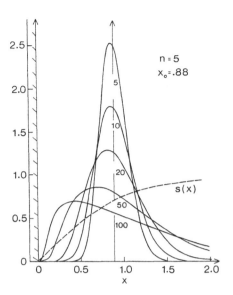

Figure 7: An illustration from hypothetical dispersion near a boundary.

Then we only transform probability back to x:

$$p_x(x) = p_y(y(x)) \mid \frac{dy}{dx} \mid = \frac{s^{-1}(x)}{\sqrt{2\pi N}} e^{-(\int_{x_o}^x s^{-1}(\hat{x})d\hat{x})^2/2N} \tag{33}$$

after N steps. Anyway, if there's not a mistake, then this is a solution. It is not Gaussian. The nice thing about an analytical expression is that we can readily examine its properties.

Consider the case of dispersion near a boundary. Set the boundary at $x = 0$. What is $s(x)$? Here let me not pretend to expertise in boundary layer turbulence but only pick a functional form for purpose of illustration. Say $s(x) = a \sinh x$. Then after N steps for a particle launched at $x_o = 0.88$,

$$p(x; N) = \frac{\coth x}{a\sqrt{2\pi N}} (\sinh x)^{-(\ln \sinh x)/2Na^2}$$

This is graphed in Figure 7.

This first illustration may seem encouraging. The picture looks somewhat realistic. Carried a little farther however, a defect becomes apparent. $p(x)$ becomes increasingly concentrated nearer to $x = 0$. For the present case of a single barrier, it may not be obvious that that is wrong. Consider instead a particle trapped between barriers at $x = -1$ and $x = 1$. Think of this as turbulence in a channel. Taking for illustration $s = a(1 - x^2)$, after N steps,

$$p(x; N) = \frac{(1 - x^2)^{-1}}{a\sqrt{2\pi N}} e^{-\frac{1}{8Na^2}(\ln\frac{(1+x)(1-x_o)}{(1-x)(1+x_o)})} \tag{34}$$

Here we could say on physical grounds that, no matter what the nature of the turbulence, after a very long time one should expect uniform $p = 1/2$ across the channel. With the analytical solution in hand, one can examine directly the limit of $N \to \infty$. It is not $1/2$. Rather, p approaches singularity at both $x = -1$ and $x = 1$, regardless of x_o. Perhaps the purported solution at Eq 33 is wrong? Or the evaluation at Eq 34 is wrong? No. These were tested against numerical realizations of thousands of walks for thousands of steps and they seem to work out OK. Neither is the problem only predicted singularities in p, which may be avoided by adding a small, positive bias to s; still p does not asymptote to its correct (*i.e.*, uniform) limit. The problem really is with trying to use *simple* random walks as models for inhomogenous turbulence. Okubo (1986) illustrates some alternatives. There are other 'fixes' possible: one may use uncentered (i.e. non-zero mean) probability distributions for the $\triangle x$ such that $\langle x \rangle$ depends upon ds/dx, for example. However, this prevents the kind of analytical solution (Eq 33) so far as I am able to see.

This discussion was intended to suggest possible promise but also pitfalls in thinking based on random walks. Clearly the problem of SGS for inhomogenous flows, especially near boundaries, is not so easily 'solved'. Here it may be appropriate to remind that these lectures are by no means complete with respect to SGS. Boundary layer turbulence is an enormously researched subject, whose size alone sets it outside the scope of these lectures (avoiding the question of author's knowledgeability).

3.6 Horizontal or isopycnal?

The question which probably stirred more debate over breakfasts or coffees during the School is: is this a question? *If* it a question, then it is an important question for SGS.

Assuming a Fickian form (of course!), then given that a diffusivity tensor $\mathbf{K}$ is highly anisotropic between its 'vertical' and 'horizontal' components, it may be very important to know carefully the orientation of those principal axes of $\mathbf{K}$. Should 'vertical' and 'horizontal' refer to geopotential coordinates or to a coordinate system that recognizes the mean orientation of isopycnal surfaces? Another question to which no answer will be offered here. Rather, what may be important in this short section is only to recognize that this question *is a question*.

Sometimes an idea becomes very popular, so much so that we begin to take it as a given rather than a hypothesis. When trying to organize these lectures, I meant to prepare a 'standard' discussion on why 'horizontal' mixing should really refer to 'isopycnal' mixing. Further complications would ensue with respect to the existence of mean isopycnal surfaces (McDougall, 1987) and with respect to transport processes *in* or *on* these surfaces – if they are surfaces. These are complications though meant to follow the **main remark:** eddy transport should **not** lie principally along equigeopotential surfaces.

Why not?

Disallowing appeal to 'higher authority', two lines of argument may be offered. The first concerns gravitational potential energy

$$P = \int_V \rho \Psi$$

where $\int_V$ is over the volume of ocean, omitting contribution from displacement of the free surface, ρ is *in situ* density, and Ψ is gravitational potential. Setting aside the difficult question of defining potential density σ of seawater, the energetic argument is that stirring which only interchanges water parcels of the same σ does not affect P since the field of ρ is unchanged. Hence work against gravity does not inhibit such eddy stirring.

However, if all we sought to do was to avoid working against gravity, another choice is available: interchange parcels along constant Ψ. (In fact, since isobaric surfaces do not coincide with Ψ and taking compressibility into account, one might not strictly interchange along Ψ. Let us only set aside this complication with the previous.) Apparently the energy argument no more favors isopycnal than geopotential stirring. Or we may take energy a step further. Suppose the principal axis of $\mathbf{K}$ lies somewhere between the isopycnal and geopotential surfaces, i.e. in the 'wedge of instability' from baroclinic instability theory. Now the interchange of parcels *releases* P, i.e. gravity works for the eddies[4]. This idea does not require us

---

[4]See also the discussion by Olbers in this volume concerning this point

to leap to a more formal statement about the role of baroclinic instability in eddy dynamics. It is closer to a 'psychological' view of eddies: they may be there for any number of reasons but they are also 'opportunistic' and will draw upon available gravitational energy where possible.

An objection (several surely!) can be raised against this 'psychological' view. If eddies are systematically releasing P, i.e. letting the center of mass of the ocean fall, what keeps the ocean up? Various influences work to increase P, such as Ekman pumping from windstress curl in presence of buoyancy differences or turbulent mixing near the benthic boundary. The question calls for more quantitative answer than will be attempted here.

A second argument against stirring along geopotentials is the implied transport of properties through isopycnal surfaces, i.e. diapycnal transport. Let the angle between an isopycnal and a geopotential surface be α. Oceanic values for α will be small in most regions, say 10^{-3} to 10^{-4} radians. Suppose a principal component of $\mathbf{K}$ really is along geopotentials, with a value of, say, $10^3 m^2 s^{-1}$. Denote this component K_2. In a coordinate system aligned with mean isopycnals, an orthopycnal component $\hat{K}_{11}$ is given from Eq 35:

$$\hat{K}_{11} \approx K_1 + K_2\alpha^2 \tag{35}$$

where K_1 represents 'vertical' diffusivity, perhaps from internal gravity wave breaking or some other small scale process. Smallness of α was used to simplify Eq 35. Concern arises from inferred values of $\hat{K}_{11}$ based upon microstructure profiler observations of dissipation rates for temperature and velocity variances in the upper ocean. Inferred $\hat{K}_{11}$ take values of $O(10^{-5}m^2s^{-1})$. With upper ocean α of 10^{-4}, we have already $K_2\alpha^2 \approx 10^{-5}$. So that seems marginally acceptable against inferred $\hat{K}_{11}$ but threatens somewhat the possible role of K_1, recognizing however that this is only a discussion to order of magnitude.

It may be interesting here to look at the 'vertical diffusivity' maps obtained from inverse calculations by Olbers *et al.* (1985) and ask how well these might be read as maps of $K_2\alpha^2$. I think that both the horizontal structure and the depth dependence are not inconsistent with this view, respecting the uncertainty of the calculation. On the other hand, if we think of some internal wave breaking K_1 and imagine that related to a universal internal wave spectrum, then Olbers's maps may be more difficult to rationalize.

3.7 SGS within partly resolved eddy fields

As ever more computing power becomes available, eddy resolving prognostic ocean models become feasible on ever larger scales. Perhaps SGS theoreticians and sundry 'hand-wavers' can be put out of business by taking over more of the formerly-SGS into resolved scales. On the contrary, this may provide new opportunities for SGS research. (The last apatosaurus may have entertained analogous thoughts.)

First let me elaborate only briefly the defense of SGS. When resolution is sufficient, hence explicit diffusion can be made small enough, that spontaneous eddies appear in a model, one should ask "Ah, but are these the 'right' eddies?" Concern is that they are not faithfully dynamically coupled to smaller scale eddies which are not yet resolved. A computer generation later, smaller eddies are resolved and the question moves to yet smaller scales. Presumably (but without guarantee!) the process converges at least with respect to mesoscale eddy resolution, leaving scales of internal wave breaking, etc. to parameterize. In fact this is terribly ambitious, whatever faith one may have in computing technology. (The last apatosaurus might also have regarded it as inelegant.) The bottom line is: Whatever the computing power available, we play a zero-sum game between eddy resolution on one hand and the desire to integrate over longer, climatologically interesting time scales, assimilate data in more sophisticated ways, or make wider exploration of parameter spaces on the other hand. Unless we really care about the details of all those eddies, they are only 'baggage' along the way to some other question that we really do care about. Expensive 'baggage'! For this reason one may be motivated to try to refine the SGS representation of smaller eddies within a field in which the larger eddy scales are resolved.

Whatever basis one supposed for Fickian representation previously, that basis becomes less tenable when one tries to represent eddies within eddies. From a dynamical systems view, the problem is to project a system of very many degrees of freedom (say 10^8 for the global ocean eddy field) onto a system of many fewer, but still many, degrees of freedom (say 10^5 in an ocean model).

Some progress can be made if we retreat to the simplifying case of statistically homogenous flows. We have in mind to treat the larger scales of flow explicitly and the smaller scales statistically. Thus each realization of the large scale flow is a strongly inhomogenous environment for the small scales. However, we might average over realizations of the large scale so that overall the problem is statistically homogenous. Spatially Fourier transformed, the situation may be as sketched in Figure 8.

Statistical closure theories such in section 3.2.2 are one approach, seen for example in Leith (1971).

A numerical empirical example may provide a more direct illustration. Figure 8b is from an integration of 2D turbulence, Eq 2, without β. The computation was run to forced-dissipative statistical stationarity with a truncation at $k_* = 60$. An energy transfer function $\Upsilon(k)$ was evaluated as the rate at which all resolved interactions act toward increasing or decreasing energy at k. What we wish we could know would be the correction for interactions with waves $k > k_*$ but these are not available. Instead we may see what happens if we successively eliminate portions of the resolved band $k < k_*$. k_* was reduced to $55, 50, 45, 40$, and in each case the difference of Υ was calculated with respect to the $k_* = 60$ case. These $\triangle\Upsilon$ are shown in Figure 8b. The figure is more than a decade old; such calculations could be carried out for very much greater k_* today.

What we see in Figure 8b is the error due to different k_* relative to the case when $k_* = 60$. What we *wish* we could see is the error when $k_* = 60$ relative to $k_* \to \infty$. Choices are either of theoretical sort, after Leith, say, or to hazard an extrapolation from numerical empiricism, say by guessing that $k_* = 40$ relative to 60 should be 'similar' (in some sense) to $k_* = 60$ relative to ∞. Both approaches are frought with unreliable guessing. However, let us imagine we have guessed correctly and that we have 'true' $\triangle\Upsilon$ for some actual problem of interest. We still need to think of an SGS scheme to 'correct' our computed Υ for $\triangle\Upsilon$. In fact one can think of a number of schemes to accomplish this. Each will produce a different flow evolution; which, if any, is 'right'?

The scheme closest in spirit to our previous discussions is to devise a kind of spectral eddy viscosity which need bear no relation to a Fickian form. This scheme is most suited to models integrated in spectral representation. The energy spectrum $E(k)$ for such a model might evolve according to

$$\frac{\partial}{\partial t}E(k) = \Upsilon(k) + F(k) - 2\nu(k)E(k) - \triangle\Upsilon(k) \qquad (36)$$

where $\Upsilon(k)$ are the nonlinear transfers explicitly computed by the model, $F(k)$ represents any external forcing, and $\nu(k)$ is the Fourier representation of any explicit, linear dissipation term in the model. If the model has a Fickian term $\nu_o\nabla^2$

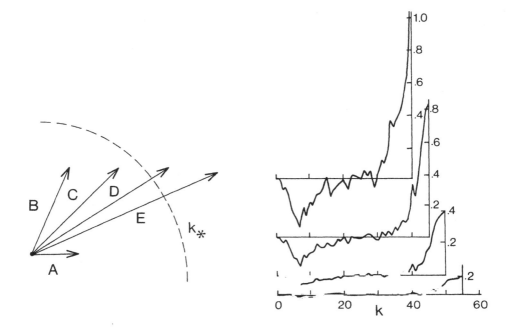

Figure 8:

a: The effect of an isotropic truncation at k_* is to prevent certain wavevector triad interactions. Suppose we are interested in evolution at wavevector A. Triad ABC is resolved. Two distinct classes of triads that are eliminated are ACD and ADE. The goal of an SGS scheme is to provide new terms at A to compensate the loss of ACD and ADE.

b: Differences $\triangle \Upsilon(k)$ in energy transfer are graphed as k_* is reduced from 60 to 55, 50, 45 and 40. Discussion is in the text.

then $\nu(k) = \nu_o k^2$, for example. As well $\triangle\Upsilon(k)$ is subtracted in Eq 36, indicating our intent to somehow 'correct' the model. Hypothesize that we know $\triangle\Upsilon(k)$; the question is what explicit change to the model will be made so that the model itself realizes the 'true' energy equation (36). One choice is to revise the dissipation term, replacing $\nu(k)$ by $\nu(k) + \nu_*(k)$ where

$$\nu_*(k) = \frac{\triangle\Upsilon(k)}{2E(k)} \tag{37}$$

Looking back to Figure 8b and imagining dividing by $2E(k)$, where all we need consider here is that $E(k)$ will be strongly decreasing with increasing k at higher k, we observe that ν_* will have a very sharp, positive cusp for $k \rightarrow k_*$. Modellers sometimes represent this by $\nu_*(k) = \nu_n k^n$ with some large value for n.

3.7.1 Biharmonic (∇^4)

In spectral models one is at liberty to assign any desired shape to $\nu_*(k)$. For more usual finite difference models, larger n will require a larger stencil, so one is practically limited to $n = 4$, perhaps, i.e. to an operator $\nu_4\nabla^4$. This is not a form derived from any theory, so far as I know. It is only 'better' than Fickian ∇^2 insofar as k^4 is more strongly cusped hence *somewhat* more like a plausible $\nu_*(k)$. By weighting dissipation more strongly to small scales, ∇^4 is more dissipative of vorticity variance relative to dissipation of velocity variance, this being a theoretical attribute of 2D turbulence. However, it should be emphasized that ∇^4 is *not* a theoretically derived operator. Arguments in its favor are two:

- It is practically implementable in usual finite difference models.

- It is (arguably) 'better' than ordinary ∇^2 (which may be awful).

A disadvantage of ∇^4 is that one then needs auxiliary boundary conditions for well-posedness. Lacking a theoretical derivation for ∇^4, one is quite unguided as to what sort of boundary conditions to impose. Here the fallacy is clear: ∇^4 or any other ν_* was only suggested above from statistically homogenous considerations. Approaching physical barriers, eddy transports must vanish by vanishing normal component of eddy flow. We would go back to the problem of grossly inhomogenous SGS near barriers, perhaps one of the worst of our unanswered questions.

Another caution is that the motivations listed above for ∇^4 were based on the equation for 2D vorticity advection. If we only considered a tracer advection in the same 2D vorticity field, we would get a different operator, say $\kappa_*(k)$, for the tracer. This just recalls a previous comment that vorticity and passive tracer, while both advected properties, behave differently. ∇^4 for one may not be such a good idea for the other. In fact though ∇^4 is probably sufficiently far from either $\nu_*(k)$ or $\kappa_*(k)$ that it is only arguably 'better' than ∇^2 for both!

An item that we have been overlooking from Figure 8b is the occurence of $\triangle \Upsilon < 0$ at lower wavenumbers. In such ν_* formulations this is usually omitted. What should be done?

3.7.2 Anticipated Potential Vorticity

This is quite a different and intriguing idea suggested by Sadourny and Basdevant (1985). An equation for advection of potential vorticity q would be written

$$\frac{\partial}{\partial t}q + J(\psi, q) = J(\psi, \theta \nabla^\lambda J(\psi, q)) \qquad (38)$$

where it is the nested Jacobian term that provides the SGS. θ and λ are parameters to adjust. Here I will not try to reproduce the detailed arguments of Sadourny and Basdevant, to whom the reader is referred. Perhaps the main property to observe from Eq 38 is that it is conservative of total energy, this following from $\int_D \psi \frac{\partial}{\partial t}q = 0$ where $\int_D$ denotes integration over the flow domain, assuming the domain is enclosed by fixed solid boundaries or periodic boundary conditions. Conservation is assured from $\int_D \psi J(\psi, \text{anything}) = 0$. But we don't want just 'anything'. The nested Jacobian will be chosen to be strictly dissipative of $\int_D q^2$, depending upon θ and λ, such dissipation being in conformance with theories of quasigeostrophic turbulence. This dissipative property is most easily seen in the case that one chooses $\lambda = 0$. Then

$$\int_D q\frac{\partial}{\partial t}q = \theta \int_D q J(\psi, J(\psi, q)) = -\theta \int_D [J(\psi, q)]^2$$

where a term $\int_D J(\psi, q^2)$ vanishes by specification of boundary conditions on D. Non-zero λ may be chosen to modify the scale selectivity of this SGS operator. Another way to see this dissipativity is that the nested Jacobian off-centers the time differencing operator. (Effectively the q advected by $J(\psi, q)$ is a little in advance of the current time, hence the label "anticipated potential vorticity" or APV.) Here a caution mentioned from the outset should be recalled: It would have been easy, possibly inadvertently so, to bury this SGS scheme inside 'the timestepping algorithm' never to be seen by unwary eye! Such 'technical detail' of numerical method is no place to hide the SGS, especially if one were then to represent a model as 'free' from SGS. Rather, as Sadourny and Basdevant have done, that SGS should be explicit and, hopefully, be explicitly motivated from mathematical - physical reasoning.

The APV is clearly attractive from the view of its conservation and dissipation properties, at least with respect to modeling quasigeostrophic dynamics. Whether or how APV can be extended to primitive equation dynamics remains to be seen. The APV which can be applied to vorticity advection can be applied directly to tracer advection without naively imposing the same effective ν_* on tracer as on vorticity. The different advective dynamics of the tracer field relative to the vorticity field means that APV will work differently. But is it right?

The effect of an APV operator on the vorticity field has been explored in numerical experiments by Vallis and Hua (1988) and compares qualitatively with the sort of $\triangle \Upsilon(k)$ seen in Figure 8b.

One place where APV fails though is with respect to 'random Galilean invariance' as defined by Kraichnan (1971). If the motion field were subjected to Galilean translation, but the specific realization of that translation were a random variable, one should require that SGS be independent of such random translation. What we mean physically is that, if a patch of turbulence is caught up in a large scale flow, the character of the turbulence should be independent of the large scale flow as that scale becomes asymptotically large relative to the turbulent scales. APV can be made to be strictly Galilean invariant in the sense that a linear trend in ψ can be set aside in Eq 38 and treated separately. However, the 'sweeping' of small scale structures by larger but still finite scale flow does appear to become confused under APV. So the method isn't perfect, to no one's surprise, but certainly promising.

One reputed disadvantage of APV is that it appears to be computationally intensive. The machine cost to advance equations such as Eq 38 is primarily the cost to evaluate a Jacobian. Nesting in Eq 38 appears to require that we evaluate not only $J(\psi, q)$ but also $J(\psi, J(\psi, q))$, calculating two Jacobians instead of one. However, this penalty can be evaded. If, as is often the case, one employs a leapfrog timestep, then one can readily extrapolate to a vorticity slightly advanced from the current time. Although strictly definite dissipation of q^2 is not guaranteed, this 'extrapolated' vorticity method seems operationally equivalent to APV at effectively no cost over the basic timestepping on $J(\psi, q)$.

There is yet one unsettling thought relative to all of the methods of SGS discussed thus far. *None* of them are right. They can't be for the simple reason that they are all deterministic.

3.7.3 Eddy noise

This is only a comment before moving to smaller scale 'vertical' mixing processes. Concern is whether *any* of the deterministic SGS can be right. We admit that our finite computing resources will not permit us to resolve *fully* the flow fields. Neither would we have the observations available to initialize a hypothetically fully resolved model, nor to provide the faithful details of external forcing, if our goal were to forecast or hindcast the actual ocean. There are always things we can't observe or can't compute or both. What to do about those things?

SGS operators such as ∇^2 or ∇^4 act purely as sinks for getting rid of variance of resolved flows. If one took quite seriously the suggestion at Eq 37 and supposed $\triangle\Upsilon(k)$ resembled Figure 8b, including $\triangle\Upsilon(k) < 0$ portions, then one's $\nu_*(k)$ would act both as a source and a sink for resolved variance. $\nu_*(k) < 0$ would appear as an instability of the larger scale flow, either a linear instability if one simply prescribed $\nu_*(k)$ from the outset or a kind of nonlinear instability if $\nu_*(k)$ responded to the evolving spectrum of resolved motion. APV also will act both as a source and a sink for different portions of the resolved spectrum. Importantly, in every case the SGS acts upon the resolved motion in a way that is fully determined by the resolved motion.

This picture is *dynamically inconsistent*. Recall the discussion from early in this section 3.7. We viewed the problem of ocean modelling as one of projecting an ocean of 10^8 (say) degrees of freedom onto an ocean model of 10^5 (say) degrees of freedom. So what have we done about those 10^3 degrees of freedom which are omitted for every one retained? Effectively we have said that the 10^3 are fully determined by the one! Nonsense. From a more statistical mechanical perspective, we might talk about our ocean model (10^5) immersed in a 'thermal bath' (10^8). The model can lose energy to the bath, perhaps represented by ordinary eddy viscosity, but the model will also receive energy from the bath effectively by a kind of random bombardment from 'molecules' in the bath. Lest one too quickly imagine an ocean model simply buffeted by a kind of Brownian motion, we should remain cautious that those unseen 'molecules' are actually smaller scale eddies with very finite length and time scales. Rose (1977) makes a brave effort at some of these issues.

When we saw $\Delta \Upsilon < 0$ in portions of Figure 8b, one could say "Ah, there's the 'thermal agitation'." However, even in portions $\Delta \Upsilon > 0$, one should imagine 'thermal agitation' – indeed increasing agitation – as $k \to k_*$. For another dynamics than 2D vorticity advection, one may have $\Delta \Upsilon > 0$ everywhere; still the argument would apply that all those scales should be subject to *some* amount of 'thermal agitation' even if that is overcome by 'thermal damping'. The challenge to quantitative scientists is to say *how much* agitation at what length scales and on what time scales. This is another of our terrible Unanswered Questions. Haphazard guesses are likely to fall quite wide of the mark. A wary reading of Appendix A may also be in order. On the other hand, saying that this agitation stuff is too difficult so we should leave it out entirely is surely an extremum among haphazard guesses.

A closing aside concerns 'predictability'. Although we may not be approaching ocean models from the view 'predicting the weather in the sea', still consideration of limits of predictability is appropriate. Even if we were given True Knowledge of ocean dynamics, and perfect numerical discretization thereof in terms of the finite degrees of freedom in a model, and also given perfect initial observations of those finite degrees of freedom as well as their perfect subsequent forcing, still we must expect that any uncertainty at all about the degrees of freedom not modeled or not observed will come in time to contaminate the model results. How long it takes such uncertainty to grow up to some gross contamination level is called the predictability time. Actual errors that we always must make, whether errors of observation, of forcing, or of defective modeling, will reduce the predictability time. The danger in running a deterministic model is that one can overlook limits to predictability – or at least not know where those limits are. "But I resubmitted my job six times and got the same answers each time, so it's right." The point would be that including random agitation (hopefully more or less the 'right' kind of agitation) tends to dissuade one from overstepping predictability. To the extent that our goal may be to model longer term statistical behavior of the ocean, perhaps with respect to climate issues, the modelled agitation will affect those statistics.

4 QUASI-VERTICAL MIXING

As we run out of time and space for these lectures, discussion of the smaller scale processes leading to vertical transport will be more brief. We will not try to distinguish between 'vertical' in the sense of normal to equigeopotential from diapycnal (*i.e.*, normal to mean isopycnal). Recall though from section 3.6 that diffusivity in the 'horizontal' (equigeopotential) can already induce a diapycnal component of transport.

4.1 Shear dispersion

First, there is a direct connection between small scale vertical mixing and the larger scale horizontal mixing. This is due to 'shear dispersion', as discussed by Kullenberg (1972) and Young *et al.* (1982). For a careful analysis, the reader is referred to Young *et al.*; here I illustrate a special case in order simply to fix the idea and derive an especially useful result.

Suppose that some process, yet to be determined, is causing vertical mixing of a sort that can be described by a Fickian coefficient K_v. Let there be a vertical shear in the water column which, for simplicity, we take to be uniform in z and oscillatory in t. We may think of inertial oscillations where we are considering vertical scales small compared with vertical wavelength. In a vertical $(y - z)$ plane, the velocity is

$$V = Sz \cos \omega t$$

where S is the fixed shear amplitude. A passive tracer ϕ evolves according to

$$\frac{\partial}{\partial t}\phi + (Sz \cos \omega t)\frac{\partial}{\partial y}\phi = K_v \frac{\partial^2}{\partial z^2}\phi \tag{39}$$

For initial conditions let $\phi_o = a_o \cos(ky + lz)$. One can consider more general initial conditions. Since Eq 39 is linear in ϕ, only suppose that we have decomposed ϕ_o into Fourier harmonics with the intent later to reconstruct ϕ by superposition. Now let us see if a solution to Eq 39 can be found in the form

$$\phi = a(t) \cos \theta$$

where

$$\theta = k(y - \frac{Sz}{\omega} \sin \omega t) + lz$$

Substituting in Eq 39, obtain

$$\frac{1}{a}\frac{d}{dt}a = -K_v(l - \frac{kS}{\omega}\sin\omega t)^2$$

If we assume that diffusion is sufficiently weak that $\frac{1}{a}\frac{d}{dt}a << \omega$, then the slow decay of a is approximated by

$$\frac{1}{a}\frac{d}{dt}a \approx -K_v(l^2 + \frac{1}{2}\frac{S^2}{\omega^2}k^2)$$

It is as if the explicit decay from vertical diffusion (i.e. $K_v l^2$) is supplemented by $\frac{1}{2}K_v\frac{S^2}{\omega^2}k^2$ which is just equivalent to a horizontal diffusivity

$$K_h = \frac{1}{2}\frac{S^2}{\omega^2}K_v \qquad (40)$$

Oceanic S^2 tends to be limited by the stability frequency N^2, while lower frequencies near the inertial $\omega^2 \approx f^2$ may dominate so that in a 'rough' way

$$K_h \approx \frac{N^2}{f^2}K_v$$

where oceanic N^2/f^2 may take values of order 10^3.

Here it should be stressed that this K_h is not the mesoscale eddy diffusivity variously discussed in section 3; rather it is a smaller scale diffusion acting at the high wavenumber end of any kind of mesoscale eddy stirring process. Now, if only we knew something about K_v!

4.2   Internal wave breaking

The cartoon at Figure 1 suggests a host of small scale processes which may contribute to vertical transport, whether or not they support Fickian representations. Here we will set aside many of these processes, omitting the separate roles of heat and salt and effects of nonlinearity in the equation of state, for example.

As a conceptual overview, one may imagine internal inertial-gravity waves generated on the boundaries of the ocean, by variable windstress, by variable air-sea buoyancy flux, by exchange with surface gravity waves, by radiation from mixed layer turbulence, by tidal flow over bottom topography and/or by other mechanisms. So generated, the internal waves radiate into the interior. Along the way, they are modified by wave-wave interactions, by possible critical layer encounters in the ambient shear field, by time-dependent caustic focusing, by radiation stress

/ shear interaction with mesoscale eddies and/or by other mechanisms. Somehow all of this may lead to the waves giving up energy to small scale turbulence which may sustain some K_v. As well there may be some turbulence generated directly by unstable mean shears; however, except in special places such as the equatorial undercurrents, it appears that most mean shears are quite stablized by the mean stratification.

The whole problem area of internal wave energetics and the transfer of energy through internal waves down to 'turbulence' scales is enormously difficult and theories on the matter are quite controversial. Maybe one can circumvent such difficulties by making many direct measurements of K_v under a wide variety of circumstances and then discover empirically how to characterize the observed K_v in terms of large scale environmental parameters such as would be available to an ocean model. In part this approach is being pursued using microstructure profilers which observe very small scale fluctuations of temperature and velocity. The question then becomes how to estimate K_v from fluctuation statistics of temperature and velocity.

Osborn and Cox (1972) have attempted to infer K_v from profiles of temperature fluctuation, arguing that conductive loss of temperature variance at rate $\kappa \mid \nabla T' \mid^2$ is compensated by production through the product of flux $K_v \partial \bar{T}/\partial z$ by the vertical gradient $\partial \bar{T}/\partial z$, hence

$$K_v = \kappa \mid \nabla T' \mid^2 /(\partial \bar{T}/\partial z)^2 \qquad (41)$$

where κ is the thermal conductivity of sea water and T' is temperature deviation from a suitably defined mean $\bar{T}$ profile. $\mid \nabla T' \mid^2$ is usually estimated from a component derivative with some assumption about dissipation isotropy or lack thereof. The dimensionless ratio $\mid \nabla T' \mid^2 /(\partial \bar{T}/\partial z)^2$ has become known as the Cox number.

A suggestion by Lilly et al. (1974) in relation to stratospheric observations is

$$K_v = E\epsilon_o N^{-2} \qquad (42)$$

where ϵ_o is the viscous dissipation of kinetic energy, N the stability frequency, and E an efficiency factor usually supposed to lie between 0.1 and 0.3. The interpretation of Eq 42 is that, of the total kinetic energy lost to viscous dissipation, an additional fraction $E\epsilon_o$ is expended in work against gravity at rate $K_v N^2$ due to vertical mass transport.

Both relations Eq 41 and 42 are in use by oceanographers to 'measure' K_v. As with each of our other discussions, there are questions about precision. As mentioned, uncertain assumptions about dissipation scale isotropy must be made by either method. As well, both $|\nabla T'|^2$ and ϵ_o are intermittent so that inferences of true mean values from sample means are sensitive to uncertain assumptions about underlying long-tailed distribution functions. Nonetheless these are only the kinds of imprecisions that we seem to have to live with at the SGS. However, I would like to pose a more distressing question:

Is it clear that $|\nabla T'|^2$ or ϵ_o actually have anything at all to do with K_v?

The motivation behind this question is partly to remark that it *is a question*. Too often, microstructure observations are discussed as though they *are* measurements of K_v, subject only to usual kinds of imprecision. That need not be the case at all. Recall the scenario where internal waves are generated at boundaries and radiate into the ocean interior. *At the boundaries* temperature and velocity fluctuations may be impressed upon the ocean, including vertically displacing the center of mass of the ocean. Subsequently, as the waves radiate into the interior, breaking up and eventually being dissipated at small scales, there is no compelling reason why the ensuing 'turbulence' performs any work against gravity. Dissipation of $|\nabla T'|^2$ may reflect only the wave radiation of $(T')^2$ in from boundaries with no $K_v(\partial \bar{T}/\partial z)^2$ production at all (except at the boundary sites which acted as sources for the wave radiation). The stationary balance underlying Eq 41 (here listing only vertical derivatives of mean quantities) is

$$K_v(\frac{\partial \bar{T}}{\partial z})^2 = \kappa |\nabla \bar{T}'|^2 + \frac{\partial}{\partial z}\overline{w'T'^2} \tag{43}$$

The usual assumption is that $\frac{\partial}{\partial z}\overline{w'T'^2}$ may be omitted. However, 'back of the envelope' calculation readily shows that this term may be at least as large as any other term in the balance, leaving even the sign of K_v indeterminate.

Likewise ϵ_o need not imply $K_v N^2$. Why convert some kinetic to potential energy? Wave radiation carries both kinetic energy and available potential energy (as T'^2). In the process of breaking waves down to dissipation, one could as well imagine converting potential to kinetic energy, thereby inducing negative K_v – referring back to section 3.3.

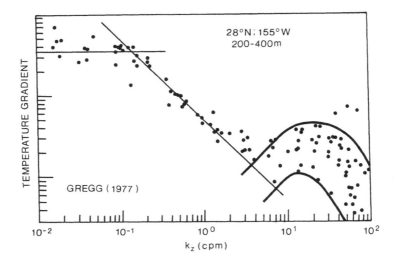

Figure 9: Vertical wavenumber spectra of temperature gradient, after Gregg (1977).

It is not the intent here to say whether K_v by internal wave breaking is positive, negative or otherwise. The main point of this section is only to caution that microstructure dissipation measurements don't *necessarily* have anything to do with K_v. What then, if anything, can be said?

4.3 Buoyant turbulence

Velocity or temperature variance dissipation rates are only moments of their respective spectra. Perhaps clues to the underlying processes can be found in spectral distributions. There are certain striking and reproducible similarities between vertical wavenumber spectra of temperature gradient ($\partial T'/\partial z$) reported by Gregg (1977) and vertical wavenumber spectra of shear ($\partial v/\partial z$) reported by Gargett *et al.* (1981), as shown in Figures 9 and 10.

Is there a dynamical theory of internal wave breakdown / buoyancy modified turbulence which is consistent with these spectra and which serves to tie K_v to the dissipation moments of the spectra? It turns out that there is. Or rather, there are. At least two theories seem to be completely consistent with observed spectra both of velocity and of temperature; unfortunately they give almost diametrically opposite predictions for the corresponding K_v!

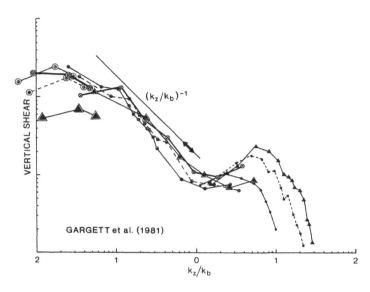

Figure 10: Vertical wavenumber spectra of shear, after Gargett *et al.* (1981).

Space here will not permit a full discussion so we will look only at some results, with the interested reader referred to the original sources for more detail. A theory that has had the widest consequence dates from Lumley (1964), wherein attention was focused upon the kinetic energy spectrum. The approach was to extend the kind of argument developed by Kolmogorov (1941) for high Reynolds number, neutrally stratified turbulence to cases where stable stratification was present. Central to Kolmogorov's theory is the role of kinetic energy transfer ϵ from large eddies to ever smaller eddies. For neutrally stratified turbulence at scales smaller than those directly forced by mean shear and larger than the scales influenced directly by viscosity, ϵ is independent of scale. This leads to the $k^{-5/3}$ turbulent subrange. In the presence of stable stratification, ϵ cannot be assumed to be independent of scale because the turbulence is thought to lose kinetic energy by conversion to potential energy via the buoyancy flux, here denoted $b(k)$. At scales not subject to direct (large scale) forcing nor to direct (small scale) viscous-diffusive decay, energy conservation requires $\partial\epsilon/\partial k = b(k)$, where negative (downward) transport of buoyancy (density defect) has the sense of converting kinetic to potential energy. With further assumptions concerning $b(k)$, Lumley obtained a kinetic energy spectrum

$$V(k) \approx N^2(k^{-3} + k_b^{-4/3}k^{-5/3}) \tag{44}$$

where $k_b = (N^3/\epsilon_o)^{1/2}$ is a 'buoyancy wavenumber' and ϵ_o is the viscous dissipation of kinetic energy.

There are a number of *caveats* associated with Eq 44 which are discussed in the original work of Lumley. What we should note is that Eq 44 predicts much more than only the power law dependences of certain subranges. It predicts also the amplitude, as on the subrange $V(k) \approx N^2k^{-3}$, and the location $k \approx k_b$ for a change of subrange, where $\approx$ recognize there are uncertain constants of order unity in such calculations. Figure 10, published 17 years after Lumley's theory, appears to confirm Eq 44 in considerable detail[5].

Encouraged that the theory at Eq 44 appears to get the kinetic energy spectrum right (at least the vertical wavenumber component thereof), we can ask what the theory implies for K_v. For this, we can examine the kinetic energy transfer rate associated with Eq 44:

$$\epsilon(k) = \epsilon_o(1 + (k_b/k)^{4/3})^{3/2} \tag{45}$$

One sees that $\epsilon(k)$ diverges toward low wavenumber as k^{-2}, which can't go on forever. So a lower wavenumber k_l must be imposed. Presumably this occurs in Fig 10 where the k^{-1} range rolls off toward lower wavenumber; hence $k_b/k_l \approx 10$. In the stratosphere, a corresponding k^{-1} subrange extends further(Dewan and Good, 1986), perhaps to $k_b/k_l \approx 100$. It is the loss of kinetic energy $\triangle\epsilon = \epsilon(k_l) - \epsilon_o$ that would appear as buoyancy flux:

$$\triangle\epsilon \approx (k_b/k_l)^2\epsilon_o \approx K_vN^2 \tag{46}$$

Now we see a problem. If this theory is right, leading to Eq 46, then the efficiency E in Eq 42 should be $E \approx (k_b/k_l)^2 \approx 10^2$ to 10^4 rather than 0.1 to 0.3. That's some wide discrepancy!

There is a further problem. *If* all that kinetic energy were being exchanged to available potential energy (APE) – appearing as temperature variance – then, over time average, the dissipation rate χ_o for temperature variance (expressed as APE) should exceed ϵ_o by roughly $(k_b/k_l)^2$. In fact observations by Oakey (1982)

[5]Because Fig 10 is a shear spectrum, the velocity spectrum in Eq 44 needs be multiplied by k^2 for comparison.

or Gregg (1987) contradict this, obtaining average χ_o (expressed as APE) rather *smaller* than ϵ_o.

So the theory is disproved? No. As shown by Weinstock (1985), it is not necessary for APE generated through $b(k)$ to be transfered on out to higher wavenumbers, ultimately to dissipation as χ_o. If instead most of that APE were transfered *backwards* from high wavenumber toward lower wavenumber, then Weinstock shows that just the arguments posed by Lumley for the kinetic energy spectrum can be carried over to the temperature (APE) spectrum with results that are consistent–within observational uncertainty – with stratospheric observations (Dalaudier and Sidi, 1987) or with Fig 9. Thus, observed spectra both of velocity and of dissipation appear to be completely consistent with efficiency E exceeding unity by *orders of magnitude!* More than simple consistency, these theories appear to *demand* such large E. On the other side, conventional usage as well as inferences from property distributions (Gargett, 1984; Ledwell *et al.*, 1986) seem to fall on the side of small E. What to do? The practical recourse may be to abandon theoretical underpinning and simply assert some rules such as Eq 41 or 42. Only one might also feel some discomfort or wariness at this.

There is another theory, perhaps better termed a conjecture, for internal waves / buoyant turbulence. The premise is almost diametrically opposite to that of Lumley and Weinstock and to a good deal of 'conventional wisdom'. Suppose that the rate of work against gravity is entirely insignificant with respect to the energetics of stratified turbulence. More precisely, suppose $b(k)$ is everywhere so small that $\epsilon(k)$ may be considered to be nondivergent in k. Hypothesize as well that properties of the waves / turbulence are determined locally in wavenumber, i.e. by ϵ which is only weakly (if at all) a function of k. Are we back at Kolmogorov and compelled to say that stable stratification should have no influence? No. The argument is that internal wave tendencies in buoyant turbulence are influential mainly through the expression $\theta_{\mathbf{kpq}}$ from Eq 22. Abbreviating (brutally!) a theoretical approach such as section 3.2.2, and imposing localness in k, the suggestion is to 'approximate' $\theta_{\mathbf{kpq}}$ by

$$\theta(k) \approx \frac{\mu}{N^2 + \mu^2}$$

where 'local cascade scaling'

$$\mu(k) \approx \epsilon^{1/3} k^{2/3}$$

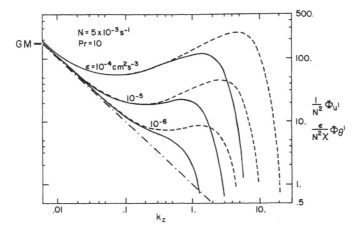

Figure 11:

Shear (solid) and temperature gradient (dashed) spectra are predicted for three values of ϵ_o. The theory here includes direct dissipation due to viscosity ν and thermal conductivity κ at a Prandtl number for sea water near $Pr = \nu/\kappa \approx 10$.

is assumed. Supposing $b(k)$ negligible, resulting predictions both for the shear spectrum and for the temperature gradient spectrum are shown in Fig 11. Details of the calculation are omitted here with an interested reader referred to Holloway (1983).

The shear spectrum in Fig 11, predicated upon the *energetic insignificance* of $b(k)$ turns out to be just the shear $k^2V(k)$ from Lumley at Eq 44, which was predicated upon *energetic dominance* by $b(k)$. The two predictions are not only the same but they fall on top of observations such as Fig 10 within experimental uncertainty. This is bad news. From observations of velocity spectra (or moments thereof, such as ϵ_o) we can't decide whether $b(k)$ is dominant, leading to efficiency $E >> 1$ in Eq 42, or whether $b(k)$ may be insignificantly small – hence $E << 1$. Is a value as large as $E = 0.3$ allowable under Holloway's 'theory'? Maybe, but only as an uncomfortable upper bound.

Do we learn more by considering the temperature spectra? Not much. The arguments by Weinstock, predicated upon energetic dominance by $b(k)$ following Lumley, yield a predicted temperature spectrum only slightly different[6] from that

---

[6]Weinstock would expect a slight spectral gap near k_b, the existence of which may be difficult to

predicted by Holloway (Fig 11). Again the spectra (or moments thereof as considered at Eq 41) do not permit us to assign a value to K_v except either *choosing* a theory, hence *choosing* either very large or very small K_v (in the sense $E >> 1$ or $E << 1$) or else setting aside these contending theories and *choosing* something else. Like $E = 0.3$.

It may be noteworthy before closing this portion that there is another major difference between Weinstock's account and Holloway's although the spectra are much the same. Under Weinstock we expect temperature variance or APE to cascade *backwards* from scales near k_b out to lower k, whereas Holloway imagines a simple *forward* cascade of APE from large scales down to small scales with little modification along the way. While direct measurement of such cascades is quite difficult, perhaps the sign of the transfer can be determined from some budget calculation. *If* this were resolved, then the surviving account might be thought more reliable with respect to estimating K_v.

4.3.1   More numerical empiricism

Daunted in our theoretical initiatives, can we turn to direct numerical simulation as an approach to parameterizing stably stratified turbulence? Surely we can, but not with the relative ease by which computers can be loosed to explore the larger scale quasi-horizontal stirring. Here, as previously, we are concerned with interactions across a wide range of scales– indeed wider perhaps than at our concerns with quasi-horizontal stirring. What's worse, we have no convincing reason (so far as I am aware) as to why we might neglect or under-represent some dimension of space. Thus three-dimensionality together with interest across a wide range of scales appear to rule out (within practical bounds of currently available computing resource) the most straightforward numerical empirical approach.

What to do?

First, we can hope that three-dimensionality isn't crucial. All variation is restricted to lie in a vertical plane although there may be a component of flow normal to the plane so long as that normal component of flow is independent of the normal coordinate. Earth's rotation and a geostrophic component of flow can be retained. What we gain is the power to resolve interactions across a tolerably wide range, say

---

establish with confidence from observed spectra.

two to three decades in wavenumber. What we lose certainly is that the character of small scale, nearly inertial turbulence is grossly corrupted by the imposition of two-dimensionality. What else we lose we may not know! We can say that whatever solutions we obtain under the 2D constraint, these are a valid subset of possible 3D solutions. However, they may be unstable to 3D, hence unrepresentative of reality. Anyway there should be no harm in looking.

Only as a visual reference, Fig 12 shows snapshots of isopycnal surfaces from four cases in which a background random agitation has built up more and more excitation until the waves begin overturning. The strength of the disturbances is measured by a kind of Froude number given by the ratio of r.m.s. vertical shear to background stability: $F = (\partial v/\partial z)_{rms}/N$.

Of interest are quantitative diagnostics including energy spectra, nonlinear transfer functions and, perhaps most importantly, the vertical buoyancy flux. A synopsis of these diagnostics is shown in the top portion of Fig 13 for a case $F \approx 1$ with earth's rotation set to zero. Energy spectra and transfer functions, both for kinetic (KE) and available potential (PE) energies, are plotted against $\ln k$, where k is magnitude wavenumber. The buoyancy flux cross-spectra $b(k)$ are plotted against $\ln k$ and as marginal spectra against $\ln k_y$ and $\ln k_z$. Negative (downwards) $b(k)$ has the sense of converting KE to PE, corresponding to down-gradient transport of buoyancy or uplifting of mass. A vertical scale of overturning in the sense of Thorpe (1977) is marked Th.

Perhaps the most striking feature of Fig 13 is the numerical experimental result that downward mixing of buoyancy occurs on scales larger than overturning while the buoyancy transport at overturning and shorter scales is systematically upwards, i.e. counter-gradient or 'restratifying'. Here we see most clearly another illustration of negative diffusion. We should not be surprised though since the advected stuff is the dynamically active buoyancy. Neither does it just happen to go this way in this particular experiment. Over many, many experiments with different kinds of forcing, at different F, with different explicit dissipation coefficients, it is really a generic feature of these numerical experiments that – over long time average – the buoyancy transport is downwards at large scales and systematically upwards (counter-gradient) at small scales. Often this statement is met with disbelief – or the suspicion that I got a sign switched. (Not so; I checked.) One particular concern must be for the role of two-dimensionality.

Figure 12: Snapshots of the density field at Froude numbers 0.1, 0.3, 1. and 3.

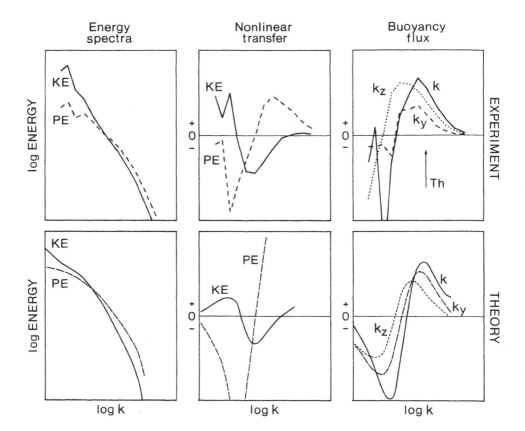

Figure 13: Numerical experiment and closure theory are compared.

Top: Numerical experimental results near $F \approx 1$, without earth's rotation, are shown. Panels include kinetic and potential energy spectra and nonlinear transfers of KE and PE as functions of total wavenumber. Buoyancy flux cross-spectra are shown as functions of total, horizontal and vertical wavenumbers.

Bottom: Closure theoretical results are plotted in the same format as the numerical experimental results.

Do these peculiar things happen also in 3D?

One can just go ahead and solve the 3D, stably stratified, Navier - Stokes - Boussinesq equations, with uniform backgound rotation if desired. A pioneering effort in this direction was by Riley *et al.* (1981). Using spectral transform techniques with collocation grid of dimensions $32 \times 32 \times 32$, available computational resource limited such early efforts to experiments on decay from prescribed initial conditions. Results are strongly dependent upon what one prescribes for initial conditions. However, the question that we might like to ask is: given some prescription of external forcing, presumably at large scales, then under conditions of statistical stationarity are we able to characterize the stably stratified turbulence? This requires a much greater extent of computations in order to achieve stationarity and to obtain confident statistical measures. However, there is uncomfortably strong dependence upon the nature of prescribed forcing and it is not possible to enforce very much separation between scales of forcing and of dissipation (assuming Laplacian diffusion for explicit dissipation operator). These are limitations we simply live with while awaiting ever bigger, faster computers. Meanwhile, some representative results for statistically stationary buoyancy flux in 3D stratified turbulence are shown in Fig 14.

For very small amplitude motion, the buoyancy flux is strongly oscillatory in time so that very long averaging periods are required. The results shown at $F = .1$ should be considered to be not significantly different from zero. However, when the level of excitation is increased toward $F = 1$, averages soon stabilize and we are struck! *This looks just like 2D!* To be honest, I've rigged it somewhat – but only slightly. A lot depends on how you force. The case shown used random agitation both of velocity and of buoyancy fields at large scales. Then one must decide how strongly to agitate each kind of field. Here I have injected KE over PE in the ratio 2:1. If you make the ratio bigger, you get more negative $b(k)$. If smaller, more positive $b(k)$. Whether the integral of b over k adds up to overall positive or negative K_v depends upon the energy input ratio and the overall excitation level with higher levels favoring negative b, hence positive K_v. An alternative method of forcing is to stipulate that larger scale waves will propagate strictly as neutral waves while the rest of the spectrum is coupled to them. This seems to favor negative K_v. It is worrisome: at least at the level of these numerical experiments, there is little evidence of 'universal' (hence parameterizable?) behavior of small scale

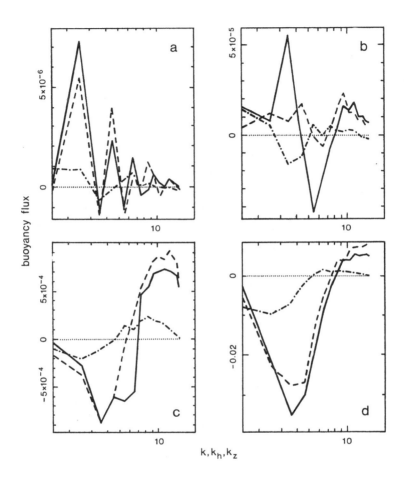

Figure 14: Buoyancy flux spectra from 3D numerical experiments

Buoyancy flux cross-spectra are plotted against total wavenumber k (solid), magnitude horizontal wavenumber k_h (dashed) and vertical wavenumber k_z (dot-dash) for four cases: $F = .1$, $F = .3$, $F = 1.$, and $F = 3.$, where $F = (\partial v_h / \partial z)_{rms}/N$. Forcing is by large scale random excitation such that the rate of injection of KE is twice the rate of PE. Prandtl number $Pr = \nu/\kappa = 1$.

turbulence. Instead, the nature of small scales seems to be determined by details of large scale forcing. However, it is altogether plausible that we are just plagued by computational limitation on the range of resolvable scales.

Although not yet a parameterization of K_v, there is an aside to mention in closing this discussion. One of the more robust features of a wide range of 3D numerical experiments on stably stratified turbulence is that the buoyancy flux tends to be more negative at low k and more positive at high k, whatever its overall sign. Why? Here is only a speculation: I think that there is almost a limitation in kinematics alone that makes less efficient the transfer of KE from low to high wavenumber due to the fact that the stuff being transferred is an incompressible velocity field. This inhibition is acute in the limit of the 2D restriction considered earlier. In 3D, I think that both KE and available PE (buoyancy variance) are transferred *forward* from low to high wavenumber – in contrast to Weinstock's backward transfer of PE. Indeed, I think that the PE transfers forward *more readily* than KE since PE transfer is not compelled to maintain an incompressible vector field representation. Consequently the larger scales are characteristically left with too much KE relative to PE while small scales tend to receive more PE relative to their KE. So they each drive buoyancy fluxes to relieve these imbalances: $b < 0$ at larger scales, $b > 0$ at smaller scales. Now, to somehow quantify and make this respectable ...(lotsa luck!)

4.3.2   More closure theory

A complete closure theoretical treatment, perhaps along lines such as section 3.2.2, for the 3D stratified turbulence problem, including earth's rotation and geostrophic modes, has proven too difficult to date – so far as I know. It remains an attractive possibility, especially if one can assume axisymmetry about the vertical for statistical quantities whereas no such axisymmetry need occur (nor is likely to occur) in the realizations of the dynamical fields. For the moment, let us only retreat to the 2D problem – as in Fig 12 – for a brief comment.

Quite apart from practical concerns, the numerical results seen, for example, in the top panel of Fig 13, stand as a challenge to statistical fluid theory. Especially when some of the results are not simply intuited, as perhaps in the shape of $b(k)$, one may look for opportunities to test one's statistical theory. Here it is not so great an extension to reproduce work that lead previously to Eq 12 through 23, but here for stratified flow in vertical plane constraint, omitting earth's rotation. Results are

$$\frac{1}{2}\frac{\partial}{\partial t}V_\mathbf{k} - b_\mathbf{k} + \sum_\Delta \theta_\mathbf{kpq} b_\mathbf{kpq} V_\mathbf{q}(V_\mathbf{k} - V_\mathbf{p}) = F_V - D_V \tag{47}$$

$$\frac{1}{2}\frac{\partial}{\partial t}R_\mathbf{k} + b_\mathbf{k} + \sum_\Delta \theta_\mathbf{kpq} c_\mathbf{kpq} V_\mathbf{q}(R_\mathbf{k} - R_\mathbf{p}) = F_R - D_R \tag{48}$$

$$\frac{\partial}{\partial t}b_\mathbf{k} + \omega_\mathbf{k}^2(V_\mathbf{k} - R_\mathbf{k}) + 2\mu_\mathbf{k} b_\mathbf{k} = F_b - D_b \tag{49}$$

where $V_\mathbf{k}$ is velocity variance, $R_\mathbf{k}$ is buoyancy variance, $b_\mathbf{k}$ is buoyancy flux, F and D represent explicit forcing and dissipation operators, $\omega_\mathbf{k}^2 = k_h^2/k^2$ and $b_\mathbf{kpq}$, $c_\mathbf{kpq}$ and $\mu_\mathbf{k}$ are just as given in section 3.2.2.

$\theta_\mathbf{kpq}$ is as given by Eq 22 but now I am obliged to take a shortcut. Rather than $\omega_\mathbf{kpq} = \omega_\mathbf{k} + \omega_\mathbf{p} + \omega_\mathbf{q}$, I have taken $\omega_\mathbf{kpq}^2 = \omega_\mathbf{k}^2 + \omega_\mathbf{p}^2 + \omega_\mathbf{q}^2$. The reason is that, by addressing the problem in velocity and buoyancy variables, we do not easily keep track of complex wave amplitudes. On the other hand, equations for velocity and buoyancy statistics are simplified. What I give up is the ability to recognize actual three wave resonances which may be expected to become more important at smaller amplitudes. On the other side, Carnevale and Frederiksen (1983) approached the closure from wave amplitudes but found it too difficult to admit non-zero buoyancy flux. Perhaps a conservation of difficulty principle is at work. Because we are here focused upon the buoyancy flux, I have gone this alternative route.

One other point to mention is that Eq 49 is here caused to be random Galilean invariant at the term $2\mu_\mathbf{k} b_\mathbf{k}$ in order to improve the estimation of high wavenumber $b_\mathbf{k}$. This was not done at Eq 14 where the consequences were not of direct concern.

An evaluation of these equations is shown in the lower panel of Fig 13. The point here is not to make detailed intercomparison. Numerical values are not included in the figure and the parameter regimes for the numerical experiment and for the closure evaluation were not identical. Nonetheless, some problems appear – as when closure appears to predict too strong PE transfer. I have not figured out why. On the other hand, it might be viewed as encouraging that the buoyancy flux spectra are so similar between theory and experiment – especially because this buoyancy flux may be surprising or unexpected from a more 'hand waving' approach. With greater effort the quantitative skill from this comparison could surely be refined. Only one must ask if it would be worthwhile given that we here test 2D theory against 2D experiments while the reality is fundamentally 3D.

5 BENTHIC BOUNDARY PROCESSES

Time and space for these lectures have fully run out. This last part is reduced to only a couple of comments, which are by no means intended to reflect the unimportance of the subject. On the contrary, if the processes of vertical mixing described in the previous section prove too feeble, then boundary-induced mixing may be what is left to do the work.

Remark: One of the most widely cited papers with respect to ocean mixing is the 'Abyssal recipes' article of Munk (1966), who inferred that an effective vertical diffusivity over the depth range 1 km to 4 km and averaged over the extent of the ocean *must* take a value near $1 cm^2 s^{-1}$ (– perhaps accounting for some reluctance among oceanographers to give up c.g.s. units![7]) This has led to a number of attempts to look into the interior of the ocean to see if one can observe Munk's $1 cm^2 s^{-1}$, perhaps by observational techniques based upon Eq 41 or 42, usually observing at depths shallower than 1 km, those being more accessible. Such efforts are essential to quantitative understanding of the oceans. It is interesting though to read in 'Abyssal recipes' where Munk frankly – and with characteristic bravery! – speculates on how $1 cm^2 s^{-1}$ comes about. Ranked according to their 'strangeness', internal wave breaking is set next to highest (most strange) while benthic boundary mixing heads the list for least strange. So that speculation, c. 1966, is that K_v is interior to the ocean – but only barely.

Substituting a cartoon for words, Fig 15 suggests such a scenario in which fluid from the nearly laminar ocean interior makes occasional encounters with the benthic boundary, there undergoing vigorous mixing before reentering the quieter interior.

Quantitative questions will concern how frequently any fluid element is likely to make such benthic encounters and whether there is sufficient energy available in the benthic boundary region to sustain significant vertical mixing. Finally, how will such processes be implemented in an ocean model SGS scheme?

There are two comments, drawing upon material discussed in previous sections.

First, as regards horizontal stirring, we must suppose that fluid elements do contact the boundary region sufficiently often – without being quite sure what we mean by 'sufficiently'. A concern that may not have received enough attention is that sloping boundaries impress strong potential vorticity gradients upon components

---

[7]See also the discussion by Olbers in this volume concerning the ill-conditionedness of this celebrated value.

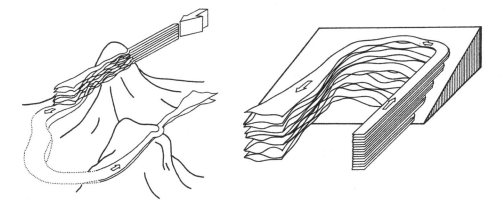

Figure 15: Interior mixing induced by topographic encounters. After Armi (1978).

of motion that are coherent over significant vertical extent. Therefore eddy dispersal in the cross-isobath direction will be inhibited analogously the suppression of meridional diffusivity on the β-plane (sect 3.2). As well, statistical inhomogeneity (sect 3.5) near boundaries is a difficult matter. These concerns arise as we assess how effectively the ocean interior communicates with its boundary regions.

The second comment draws upon internal wave dynamics and will serve to close these lectures. Is there sufficient energy in the benthic boundary region to sustain significant mixing? What are the sources for such energy? A usual consideration for energetics is shear instability resulting from mean and mesoscale eddy circulation above the benthic boundary layer. Is that energetic source sufficient (Garrett, 1979)? Another source candidate that is attracting a good deal of recent attention is internal wave breaking due to near-critical boundary reflections (Eriksen, 1982).

Among the many fascinating properties of internal inertial-gravity waves is that they do not make specular reflections from barriers, i.e. they are unlike light or sound or other waves for which the angle of reflection relative to the normal to a barrier is equal to the angle of incidence. Instead, the reflected internal wave makes an equal angle relative to vertical (defined by gravity). This is because the frequency of internal gravity waves is determined by the propagation angle relative to gravity and frequency must be a preserved property upon reflection. The other thing that must be preserved is the component of wavevector lying along the boundary surface.

More thorough discussion can be read in Eriksen or in some number of texts.

As an extreme case, suppose that the angle of an incident wave relative to vertical is the same as the angle of a barrier relative to vertical. The would-be reflected wave lies entirely on the surface of the barrier so that all of the incident energy is deposited singularly at the barrier interface. This defines critical reflection. Of concern to us will be the many waves that make nearly critical reflection, including the extreme case at criticality. A number of factors combine to enhance wave breaking near the boundary.

- The flux of energy through a 'ray tube' is $c_g E$ where c_g is wave group speed and E is wave energy density. Near criticality, the vertical wavenumber k_z of the reflected wave will be much greater than k_z of the incident wave. $c_g \propto k_z^{-1}$ will thus be much reduced upon reflection.

- The cross-sectional area of a 'ray tube' is also reduced in proportion to k_z^{-1}. Therefore energy conservation requires that reflected wave energy density be increased proportionally to k_z^2.

- Supposing that wave stability is most sensitive to vertical shear in the wave, shear variance density will scale as $k_z^2 E$, so that reflected wave shear variance increases as k_z^4.

A numerical illustration of an internal wave packet approaching a sloping planar boundary is given in Fig 16. The collapse of the packet with corresponding increase of energy density is evident. Oceanic interest will concern cases where a broad band spectrum of internal waves fill up the ocean interior and encounter continuously varying topography, bringing more or less of the incident spectrum to near criticality depending upon the convexity of topography, for example. How significant this is with respect to other sources of benthic boundary mixing remains to be assessed. Likewise its inclusion in numerical ocean models. And, since the student has come to recognize my penchant for misgivings, let me end on one such:

Recall from our discussion of internal wave breaking that the sense of resulting buoyancy flux (as also a passive chemical tracer that may be compelled to follow the buoyancy) depends upon the ratios of kinetic to available potential energies as supplied to, or by, the larger scale waves. Incident internal waves that make nearly critical reflections with the

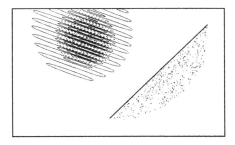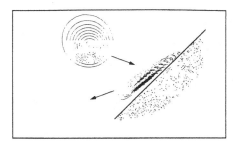

Figure 16: Internal wave packet reflecting from a sloping boundary.

boundary will supply both kinetic and potential energies. Can we be so sure that breaking these waves will induce down-gradient transports? What if all kinds of wave breaking occurs near boundaries and all that does is to un-mix or restratify the ocean? (I better stop.)

A conclusion

There isn't a conclusion to these lectures. I hope the picture presented will not be regarded as discouraging. It might have been pleasant, in a relaxing, easy way, to limit discussion on SGS to a 'how to tune up your eddy viscosity coefficients' cookbook, but I doubt that either the organizers of this School or the students would accept that. They shouldn't. As uncomfortable as it may be to admit that we don't know what we are doing, it is so much worse to suppose that we do know, taking care not to ask dangerous questions. Here the goal has been to emphasize those questions. But that is not to say that truly we don't know anything. A number of research directions have been indicated and many, many more have not been mentioned. (Indeed, looking back over these lectures as they are prepared for the volume, I wonder if they resemble an 'SGS garage sale'.) Certainly challenging opportunities abound. Finally, for lectures that opened on a Chinese fortune cookie, we might hope to close on a higher note. Perhaps it is just at the SGS that we can best be reminded:

Things should be made as simple as possible, but not simpler.
A. Einstein

Acknowledgements

It is a pleasure to thank David Anderson and Jurgen Willebrand for having caused this School to happen. The enthusiasm and critical spirit of students and faculty contributed a great deal to these lectures. I'm grateful to Patricia Kimber for assistance with figures and to colleagues Dave Ramsden, Jane Eert and Tom Keffer on whose research I have drawn for illustrations. Thoughtful criticism from Chris Garrett, Trevor McDougall, Robert W. Stewart, Geoff Vallis and George Veronis have been most helpful in revising these notes. Research here reported has been supported in parts by the Office of Naval Research and by the National Science Foundation.

Appendix A:
Wherein the *K*ey to the *S*ecret of the *U*niverse is *R*evealed

This material was not presented as a part of lectures but arose separately in discussions with students. When included in a draft for these lectures, this appendix drew some howls of protest. I think that it was seen as only a 'dumb joke', an effort to be 'cute' or 'silly' or something. Why not omit? Upon very earnest reconsideration, I have wished to retain the appendix – not as a silly, dumb joke – but as a point upon which I feel very strongly. I hope it may be a discussion of value to others. It can also be remarked that the calculation which follows could be recast in a much more transparent way. That will not be my point; rather I emphasize the ease of obfuscation – inadvertent or deliberate. But to say more here will give away the punch line.

There is a method of such awesome power that I had previously thought that its Terrible Truth should remain hidden. Such a Terrible Truth would emerge in time though, so perhaps this is the time.

As we've seen throughout the consideration of SGS, there is uncomfortable dependence upon 'hand-waving' and guessing. If only one could proceed in a more systematic way! One can.

Consider any question at all that can be posed in the form

$$A =$$

where A is a symbolic (implicit) representation of some Answer that we seek, and our goal is to provide an explicit (computable) right hand side. Unable to deduce a RHS, we may be obliged to Guess, say $A = G$. However, there is no assurance that $A = G$ is an equality. We might happen to get lucky but still would endure derision from our mathematically sophisticated colleagues. So let's fix that:

$$A = G - \epsilon(G - A) \tag{50}$$

where ϵ will be called a 'formal ordering parameter' with the value $\epsilon = 1$. We have re-established equality but now the RHS is not explicit due to the occurence of A. Let us substitute Eq 50 for A and collect terms:

$$A = G - \epsilon(G - (G - \epsilon(G - A))) = G - \epsilon^2(G - A) \tag{51}$$

The remarkable thing is that terms of $O(\epsilon)$ have dropped out from Eq 51. But A still occurs on the RHS. So we repeat the step n times to

$$A = G - \epsilon^{n+1}(G - A) \tag{52}$$

We need a name for this procedure. Call it a 'formal perturbation series' where 'formal' recognizes that ϵ is not a small parameter. In fact ϵ is no larger than unity so it is not very big and people do sometimes expand in parameters much larger than that! We observe again the most remarkable property that, at each level $n + 1$ in our iterated procedure, the collection of all terms through ϵ^n keeps coming up the same. There may be some sophisticated name for this property, else we might think of one. Now we admit that we can't go on iterating forever. So we carry out our procedure to some very high order, noting the remarkable property mentioned above (to be named), then finally say that we will retain terms only through order n for some arbitrarily large n. And we have the Answer:

$$A = G \tag{53}$$

So my first Guess, indeed any old Guess at all, turned out to be Right. Well, it is not *exactly* Right, but many things aren't. The important point is that I didn't *simply* guess; I *sytematically derived* my Guess. Unfortunately I may not have escaped the derision from my mathematically sophisticated colleagues – nor even from some who aren't so mathematically sophisticated. But it's OK because that too can be fixed. All we have to do is, at different orders in the procedure, add some complicated identity across the equation. It helps also to pick a complicated-looking G. Then it helps to perform some successive integrations by parts on terms within G, recombining some terms, expanding others, and so on. It is especially helpful to keep a little 'extra' in the way that you originally set out A. Later you can very carefully expand 'extra' in multiple perturbation parameters, showing in thorough detail that you are justified to neglect 'extra' after all. Colleagues will be impressed at your diligence. (Only some care is required to choose 'extra' that really doesn't matter very much, lest some unsophisticated colleague look and see that you are wrong.)

Clearly this invites outright charlatanism. One hopes the incidences of such are few in our science. The greater danger is that we may inadvertently do these things to ourselves and then burden our colleagues with the resulting Garbage. It is a caution with respect both to our own research and to things we may read in the Literature.

Appendix B: The eddy stress in sheared 2D turbulence

This material was not presented during lectures nor in an earlier draft of these lecture notes. Rather there was only 'it can be shown'. In revision, this Appendix is included in response to a number of queries from readers of the draft notes. Neither is the calculation trivial. The possibility of elementary blunder on my part is always a consideration.

Imagine a field of quasigeostrophic turbulence embedded in a large scale, horizontal shear $U(y)$ where U is eastward and y is a Cartesian representation of latitude. For tractability, consider barotropic motion assuming flat, rigid top and bottom boundaries. One may retain β without too great a burden or else consider only the f-plane. Among the more grievous omissions are the role of vertical shear, hence possible baroclinic exchanges, and the role of bottom topography. However, since even the simplest case seems theoretically ambiguous with respect to the *sign* of the expected eddy stress, we start with this.

Assume that one has two spatial scales of variability. For any two coordinate locations x_1 and x_2, define sum and difference variables $X = \frac{1}{2}(x_1 + x_2)$ and $r = x_1 - x_2$. Expressing all second order correlations (such as energy or eddy stress) in terms of X and r, Fourier transform on the 'fast' variable r to obtain an equation for evolution of an enstrophy spectrum $Z_k(X, t)$:

$$(\frac{\partial}{\partial t} + \mathbf{c} \cdot \partial_{\mathbf{X}} + \mathbf{R} \cdot \partial_{\mathbf{k}} + 2\nu_k)Z_{\mathbf{k}}(\mathbf{X}, t) = F_{\mathbf{k}}(\mathbf{X}, t) + T_{\mathbf{k}}(\mathbf{X}, t) \qquad (54)$$

where

$$\mathbf{c} = [U + (\beta - \partial_{YY}U)\frac{k_x^2 - k_y^2}{k^4}, 2(\beta - \partial_{YY}U)\frac{k_x k_y}{k^4}] \qquad (55)$$

is the group velocity of advected Rossby wave packets,

$$\mathbf{R} = [0, -k_x \partial_Y U] \qquad (56)$$

is shear-induced wave refraction, $\partial_{\mathbf{X}}$ and $\partial_{\mathbf{k}}$ are gradient operators in $\mathbf{X}$ and $\mathbf{k}$, respectively, $F_{\mathbf{k}}(\mathbf{X}, t)$ represents external forcing such as windstress curl, and $T_{\mathbf{k}}(\mathbf{X}, t)$ represents any enstrophy tendency due to eddy-eddy interactions. Spectra Z, F and T are said to have parametric dependence upon the 'slow' spatial variable $\mathbf{X} = (X, Y)$. We have not been obliged to make a similar distinction between 'fast' and 'slow' time scales although it may be recalled from Sect 3.2.2 that theoretical evaluation of the eddy term T depends upon slow evolution of Z relative to a fast eddy time scale.

In these notes, no effort is made to give the derivation of the left side of Eq 54 for which further discussion may be found in McWilliams (1976) or Muller (1978). One may also view the left side of Eq 54 as a statement of wave action conservation in shear flow, c.f. Bretherton and Garrett (1968), where action is

$$A_{\mathbf{k}} = \frac{E_{\mathbf{k}}}{|\sigma_{\mathbf{k}}|} = \frac{\frac{1}{2}Z_{\mathbf{k}}/k^2}{(\beta - \partial_{YY}U)k_x/k^2} \tag{57}$$

Here $E_{\mathbf{k}}$ is wave energy (kinetic in this case) and $\sigma_{\mathbf{k}}$ is the intrinsic phase speed. k_x is not affected by the mean flow $U(y)$.

The subject of this Appendix is the derivation of consequences from Eq 54. Assuming that the eddy statistics do not depart too greatly from isotropy, adopt an approximate representation

$$2\pi Z_{\mathbf{k}} = \sum_{n=-\infty}^{\infty} Z_n(k)e^{in\phi_{\mathbf{k}}} \approx Z(k)\delta k(1 - P(k)\cos 2\phi_{\mathbf{k}} - Q(k)\sin 2\phi_{\mathbf{k}}) \tag{58}$$

where now $Z(k)$ is the enstrophy spectral density such that $Z(k)\delta k$ is the enstrophy within a band of width δk about k. $\phi_{\mathbf{k}}$ is the angle between $\mathbf{k}$ and the k_x-axis. Odd powers of n drop out of Eq 58 by a symmetry $Z_{\mathbf{k}} = Z_{-\mathbf{k}}$ due to reality of the vorticity field. Thus in Eq 58 we retain only the lowest order departure from isotropy, assuming $|P|$ and $|Q|$ remain small compared with unity.

The objective is to obtain the eddy stress

$$\langle \mathbf{u}\mathbf{u} \rangle = \int dk \begin{pmatrix} 1 + P(k) & Q(k) \\ Q(k) & 1 - P(K) \end{pmatrix} \frac{Z(k)}{2k^2} \tag{59}$$

Substituting from Eq 58 into Eq 54 then performing $\oint d\phi_{\mathbf{k}}$, $\oint d\phi_{\mathbf{k}}\sin 2\phi_{\mathbf{k}}$ and $\oint d\phi_{\mathbf{k}}\cos 2\phi_{\mathbf{k}}$ on the results, obtain

$$(\frac{\partial}{\partial t} + c_y\partial_Y + 2\nu_k)Z(k) + \frac{1}{4}\frac{\partial U}{\partial Y}\partial_k(kZ(k)Q(k)) = F(k) + T(k) \tag{60}$$

$$(\frac{\partial}{\partial t} + c_y\partial_Y + 2\nu_k)\frac{Z(k)Q(k)}{k^2} + \frac{\partial U}{\partial Y}\frac{Z(k)P(k)}{k^2} + \frac{1}{2}\frac{\partial U}{\partial Y}\partial_k\frac{Z}{k} = T_Q(k) \tag{61}$$

$$(\frac{\partial}{\partial t} + c_y\partial_Y + 2\nu_k)\frac{Z(k)P(k)}{k^2} - \frac{\partial U}{\partial Y}\frac{Z(k)Q(k)}{k^2} = T_P(k) \tag{62}$$

where only the isotropic part $F(k)$ of forcing is listed and T, T_Q and T_P represent eddy interactions which have not yet been discussed.

Apart from the uncertain role of eddy interactions, a couple of remarks are immediate. Integrating Eq 60 over all k, and assuming that eddy variance (hence kZQ) sufficiently nearly vanishes outside some finite wavenumber band, we observe that the mean flow interaction conserves eddy enstrophy. Similarly integrating Eq 61 and 62 over all k and here requiring that Z/k sufficiently nearly vanish outside the finite eddy band, we see that the mean flow generation term $\frac{1}{2}\frac{\partial U}{\partial Y}\partial_k \frac{Z}{k}$ results in no net stress generation at all! This is the basis for the claim that the negative viscosity interpretation of Figure 5 is mistaken. What one tends to 'see' in the figure are flow lines tilting SW-NE. These are associated with the higher wavenumber portion of the eddy spectrum for which $\partial_k \frac{Z}{k} < 0$, tending to generate $Q > 0$. The lower wavenumber portion for which $\partial_k \frac{Z}{k} > 0$ tends to generate $Q < 0$. A calculation is necessary to decide which sign of Q wins; the purpose of this Appendix is to examine that calculation. What we appear to find is that overall generation of ZQ/k^2 is zero. The 'correct' interpretation of Figure 5 should be *zero eddy viscosity*.

Of course the story is not over. We have been compelled to assume a spectral gap or, equivalently, that eddies occupy a wavenumber spectrum of finite width. Calculation helps us now to quantify this requirement. The condition $\partial_k \frac{Z}{k} > 0$ at low wavenumbers corresponds to a kinetic energy spectrum $E(k)$ climbing at least as rapidly as k^3. This may be a severe demand for actual oceanic eddy spectra. The result may be to weaken or eliminate the low wavenumber contributions to $Q < 0$ in actuality. However, one should not yet conclude that eddy viscosity ought to be negative since, in the absence of a gap separating low wavenumber 'eddies' from large scale 'mean' shear, we had no basis for a Fickian eddy viscosity.

Before returning to eddy - eddy interactions terms on the right sides of Eq 60, 61 and 62, let us briefly consider the role of dissipative terms $2\nu_k$ on the left sides.[8] We might also set aside Eq 60 by saying that $Z(k)$ will be given on account of $F(k)$. Then statistically stationary and homogenous solution of Eq 61 and 62 yields

$$\frac{Z(k)Q(k)}{k^2} = -\frac{\nu_k \partial U/\partial k}{4\nu_k^2 + (\partial U/\partial k)^2}\partial_k \frac{Z(k)}{k} + \frac{2\nu_k T_Q(k)}{4\nu_k^2 + (\partial U/\partial k)^2} - \frac{(\partial U/\partial k)T_P(k)}{4\nu_k^2 + (\partial U/\partial k)^2} \quad (63)$$

[8]After all, at our most naive, we might try to toss the eddy - eddy interactions over onto the left sides by simply calling them 'eddy viscosities', *i.e.*, saying $\nu_k = \nu_{eddy}k^2$, within this overall discussion of eddy viscosity.

If now we took ν_k to be a simple constant (independent of k), such as a Rayleigh drag, we would again see that the apparent source term (first term on right) vanishes when integrated over a range of k containing a band-limited eddy spectrum. If we suppose ν_k is some increasing function of increasing k, then the situation is more ambiguous since ν_k tends to cut off contributions from the source term at both small and large wavenumber, depending upon the magnitude of ν_k relative to $\partial U/\partial k$. It is noteworthy that ν_k does not directly affect the relative anisotropies P and Q. This can be seen, *e.g.*, by subtracting the product of $Q(k)$ by Eq 60 from the product of k^2 by Eq 61. ν_k does not induce a return-to-isotropy but only tends to decay products such as ZQ.

Finally, eddy terms $T_Q(k)$ and $T_P(k)$ present a research challenge which lies outside the scope of this Appendix. Major work in the treatment of anisotropic 2D turbulence is by Herring (1977) with discussion of the extension to geophysical influences such as β included in Holloway (1986). Brief comments here will only be of a qualitative nature. T_Q and T_P are found to contain expressions for nonlinear transfer of anisotropy among different scales of motion as well as exhibiting a return-to-isotropy tendency. In part, the transfer expressions tend to compete with the return tendency so that, to a rough approximation, the overall influence resembles a weaker return-to-isotropy. A more quantitative characterization of the return tendency depends upon quantitative details of $Z(k)$; however, the return tendency usually appears as an increasing function of wavenumber. Since this is a direct decay of Q or P, it further weakens the high wavenumber portion of a product such as ZQ at Eq 63, tending to enhance to overall positivity of eddy viscosity.

Of geophysical interest is a positive bias that tends to occur in T_P when β is non-zero. This is a tendency to enhance E-W motion relative to N-S, as observed, *e.g.*, in numerical experiments of Rhines (1975). Under the influence of $\partial U/\partial k$, some of the resulting P is tipped into Q with the sense from Eq 63 that $\partial U/\partial k > 0$ induces $Q < 0$, further enhancing overall positivity of eddy viscosity.

This Appendix, along with discussion under Sect. 3.3.4, should be read as speculative research suggestions. The summary – to my mind – is that *if* an eddy spectrum were band limited in the presence of a large scale mean shear, then positive eddy viscosity would result. *But* if there is not a sufficient spectral gap separating 'eddy' from 'mean', then positivity quickly collapses along with the basis for having assumed an eddy viscosity formulation in the first place.

REFERENCES

Armi, L., 1978, 'Some evidence for boundary mixing in the deep ocean', *J. Geophys. Res.*, **83**, 1971-79.

Babiano, A., C. Basdevant, B. Legras and R. Sadourny, 1987; 'Vorticity and passive scalar dynamics in two-dimensional turbulence', *J. Fluid Mech.*, **183**, 379-397.

Carnevale, G. F. and J. S. Frederiksen, 1983; 'A statistical dynamical theory of strongly nonlinear internal gravity waves', *Geophys. Astrophys. Fluid Dyn.*, **23**, 175-207.

DeSzoeke, R. A. and M. D. Levine, 1981; 'The advective flux of heat by mean geostrophic motions in the Southern Ocean', *Deep-Sea Res.*, **28**, 1057-85.

Dewan, E. M. and R. E. Good, 1986; 'Saturation and universal spectrum for vertical profiles of horizontal scalar winds in the atmosphere', *J. Geophys. Res.*, **91**, 2742-48.

Eriksen, C. C., 1985; 'Implications of ocean bottom reflection for internal wave spectra and mixing', *J. Phys. Oceanogr.*, **15**, 1145-56.

Freeland, H., 1987; 'Oceanic eddy transports and satellite altimetry', *Nature*, **326**, 524.

Gargett, A. E., 1984; 'Vertical eddy viscosity in the ocean interior', *J. Mar. Res.*, **42**, 359-393.

Gargett, A. E., P. J. Hendricks, T. B. Sanford, T. R. Osborn and A. J. Williams III, 1981; 'A composite spectrum of vertical shear in the upper ocean', *J. Phys. Oceanogr.*, **11**, 71.

Garrett, C., 1979; 'Comments on 'Some evidence for boundary mixing in the deep ocean' by Laurence Armi', *J. Geophys. Res.*, **84**,

Gill, A. E., 1982;' Atmosphere-Ocean Dynamics', Academic Press; 662 pp.

Gregg, M. C., 1977;'A comparison of fine-structure spectra from the main thermocline', *J. Phys. Oceanogr.*, **7**, 33-40.

Gregg, M. C., 1987;'Diapycnal mixing in the thermocline: a review', *J. Geophys. Res.*, **92**, 5249-86.

Herring, J. R., 1977; 'Theory of two-dimensional anisotropic turbulence', *J. Atmos. Sci.*, **34** 1731-50.

Holloway, G., 1983;'A conjecture relating oceanic internal waves and small scale processes', *Atmos.-Ocean*, **21**, 107-122.

Holloway, G., 1986;'Eddies, waves, circulation and mixing: statistical geofluid mechanics', *Ann. Rev. Fluid Mech.*, **18**, 91-147.

Keffer, T. and G. Holloway, 1988;'Estimating Southern Ocean eddy flux of heat and salt from satellite altimetry', *Nature*, **332**, 624-626.

Kolmogorov, A. N., 1941;'The local structure of turbulence in an incompressible viscous fluid for very large Reynolds numbers', *C. R. Akad. Nauk, SSSR*, **30**, 301-305.

Kraichnan, R., 1959;'The structure of isotropic turbulence at very high Reynolds numbers',*J. Fluid Mech.*,**5**, 497-543.

Kraichnan, R., 1971; 'An almost-Markovian Galilean-invariant turbulence model',*J. Fluid Mech.*,**47**, 512-524.

Kullenberg, G., ed., 1982; Pollutant Transfer and Transport in the Sea; Boca Raton: CRC Press.

Ledwell, J. R., A. J. Watson and W. S. Broecker, 1986; 'A deliberate tracer experiment in the Santa Monica Basin', *Nature*,**323**, 322-324.

Legras, B., 1980;'Turbulent phase shift of Rossby waves', *Geophys. Astrophys. Fluid Dyn.*,**15**, 253-281.

Leith, C. E., 1971; 'Atmospheric predictability and two-dimensional turbulence', *J. Atmos. Sci.*, **28**, 145-161.

Lesieur, M., 1987; Turbulence in Fluids, Dordrecht: Martin Nijhoff Publ., 286 pp.

Leslie, D. C., 1973; Developments in the Theory of Turbulence, Oxford: Clarendon, 368 pp.

Levitus, S., 1982;'Climatological Atlas of the World Ocean', NOAA Prof. Paper 13, Washington, D. C.

Lilly, D. K., D. E. Waco and S. I. Adelfang, 1974; 'Stratospheric mixing estimated from high-altitude turbulence measurements',*J. Applied Meteorology*,**13**, 488-493.

Lumley, J. L., 1964; 'The spectrum of nearly inertial turbulence in a stably stratified fluid',*J. Atmos. Sci.*,**21**, 99-102.

McDougall, T. J., 1987; 'Neutral surfaces',*J. Phys. Oceanogr.*, **17**, 1950-64.

McWilliams, J. C., 1976; 'Large-scale inhomogeneities and mesoscale ocean waves: a single, stable wave field',*J. Marine Res.*,**34**, 423-456.

McWilliams, J. C., 1984; 'The emergence of isolated coherent vortices in turbulent flow',*J.Fluid Mech.*,**146**, 21-43.

Muller, P., 1978;'On the parameterization of eddy-mean flow interaction in the ocean',*Dyn. Atmos. Oceans*,**2**, 383-408.

Muller, P., R. Lien and R. Williams, 1988; 'Estimates of potential vorticity of small scales in the ocean',*J. Phys. Oceanogr.*,**18**, 401-416.

Munk, W., 1966; 'Abyssal recipes',*Deep-Sea Res.*, **13**, 707-730.

Oakey, N. S., 1982; 'Determination of the rate of dissipation of turbulent energy from simultaneous temperature and velocity shear microstructure measurements',*J. Phys. Oceanogr.*,**12**, 256-271.

Ogura, Y., 1963; 'A consequence of the zero-fourth-cumulant approximation in the decay of isotropic turbulence',*J. Fluid Mech.*, **16**, 33-40.

Okubo, A., 1986; 'Dynamical aspects of animal grouping: swarms, schools, flocks, and herds',*Adv. Biophys.*,**22**, 1-94.

Osborn, T. R. and C. S. Cox, 1972; 'Oceanic fine structure', *Geophys. Fluid Dyn.*,**3**, 321-345.

Olbers, D. J., M. Wenzel and J. Willebrand, 1985; 'The inference of North Atlantic circulation patterns from climatological hydrographic data',*Rev. Geophysics,*23, 313-356.

Pouquet, A., M. Lesieur, J. C. Andre and C. Basdevant, 1975; 'Evolution of high Reynolds number two-dimensional turbulence', *J. Fluid Mech.,*72, 305-319.

Rhines, P. B., 1975; 'Waves and turbulence on a beta-plane', *J. Fluid Mech.,*69, 417-443.

Richardson, L. F. and H. Stommel, 1948; 'Note on eddy diffusion the sea',*J. Meteorol.,*5, 238-240.

Riley, J.J., R. W. Metcalfe and M.A. Weissman; 'Direct numerical simulations of homogenous turbulence in density-stratified fluids', in Nonlinear Properties of Internal Waves B. J. West, ed., New York: Amer. Inst. Phys.

Rose, H. A., 1977; 'Eddy viscosity, eddy noise and subgrid-scale modelling',*J. Fluid Mech.,*81 719-734.

Sadourny, R. and C. Basdevant, 1985; 'Parameterization of subgrid-scale barotropic and baroclinic eddies in quasi-geostrophic models: Anticipated potential vorticity method'., *J. Atmos. Sci.,*42, 1353-63.

Salmon, R., G. Holloway and M. C. Hendershott, 1976; 'The equilibrium statistical mechanics of simple quasi-geostrophic models', *J. Fluid Mech.,*75, 691-703.

Starr, V. P., 1968; The Physics of Negative Viscosity Phenomena, McGraw-Hill, New York.

Stewart, R. W. and R. E. Thomson, 1977; 'Re-examination of vorticity transfer theory',*Proc. Roy. Soc. Lond.,*A 354, 1-8.

Taylor, G. I., 1921; 'Diffusion by continuous movements', *Proc. Roy. Soc. Lond.,*20 196-212.

Thorpe, S. A., 1977; 'Turbulence and mixing in a Scottish Loch', *Proc. Roy. Soc. Lond.,*A 286, 125-181.

Vallis, G. K. and B.-L. Hua, 1988; 'Eddy viscosity of the anticipated vorticity method',*J. Atmos. Sci.,*45, 617-627.

Weinstock, J., 1985; 'On the theory of temperature spectra in a stably stratified fluid',*J. Phys. Oceanogr.,*15, 475-477.

Young, W. R., P. B. Rhines and C. J. R. Garrett, 1982; 'Shear-flow dispersion, internal waves and horizontal mixing in the ocean', *J. Phys. Oceanogr.,*12, 515-527.

600

Inversion, 163
        -, linear, 428
        -, nonlinear, 433
        -, total, 191,435,436
Isentropic coordinate, 261
Isopycnal, 56
        - coordinates, 488ff.
        - diffusion, 129,134,135
        - mixing, 397ff.,454,456,457,461,551

Kalman filter, 235ff.
Kelvin wave, 288,295
Krigging, 53
Krypton-85, 351,362ff.,378

L1-norm, 52
L2-norm, 49,51
Lagrange function, 290
        - multipliers, 49,52,65,83,263,288,290,291
Lagrangian correlation time, 520
        - dispersion, 521
        - function, 262
        - interpolation, 314
Laplace's equation, 54
Laplace-Poisson equation, 14
Leap-frog scheme, 266
Least squares, 28,34,36,44,52,53,237ff.,258,260
        -, separable, 436,437
        -, total, 433ff.
        -, weighted, 20
Level-of-no-motion, 56,96,102,118
Linear maximal sequence, 160,161
Linear programming, 50,51,53,54,412,436,451ff.
Linear regression, 44,258

Mathematical programming, 52
Matrix condition, 90,118,120
Matrix inversion, 81
Maximum entropy, 530
Maximum likelihood estimate, 33
Maximum likelihood solution, 50,337
Measurement errors, 268,272,273
Mediterranean Sea, 444
Mesoscale fluctuation, 166
Mesoscale mapping, 173
Metric, 80,83
Minimization, unconstrained, 261,262,263,276
Minimum-mean-square estimation, 164
Mixed layer, 261
Mixing, 349,369ff.,455,462,487ff.
        - length, 536
        - parameters, 97,101,104,122

Printed in the United States
By Bookmasters